Biology of Bryozoans

CONTRIBUTORS

GENEVIÈVE BOBIN

JACK D. FARMER

ÅKE FRANZÉN

DENNIS P. GORDON

DIETHARDT JEBRAM

GENEVIÈVE LUTAUD

CLAUS NIELSEN

J. S. RYLAND

PHILIP A. SANDBERG

THOMAS J. M. SCHOPF

LARS SILÉN

DOROTHY F. SOULE

JOHN D. SOULE

ROLF STRÖM

JUDITH E. WINSTON

ROBERT M. WOOLLACOTT

RUSSEL L. ZIMMER

Biology of Bryozoans

Edited by

ROBERT M. WOOLLACOTT

Museum of Comparative Zoology
Harvard University
Cambridge, Massachusetts

RUSSEL L. ZIMMER

Department of Biological Sciences
and Allan Hancock Foundation
University of Southern California
Los Angeles, California

ACADEMIC PRESS New York San Francisco London 1977

A Subsidiary of Harcourt Brace Jovanovich, Publishers

ACADEMIC PRESS, INC.
111 Fifth Avenue, New York, New York 10003

United Kingdom Edition published by
ACADEMIC PRESS, INC. (LONDON) LTD.
24/28 Oval Road, London NW1

Library of Congress Cataloging in Publication Data

Main entry under title:

Biology of bryozoans.

Bibliography: p.
Includes index.
1. Polyzoa. I. Woollacott, Robert M.
II. Zimmer, Russel L.
QL396.B56 594′.7 76-19494
ISBN 0-12-763150-X

PRINTED IN THE UNITED STATES OF AMERICA

The editors dedicate their efforts in the organization and production of this book to the memory of David Robert Jahn (1950–1976). David will be remembered not only as a promising graduate student fascinated with bryozoan biology but as a valued colleague and friend.

Contents

5 Ultrastructure, Mineralogy, and Development of Bryozoan Skeletons

PHILIP A. SANDBERG

6 Polymorphism

LARS SILÉN

7 Feeding in Marine Bryozoans

JUDITH E. WINSTON

8 Experimental Techniques and Culture Methods

DIETHARDT JEBRAM

List of Contributors

Numbers in parentheses indicate the pages on which the authors' contributions begin.

GENEVIÈVE BOBIN (307), Ecole Pratique des Hautes Etudes, Histologie et Cytologie des Invertébrés Marins, Laboratoire de Cytologie, Université de Paris, Paris, France

JACK D. FARMER (487), Department of Geology, University of California, Davis, California

ÅKE FRANZÉN (1), Institute of Zoology, University of Uppsala, Uppsala, Sweden

DENNIS P. GORDON* (335), Department of Biology, Dalhousie University, Halifax, Nova Scotia, Canada

DIETHARDT JEBRAM (273), Zoologisches Institut der Technischen Universität, Braunschweig, West Germany

GENEVIÈVE LUTAUD (377), Histologie et Cytologie des Invertébrés Marins, Laboratoire de Cytologie, Université de Paris, Paris, France

CLAUS NIELSEN (519), Marine Biological Laboratory, University of Copenhagen, Helsingør, Denmark

J. S. RYLAND (411), Department of Zoology, University College of Swansea, Wales, United Kingdom

PHILIP A. SANDBERG (143), Department of Geology, University of Illinois, Urbana, Illinois

THOMAS J. M. SCHOPF (459), Department of Geophysical Sciences, University of Chicago, Chicago, Illinois, and Marine Biological Laboratory, Woods Hole, Massachusetts

LARS SILÉN (183), Department of Zoology, University of Stockholm, Stockholm, Sweden

DOROTHY F. SOULE (437), Allan Hancock Foundation, University of Southern California, Los Angeles, California

JOHN D. SOULE (437), Allan Hancock Foundation and School of Dentistry, University of Southern California, Los Angeles, California

*Present address: Marine Research Laboratory, Northland, New Zealand.

ROLF STRÖM (23), Institute of Zoology, University of Uppsala, Uppsala, Sweden

JUDITH E. WINSTON* (233), Department of Geophysical Sciences, University of Chicago, Chicago, Illinois

ROBERT M. WOOLLACOTT (57, 91), Museum of Comparative Zoology, Harvard University, Cambridge, Massachusetts

RUSSEL L. ZIMMER (57, 91), Department of Biological Sciences and Allan Hancock Foundation, University of Southern California, Los Angeles, California

*Present address: Department of Paleobiology and Fort Pierce Bureau, U.S. National Museum, Smithsonian Institution, Fort Pierce, Florida

Preface

For curiosity and beauty, I have not, among all the plants or vegetables I have yet observed, seen any one comparable to this seaweed.[1]

The above quote correctly suggests that bryozoans are attractive, rich in enigmatic features, and provide an opportunity for gross misinterpretation. That bryozoans are indeed colonial animals and not plants is now beyond dispute, but the number of gaps in our knowledge is rivaled only by the areas in which generalizations are based on one or at most a few examples. All too often when there are comparative observations they are in conflict.

This volume was conceived not so much to answer all of these deficiencies (for such would be impossible), but to provide a forum for the statement of problems, an indication of our current state of the art, and directions in which future studies might be profitably oriented. This book is obviously intended for students and specialists in the field. However, by providing current reviews of some of the unusual and even unique systems and aspects of bryozoans which may reveal information otherwise unavailable on basic problems of fundamental interest to biologists and paleontologists, we hope it will also be useful to a broader audience.

The early history of our knowledge of this phylum, to a great degree, featured studies by lay naturalists, but at the turn of the twentieth century the field was benefited by a small cadre of predominately European specialists who are still recognized as giants of bryozoology. Recently, interest in the phylum has been considerably more catholic, and experts can be found in countries throughout the world. This renaissance of interest is reflected by the formation of the International Bryozoology Association, which has been instrumental in bringing paleontologists and neontologists to a greater understanding of each other's field of interest. The Association sponsors triannual meetings, the proceedings of which provide a varied menu of current studies. Additionally, the membership roster of the Association constitutes a convenient listing of most specialists.

[1] An observation by Hooker on the frondose bryozoan *Flustra foliacea* as quoted by D. Landsborough (1852). "A Popular History of British Zoophytes, or Corallines," pp. 346–347. Reeve and Co., London.

This book is not comprehensive in its coverage, but focuses attention on topics which have not been covered elsewhere recently and which we felt have one or both of the following attributes: they are presently being extensively studied so that a current review and synthesis are appropriate, or they are aspects of bryozoan biology which should be brought to the attention of a broad audience but have not received sufficient emphasis by bryozoologists. The selection of chapter topics was further restricted by the availability of current experts. For example, it would have been extremely beneficial to have a chapter on the musculature of bryozoans as a striking range of variation is found in the phylum including the fastest and among the smallest fibers, striated and smooth myoepithelia, and supercontracting muscles. Similarly, it would have been useful to have a chapter on the biological associations and ecology of bryozoans; fortunately, Ryland (1976)[2] has provided a thorough review of these topics. These examples reveal a greater problem: the surprisingly large deficiencies in our understanding of the basic biology of this conspicuous and interesting group of colonial invertebrates.

The volume contains no introduction to the phylum because useful accounts are found—at least in a limited sense—in every invertebrate textbook. The compendia by Brien (1960)[3] and Hyman (1959)[4] provide more detailed sources of information. The highly readable and informative book by Ryland (1970)[5] also gives a concise introduction to the phylum as well as correlating paleontological information with that derived from studies of the living animals. The bryozoan section of the "Treatise on Invertebrate Paleontology,"[6] first printed in 1953 and currently under extensive revision, is the authoritative work on the very rich fossil record (there are 3 to 4 fossil species for each of the 4000 or so living ones). The revised treatise will also prove a valuable resource for information on recent forms and on systematics. In view of these considerations, no specific attention is given here to the fossil history or to systematics. However, a classification of all genera cited in the text is provided in the Appendix. Even the casual reader will note that the cheilostome and ctenostome gymnolaemates receive disproportionate attention with a shortchanging of the stenolaemates and phylactolaemates. This imbalance is paralleled in part by the distribution of specialists and by the biases in interests of the coeditors.

[2] Ryland, J. S. (1976). Physiology and ecology of marine bryozoans. *Adv. Mar. Biol.* **14**, 285–443.

[3] Brien, P. (1960). Classe de Bryozoaires. *In* "Traité de Zoologie" (P.-P. Grassé, ed.), Vol. 5, fascile 2, pp. 1053–1355. Masson, Paris.

[4] Hyman, H. L. (1959). "The Invertebrates: Smaller Coelomate Groups." McGraw-Hill, New York.

[5] Ryland, J. S. (1970). "Bryozoans." Hutchinson Univ. Library, London.

[6] Bassler, R. S. (1953). "Treatise on Invertebrate Paleontology" (R. C. Moore, ed.), Part G, Bryozoa. Univ. of Kansas Press and Geological Society of America, Lawrence, Kansas.

The reader is cautioned that the use of terms and organization of taxa, while consistent within each chapter, are not uniform throughout the volume. Because various contributors have conflicting interpretations of the constitution or significance of a given structure or taxon and feel strongly that their interpretation is preferable to alternative ones, it was deemed impractical and undesirable to impose a rigid "glossary" or classification. Even the term Bryozoa is used in different contexts within the book, reflecting a continuing controversy that is partially national in scope. The majority of bryozoologists accept the term Bryozoa as a phylum name to indicate only the Ectoprocta (and not these plus the Entoprocta as in earlier usage), but some specialists insist on using Ectoprocta to avoid any confusion. This usage has certain merit not only because of the past breadth of the term Bryozoa but also because there are several who conclude—in contrast to generally accepted dogma—that the entoprocts and ectoprocts are intimately related and consequently use Bryozoa in its earlier broader composite meaning. As well as with the phylum name, there are disagreements concerning the composition of even the highest taxonomic subdivisions. There is growing acceptance of the division of the phylum into three classes: the Phylactolaemata, the Gymnolaemata, and the Stenolaemata, but some retain a classification in which the latter two classes are united as a single unit, the Gymnolaemata, with rank equal to the Phylactolaemata. This expanded group is then frequently subdivided as the Eurystomata and Cyclostomata.

As with any multiauthored work, the majority of manuscripts were solicited, completed, and accepted for publication considerably before the book went to press. In most cases there has been a revision of the chapters to accommodate data to 1977. We thank the contributors for their patience and for their efforts to present thorough and up-to-date chapters. It is hoped they will accept our apologies for delays in preparation of the final volume.

Production of this book has been facilitated by numerous persons. Their combined suggestions and aid have greatly eased the editors' burden. We are especially grateful to Mrs. Cheryl Zimmer for her major efforts in translating the contribution from one of the foreign authors, help with proofreading, preparation of the index, and, especially, for her cheerful moral support throughout the duration of this project.

Robert M. Woollacott
Russel L. Zimmer

1

Gametogenesis of Bryozoans

ÅKE FRANZÉN

I. MORPHOLOGY OF THE OVA AND OOGENESIS

The germ cells are recognizable at an early stage of zooid budding by their larger size and larger nuclei with a conspicuous nucleolus. Most gymnolaemate zooids are hermaphroditic. The ovary is often more distally situated in the body cavity than the testis and the two sorts of gonads do not ripen simultaneously.

In some Gymnolaemata it seems to be a rule that the ovary develops at the wall toward which the stomach is directed. Prouho (1892) describes a special type of protandry in *Alcyonidium duplex* Prouho. In this species the male zooid degenerates after producing testes and in the same zooecium a female zooid regenerates and develops an ovary. Silén (1966) reports that there is a strong tendency toward protandric hermaphroditism in *Electra posidoniae* Gautier, but in *Electra crustulenta* (Pallas) spermatid cytophores, mature sperm and ova are present simultaneously.

Silén (1945) has made a detailed study of the development and the oogenesis in some brooding gymnolaemates, especially *Callopora dumerilii* Aud., and reports that both oocytes and eggs have a loose consistency and are capable of changing form. The elongated egg in this species may sometimes be flattened, owing to the pressure of the moving stomach, but the egg immediately resumes its original shape when the pressure ceases. The egg is large with a length of about 200 μm, and the total length of the zooid averages 450 μm. When it is transferred to the ovicell it runs as a thin cord through the opening into the ovicell and then rapidly resumes its normal shape.

Generally a second oocyte appears in the ovary only when the first egg has almost reached its final size and has freed itself from the ovarian wall. Movement of the alimentary canal effects the carrying of the egg to the distal part of the ovary and to the ovicell. Generally, there is also only one egg in the ovicell at a time.

In *Electra posidoniae* 8–9 eggs are generally formed, but they may number up to 20 (Silén, 1966). Their form is irregular. The majority of gymnolaemates brood their embryos, and only one oocyte at a time grows and is extruded out of the peritoneum. Silén (1944) reported that in *Labiostomella gisleni* each ovary develops about 100 oocytes, about 10 reach the distal part of the zooid, but only one develops.

In *Membranipora membranacea* (L.) a large number of small eggs are rapidly emitted (Silén, 1945); also in *Electra pilosa* (L.) the number of eggs produced in one reproductive period is large. A low reproduction occurs in a female zooid of *Synnotum aegypticum* (Aud.) where the polypide degenerates after it has developed a single larva (Marcus, 1941). The eggs of brooding bryozoans are relatively large and pigmented with yellow, red, or orange carotenoids (Ryland, 1958). Mature eggs commonly measure about 200 μm in diameter. In nonbrooding species the eggs are small and generally more or less colorless. The oocyte may be nourished by the fusion with another oocyte, or by the yolk stored in peritoneal cells. The gonozooid in Cyclostomata contains a female polypide, differing in development from species to species. Generally only one egg develops in each gonozooid in the cyclostome colony.

Silén (1945) reports that in specimens of *Callopora* kept in aquaria, about two weeks passes from the first appearance of the oocyte until it leaves the body cavity and that its development in the ovicell takes about the same time.

The eggs of nonbrooding gymnolaematous bryozoans are delivered into the seawater where they rapidly develop into planktotrophic pelagic larvae. These larvae have an intestine and are capable of nourishing themselves, and only a small amount of stored nutrition in the egg is needed. In most species, however, the amount of yolk in the egg has increased, and the embryo develops in the ovicell for a comparatively long time, but needing no further supply of food.

Most authors believed that autogamous fertilization is the rule among bryozoans. Perhaps the main reason for this conclusion is the simultaneous appearance in many species of morphologically mature eggs and spermatozoa within one zooid. In some species a spermatozoon has already entered the immature oocyte. In cross-fertilization there is some difficulty in explaining how the spermatozoa from one zooid can enter into the coelom to fertilize the eggs.

Silén (1966) discovered how cross-fertilization in bryozoans takes place. The spermatozoa in *Electra* are discharged through a minute pore in each of the two most dorsal tentacles. After this they drift away through the water and are carried away from the parent zooid by water currents. They are then sucked in by the tentacle crown of another zooid and become attached to the tentacles. When this zooid discharges ova, the spermatozoa detach and fertilization occurs. Silén also reports that spermatozoa are seen to move swiftly around the ova and soon disappear, but the entering of the spermatozoon into the ovum could not be discerned.

II. MORPHOLOGY OF THE SPERMATOZOON AND SPERMIOGENESIS

The germ cells differentiate from the peritoneum in the developing zooid. The testis of Gymnolaemata is usually located proximally in the zooid on the peritoneum of the body wall or on the funiculus. At sexual maturity the testis subdivides into groups of cells consisting of different stages of spermatogenesis. Spermatogonia produced from primitive germ cells in the testis also migrate into the coelomic cavity to produce morulae of spermatocytes clustered around a cytoplasmic mass or cytophore. The two divisions of meiosis produce four daughter cells or spermatids. The spermatids undergo a complex transformation involving most or all organelles. The spermatogenesis, or transformation of the spermatid into the long, slender spermatozoon, specialized for swimming and penetrating the ovum, has been the subject of rather few studies in Bryozoa.

Retzius (1904, 1906, 1910), in his splendid monograph, gave descriptions of the spermatozoa of *Flustra foliacea* (L.), *Scrupocellaria reptans* (L.), *Triticella korenii* G.O. Sars, and *Plumatella fungosa* Pallas. Discussing spermiogenesis of *Triticella,* Retzius reported the occurrence in certain stages of large spheres around the base of the tail or flagellum. These spheres were obviously the mitochondria.

At the level of light microscopy spermiogenesis and particularly the transformation of mitochondria have been described and illustrated by Franzén (1956) in some species, viz., *Tubulipora liliacea* (Pallas), *Membranipora membranacea* (L.), *Flustra foliacea* (L.), *Triticella koreni* G.O. Sars, *Alcyonidium gelatinosum* (L.), and *Bowerbankia* sp.

Comparing the different species within Gymnolaemata, all the species examined seem to exhibit during spermiogenesis certain primitive features, and genesis on the whole follows a similar course. Yet the Cyclostomata can be distinguished here with a fair amount of justification.

A. Stenolaemata

The following description of spermiogenesis in Cyclostomata refers to an examination of *Tubulipora liliacea* (Pallas) by Franzén (1956). At the division of a primary spermatocyte the daughter cells remain connected, and this applies also to the next division, with the result that four spermatid nuclei are situated together in a cytoplasmic mass or cytophore (Fig. 1a–c). Borg (1926), who described the spermatozoon of *Crisiella producta* (Smitt), has made the same observation. In light of the observations by Fawcett (1972) on intercellular bridges among male germ cells, and his assumption that the theoretical number of conjoined spermatids would be at least 256, it is of interest to note that a characteristic feature of spermiogenesis in Cyclostomata seems to be that only the four spermatids derived from a primary spermatocyte differentiate together within a syncytium and maintain synchrony. Fawcett considers it plausible to explain the existence of intercellular bridges during spermiogenesis as a means of maintaining synchrony. Four flagella with four nuclei grow out from the syncytium (Fig. 1d). During transformation the chromatin is concentrated in the anterior part of the nucleus. The anterior pole of the spermatid nucleus becomes increasingly dense, whereas the posterior portion of the nucleus is hardly visible in the phase contrast microscope. The nucleus gradually assumes conical shape (Fig. 1e) and becomes more and more elongated. The syncytium becomes polarized with the nuclei at one end and the flagella growing out at the opposite pole. Four small, dark granules or mitochondria are later found around the proximal part of each flagellum. As yet there is evidence only from the light microscope that the dark granules around the bases of the flagella are mitochondria. On vital staining with Janus green they are strongly tinted; this finding may be taken as a demonstration of their mitochondrial nature.

The four mitochondria disappear, but in their place a thin layer of mitochondrial material appears around the flagella from the base of the nuclei to the syncytium membrane. These mitochondrial layers around the flagella form the midpieces of the mature spermatozoa. The cytoplasm slides backward along the flagella, which still point mainly in the same direction. The flagella are surrounded by cytoplasm with mitochondria up to the point forming the limit between midpiece and tail. At this point is a dark structure, possibly a centriole. In the tail region, the flagellum seems to obtain a cover of cytoplasm, but this cover may be of the same nature as described below for *Triticella* on the basis of electron microscopic observations. The spermatids separate from each other after reaching morphological maturity. This separation is initiated by the separation of the nuclei and the posterior part of the midpieces.

The head of the mature spermatozoon (Fig. 1h) is 8 μm long and is pointed. No acrosome has been observed at the distal part of the head. The midpiece is

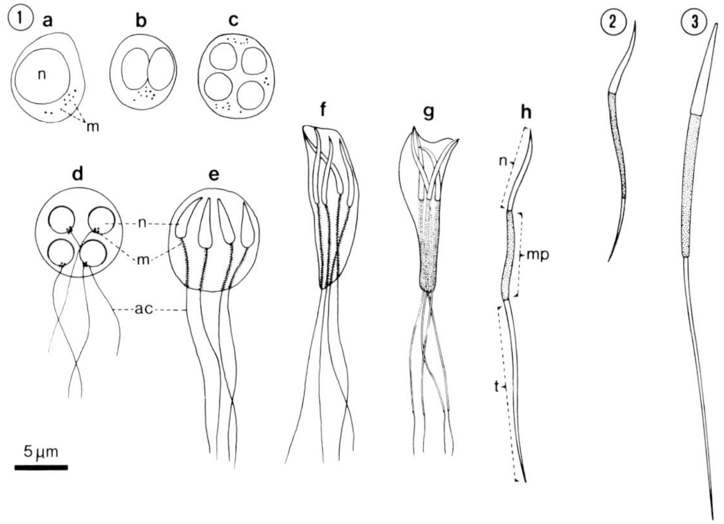

Fig. 1. Diagrammatic representation of successive stages of spermatogenesis in a cyclostome bryozoan, *Tubulipora liliacea* (Pallas). Drawings based on phase contrast microscopy. (a) Primary spermatocyte. (b) Secondary spermatocyte. (c) Four spermatid nuclei in a syncytium. (d) Spermiogenesis; chromatin condensation occurs in the anterior poles of the nuclei, n; mitochondria, m, are collected in rings around the base of each flagellum or axial filament complex, ac. (e) Nuclear condensation and elongation continue; caudal displacement of mitochondria along the flagella has started. (f) Elongation of the whole spermatid syncytium. (g) Nuclei and midpieces possess their final shape, but the tails are not yet morphologically complete, and the sperm cells are still in a syncytial arrangement. (h) Spermatozoon: n, nucleus; mp, midpiece with mitochondria; t, tail in which the axial filament complex is no longer seen in phase contrast microscope.

Fig. 2. Spermatozoon of *Berenicea patina* (Lamarck).

Fig. 3. Spermatozoon of *Diplosolen obelia* (Johnston). From Franzén (1956).

built of mitochondria around the anterior part of the flagellum and its length is about 8 μm. The midpiece is followed by a tail flagellum with a thinner terminal portion. The midpiece seems to be limited in either direction by dark granules, possibly centrioles.

In some other cyclostomes, *Diplosolen obelia* (Johnston) in Fig. 3, and *Berenicea patina* (Lamarck) in Fig. 2, the spermatozoa are of the same type as in *Tubulipora.* However, the size of the spermatozoa differs. The length of the morphologically mature spermatozoon is 45 μm in *Diplosolen obelia* as compared to only 22 μm in *Berenicea patina.*

B. Gymnolaemata

1. Phase Contrast Observations of Living Spermatozoa and Spermatids

At the level of light microscopy the spermatozoa of this class are of a special modified type with lancelike heads, and an elongate mitochondrial midpiece. All species examined are found to exhibit primitive features during their spermiogenesis, and genesis on the whole follows a similar course. On comparison with the Cyclostomata a tendency may be found in the Gymnolaemata toward an increase in the content of cytoplasm in the midpiece and tail (Figs. 4–14). An effect of this morphology may be that the spermatozoa swim in a characteristic way with slightly undulating midpieces and tails.

Mitochondria are seen in the cytoplasm of young spermatids with a spherical nucleus. The number of mitochondria decreases while at the same time those remaining become larger (Fig. 14a–c). When only four large mitochondria remain these form a regular ring around the point of origin of the growing flagellum. This stage is followed by an additional concentration of the mitochondrial material so that succeeding stages exhibit two large masses near one another on either side of the flagellum (Fig. 14c). These two masses seem to be stretched, and they first assume a droplike shape (Fig. 14d). Later on they form two dark rods and elongate in the posterior direction (Fig. 14e,f). Gradually they become more or less indistinguishable from each other under the light microscope. However, it may be stressed that in many species it is possible to separate the two dark mitochondrial rods from each other and from the flagellum by pressing the coverslip.

In *Triticella*, the flagellum of the spermatid attains a length of about 60 μm before the mitochondrial masses emerge from the rounded cell and form a peripheral cover to the axoneme. During the backward movement of the cytoplasm the shape of the cell becomes more or less irregular, but the lumps of cytoplasm with Golgi material which appear here and there disappear at the end of spermiogenesis.

Spermatids can occasionally be seen in which the flagellum has a distal, tubelike cover before the formation of the midpiece is complete (Fig. 14f).

Initially, the chromatin is evenly distributed in the nucleus of the spermatid, but it is soon concentrated toward the anterior periphery like a cone. Later, the nuclear shape becomes increasingly pointedly conical (Fig. 14d–f), and a simultaneous concentration of the chromatin results in the appearance of the nucleus of the mature *Triticella* spermatozoon as a dark, awl-shaped structure (Fig. 14g). The mature spermatozoon of *Triticella* is filiform, and in the light microscope a head about 16 μm long, a midpiece about 50 μm long, and a broadly lancet-shaped tail about 33 μm long, can be distinguished.

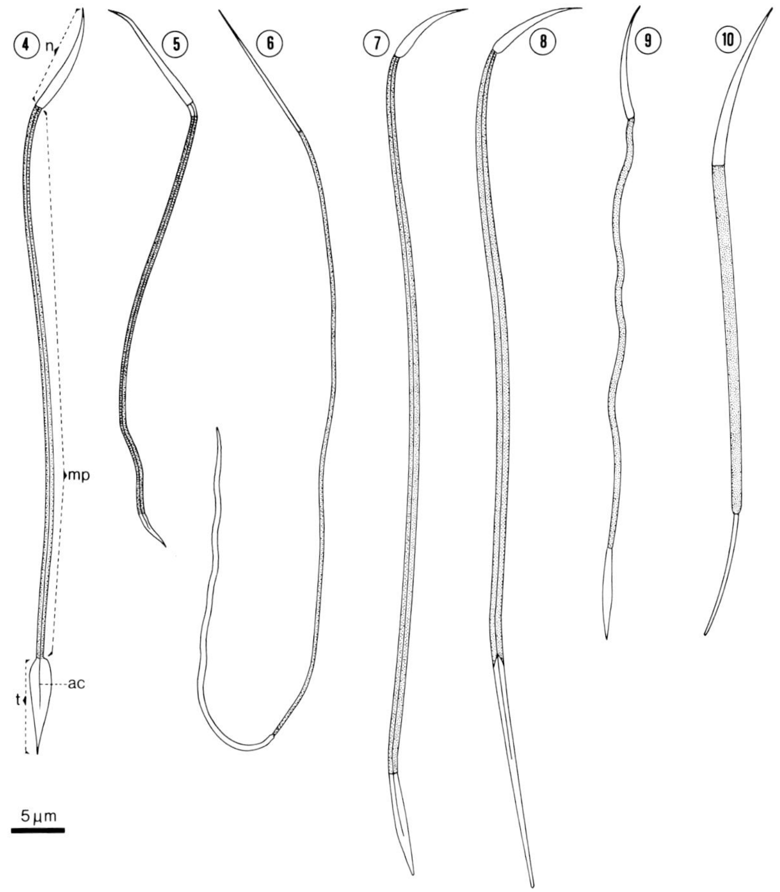

Figs. 4–10. Schematic diagrams of mature spermatozoa from different cheilostomes. Abbreviations: n, nucleus; mp, midpiece with mitochondria; t, tail; ac, axial filament complex. Figures 4 and 7–9 from Franzén (1956).

Fig. 4. *Flustra foliacea* (L.).

Fig. 5. *Flustra securifrons* (Pall).

Fig. 6. *Electra pilosa* (L.).

Fig. 7. *Caberea ellisii* (Flemming).

Fig. 8. *Cryptosula pallasiana* (Moll).

Fig. 9. *Scrupocellaria scruposa* (L.).

Fig. 10. *Bicellariella ciliata* (L.).

Structures resembling centrioles appear at both ends of the midpiece in spermatozoa of some bryozoan species (Franzén, 1956). Electron microscopic studies have not confirmed their centriolar nature, and this is why they have been omitted from the figures (Figs. 4–14). Perhaps these structures seen in the light microscope may be the nuclear vesicles at the base of the elongating nucleus. These vesicles often contain electron-dense material (Fig. 17) and may appear as ring centrioles in the light microscope.

Three special features of possible functional significance are characteristic for the gymnolaemate spermatozoon. These include absence of an acrosome, two mitochondrial rods in the midpiece, and a tubelike cell membrane surrounding the axoneme in the tail. This basic pattern may be modified in different species (Figs. 4–14). It may be noted that two of the ctenostomes studied, viz., *Alcyonidium gelatinosum* (L.) (Fig. 11) and *Bowerbankia* sp. (Fig. 13), are drawn with a sperm tail lacking the tubelike cell membrane around the axoneme. However, the stages studied in both these species may represent late spermatids and morphologically immature spermatozoa.

In *Membranipora* the mature spermatozoa are aggregated in tightly adherent groups of 32 or 64, as so-called spermatozeugmata. This was first described by Bonnevie (1907) and has been confirmed by others (Franzén, 1956; Zimmer and Woollacott, 1974).

2. *Electron Microscopy of Sectioned Spermatozoa and Spermatids*

Electron microscopy has up to now yielded rather limited new information on the morphology of the bryozoan spermatozoa and spermiogenesis. There are only three electron microscopic studies on this subject. Reger (1971) reported that in *Bugula* sp. the spermiogenesis is similar to spermiogenesis in typical flagellate spermatozoa, but the mature spermatozoon lacks an acrosome and a nuclear envelope. As a third difference from the typical spermatozoon, Reger mentions that in late stages of spermiogenesis the developing flagellum with the axoneme complex recoils around the circumference of the late spermatid and becomes incorporated into the spermatid cytoplasm. However, it is possible that the flagellum later uncoils again; it is also possible that two types of spermatozoa are produced.

Reger has described and presented electron micrographs of mature *Bugula* spermatozoa characterized by a typical flagellum, with the 9 + 2 pattern of microtubules, mitochondria surrounding the basal centriole, and a head containing a nucleus but lacking an acrosome. Thus it appears quite clear that the spermatozoa in this species is of a primitive type; the only modification is that the head region is elongated and lacks an acrosome and nuclear envelope. It may also be stressed that this type of mature spermatozoon corresponds to the spermatid stages shown in Fig. 1d (*Tubulipora*) and Fig. 14b (*Triticella*). The

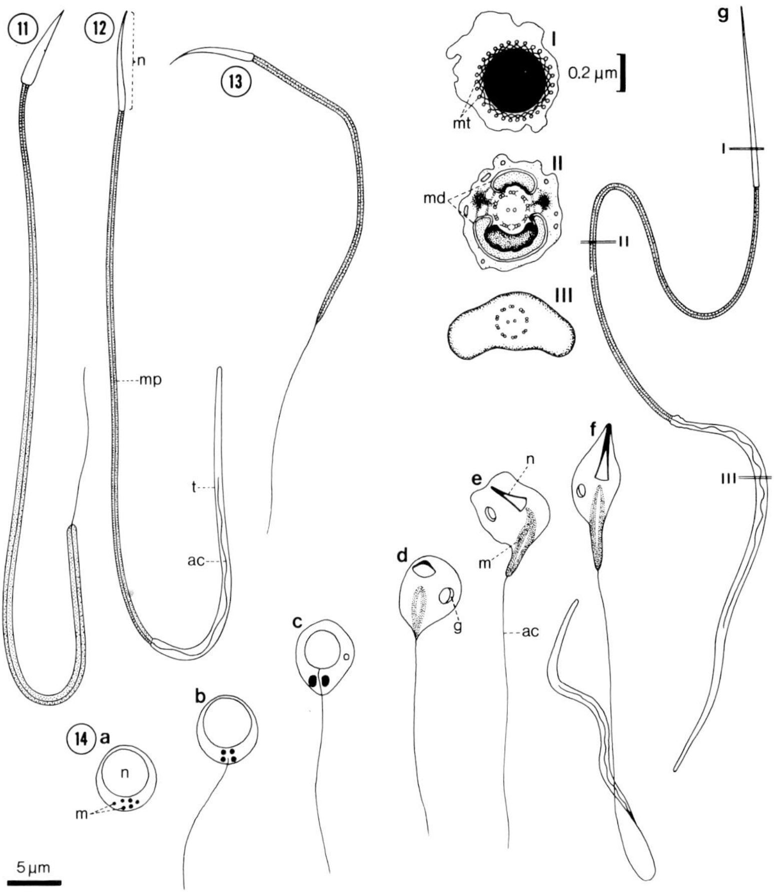

Figs. 11–14. Schematic diagrams of spermatozoa and spermiogenesis in some ctenostomes. Abbreviations: ac, axial filament complex; md, mitochondrial derivative; mt, microtubules; g, Golgi body; n, nucleus; mp, midpiece with mitochondria; t, tail. Figures 11, 13, and 14 from Franzén (1956). Figure 12 modified from Franzén (1960). The drawings of the sectional planes I, II, and III have not been previously published.

Fig. 11. Spermatozoon or late spermatid of *Alcyonidium gelatinosum* (L.).

Fig. 12. Spermatozoon of *Monobryozoon limicola* Franzén.

Fig. 13. Spermatozoon or late spermatid of *Bowerbankia* sp.

Fig. 14. Successive stages in spermiogenesis in the ctenostome *Triticella koreni* G.O. Sars, showing the formation of the midpiece from the mitochondria and the condensation and elongation of the nucleus in the head region. (a) Young spermatid. (b) Orientation of four mitochondria around the growing flagellum. (c) Two large mitochondria at the base of the flagellum. (d, e, f) Successive stages of nuclear condensation and spermatid elongation. Mitochondria form two dark rods flanking the growing axoneme. (g) Spermatozoon. The sectional planes I, II, and III correspond to the drawings I, II, and III illustrating the ultrastructure at different levels. I shows nuclear region (= head) in transverse section; II shows the midpiece, and III shows the tail with the axoneme within a broadly tube shaped cell membrane.

principal difference between these spermatid stages and the mature *Bugula* spermatozoon is that the amount of cytoplasm is reduced and the nucleus is condensed and elongated in *Bugula*. Thus *Bugula* seems, according to Reger's description, to represent a more primitive condition in sperm morphology than other bryozoans. In this connection it is surprising that in one *Bugula* species, according to Gerwerzhagen (1913), the eggs are fertilized by spermatozoa from the same zooid. The second of the three existing electron microscopic studies on bryozoan spermiogenesis is an interesting but unillustrated abstract by Zimmer and Woollacott (1974) on spermiogenesis in *Membranipora* sp. During spermiogenesis in this species the mitochondria aggregate in two composite masses adjacent to the flagellar centriole. These two mitochondrial masses later elongate as straight rods flanking the developing axoneme. Initially, the mitochondrial rods have a typical mitochondrial structure but later become reduced with loss of cristae. Thus there seems to be variability in the ultrastructure of the mitochondrial midpiece among bryozoans. It should be pointed out that despite small variations in morphology of the spermatozoa and in the spermiogenesis among the bryozoan species studied by me and others with the light microscope, all these Gymnolaemata spermatozoa may be referred to as the same general type with lancelike head and elongate mitochondrial midpiece. The investigation by Zimmer and Woollacott confirms this concept.

In order to present more data on the morphology of this typical bryozoan spermatozoon and its development, I started a fine structure study using the spermatozoon of the ctenostome bryozoan *Triticella korenii* (Figs. 15–21). Young spermatids become polarized with mitochondria, a centriole, and a growing filament at one end of the nucleus. At the opposite pole the nuclear chromatin starts to condense. Complete mitochondrial fusion results in the formation of two large masses adjacent to the base of the flagellum. Each of these large mitochondrial masses soon elongates along the developing axoneme (Fig. 15). The mitochondrial cristae become oriented more or less parallel to the axoneme.

At this stage microtubules appear near the anterior surface of the nuclear envelope where the chromatin is condensing. In the cytoplasm a Golgi body is seen but it is not in close apposition to the nuclear membrane. Early spermatid nuclei are roughly spherical and contain scattered flecks of chromatin (Fig. 16). The posterior region of the nucleus is invaginated and in this invagination a centriolar complex is developed, forming the basal body of the flagellum (Fig. 16). At a stage when the nucleus becomes conical (Fig. 17) the chromatin condensation has given rise to a very electron-dense anterior part, an intermediate granular or filamentous part, and a posterior, very translucent part forming saclike protrusions of the nuclear envelope.

Further development is characterized by condensation of the nucleus (Fig. 18). Coincident with the elongation of the nucleus, microtubules appear

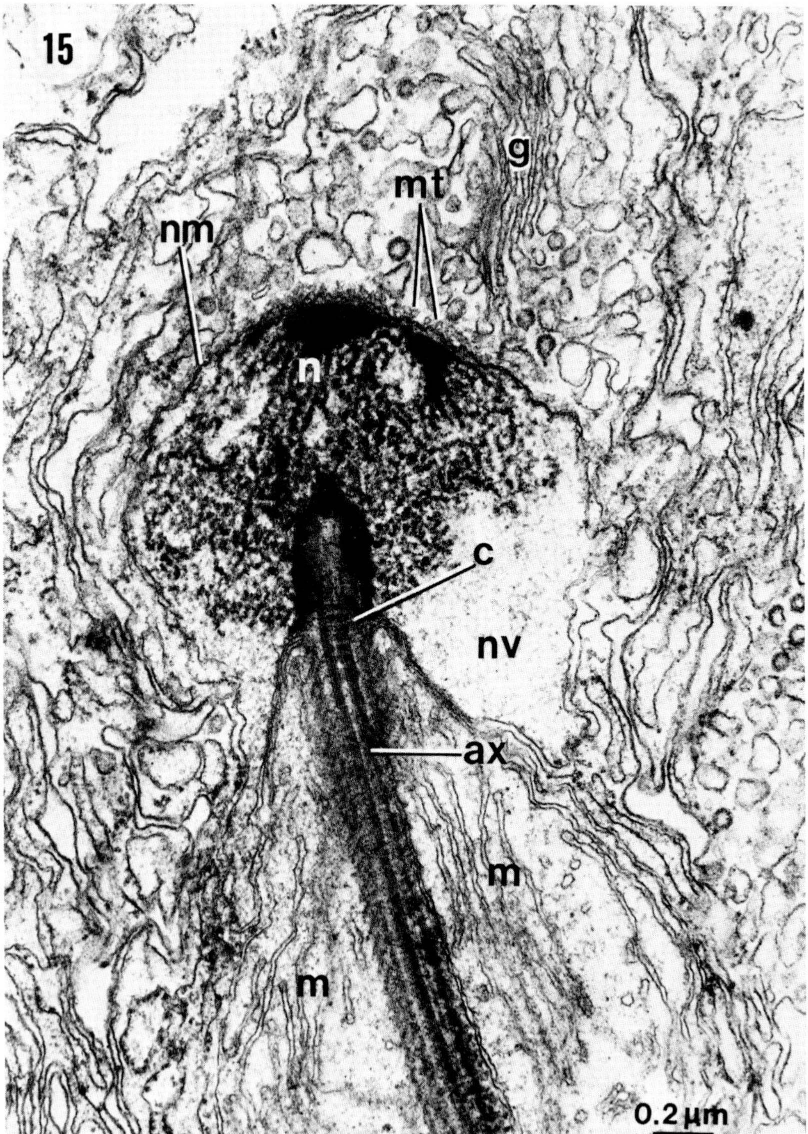

Fig. 15. Longitudinal section of a *Triticella* spermatid in an early stage of nuclear condensation. Formation of nuclear vesicles has started in the posterior pole. Along the flagellum or axoneme two large elongated mitochondrial masses have started to move in a posterior direction from the centriolar region. Abbreviations: ax, axoneme; g, Golgi body; c, centriole; m, mitochondria; mt, microtubule; n, nucleus; nm, nucleaı membrane; nv, nuclear vesicle.

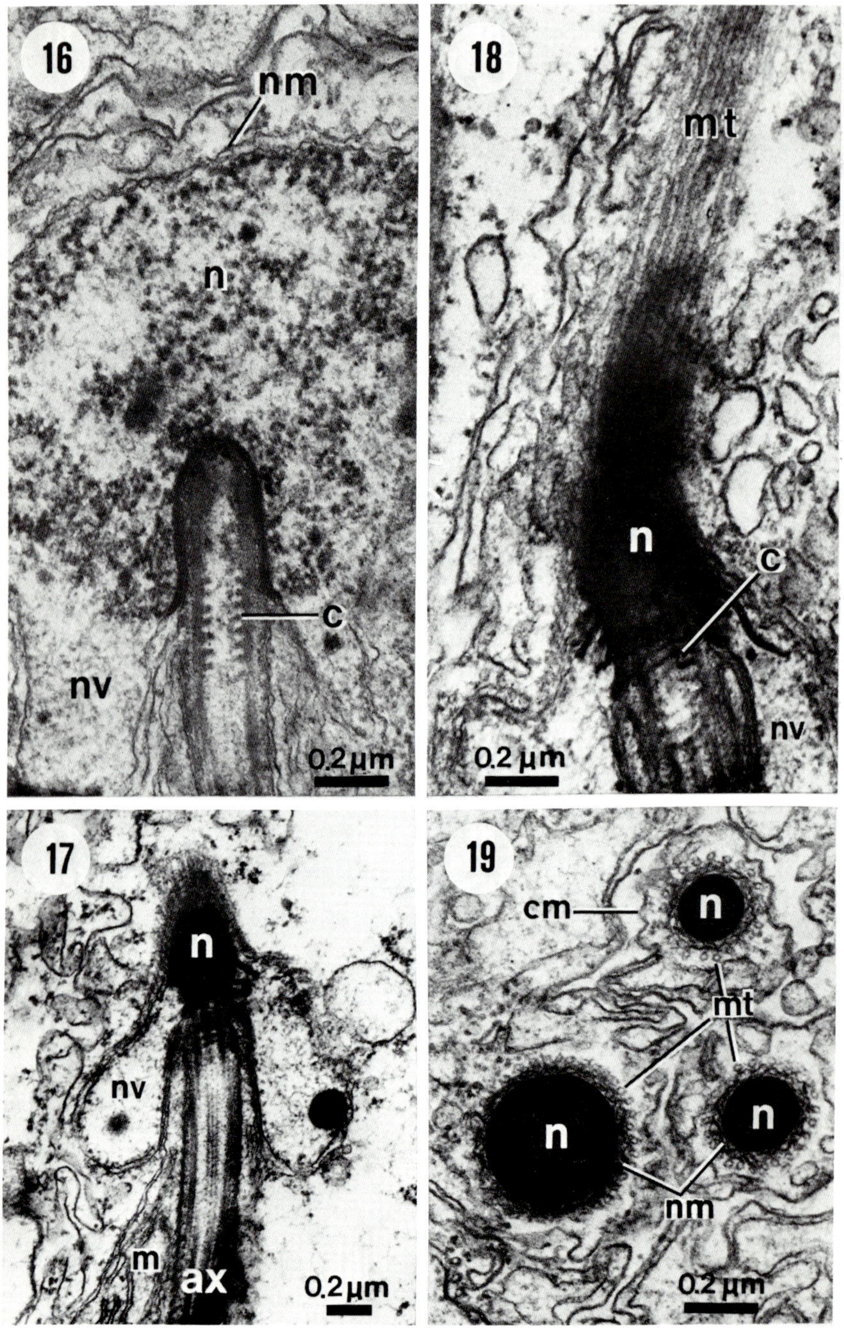
16
nm
n
c
nv
0.2 µm
18
mt
n
c
nv
0.2 µm
17
n
nv
m
ax
0.2 µm
19
cm
n
mt
n
n
nm
0.2 µm

oriented in a ring around the nucleus (Figs. 18 and 19). The microtubules are of uniform diameter throughout their length and appear to number 30 at the level indicated in Fig. 14gI. In the stage shown in Fig. 18 coarse chromatin fibrils appear in the nucleus near the centriolar invagination.

As the spermatids continue to mature the nucleus becomes more and more elongated (Figs. 14, and 17–19). I have not succeeded in finding any acrosome in the spermatids or in the mature spermatozoon. The nucleus terminates as a very thin thread. The fact that microtubules appear at the time of nuclear condensation and that their orientation is so closely allied to that of the nucleus suggests that they are in some way involved in the formation of the mature nucleus. There is, however, a great deal of evidence indicating that microtubular systems do not have an active molding function in spermiogenesis (Fawcett *et al.*, 1971).

A relatively prominent specialization along the lateral sides of the nucleus contains microtubules and a nuclear membrane (Figs. 19 and 20). The surrounding cytoplasm contains some small vesicles which seem to derive from the Golgi body (Fig. 15). One could possibly speculate that these vesicles have, or develop, an acrosomal function. The whole head is limited by the cell membrane.

The centriolar region in the basal part of the nucleus is formed at a stage when the nucleus is approximately spherical. The longitudinal section in Fig. 16 shows that the centriole at its inner surface contains about seven rings of dense material. In cross sections the centriole appears as nine dense rods, but their detailed structure has not been studied.

In early spermatids the mitochondria have a normal structure with cristae. When the two large mitochondrial masses start to elongate along the axoneme as straight rods (Figs. 16 and 21) they still have long cristae. One of the two mitochondrial rods becomes larger and this rod seems to be centered peripherally to doublets 5 and 6 (compare Afzelius, 1969). Zimmer and Woollacott (1974) have come to the conclusion that in *Membranipora* the larger mitochondrial rod is centered peripherally to doublets 4 and 5.

Fig. 16–19. Sections of spermatid nuclei from *Triticella*. Abbreviations: ax, axoneme; c, centriole; cm, cell membrane; mt, microtubule; n, nucleus; nm, nuclear membrane; nv, nuclear vesicle.

Fig. 16. Longitudinal section of early spermatid with the centriole cut along its long axis.

Fig. 17. Longitudinal section through the base of the nucleus. Note the caudal position of the nuclear vesicles containing some electron-dense material. An elongating mitochondrion is seen near the axoneme.

Fig. 18. Longitudinal section of a spermatid with elongated nucleus. Microtubules are seen cut along the surface of the nucleus. At the base of the nucleus a centriole and coarse chromatin filament are seen.

Fig. 19. Cross section of late spermatid nuclei cut at different levels.

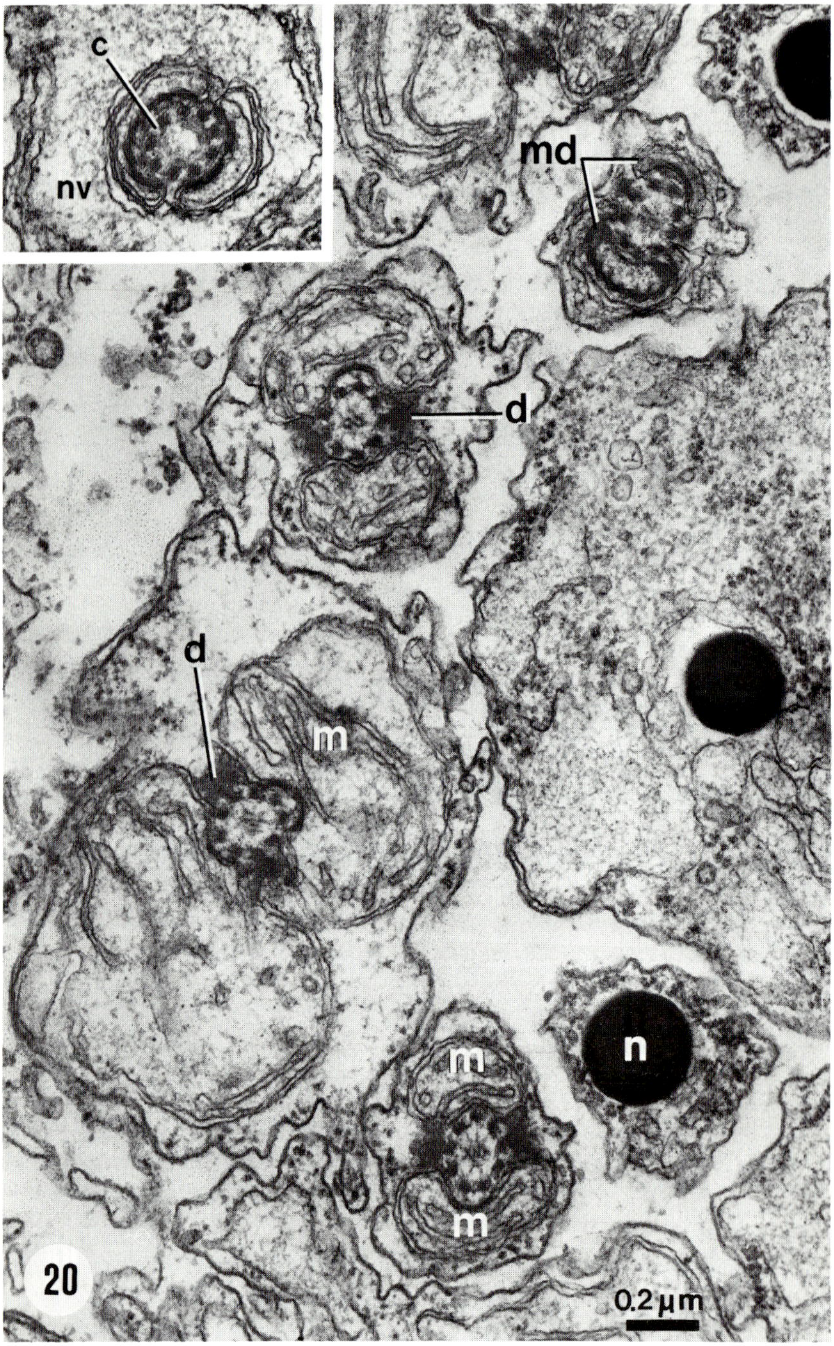
c
nv
md
d
d
m
m
n
m
20
0.2 µm

In rather early *Triticella* spermatids the crescentic rods around the axoneme are connected by electron-dense material. The amount of this dense material is reduced as spermiogenesis continues, and at the same time the mitochondria become more and more modified and cristae are lost. Perhaps this dense material is some source of nutrient material during gamete transformation.

The smaller mitochondrial rod is diametrically opposed to the larger one. In cross section the modified mitochondrial rods are crescentic both in *Membranipora* and in *Triticella.* In *Triticella* the convex face of the larger rod thickens and increases in electron density, partly depending on a complicated spiral structure in the mitochondrial membrane. In the smaller rod the concave face increases in electron density (Fig. 21). A system of membranes is seen in connection with the modified mitochondria or mitochondrial derivatives.

Cytoplasm with a few microtubules peripheral to the mitochondrial derivatives in Fig. 21 indicates that the spermatozoa shown here are not completely mature. Zimmer and Woollacott have reported tests for succinic dehydrogenase activity showing that modification of morphology and degradation of enzymatic activity are correlated.

In the mature spermatozoon the tail is about 33 μm long and of a somewhat flattened tubelike shape. The axoneme has the typical 9 + 2 structure, and the peripheral double filaments carry arms on one of their microtubules. Throughout most of its length the tail structure is constant. The most remarkable feature in the tail of bryozoan spermatozoa is that the axoneme is surrounded by a spacious tube formed by the cell membrane. This detail is clearly seen in the phase contrast microscope in most of the species studied.

A mitochondrial sheath envelops the axoneme in the midpiece of most types of modified flagellate spermatozoa. In *Triticella* and perhaps also in the gymnolaematous bryozoan *Flustra* the two mitochondrial derivatives are not in direct contact with each other but are separated by electron-dense material. The amount of this material seems to depend on the developmental stage in spermiogenesis. A similar condition with opaque material separating the mitochondria in the midpiece has been reported for Squamata spermatozoa by Furieri (1970).

Spermiogenesis and sperm morphology in *Flustra foliacea* (L.) and *Triticella korenii* G.O. Sars was studied by Franzén (1976) with the electron microscope. The head of the spermatozoon in *Flustra* consists of the elongated nucleus,

---

Fig. 20. Cross sections of *Triticella* spermatids at different levels through nuclei and developing midpieces. Elongating mitochondrial rods are flanking the axoneme. Electron-dense material is accumulated between the two mitochondrial rods in each spermatid. The inset is a cross section through the centriole in a young spermatid with a prominent nuclear vesicle corresponding to the stage seen in Fig. 17. Abbreviations: c, centriole; d, electron-dense material; m, mitochondria; md, mitochondrial derivative; n, nucleus; nv, nuclear vesicle.

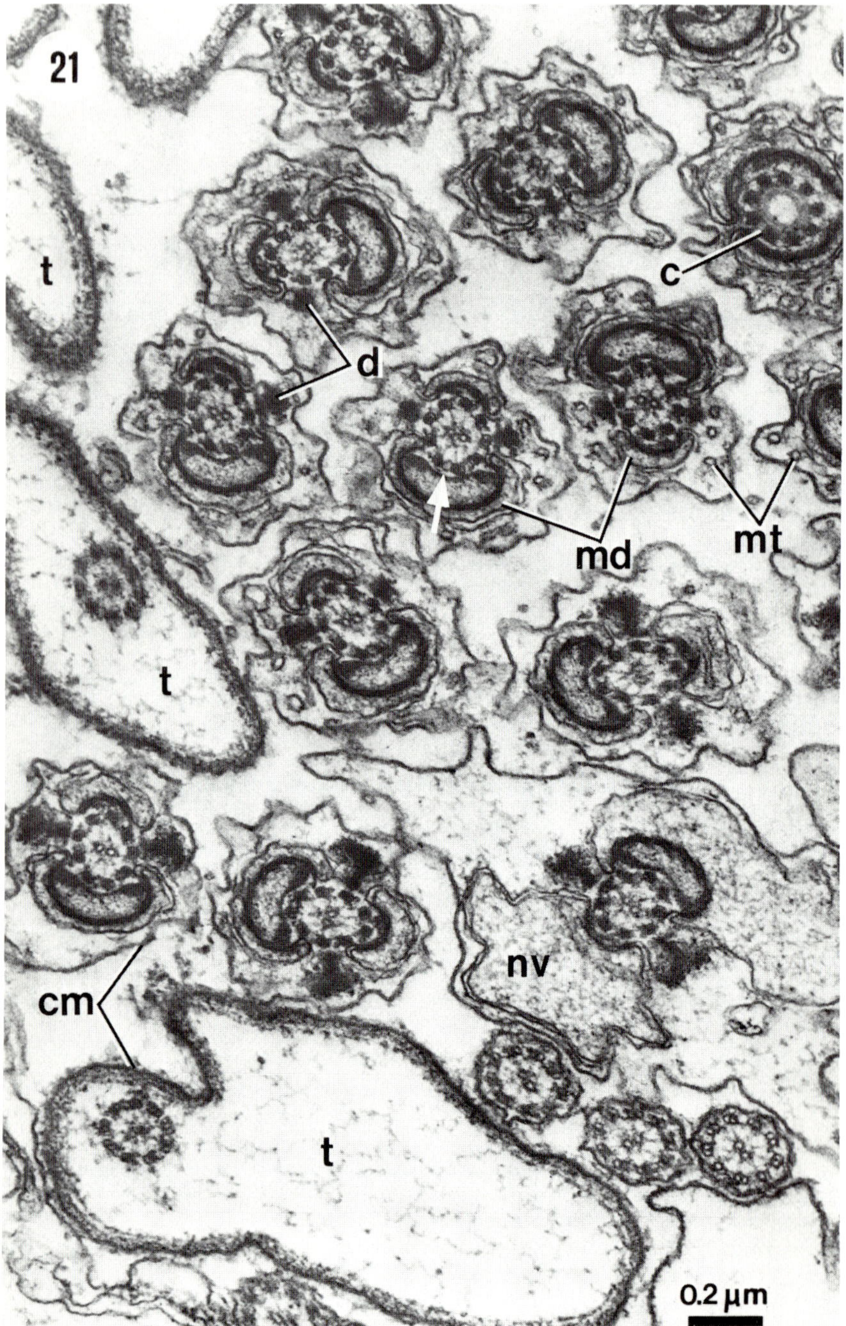
21
t
c
d
md
mt
t
cm
nv
t
0.2 µm

which during elongation is surrounded by approximately 60 microtubules. In the midpiece the mitochondrial derivatives form two straight rods flanking the axoneme (Figs. 22 and 23).

There is a definite relationship between the morphology of the spermatozoon and the biology of fertilization in the Metazoa in general (Franzén, 1956, 1970). The primitive type of spermatozoon is a small cell with a short, rounded or conical head, a midpiece containing a few mitochondria (generally four), and a tail consisting of a long flagellum. When released in water in which the animal lives, this primitive spermatozoon shows vigorous movements. This primitive spermatozoon is retained in groups and species with external fertilization in which gametes are shed directly into the water.

Animals which transfer their spermatozoa to the female either by copulation, pseudocopulation, or spermatophores possess sperm that differ from the primitive type of sperm by modifications in the mitochondrial midpiece and sometimes also in the head. The mitochondria often extend along the flagellum and are recognized and transformed in various ways.

In considering spermiogenesis and fertilization in the Bryozoa, the so-called pseudocopulation is perhaps of special interest. In pseudocopulation the gametes are shed close together in some way and often within a gelatinous cover.

The reason why we find a modified type of spermatozoon in many or most bryozoans seems to be the special type of fertilization that has been demonstrated by Silén to occur in *Electra* and probably in other gymnolaematous species as well. Fertilization takes place immediately outside the intertentacular organ in *Electra posidoniae* Gautier and in the intertentacular organ in *Electra crustulenta* (Pallas). The intertentacular organ seems to serve as a receptaculum seminalis in some of the species with this structure (Silén, 1966).

An explanation of the retention by *Bugula* of the primitive type of spermatozoon might be that the primitive biology of fertilization is retained in this species. It is, however, necessary to conduct a special examination of each separate species to elucidate the details in fertilization.

The condition that a few species within an animal group have developed modified spermatazoa, while the majority of the species have retained the

---

Fig. 21. Cross sections through midpieces and tails of spermatozoa or possibly late spermatids of *Triticella*. The mitochondria flanking the axoneme are reduced, in cross section, to crescentic mitochondrial derivatives. The larger mitochondrial rod is centered peripherally to doublets 5 and 6 (at arrows). The smaller rod is diametrically opposed. In the tail region the 9 + 2 axoneme and the cell membrane are seen. Near the bottom right corner of the figure three spermatid flagella are cut and in their axoneme it is seen that the nine peripheral double filaments carry arms (composed of dynein) on one of their microtubules. Abbreviations: c, centriole; cm, cell membrane; d, electron-dense material; md, mitochondrial derivative; mt, microtubules; nv, nuclear vesicle; t, tail.

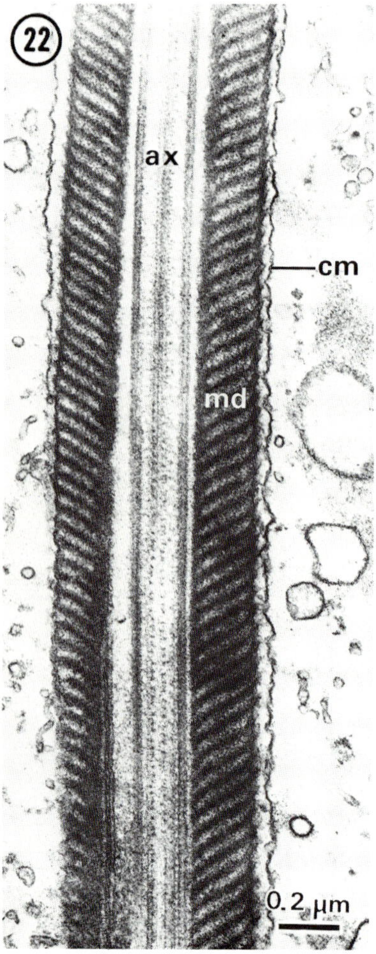

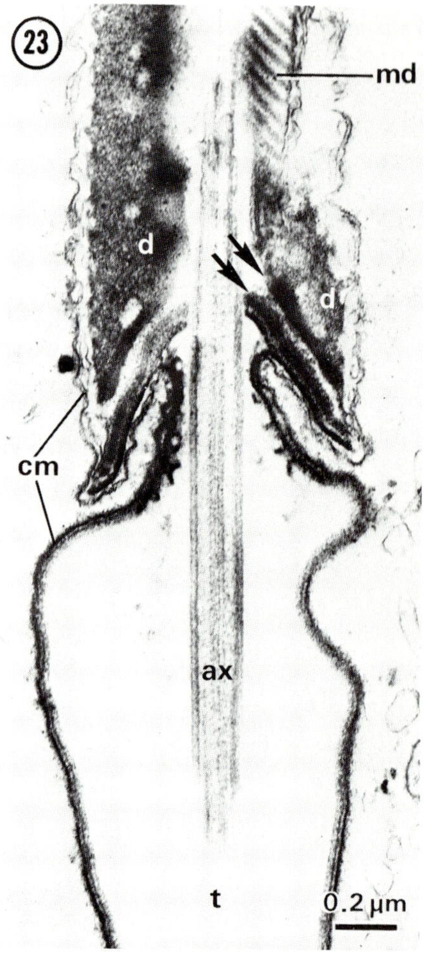

Fig. 22. Longitudinal section of midpiece of a *Flustra foliacea* spermatozoon showing the two rod-shaped mitochondrial derivatives flanking the axoneme (ax). Abbreviations: cm, cell membrane; md, mitochondrial derivatives.

Fig. 23. Longitudinal section through the junction between mitochondrial midpiece region and tail region of *Flustra foliacea* spermatozoon. In the midpiece region only a small part of the mitochondrial derivatives is seen. The granular electron-dense material (d) is seen on both sides of the axoneme. Dense material (at arrows) marks the end of the midpiece. Abbreviations: ax, axoneme; cm, cell membrane; d, electron-dense material; md, mitochondrial derivatives; t, tail.

primitive type, has been demonstrated for the polychaetous annelids (Franzén, 1975).

In the primitive type of spermatozoon the tip of the sperm head carries the acrosome, and this acrosome is the first part of the sperm to come in contact with the egg at fertilization. Sometimes the acrosome is composed of a few separate electron-dense bodies (Afzelius and Franzén, 1971). Grellet (1958) reports that the sperm head in *Alcyonidium gelatinosum* (L.) has a granular cover of Golgi material. He demonstrated this in material after fixation in Cajal and subsequent silver impregnation.

The absence of a typical acrosome thus seems to be a characteristic feature in the bryozoan spermatozoon (cf. Figs. 1–22). It may be admitted that in one species, *Monobryozoon limicola* Franzén, I earlier illustrated a sperm with an acrosome (Franzén, 1960). However, on reexamination of this material I have come to the conclusion that this was wrong. What I believed to be an acrosome in that spermatozoon was probably merely the anterior part of the cell membrane and nuclear membrane which had become free of nuclear material because of contraction.

It may be mentioned that the small vesicles I found near the nuclear envelope in *Triticella* may come from the Golgi body and correspond to a typical acrosome. Golgi bodies are found in the cytoplasm in all spermatid stages but further studies on their importance are needed.

C. Phylactolaemata

Among bryozoans the most aberrant spermatozoon is found in the class Phylactolaemata. Marcus (1934) states that in *Lophopus crystallinus* (Pallas) the mitochondria do not assemble into four spheres in the young spermatid, but instead constitute a single mass.

Figure 24a–c shows some stages in the differentiation of the spermatozoon in *Plumatella fungosa.* The stage shown in Fig. 24a is a spermatid from a cytophore of several spermatids. The nucleus is more or less spherical and is surrounded by cytoplasm with scattered mitochondria. A characteristic is that the distal end of the flagellum receives a cover of cytoplasm at an early stage, before the midpiece is formed. A similar sequence in development has been reported for *Lophopus* by Marcus (1934). Probably this cytoplasmic cover corresponds to the tubelike tail cover in the gymnolaemate spermatozoa described above. The mitochondria collect in a ringlike structure near the posterior pole of the nucleus where the flagellum grows out (Fig. 24b). Later the mitochondria extend backward along the flagellum (Fig. 24c).

The mature spermatozoon of *Plumatella* is relatively large and rich in cytoplasm (Fig. 24d). It consists of a pointedly conical head, a long midpiece with mitochondrial material, probably in a helical arrangement, and a long tail similar in structure to that in the Gymnolaemata. No acrosomal structure has been detected in the head region, but fine structural studies have yet to be made.

Early investigations, particularly by Kraepelin (1892), Retzius (1906), Braem (1908a,b), and Marcus (1941), indicated that the spermatozoon of Phylactolaemata is of a type which differs from that in Gymnolaemata. Braem described the external morphology of the mature spermatozoon in *Plumatella, Pectinatella,* and *Fredericella* and observed that these three agree in all essential points. He also made the important observation that the freshwater gymnolaemate *Paludicella* differs from freshwater phylactolaemates in sperm morphology and agrees instead with *Triticella*. Müller (1914), on the other hand, reports that he could not find any essential difference in sperm morphology between *Triticella*

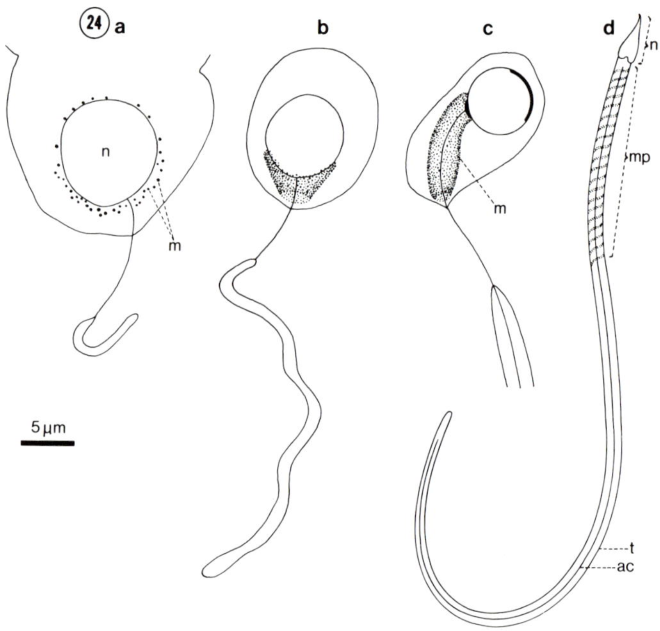

Fig. 24. Diagrammatic representation of successive stages of spermiogenesis in the Phylactolaemata (*Plumatella fungosa* Pallas). Drawings based on phase contrast microscopy of living material. (a) Young spermatid with a flagellum growing from the posterior pole. (b) Mitochondria appear tangentially to the caudal part of the nucleus. (c) Nuclear condensation begins in the anterior part of the nucleus. Mitochondria gather around the anterior part of the flagellum. (d) Spermatozoon. The nucleus has the form of a pointed cone, and the chromatin is highly condensed. The mitochondria have formed a helix in the midpiece. In the long tail region the axial filament complex is visible through the cell membrane. Abbreviations: ac, axial filament complex; m, mitochondrion; mp, midpiece; n, nucleus; t, tail. Modified from Franzén (1970).

and *Plumatella,* but this conclusion probably is based on insufficiently detailed studies.

Studies of living spermatozoa with phase contrast microscopy (Franzén, 1970) have confirmed the observations by Retzius, Braem, and others; there is no doubt that great dissimilarity exists between the Phylactolaemata, on one hand, and the Stenolaemata and the Gymnolaemata, on the other. These differences refer to the morphology of both the head and the midpiece. The functional significance of these differences is not fully understood at present. Whether the habitat is freshwater or marine does not seem to have any direct importance as the freshwater form *Paludicella* has the same type of spermatozoon as other gymnolaemata, with the exception of *Bugula.*

REFERENCES

Afzelius, B. (1959). Electron microscopy of sperm tails. *J. Biophys. Biochem. Cytol.* **5,** 269–278.

Afzelius, B., and Franzén, Å. (1971). The spermatozoon of the jellyfish *Nausithoë. J. Ustrastruct. Res.* **37,** 186–199.

Bonnevie, K. (1907). Untersuchungen über Keimzellen. II. Physiologische Polyspermie bei Bryozoen. *Jena. Z. Naturwiss.* **42,** 567–598.

Borg, F. (1926). Studies on recent cyclostomatous Bryozoa. *Zool. Bidr. Uppsala* **10,** 181–507.

Braem, F. (1908a). Die Spermatozoen der Süsswasser-Bryozoen. *Zool. Anz.* **32,** 671–673.

Braem, F. (1908b). Die Spermatozoen von *Paludicella* und *Triticella. Zool. Anz.* **33,** 380–381.

Fawcett, D. W. (1972). Observations on cell differentiation and organelle continuity in spermatogenesis. Proc. Int. Symp. Genet. Spermatozoon, Edinburgh, 1971, pp. 37–68.

Fawcett, D. W., Anderson, A., and Phillips, D. M. (1971). Morphogenetic factors influencing the shape of the sperm head. *Dev. Biol.* **26,** 220–251.

Franzén, Å. (1956). On spermiogenesis, morphology of the spermatozoon, and biology of fertilization among invertebrates. *Zool. Bidr. Uppsala* **31,** 355–482.

Franzén, Å. (1960). *Monobryozoon limicola* n.sp., a ctenostomatous bryozoan from the detritus layers on soft sediment. *Zool. Bidr. Uppsala* **33,** 135–148.

Franzén, Å. (1970). Phylogenetic aspects of the morphology of spermatozoa and spermiogenesis. *In* "Comparative Spermatology" (B. Baccetti, ed.), pp. 29–46. Academic Press, New York.

Franzén, Å. (1975). Sperm ultrastructure in some Polychaeta. *In* "The Functional Anatomy of the Spermatozoon" (B. Afzelius, ed.), pp. 267–278. Pergamon, Oxford.

Franzén, Å. (1976). On the ultrastructure of spermiogenesis of *Flustra foliacea* (L.) and *Triticella koreni* G. O. Sars (Bryozoa). *Zoon* **4,** 19–29.

Furieri, P. (1970). Sperm morphology in some reptiles: Squamata and Chelonia. *In* "Comparative Spermatology" (B. Baccetti, ed.), pp. 115–131. Academic Press, New York.

Gerwerzhagen, A. (1913). Untersuchungen an Bryozoen. *Sitzungsber, Heidelb. Akad. Wiss., Math.-Naturwiss Kl. B.* **4,** No. 9, 1–16.

Grellet, P. (1958). Histologie du testicule et spermatogénèse chez *Alcyonidium gelatinosum* (L.) (Bryozoaire ctenostome). *Bull. Lab. Marit. Dinard* **43**, 1–24.

Kraepelin, K. (1892). Die Deutschen Süsswasser-Bryozoen. II. Entwicklungsgeschichtlicher Teil. *Abh. Naturwiss. Hamburg* **12**, 1–67.

Marcus, E. (1934). Über *Lophopus crystallinus* (Pall.). *Zool. Jahrb., Abt. Anat. Ontog. Tiere* **58**, 501–606.

Marcus, E. (1941). Sobre Bryozoa do Brasil. *Univ. Sao Paulo Fac. Filos., Cienc. Let., Bol. Zool.* **22**, No. 5, 3–208.

Müller, A. (1914). Histologie des Darmtraktes und Spermatogenese der *Plumatella polymorpha* Krpl. *Festschrift Siebenburg. Ver. Naturwiss. Hermannstadt* **65**, 34–48.

Prouho, H. (1892). Contribution à l'histoire des Bryozoaires. *Arch. Zool. Exp. Gén.* [2] **10**, 557–656.

Reger, J. F. (1971). A fine structure study on spermiogenesis in the ectoproct, *Bugula* sp. *J. Submicrosc. Cytol.* **3**, 193–200.

Retzius, G. (1904). Zur Kenntnis der Spermien der Evertebraten. *Biol. Unters. [N.S.]* **11**, No. 1, 1–32.

Retzius, G. (1906). Die Spermien der Enteropneusten und der Nemertinen. Die Spermien der Turbellarien. Die Spermien der Bryozoen. *Biol. Unters. [N.S.]* **11**, Nos. 2, 3, and 4, 37–48.

Retzius, G. (1910). Zur Kenntnis der Spermien der Echinodermen. Weitere Beiträge zur Kenntnis der Spermien mit besonderer Berücksichtigung der Kernsubstanz. *Biol. Unters. [N.S.]* **15**, Nos. 2 and 3, 54–82.

Ryland, J. S. (1958). Embryo colour as a diagnostic character in Polyzoa. *Ann. Mag. Nat. Hist.* [13] **1**, 552–556.

Silén, L. (1944). Anatomy of *Labiostomella. K. Sven. Vetenskapsakad. Handl.* [3] **21**, 1–111.

Silén, L. (1945). Main features of development of the ovum, embryo, and ooecium in the ooeciferous Bryozoa Gymnolaemata. *Ark. Zool.* **35**, No. 17, 1–34.

Silén, L. (1966). On the fertilization problem in the gymnolaematous Bryozoa. *Ophelia* **3**, 113–140.

Zimmer, R. L., and Woollacott, R. M. (1974). Morphological and biochemical modifications of the spermatozoan mitochondria of *Membranipora* (Bryozoa). *J. Cell Biol.* **63**, 385a.

2

Brooding Patterns of Bryozoans

ROLF STRÖM

I. INTRODUCTION

The commonest division of the phylum Bryozoa is into three classes: Phylactolaemata, Stenolaemata, and Gymnolaemata (Ryland, 1970). The Phylactolaemata live exclusively in freshwater. It is a well-defined group, both regarding morphology and mode of reproduction. Of the Stenolaemata only one order, the Cyclostomata, has survived to the present time. This is an exclusively marine group. The Gymnolaemata are mainly marine and are usually divided into two orders: Ctenostomata and Cheilostomata. Some authors unite these into one group, the Eurystomata or Cheilo-ctenostomata (Silén, 1942). Within the Gymnolaemata there exist wide variation in both general morphology of the zooids and their manner of reproduction.

Despite the fact that most bryozoans are hermaphroditic, cross-fertilization does occur (Silén, 1966). Silén (1972) believes this is the common way of fertilization in the Gymnolaemata. Fertilization occurs the moment the oocyte leaves the coelomic cavity of the zooid either through a supraneural pore or an intertentacular organ. Thus the oocyte leaves the parent zooid before fertilization and separation of the polar bodies. Before leaving the zooid the oocytes float in the coelom, and the nucleus is at the first meiotic division of maturation (Braem, 1951; Ström, 1969).

Concerning reproduction of the Metazoa, it is generally considered that release of ova into the surrounding water and subsequent development in the plankton is the primitive pattern (Jägersten, 1972). Other primitive features include mass production of eggs and external fertilization in combination with the release of ova. The ova have small amounts of yolk, and the cleavage is total and usually equal. The larvae that develop are planktotrophic.

Secondary conditions occur when the zygote is retained at, or in, the parent animal. The number of ova is reduced and internal fertilization usually takes place. Ova with larger amounts of yolk cleave unequally or partially, and the resulting larvae are lecithotrophic. In connection with internal fertilization the sperm changes from a generalized to a specialized type (Franzén, 1956). This general pattern can also be applied to the reproductive conditions in the Bryozoa.

Brooding of lecithotrophic larvae occurs in most bryozoans. Groups solely with this type of brooding are the Phylactolaemata and the recent group within the Stenolaemata, i.e., Cyclostomata. Brooding also dominates in the Gymnolaemata, but development in seawater occurs in some species within both the Ctenostomata and the Cheilostomata. These forms have a pelagic, planktotrophic larva known as a cyphonautes. It is, therefore, most logical to begin with the group Gymnolaemata.

II. GYMNOLAEMATA

A. Embryology

The embryonic development has been described by many authors, e.g., Barrois (1877), Prouho (1890, 1892), Calvet (1900), Pace (1905), Marcus (1926, 1938, 1941), Côrrea (1948), Mawatari (1952), and Ström (1969). The early development is largely uniform both in forms with release of ova into the surrounding water and in those which brood their embryos. The development of *Triticella koreni* may serve as an example (Ström, 1969). When the oocyte leaves the zooid, the nucleus is in the first metaphase of meiosis. Fertilization probably occurs at egg emergence, whereupon the membrane starts to lift up irregularly from the surface (Plate I, A,B). The polar bodies then separate (Plate I, B), and cleavages which are total and equal begin (Plate I, C–E). The cleavages are radial

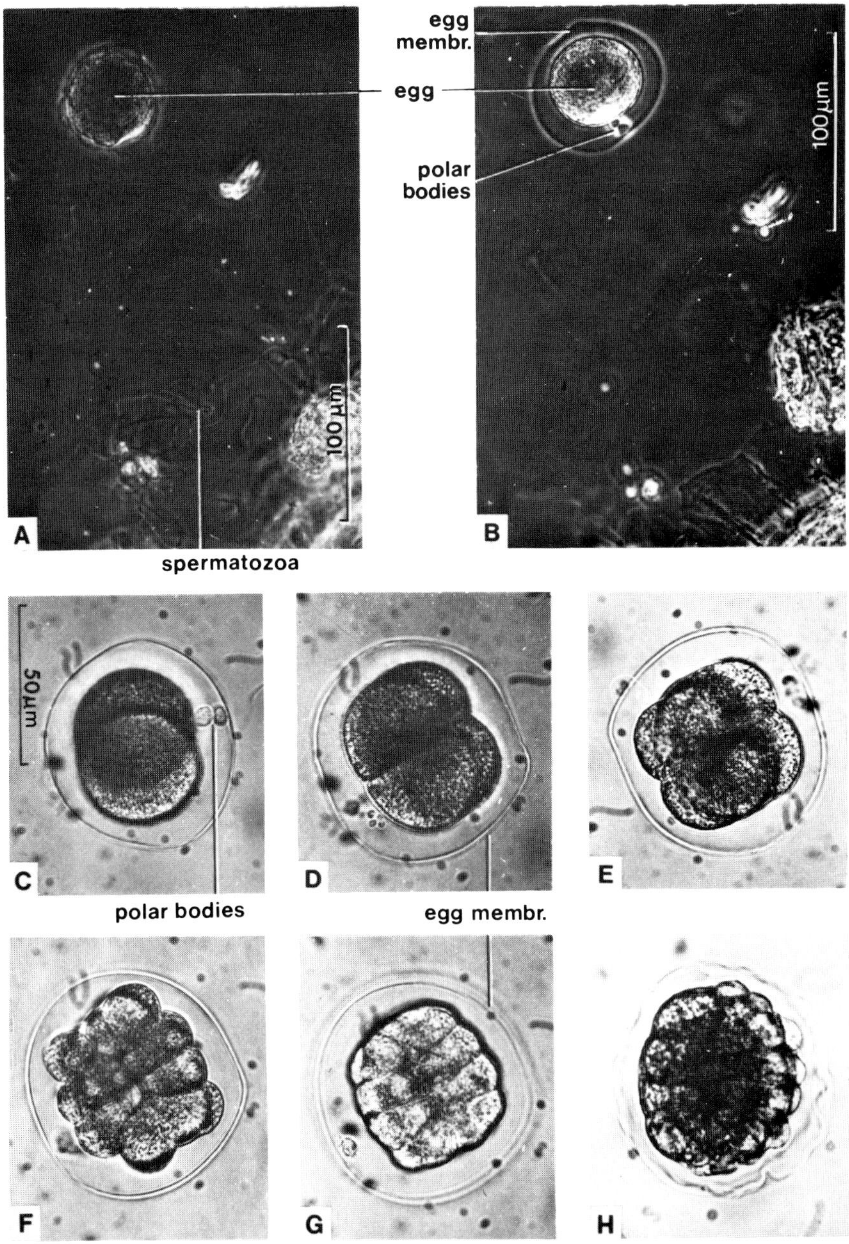

Plate I. Maturation division and embryos in *Triticella koreni.* Photographed alive by Ström (1969). (A) Egg just leaving the zooid and the egg membrane lifting up, irregularly (phase contrast); (B) membrane entirely uplifted, the polar bodies are separated (phase contrast); (C) 2-cell stage; (D) 4-cell stage; (E) 8-cell stage; (F) 16-cell stage; (G) 32-cell stage, morula; (H) 64-cell stage, blastula.

and, in the 64-cell stage, result in a blastula (Plate I, F–H). Gastrulation has been described by Prouho (1892) in embryos with small amounts of yolk, and by Pace (1906) and Côrrea (1948) in embryos richer in yolk.

In some forms with yolk-rich eggs the cleavages may be somewhat unequal and result in larger cells at the vegetal pole. This is the case with *Flustrellidra hispida* (Bassler) (Pace, 1906). Four cells at the vegetal pole are larger and richer in yolk (Fig. 1a,b). Gastrulation begins when these four cells divide off their inner halves into the blastocoel. The process is the same in embryos with less yolk. In the 64-cell stage the four cells are divided horizontally (Figs. 1c, 2a, and 3a). The inner cells continue to divide as they move into the blastocoel and eventually form a compact entoderm (Fig. 1d). Prouho (1892) has observed in *Alcyonidium albidum* a cell group in front of the stomodaeum which developed into muscle tissue in the larva (Fig. 3b). A special ectodermal ring of cells with very long cilia forms the corona, which is the larva's swimming organ (Figs. 2b and 4a). In planktotrophic larvae, and even some lecithotrophic ones, the corona has a limited extent and is situated orally (Fig. 3c). In most lecithotrophic larvae it is more extensive and in some, e.g., larvae of *Bugula neritina* described by Woollacott and Zimmer (1971), it covers most of the larva's surface (Fig. 5).

At the same time as the corona is being formed, certain of the ectodermal cells are enlarged at the animal pole and form an apical organ, which in some species possesses long sensory cilia (Figs. 3c and 4b). When the corona and apical organ are fully developed the embryo in some forms now escapes from the egg membrane (Fig. 3c). In other forms the embryo remains for a longer period of time in the egg membrane. In nonbrooding forms or those in which the egg

---

Fig. 1. Early development of *Flustrellidra hispida.* Redrawn from Pace (1906). (a) 32-cell stage from below; (b) 32-cell stage, median section; (c) 64-cell stage, median section, the four vegetative cells have divided horizontally; (d) gastrula, median section, with entoderm cells.

Fig. 2. Embryo stages of *Bugula flabellata.* Modified from Corrêa (1948). (a) Median section of a 64-cell stage; (b) median section through a young larva. Abbreviations: ao, apical organ; as, adhesive sac; c, corona; nm, neuromusculature; po, pyriform organ.

Fig. 3. Embryo stages of *Alcyonidium albidum.* Redrawn from Prouho (1892). (a) 64-cell stage; (b) embryo after gastrulation: ao, apical organ; c, corona; en, entoderm, m, mesoderm; (c) embryo leaves the embryo membrane: mb, membrane; (d) young larva with an organization like a cyphonautes with a functional alimentary canal; nm, neuromusculature; po, pyriform organ; s, stomach, v, vestibule.

Fig. 4. Embryo stages of *Electra pilosa.* Redrawn from Prouho (1892). (a) Young embryo: ao, apical organ; c, corona; mb, membrane; (b) embryo after leaving the membrane; en, entoderm; m, mesoderm; (c) young cyphonautes: as, adhesive sac; oe, oesophagus; po, pyriform organ; r, rectum; s, stomach; sh, shell; v, vestibule.

Fig. 5. Lecithotrophic larva of *Bugula neritina.* Redrawn from Woollacott and Zimmer (1971). Abbreviations: ao, apical organ; as, adhesive sac; c, corona; ct, ciliary tuft; nm, neuromusculature; po, pyriform organ.

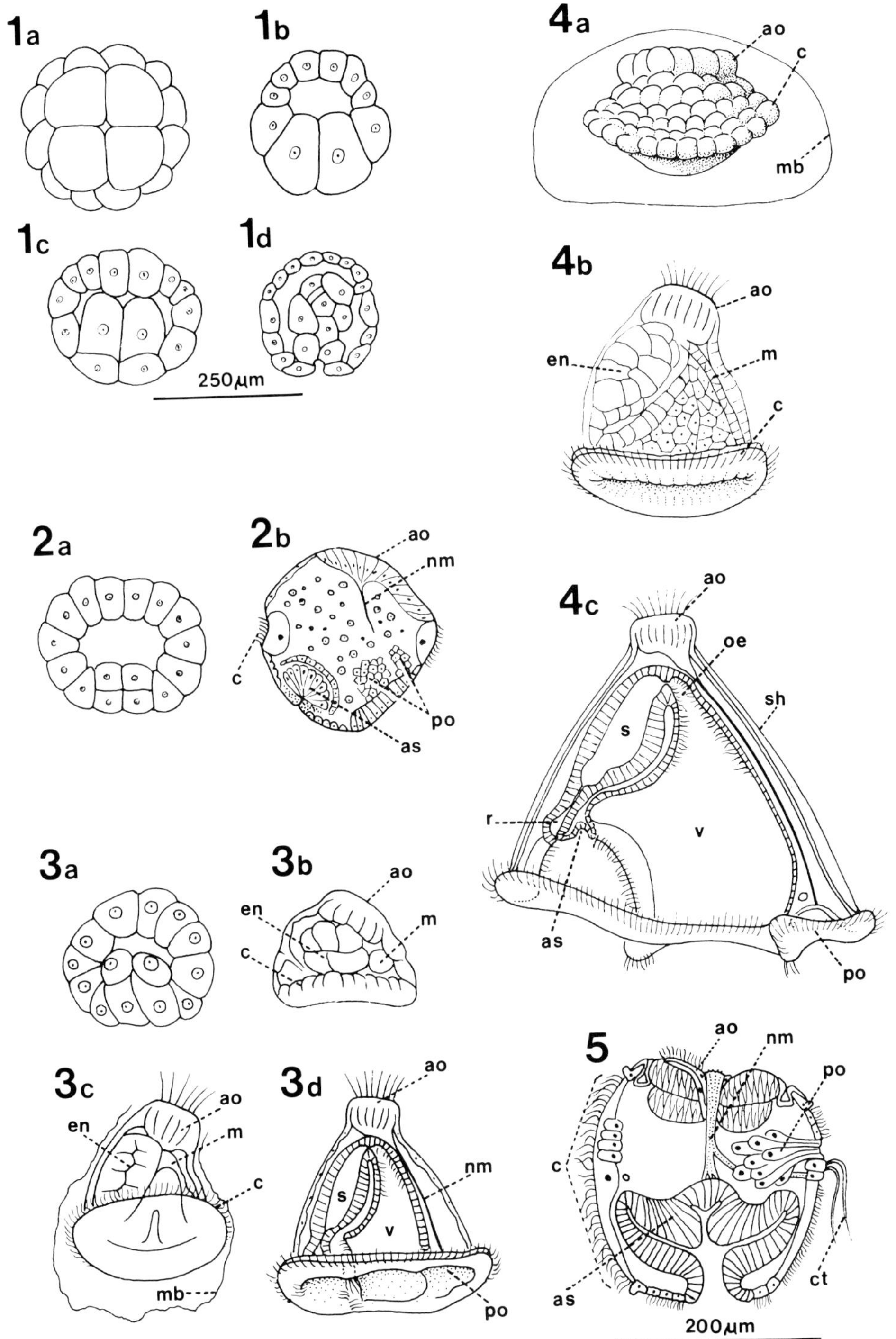

1a
1b
1c
1d
250µm
2a
2b
ao
nm
c
po
as
3a
3b
ao
en
m
c
3c
en
ao
m
c
mb
3d
ao
s
nm
v
po
4a
ao
c
mb
4b
ao
en
m
c
4c
ao
oe
sh
s
r
v
as
po
5
ao
nm
po
c
ct
as
200µm

membrane adheres to the outside of the parent zooid, the embryo after escaping the egg membrane, swims freely, and the rest of its development to a larva takes place in the water. In forms which brood, the embryo remains in the brood chamber until the larva is fully developed. The embryo usually remains in the brood chamber for about 14 days (Silén, 1945).

Just before or after the escape of the embryo from the egg membrane an ectodermal invagination is formed near the oral pole in front of the stomodaeum (Figs. 2b and 3d). This is known as the pyriform organ and contains glandular cells which form an aboral or "superior" glandular field and ciliated cells which form an elongated ciliated groove orally. In most larvae, especially long cilia, the vibratile plume or ciliary tuft, are formed near the aboral end of the organ (Fig. 5). The function of the pyriform organ has long been discussed. It is probably a sensory and locomotory organ used when the larva is seeking a suitable substrate for attachment (Jägersten, 1972). A strand of both muscular and neural tissue (Figs. 3d and 4b,c) develops between the apical organ and the pyriform organ.

An ectodermal invagination of very tall, secretion-filled cells is formed posteriorly at the oral side (Figs. 2b and 4c). This invagination forms the adhesive sac (Fig. 5). The adhesive sac is everted at metamorphosis and attaches the larva to the substrate.

The embryos that continue their development in the water usually grow into planktotrophic cyphonautes larvae with a functional digestive tract (Figs. 3d and 4c). In some species the differentiation of the gut is incomplete and only rudiments of an alimentary canal are formed. In typical brooding forms, the larvae are lecithotrophic and lack a functional digestive tract (Fig. 5). The morphology of the mature larva and different types of larvae are discussed in Chapter 3.

B. Brooding in Ctenostomata

Among the ctenostomes only three species have been described as undergoing larval development in the water and presumably possessing planktotrophic larvae: *Alcyonidium albidum* within Carnosa (Prouho, 1892); *Hypophorella expansa* (Ehlers) within Stolonifera (Prouho, 1892), and *Farella repens* (Farre) within Stolonifera (Marcus, 1926). No form of brooding seems to occur in these species and the larvae are of the cyphonautes type but are not typically developed. The combination of development in the water and planktotrophic cyphonautes larvae otherwise occurs in the Bryozoa in only a few species of cheilostomes.

In some other species of ctenostomes a simple form of brooding takes place. The simplest is in *Triticella koreni* where only a short period of the embryonic development occurs in association with the parent animal (Ström, 1969). When the egg has left the coelomic cavity the embryonic membrane adheres to the parent zooid. About twenty embryos can be attached at the same time (Fig. 6), but there are generally fewer. *Triticella koreni* lives epizooically on the crustacean *Calocaris macandreae,* which burrows in clay bottoms, and consequently the eggs which do not fasten to the parent probably die in the clay. Only the earliest development takes place within the embryonic membrane. After gastrulation, when the embryo has become ciliated and is capable of movement, the embryo escapes. The rest of the development to a larva takes place in the water. The larva is probably lecithotrophic, but has rudiments of an alimentary canal and an appearance similar to that of planktotrophic larva. Consequently, the brooding is the shortest possible for the embryos to survive. In addition, reproduction of the bryozoan is synchronized with molting in *Calocaris,* which enables it to colonize the new shell, maintaining the species.

A similar form of brooding was described by Braem (1896) in *Paludicella articulata,* which is a freshwater species belonging to Carnosa, superfamily Paludicelloidea. The cleavages commence when the eggs leave the parent animal. The uplifted membrane sometimes adheres to the cuticule of the parent zooid. It takes three days for the embryo to develop into a gastrula, whereupon it emerges from the egg membrane. Braem presumed the larva to be fully developed at this stage, but judging from his illustration (Braem, 1896; Fig. 4), it is still only partially developed.

The commonest form of brooding in the Ctenostomata is that in which the embryos are retained in the vestibule or in the tentacle sheath. The tentacle sheath is the part of the animal's body wall that encircles the tentacles when the polypide is withdrawn. On strong retraction of the polypide, an additional portion of the body wall is introverted, establishing a space called the vestibule. This is separated from the cavity of the tentacle sheath by an atrial sphincter or diaphragm.

There are transitional forms between those which have embryos fastened externally and those with brooding within the introvert. One of these is *Bulbella abscondita* (Braem), which is closely related to *Paludicella.* The egg is released through an intertentacular organ and adheres to the outside of the polypide below the diaphragm (Fig. 7a,b). When the polypide is extended the embryos are outside the zooid, but when it retracts they are pulled into the vestibule (Braem, 1951) (Fig. 7c). Braem considered that, phylogenetically, this situation is more primitive than that in which the embryo sinks into the body wall (see below). The embryos are usually four in number, maximally six. Braem also observed that if the eggs were separated from the parent zooid they failed to develop. The larva of *Bulbella,* although lecithotrophic, has a relatively primitive appearance (cf. Jägersten, 1972) with a subequatorial corona and a rudimentary digestive tract. Brooding in *Bulbella* might be indicative of how brooding in the introvert originated.

6

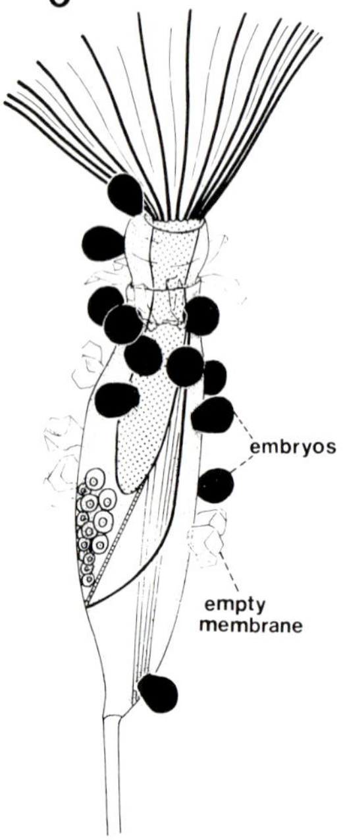

7

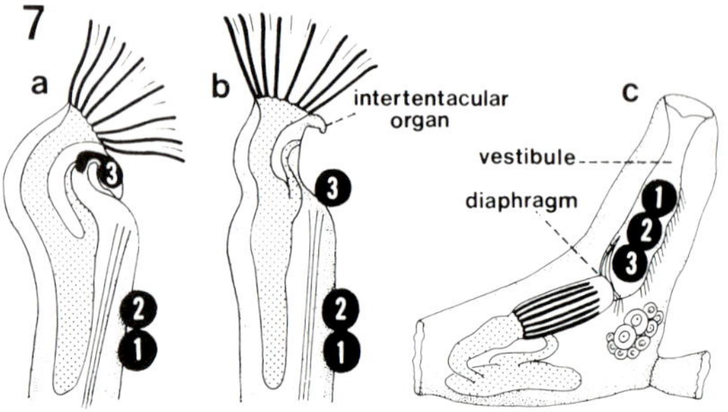

8

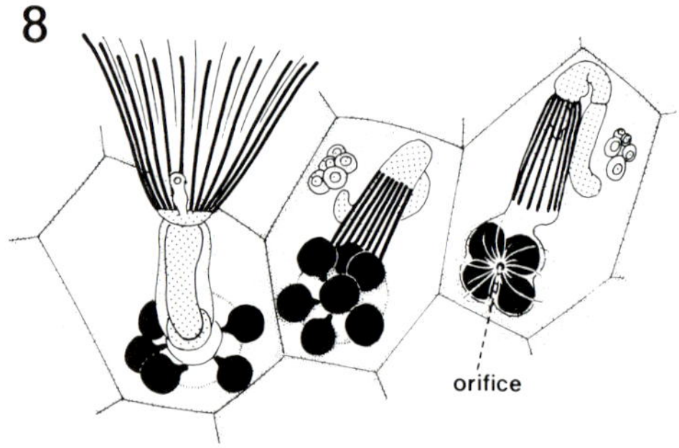

Fig. 6. Fertile zooid of *Triticella koreni* with embryos adhered to the outside.

Fig. 7. Emergence of eggs and adhesion to the body wall of *Bulbella abscondita*. Redrawn from Braem (1951). (a) Emergence of third egg; (b) embryos at the outside of the polypide; (c) embryos in the vestibule when the polypide is retracted.

Fig. 8. *Alcyonidium duplex* with embryos adhered to the diaphragm. Redrawn from Prouho (1892).

A similar reproductive pattern is found in *Alcyonidium duplex,* belonging to Carnosa, superfamily Halcyonelloidea (Prouho, 1892). After the egg is released through the intertentacular organ it adheres to what Prouho calls the diaphragm, i.e., the boundary between the tentacle sheath and the vestibule. The embryos are outside the zooid when the polypide is extended, but are drawn into the vestibule when the polypide is retracted (Fig. 8). *Alcyonidium albidum* has development in the water, although other species in the genus have development in the introvert (e.g., *A. polyoum*). The lecithotrophic larva of *A. duplex* has a subequatorial corona and a well-defined apical organ.

Most of the species within Ctenostomata, for which reproduction is known, have brooding within the introvert, that is to say, either in the vestibule, in the tentacle sheath, or in a brood pouch between them. *Flustrellidra hispida* within

Carnosa has been thoroughly investigated by Prouho (1890) and Pace (1906). They demonstrated that the embryos develop in the tentacle sheath. Up to five embryos develop simultaneously while the polypide degenerates. The fully developed larvae finally emerge through the orifice. The larvae are cyphonautes-like, but are lecithotrophic. *Pherusella tubulosa* (Ell. Soll.), closely related to *Flustrellidra,* has the same form of brooding and similar larvae (Prouho, 1892).

As mentioned earlier, there is a similar type of brooding within the genus *Alcyonidium.* The species investigated in this respect are *A. polyoum* and *A. variegatum. Alcyonidium polyoum* has been studied by, among others, Prouho (1892), Seeliger (1906), and Matricon (1963). Prouho and Seeliger considered that development took place in the tentacle sheath. Matricon later demonstrated that the embryos develop in a "brood pouch" (poche incubatrice) between the vestibule and the degenerated tentacle sheath (Fig. 9). The polypide degenerates first, whereupon the brood pouch develops. It eventually contains 4–6 embryos which probably enter the pouch via a ciliated funnel or intertentacular organ (Fig. 9). Here they develop into lecithotrophic larvae, which swarm out through the peristome when they reach maturity.

In the suborder Stolonifera the embryo develops within the introvert, while the polypide degenerates. This situation occurs mainly within the superfamily Vesicularioidea. Calvet (1900) noted brooding in the tentacle sheath of *Amathia lendigera* (L.), *Vesicularia spinosa* (L.), and *Bowerbankia pustulosa* (Ell. Soll.).

9

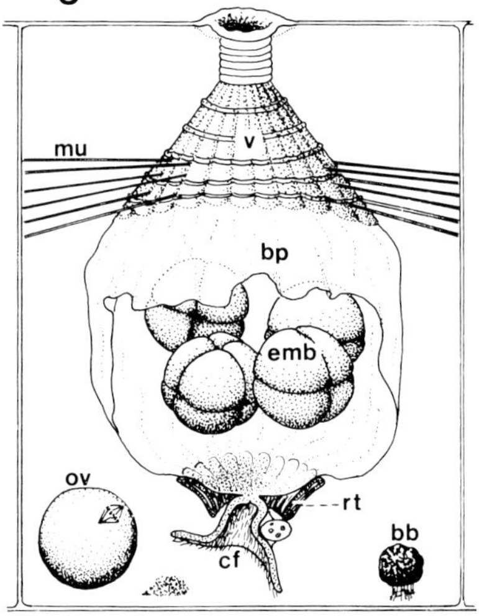

Fig. 9. Schematic figure of "brood pouch" in *Alcyonidium polyoum.* Redrawn from Matricon (1963). Abbreviations: bb, brown body; bp, brood pouch; cf, ciliary funnel; emb, embryo; mu, muscles; ov, ovum; rt, rudimentary tentacle sheath; v, vestibule.

Zirpolo (1933) reported a similar brooding pattern in *Zoobothryon verticillatum* (Della Chiaje) = *Z. pellucidum* (Vigelius). During its development the embryo is pushed progressively upward in the zooid; the larva will be positioned in a swelling just under the orifice before it leaves the zooid. Waters (1914) has interpreted this, probably in error, to mean that the embryo rests in an outer embryo sac. Braem (1951) showed that the embryo develops in the vestibule in *Bowerbankia gracilis* (Leidy) = *B. caudata* (Hincks). The polypide degenerates at the same time as the egg grows larger (Fig. 10a). The egg is pressed through a pore in the rudiment of the tentacle sheath of the developing replacement polypide and continues through the diaphragm to the vestibule (Fig. 10b). The larva remains in this position until it is fully developed, whereupon it emerges through the orifice. In these species only one larva develops at a time. Additional investigation of this form of brooding is needed to enable comparison with that described in *Alcyonidium polyoum* (Matricon, 1963).

According to Soule (1950a) and Bobin and Prenant (1954), the burrowing form *Terebripora comma* (Soule) undergoes brooding in the tentacle sheath.

Brooding in the introvert is also found in the other superfamily within Stolonifera, the Valkerioidea. According to Joliet (1877), *Valkeria uva* has brooding in the tentacle sheath. A single larva is developed which hatches when it is mature. Banta (1968) demonstrated brooding in the tentacle sheath in

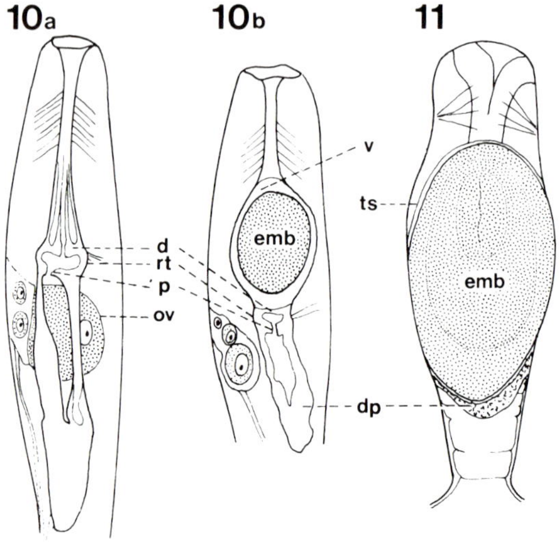

Fig. 10. Egg entry into the vestibule of *Bowerbankia gracilis.* Redrawn from Braem (1951). (a) Stage before egg entry: d, diaphragm; ov, ovum; p, pore in the tentacle sheath; rt, rudimentary tentacle sheath; (b) embryo in the vestibule: dp, degenerated polypide; emb, embryo; v, vestibule.

Fig. 11. Brooding in the tentacle sheath of *Mimosella cookae.* Modified from Banta (1968). Abbreviations: emb, embryo; ts, tentacle sheath.

Mimosella cookae (Banta). Only one embryo develops and it fills most of the zooid (Fig. 11). Within Valkerioidea there are also forms with release of eggs into the water (*Farella, Hypophorella*) and others with a simple form of brooding (*Triticella*).

A somewhat different form of brooding in Ctenostomata is that in which the embryo is found in a sac or pouch attached to the body wall so it protrudes into the coelomic cavity. Examples of how the egg leaves the zooid and sinks into the body wall were given by Braem (1951) for *Victorella (Tanganella) muelleri* (Kraepelin), which is closely related to *Paludicella.* The oocyte emerges via a supraneural pore near the anus as the polypide bends ventrally and adheres to a protuberance of the body wall (Fig. 12a–d). This thickening of the wall is below the diaphragm on the anal side and has muscle fibers fastened to it. The zygote sinks into this thickened part (Fig. 12e–g) within a few minutes. Presumably fertilization and maturation divisions occur during this time. When the polypide is retracted the embryo will be placed in the vestibule as in *Bulbella,* with the difference that in *Victorella* the embryo is placed in the invaginated wall. Several embryos, normally three, may be placed in a row, with the youngest furthest in, when the polypide is retracted (Fig. 12h). The polypide subsequently degenerates, later to grow out again after the larvae have left their embryo sacs. The larva emerges from its sac when the enveloping epithelium bursts.

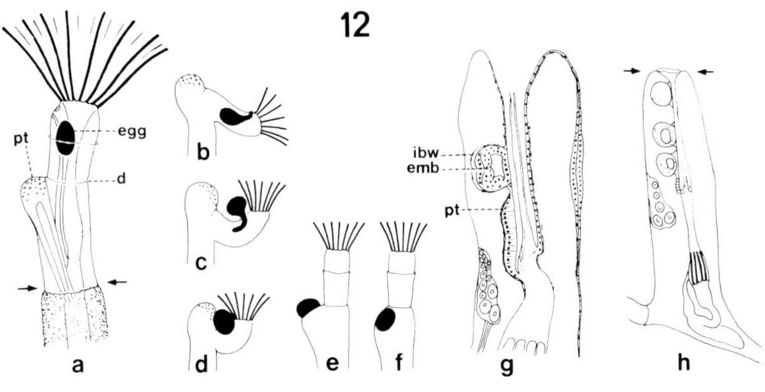

Fig. 12. Emergence of the egg in *Victorella muelleri* and the zygote sinking down into the body wall. Redrawn from Braem (1951). (a) An egg in the upper part of the extended polypide: d, diaphragm; pt, protuberance; (b) the polypide bends orally and the egg begins to emerge; (c) egg emerging through the supraneural pore; (d) egg resting between the protuberance and the curved polypide; (e) egg sinks into the body wall; (f) egg or zygote has sunk into the body wall; (g) embryo lying in the invaginated body wall of the vestibule when the polypide is retracted: emb, embryo; ibw, invaginated body wall; (h) three embryos in the vestibule.

A similar process takes place in the genus *Nolella,* also belonging to the Paludicelloidea. Brooding in *Nolella (Cylindroecium) dilatata* (Hincks) was described by Prouho (1892). According to Prouho the egg adheres to the inside of the coelomic wall and later emerges to the exterior through a rupture in the body wall (Fig. 13a–c). This may, however, be another example in which embryo sacs have been invaginated from the outside, as in *Nolella papuensis* (Busk) studied by Harmer (1915) and Silén (1944). Harmer described the process as being similar to that in *N. dilatata* but his illustrations show that the large egg is enveloped in a membrane that may be an invaginated body wall (Fig. 14a), and he has even sketched zooids with eggs adhered to the outside (Fig. 14b). Silén later clearly showed that the embryo is enclosed in an embryo sac in *N. papuensis.* Its embryo is enveloped in a fairly thick invaginated epithelium (Fig. 15). The polypide degenerates during the brooding. The embryo and the embryo sac are situated on the anal side as in *Victorella muelleri.* The larva probably emerges through the invagination point, and the epithelium is then possibly extended, as observed by Prouho. Silén (1944) considered that Prouho and Harmer had misinterpreted their observations, as from the outside it appears as if the embryo floats freely in the coelom. The smaller eggs observed by them in the coelom were probably unfertilized eggs. Each sac contains only one egg which has probably been transported there from the outside. It is possible that several embryos can develop simultaneously as has been observed in *N. dilatata* (Prouho, 1892) and in *N. gigantea* (Marcus, 1938).

Another species with this type of brooding is *Sundanella sibogae* (Harmer), which is closely related to *Nolella* (Braem, 1940). It has an embryo sac in the form of an invagination of the body wall on the anal side (Fig. 16a). There are two cell layers that are invaginated, the ectodermal layer and the mesodermal coelomic lining. The embryo and the embryo sac gradually fill up the entire zooid as the polypide degenerates (Fig. 16b). Braem did not observe how the egg entered the embryo sac, but it probably became embodied in the wall from the outside, as in the case with *V. muelleri.* The larva probably emerges through the outer part of the embryo sac. As the development in Phylactolaemata is suggestive of that in *Sundanella,* Silén (1944) considered that there may be a relationship here, but Jebram (1973) held that there is no such connection.

In the burrowing form *Penetrantia* the embryo is situated between two layers of cuticle (Silén, 1947). In *P. densa* (Silén) there is a small opening (Fig. 17a). Silén (1948) later described an outer embryo sac in *P. concharum* (Silén) that appears to be closed (Fig. 17b), and Soule (1950b) has observed the same in *P. siléni* (Soule). The outer cuticle could possibly correspond to a thickened embryonic membrane.

The descriptions above show that Ctenostomata have several modes of brooding as well as larval development. The forms of brooding range from simple adhesion of the zygotes on the outside of the zooid to brooding in the introvert

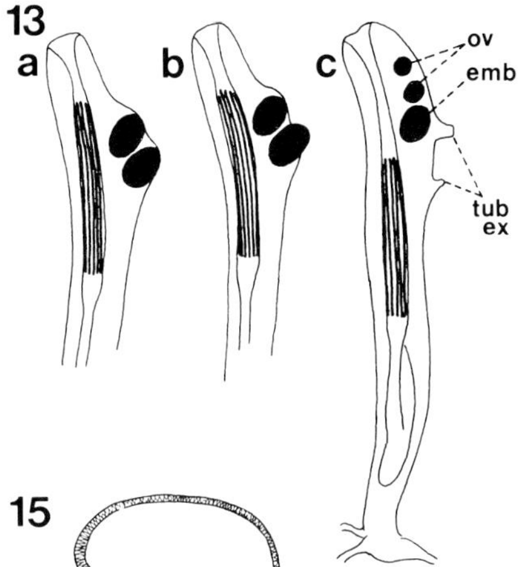

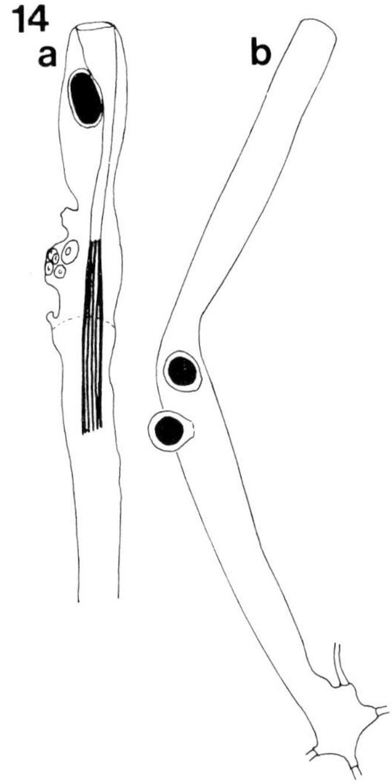

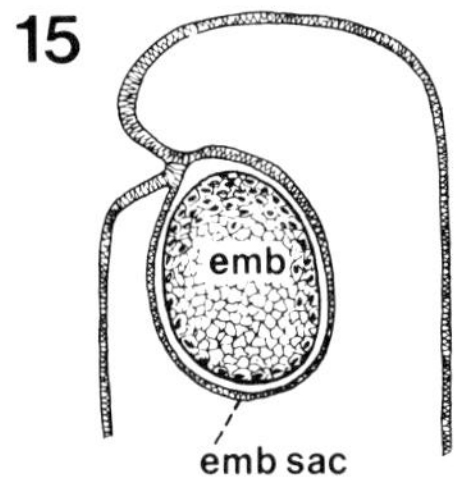

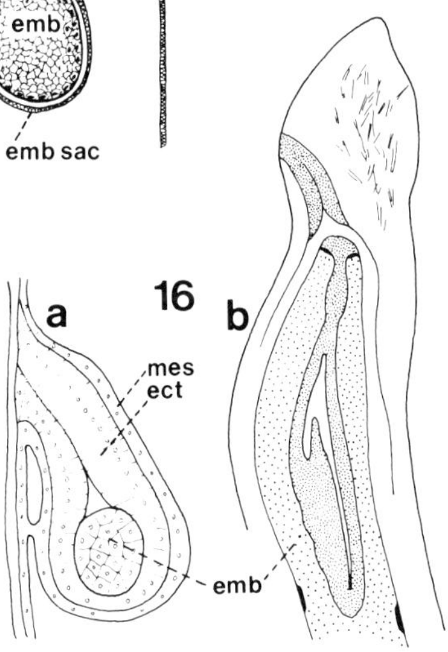

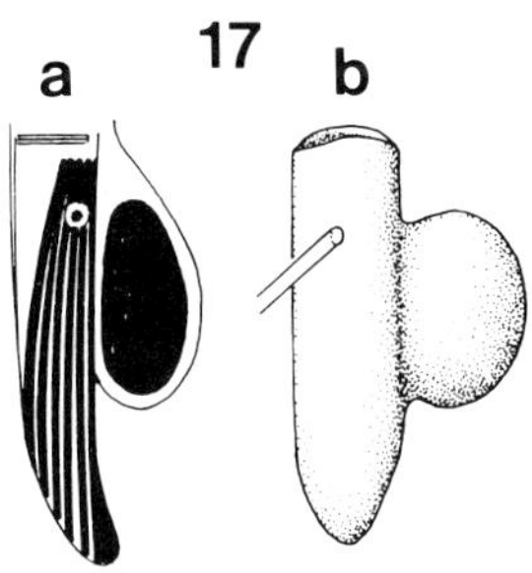

Fig. 13. Brooding in *Nolella dilatata.* Redrawn from Prouho (1892). (a) Eggs or embryos that appear to be in the coelom; (b) an embryo leaving the zooid; (c) tubelike exits: emb, embryo; ov, ovum; tub ex, tubelike exit.

Fig. 14. Brooding in *Nolella papuensis.* Redrawn from Harmer (1915). (a) Embryo that appears to be in the coelom and tubelike exits; (b) embryos adhered to the outside of the zooid.

Fig. 15. Embryo sac in *Nolella papuensis.* Redrawn from Silén (1944). Abbreviations: emb, embryo; emb sac, embryo sac.

Fig. 16. Embryo sac in *Sundanella sibogae.* Redrawn from Braem (1940). (a) Early stage of embryo sac invaginated from the outside: ect, ectoderm; emb, embryo; mes, mesoderm; (b) late stage of embryo sac and embryo.

Fig. 17. Outer embryo sac between two cuticles in *Penetrantia.* Redrawn from Silén (1947, 1948). (a) Outer embryo sac of *P. densa* with small opening; (b) outer embryo sac of *P. concharum.*

and embryonic development in an invagination of the body wall. Specially constructed brood chambers are not present. The most advanced type of brooding is probably where the development leads to a single embryo in the introvert, as found in the Vesicularioidea. The larva of this type also appears to be the most specialized in comparison with the planktotrophic type.

C. Brooding in Cheilostomata

The cheilostomes also contain forms which release eggs into the water and produce a planktotrophic larva, the cyphonautes. This primitive pattern of development has evidently been retained by certain forms within the Malacostegioidea [e.g., *Membranipora membranacea* (L.) and *Electra pilosa* (L.)] (Jägersten, 1972).

A very simple form of brooding is found in *Tendra zostericola* (Nordm.) (Paltschikova-Ostroumova, 1925). Overarching spines from the edge of the cystid form a roof over the frontal membrane. Several embryos lie in the space between these spines and the frontal membrane. A brood chamber of this sort is usually called an acanthostegous ovicell (Hyman, 1959).

Simple adherence of the zygotes on the outside of the zooid may possibly occur in some primitive forms of Anasca. Osburn (1912) and Waters (1913) described an outer embryo sac in *Aetea anguina* (L.), but judging from their illustrations it could be an adherent embryonic membrane (Fig. 18). Marcus (1937a) has also described external embryo sacs in *A. anguina* and *A. sica* (Couch) which also probably are adherent embryonic membranes. Hastings (1943) has described an external embryo sac in *A. anguina* attached to the frontal membrane of the zooid under the orifice (Fig. 19a), although other authors stated that the embryo sac is situated at the back wall of the zooid. Hastings also observed empty sacs after the emergence of the embryo of the larva (Fig. 19b). According to unpublished material of Eggleston (Ryland, 1970) both *Aetea* and *Eucratea* have external embryo sacs. I have also occasionally seen embryos attached to the zooid of *Aetea truncata* (Landsb.). There are generally three embryos somewhere on the outside of the zooid. At present there have been too few observations to confirm that this is the normal reproduction pattern. It is, however, clear that the so-called outer embryo sac does not consist of a cell layer, but that the wall is a "cuticular" membrane, which could correspond to the embryonic membrane. More thorough investigations are necessary to solve this problem.

Another species which appears to have the embryos fastened to the outside of the zooid is *Carbasea indivisa* (Ell. Sol.) within Malacostegioidea. According to Stach (1938) the embryo lies in an embryo sac fastened to the operculum (Fig. 20a). Stach considered that the embryos are surrounded by epithelium from the outer part of the tentacle sheath (Fig. 20b).

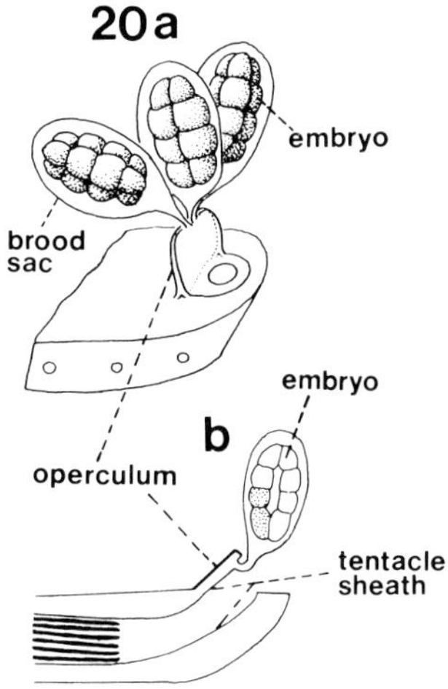

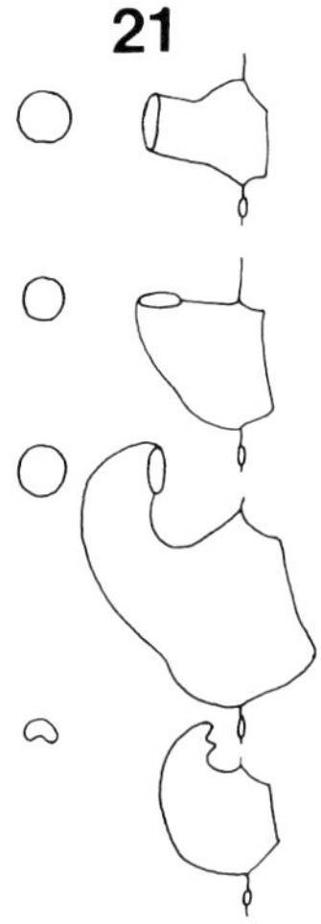

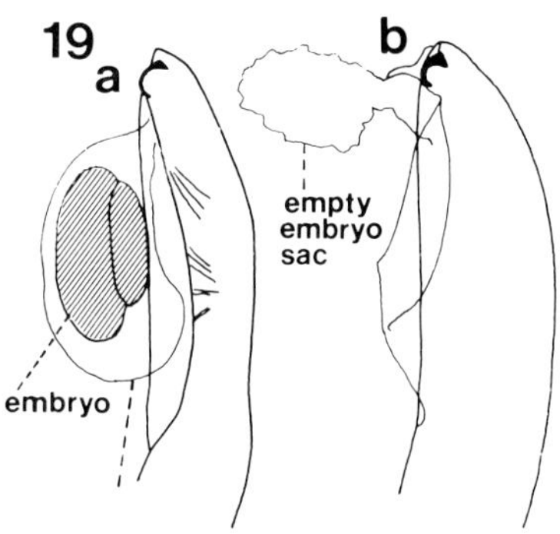

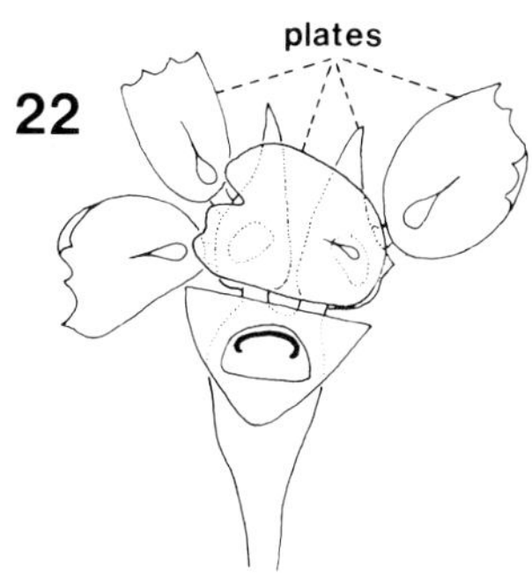

Fig. 18. *Aetea anguina,* embryo at the outside of the zooid. Redrawn from Waters (1913).

Fig. 19. Outer embryo sac of *Aetea anguina.* Redrawn from Hastings (1943). (a) Embryo sac fastened at the frontal membrane; (b) the larva has left the embryo sac.

Fig. 20. Brooding in *Carbasea indivisa.* Redrawn from Stach (1938). (a) Brood sacs fastened to the operculum; (b) schematic section showing the connection between the brood sac and the tentacle sheath.

Fig. 21. The orificial collar or peristome forming brood chamber in *Tubucellaria,* four different species. Redrawn from Marcus (1937b).

Fig. 22. Basketlike brood chamber in *Catenicula corbulifera.* Redrawn from O'Donoghue (1924).

Brooding in a tubelike extended orificial collar or peristome occurs in the genera *Tubucellaria* (Fig. 21) (Waters, 1907; Marcus, 1937b) and *Turretigera* (Levinsen, 1909), both of which are ascophorans. This form of brooding is suggestive of that observed in the vestibule of Ctenostomata. The brood chamber formed is called a peristomial ovicell (Fig. 28b) by Bassler (1953). A variant is found in *Catenicula corbulifera* (O'Donoghue). O'Donoghue (1924) demonstrated an outer basketlike brood chamber formed of plates from the front part of the zooid (Fig. 22).

An embryo sac invaginated into the coelom from the outside is also present in the Cheilostomata, as in the Ctenostomata. *Labiostomella gisleni* (Silén), which is considered among the most primitive within the group, has an embodied embryo similar to that in *Nolella* within ctenostomes. Silén (1944) demonstrated an invagination of the body wall which envelopes the embryo (Fig. 23). The epithelium is thickest on the outside, as in *Victorella.* The embryo probably sinks in from the outside, but this has not been seen. An embryo sac that invaginates into the coelom is also found in *Beania (Diachoris) magellanica* (Busk) and *Poricellaria (Diplodidymia),* both belonging to Anasca (Waters, 1912, 1913). An internal embryo sac is also found in *Watersipora cucullata* (Busk), within Ascophora (Mawatari, 1952). In *Cryptosula pallasiana* (Moll.) the embryo is developed in a brood pouch in the wall of the vestibule (Fig. 24) (Calvet, 1900), suggestive of the brooding in *Victorella.*

23

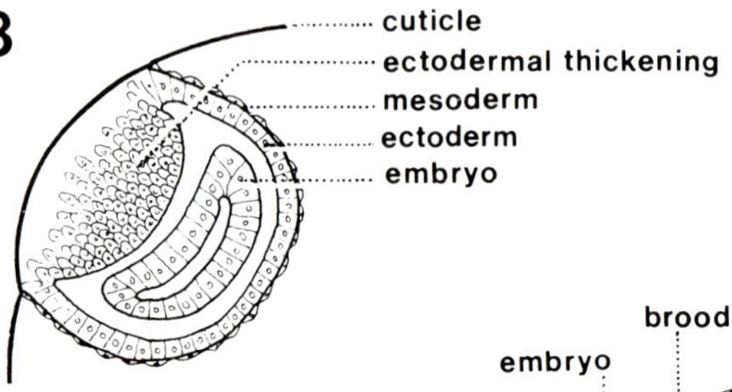

24

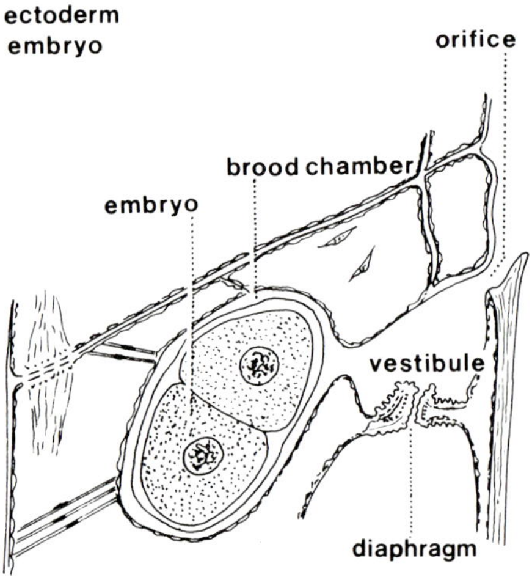

Fig. 23. Embryo sac in *Labiostomella gisleni.* Redrawn from Silén (1944).
Fig. 24. Brood chamber of *Cryptosula pallasiana.* Redrawn from Calvet (1900).

The commonest form of brooding in the Cheilostomata is the presence of a specialized brood chamber on the frontal part of the zooid. This chamber is called an ovicell or ooecium. An advantage of an ovicell is that the polypide does not have to degenerate during the development of the embryo. Silén (1944) discussed the origin of an ovicell from an embryo sac of the type formed in *Labiostomella.* His hypothesis is diagramatically shown in Fig. 25. He also noted that an ovicell develops only when a mature egg is present in the coelom; presumably there is a hormonal influence regulating its production. The transport of an egg from the coelom to the ovicell has been described by Gewerzhagen (1913) and by Silén (1945). The egg is first transported up to the supraneural pore, when the polypide is in a protruded position (Fig. 26a,b). The pore is then at the same level as the opening of the ovicell and the inner vesicle that covers the opening is drawn downward by retractor muscles. The amoeboid egg flows rapidly through the pore and then regains its spherical form (Fig. 26c–e). The entire process takes about 15 seconds. Fertilization probably occurs during this brief period when the oocyte is exposed to the ambient seawater.

Levinsen (1909) classified the different types of ovicells. This classification was later modified by Silén (1945) and by Bassler (1953). The commonest type is the hyperstomial ovicell, which is not joined to the neighboring zooid (Figs. 26–28a). This is found in *Callopora* and *Scrupocellaria* and its development and construction are described by Silén (1945). It consists of two prominences on the frontal wall, an outer and later an inner one (Fig. 27a–d_2). The outer one consists, at the start, of two protuberances that might be converted spines. Later they merge into a hood, with double walls (Fig. 27e–g). Silén called this an ectooecium. The inner prominence forms a sac which partly fills up the space in the hood; this Silén calls an endooecium (Fig. 27h_1,h_2). Some authors interpret the endooecium as being the inner wall of the ectooecium and they simply call the inner protuberance the inner vesicle. Here I have used Silén's interpretation.

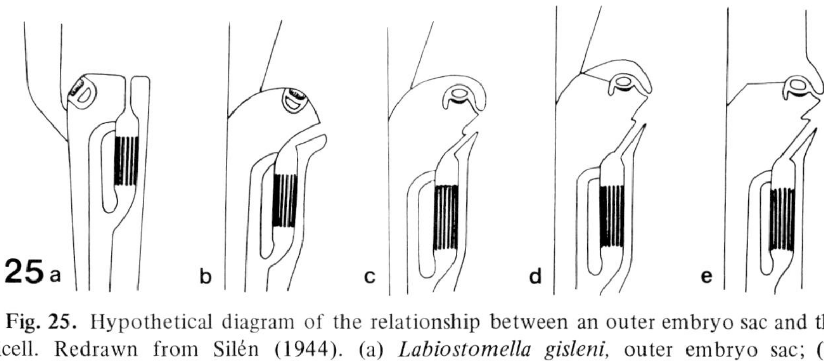

Fig. 25. Hypothetical diagram of the relationship between an outer embryo sac and the ovicell. Redrawn from Silén (1944). (a) *Labiostomella gisleni,* outer embryo sac; (b) hypothetical intermediate form; (c) ovicell at the distal part of the zooid; (d) the hyperstomial ovicell; (e) the endozooidal ovicell.

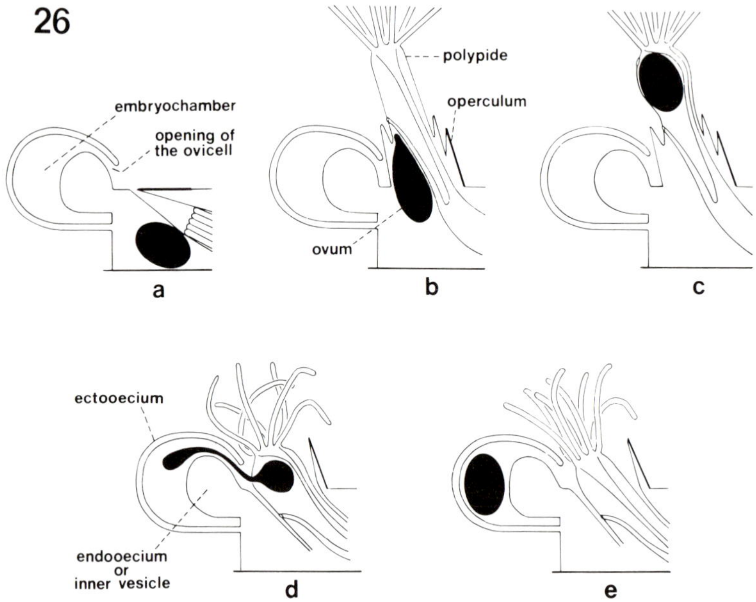

Fig. 26. The transfer of the ovum to the ovicell in *Callopora dumerilii*. Redrawn from Silén (1945). (a) The ovum at the distal end of the zooid; (b) the polypide in extended position; (c) the ovum lies near the supraneural pore; (d) the ovum flows into the ovicell; (e) the ovum or zygote lies in the embryo chamber.

The embryo will lie between the ectooecium and the endooecium. The latter can be retracted by muscle fibers. The opening to the ovicell is usually closed by the endooecium as in *Callopora* (Figs. $27h_2$ and $28a_1$). In *Scrupocellaria* that portion of the endooecium which closes the opening is more heavily chitinized (Silén, 1944). In other forms the operculum can close only the orifice or both

---

Fig. 27. Development of the hyperstomial ovicell. Redrawn from Silén (1945). (a) Double origin of the ovicell: ov, ovum; (b–c) developing ectooecium: dev ect, developing ectooecium; (d) the endooecium begins to develop, the first oocyte starts to grow: ect, ectooecium; end, endooecium; ov, ovum; (e–g) the ectooecium develops into a hood; (h) the ovicell is fully developed, the ovum is ready to enter the ovicell, the second oocyte begins to develop: emb ch, embryo chamber; (i) *Bugula neritina*, relationships between distal zooid, maternal zooid, and kenozooid forming the ooecial fold. Redrawn from Woollacott and Zimmer (1972). (j) *Bugula neritina*, diagrammatic median section of fully developed ovicell with young embryo. Redrawn from Woollacott and Zimmer (1972). Calcified layers are in black and cuticle is indicated by dots of varying sizes. Outer epithelium and peritoneum complex are indicated by dashed line, except in that region where the epithelium is hypertrophied.

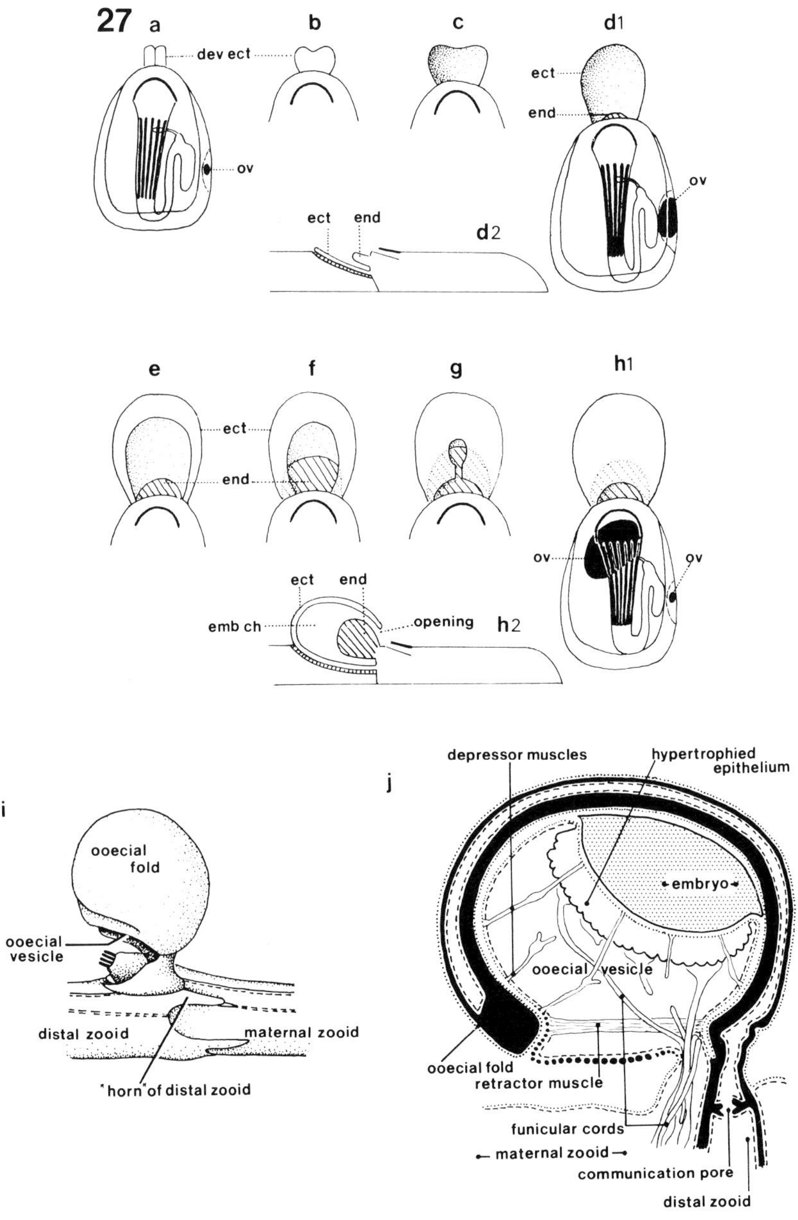
27 a
b
c
d1
dev ect
ov
ect
end
d2
ect
end
ov
e
f
g
h1
ect
end
ov
ov
ect
end
emb ch
opening
h2
i
ooecial fold
ooecial vesicle
distal zooid
maternal zooid
"horn" of distal zooid
j
depressor muscles
hypertrophied epithelium
embryo
ooecial vesicle
ooecial fold
retractor muscle
funicular cords
maternal zooid
communication pore
distal zooid

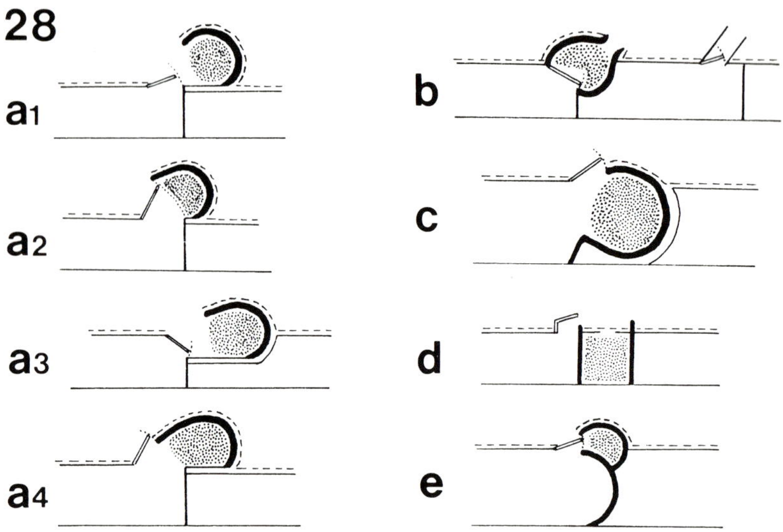

Fig. 28. Different types of ovicells and different types of closing of the ovicell. Redrawn from Bassler (1953). (a_1) Hyperstomial ovicell, the operculum closes only the orifice, the endooecium or inner vesicle closes the opening of the ovicell; (a_2) hyperstomial ovicell, the operculum closes either the orifice or the opening of the ovicell; (a_3) hyperstomial ovicell, the operculum closes either the orifice or both openings; (a_4) hyperstomial ovicell, the operculum closes both openings at the same time; (b) peristomial ovicell; (c) endozooidal ovicell; (d) endothoical ovicell; (e) recumbent ovicell.

openings at the same time (Fig. 28a_2,a_3), as in *Pentapora.* In many cases the operculum always closes both openings (Fig. 28a_4); an ovicell sealed in this way is called cleithral. Cleithral ovicells are found in *Microporella* and *Membraniporella* (Ryland, 1970).

Woollacott and Zimmer (1972) have described the brood chamber of *Bugula neritina.* It is of the hyperstomial type. The ovicell or ooecium is located distally to the orifice or aperture of the maternal zooid (Fig. 27i). Woollacott and Zimmer name the outer fold of the ovicell the ooecial fold. That corresponds to the ectooecium according to Silén (1945). This ooecial fold arises from a "horn" of the distal zooid directed proximally along the distolateral margins of the parent zooid (Fig. 27i). The ooecial fold becomes flattened and grows around the ooecial vesicle (endooecium according to Silén) to form a hood. It becomes calcified, providing protection for the embryo. The ooecial vesicle is derived from the maternal zooid as an evagination from the body wall; it is membranous. Although the ooecial fold is derived from the distal zooid it arises proximal to the ooecial vesicle (Fig. 27j). The ooecial fold is separated from the distal zooid by a calcified septum, which is perforated. The ooecial fold can thus be considered as a kenozooid. The embryo lies between the ooecial fold and the ooecial vesicle. The epithelium of the ooecial vesicle, which lies adjacent to the embryo becomes hypertrophied (Fig. 27j). Strands of funicular tissue extend from the maternal zooid through the lumen of the ooecial vesicle and contact the basal ends of the hypertrophied cells. These are considered to function in the nutrition of the early embryo. Both retractor muscles and depressor muscles extend through the cavity of the ooecial vesicle.

The *endozooidal* or endozooecial ovicell lies embedded in the base of the succeeding zooid (Fig. 28c). Such ovicells are found in *Flustra foliacea* (L.) (Fig. 29) (Calvet, 1900). Silén (1944) considered that the endozooidal ovicell may have developed from the hyperstomial ovicell (Fig. 25d).

A type of brood chamber called an endothoical ovicell is found in the family Cellariidae. This is a protuberance of the frontal part of the zooid, but is completely separated from the zooid (Fig. 28d).

Bassler (1953) named the type of ovicell in *Phyllactellipora hincksi* (Bassler) *recumbent* (Fig. 28e). It differs from the hyperstomial ovicell only in that it rests on the distal wall of its own zooid. It may be considered as a variant of the hyperstomial ovicell. The above classification may not reflect significant morphological or functional differences, but is useful in taxonomy.

The two-valved ovicell is formed by two semispherical arches growing toward each other and joining. This type of ovicell is found in *Scruparia chelata* (L.) (Fig. 30a) (Hastings, 1941), *Thalamoporella* (Harmer, 1926), and *Alycidium* (Fig. 30b) (Levinsen, 1909). These ovicells are interesting as they contain several

29

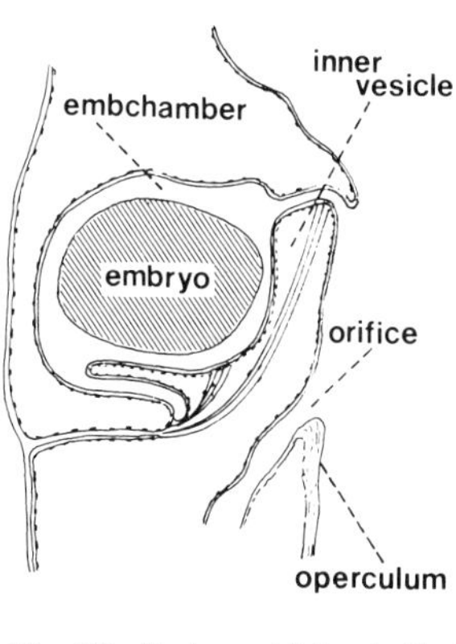

30a

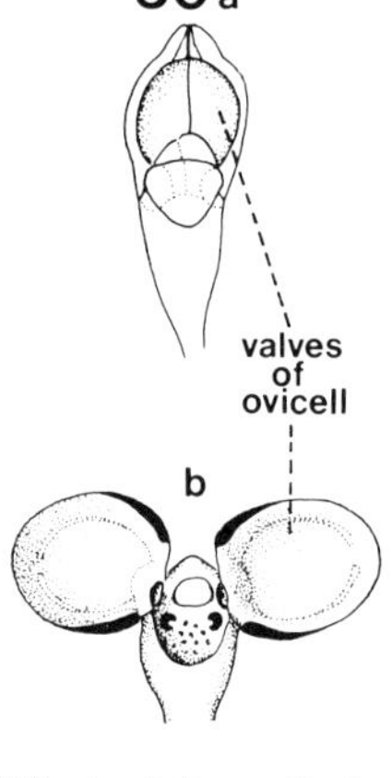

31

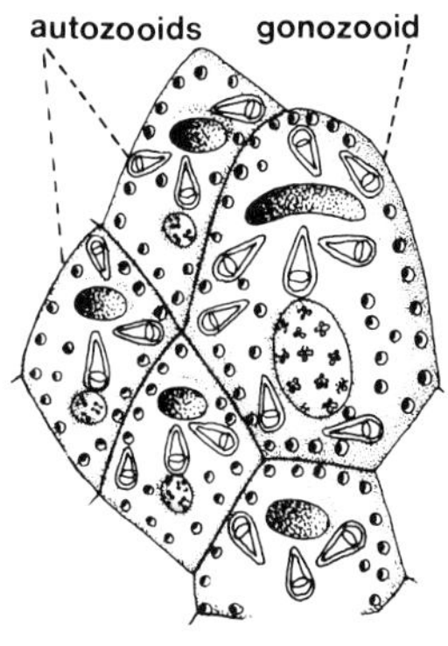

Fig. 29. Endozooidal ovicell of *Flustra foliacea.* Redrawn from Calvet (1900).

Fig. 30. Two-valved ovicells. (a) Two-valved ovicell in *Scruparia chelata*. Redrawn from Hastings (1941). (b) Two-valved ovicell in *Alcyidium*. Redrawn from Levinsen (1909).

Fig. 31. An enlarged zooid functioning as a gonozooid in *Adeonellopsis.* Redrawn from Levinsen (1909).

embryos. *Scruparia chelata* may have as many as seven embryos in an ovicell at the same time.

There are also forms within the Cheilostomata where the entire zooid functions as a brood space. These zooids could be called gonozooids. Within the family Adeonidae there are particularly large zooids with a large embryo sac containing embryos (Waters, 1912). Such enlarged zooids containing embryo sacs have been described in *Adeonellopsis* (Fig. 31) by Levinsen (1909).

The hyperstomial, endozooidal, and endothoical ovicells contain only one embryo at a time, although the two-valved ovicells contain several. The ovicells containing only one embryo can, however, contain several larvae, one after the other. *Callopora dumerilii* (Aud.) has a sequence of 3 or 4 embryos succeeding each other in the same ovicell throughout the season (Silén, 1945). Others such as *Synnotum aegypticum* (Aud.) have only one embryo per season (Marcus, 1941). The limitations in the latter cases are connected with the fact that the polypide usually degenerates during the development of each embryo.

III. STENOLAEMATA

Cyclostomata

The Cyclostomata is the only surviving order within the Stenolaemata. Reproduction is uniform within the group and development takes place in converted zooids, called gonozooids. These brood chambers have been described by Harmer (1890, 1893), Robertson (1903), Waters (1914), and Borg (1926, 1933), among others. Gonozooids are swollen protuberances with a funnellike opening which corresponds to the orifice in the normal zooid. Only a few gonozooids develop in each colony. Fertilization processes are unknown, but they are believed to be internal.

In the genus *Crisia* there are simple swollen gonozooids (Fig. 32a). *Tubulipora* also has simple swollen gonozooids but they spread out and form irregular structures (Fig. 32b). In others an extrazooidal cavity is incorporated with the originally simple gonozooid, as in *Lichenopora* (Fig. 32c) (Borg, 1926). Borg (1933) also studied *Heteropora* where the gonozooid containing one or several eggs is at first single, but later incorporates neighboring zooids, both autozooids and kenozooids, and thus forms a complex brood space (Fig. 32d).

In the Cyclostomata there is an unusual form of development called polyembryony. This phenomenon was first discovered in bryozoans by Harmer (1890, 1893) and has since been verified several times, e.g., by Robertson (1903) and Borg (1926). The processes of polyembryony may be summarized, mainly based on the investigations by Borg, as follows: Many eggs are produced but the only eggs that develop are those that adhere to a polypide bud in an incipient

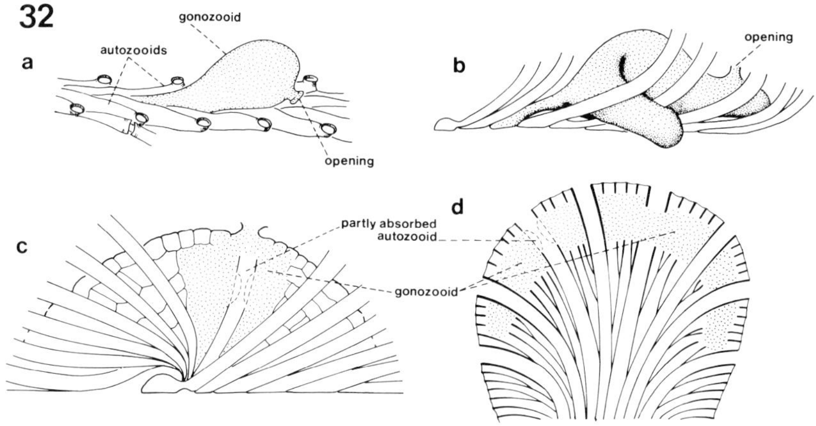

Fig. 32. Different types of gonozooids in Cyclostomata. Modified from Borg (1926, 1933). (a) Part of a colony of *Crisia denticulata* with gonozooid; (b) branched gonozooid of *Tubulipora;* (c) diagrammatic section through a fertile colony of *Lichenopora;* (d) diagrammatic section through distal end of fertile branch of *Heteropora.*

gonozooid. Generally there is only one such egg, sometimes two. The egg lies in the polypide's mesodermal layer which forms a follicle around what is now probably the zygote. Harmer (1893) and Robertson (1903) considered that the embryo developed in the tentacle sheath, but Borg showed it to be situated in what he called a membranous sac. The polypide degenerates and contributes to the embryo's nutrient requirements. The connection formed between the remnant of the polypide and the embryo is called the nutrient column. This process is uniform within the entire group. The fertilized egg cleaves a number of times to produce a primary embryo. The blastomeres will then be isolated off singly or in small groups which in turn start to divide into secondary embryos. Tertiary embryos may also develop and sometimes there may be as many as a hundred in a gonozooid. The membranous sac increases in size as the embryos grow and increase in number. They soon completely fill the zooid (Fig. 33). The secondary or tertiary embryos then develop into larvae before they leave the gonozooid. The upper part of the gonozooid usually degenerates, but later it regenerates and the gonozooid is then able to produce new embryos. The larvae are lecithotrophic and highly modified in comparison with the planktotrophic larvae. Membranous sacs, which are unique to the Cyclostomata, are not restricted to gonozooids, but are found in sterile zooids as well; in the latter they play a role in the protrusion of the tentacles.

The membranous sac cannot be considered to correspond to the embryo sac in the Gymnolaemata. It is mesodermal and envelops the entire polypide. Polyembryonic reproduction is unique to cyclostomes within the Bryozoa.

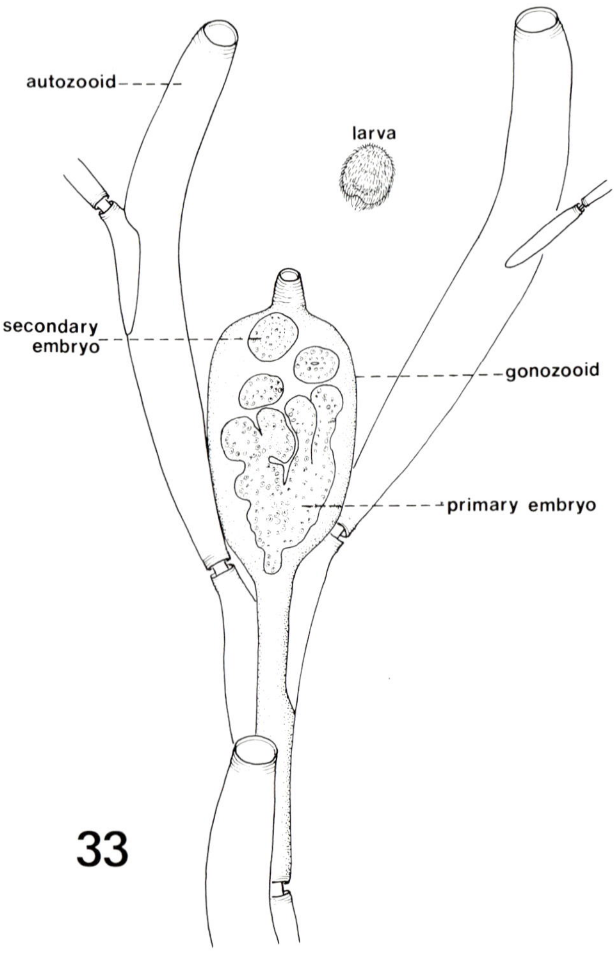

Fig. 33. Gonozooid of *Crisia* with polyembryony. Modified from Harmer (1893), Calvet (1900), Robertson (1903), and Borg (1926).

Cyclostomata are thus, with regard to reproduction, well defined from other bryozoan groups. There are forms within the Cheilostomata with special zooids that function as brood spaces, *Adeonellopsis,* for example, but this must be a parallel development. The Cyclostomata is an ancient group, and phylogenetically this unique reproduction must have appeared at the separation of the group.

IV. PHYLACTOLAEMATA

Reproduction in the Phylactolaemata has been studied most thoroughly by Davenport (1890), Kraepelin (1892), Braem (1897, 1908), Marcus (1934), and Brien (1953). The embryo develops in an embryo sac which is an invagination of the body wall into the coelom. The polypide degenerates during the growth of the embryo. The embryo sac is situated ventrally immediately above the ovarium (Fig. 34a_1,a_2). It is not known how the egg is transported into the embryo sac. The mechanism of fertilization is also unknown, but is considered to be internal. The zygote is believed to be transported directly to the embryo sac through the sac's two cell layers (Marcus, 1934). This process should be reexamined. The formation of the embryo sac is very similar to the situation in *Sundanella sibogae,* a ctenostome. First, the epidermal layer invaginates, and this leads to invagination of the mesodermal layer. Marcus (1938) suggested that the embryo sac develops under the influence of hormones from the ovary. Development is more or less uniform within the whole group; in this description I will follow that reported for *Plumatella* according to the descriptions by Braem (1897) and Brien (1953). After the initial cleavages, the embryo consists only of a single cell layer which later becomes double (Fig. 34b). It is not certain if this is a true

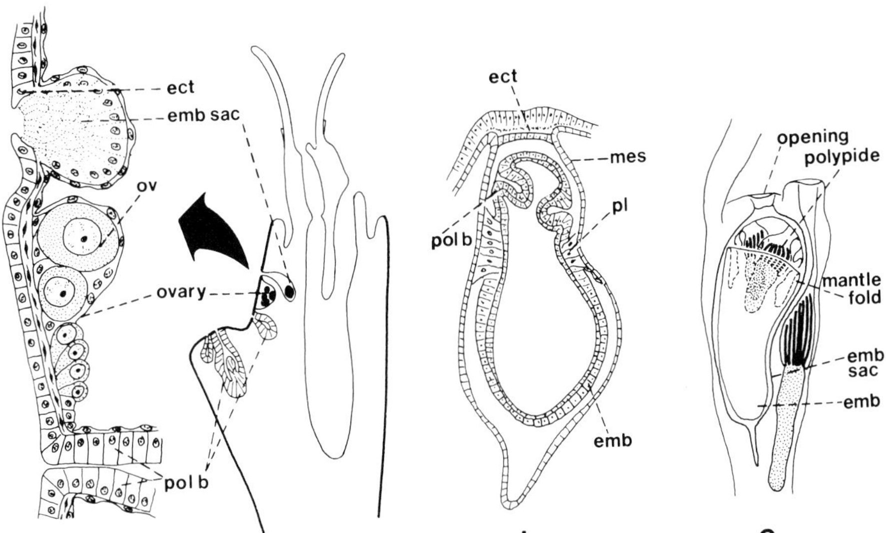

Fig. 34. The embryology of *Plumatella.* Modified from Braem (1897) and Brien (1953). (a) The site and development of the embryo sac: ect, ectoderm; emb sac, embryo sac; ov, ovum; pol b, polypide buds (b) section of embryo with placenta and developing polypide buds: mes, mesoderm; pl, placenta; (c) large embryo inside the embryo sac.

gastrulation. The cell layers are formed by cell division and rearrangement. No true entoderm is formed and no purely larval organs develop.

The inner layer of the embryo sac, i.e., the (adult) epidermis, largely disappears during the growth of the embryo. However, around the center of the embryo it forms a ring which is in contact with both the embryo and the embryo sac. This probably functions as a kind of placenta (Braem, 1897; Brien, 1953); at the same time it holds the embryo in place (Fig. 34b). In *Fredericella* this ring is placed near the top of the embryo (Braem, 1908). The embryo sac is considerably expanded during the growth of the embryo and reaches far into the coelom. Later the outer part of the embryo sac becomes somewhat thinner. The larva develops one or more polypides (usually two) and the mantle fold grows up and envelops them (Fig. 34c). At this stage the larva leaves the embryo sac through an opening which is formed in the outer part of the sac's epithelium. The larva is lecithotrophic and largely modified compared with a planktotrophic larva. It can be described as a small colony. At emergence the larva first rotates 180° so that it swims out backward. The posterior pole will then lead when swimming. Phylogenetically this reversal probably is of secondary origin (Jägersten, 1972). The folding-up of the mantle over the developing polypide buds of the larva is suggestive of the cyclostome larva. In conclusion, the Phylactolaemata have an advanced form of brooding and a highly specialized larva.

V. SUMMARIZING COMMENTS ON BROODING

The most primitive form of reproduction is the mass production of eggs and their release into the surrounding water (Jägersten, 1972). In the Bryozoa this pattern occurs in both the Ctenostomata and the Cheilostomata. External fertilization is also primitive but in bryozoans appears to take place in a somewhat altered form. It is, however, a form of external fertilization as, in those cases where the conditions are known, the egg is fertilized at the moment it leaves the zooid (Silén, 1966, 1972). In connection with the release of the egg a planktotrophic larva, a cyphonautes, is formed in these species. Jägersten (1972) maintains that the cyphonautes larva is the most primitive. Whatever the case may be, the most primitive larval form is certainly a planktotrophic larva with a complete digestive tract. Lecithotrophic larvae have developed in connection with brooding.

The simplest form of brooding is found in some ctenostomes, which still have mass production of eggs but in which the egg membrane adheres to the outside of the zooid. Only the first phase of the development takes place on the parent zooid. When ciliation is developed the larva leaves the membrane and the rest of the development takes place in the water. This form of brooding is found in *Triticella* and *Paludicella*. It may also occur in *Aetea,* within the Cheilostomata.

In the Ctenostomata the sequence can be followed from a temporary adhesion of embryos on the outside of the parent to typical brooding in the introvert, with the number of embryos produced at a time being reduced to one. A transitional stage is represented by *Bulbella* and *Alcyonidium duplex* where two to six embryos adhere to the outside of the polypide, but which are drawn into the vestibule when the polypide is retracted. The species within the Carnosa with brooding in the introvert have several embryos at the same time. However, within the Stolonifera only one embryo develops in the introvert. The larva is largely transformed. In the Cheilostomata brooding takes place in the vestibule or what corresponds to the vestibule in *Tubucellaria.*

The development of the embryo in an embryo sac formed by an invaginated body wall in ctenostomes is found in *Victorella, Nolella,* and *Sundanella.* This type of brooding is also found in the cheilostome *Labiostomella.* Embryo sacs that invaginate into the coelom are also found in *Beania, Cryptosula,* and *Watersipora* (Waters, 1912; Calvet, 1900; Mawatari, 1952). These may be regarded as invaginations in the vestibule.

According to Silén (1944) the commonest type of brood chamber in cheilostomes, the ovicell, can be traced to an embryo sac similar to that in *Labiostomella* (Figs. 25 and 35). Other types may be variants of the hyperstomial type. The egg always first leaves the zooid before it enters the ovicell.

In all phylactolaemates an embryo sac is formed which is similar to that in the ctenostome *Sundanella.* It is, however, placed opposite the anal side. The placement in relation to the polypide buds coincides with that in *Sundanella*

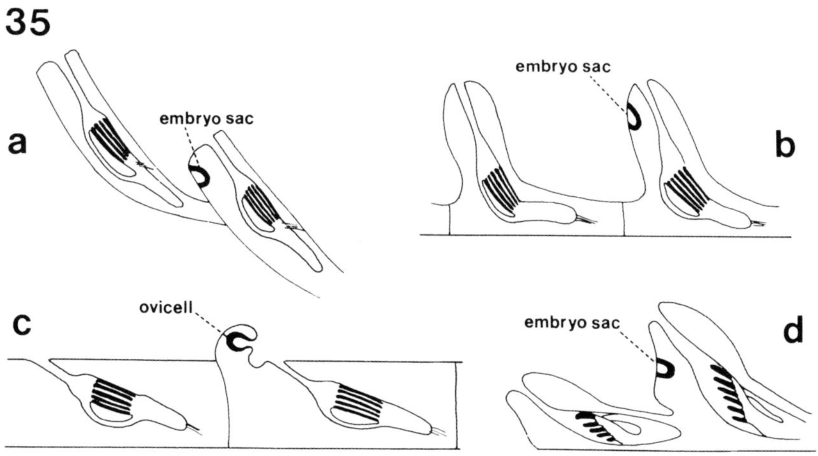

Fig. 35. Position of the brood chamber in different types of Bryozoa. Redrawn from Silén (1944). (a) *Labiostomella;* (b) *Sundanella;* (c) a cheilostome with ovicell; (d) *Plumatella.*

(Fig. 35). Fertilization in the Phylactolaemata is presumed to take place in the coelom but the actual case is unknown. It is further presumed that the zygote enters the embryo sac directly without leaving the zooid first. This process requires further investigation. In Phylactolaemata the epidermal epithelium of the embryo sac disintegrates and the embryo has a placenta. However, comparing the embryo sacs in ctenostomes and phylactolaemates, Silén considered that the similarities outnumbered the dissimilarities. He assumed that the embryo sac corresponds to a polypide bud in both cases, but this opinion is not shared by Brien (1953, 1960) or Jebram (1973). Thorough investigations of the reproduction in Phylactolaemata are needed before anything definite can be said on the relationship with the corresponding process in Ctenostomata.

Finally, Cyclostomata are a very special group as regards brooding. They have specialized zooids–gonozooids–with the unique form of reproduction called polyembryony. The embryos lie in a membranous sac the like of which is not found in any other group of Bryozoa. Fertilization has not been reported on. This way of reproduction probably appeared at the same time as when the group separated from other bryozoan groups.

VI. BREEDING SEASONS

Most of the available information on reproductive seasons of bryozoans is derived from settlement studies. Observations on the seasonal abundance of embryos of marine species have been made by Gautier (1962) and Ryland (1963, 1967). According to Gautier some species in the Mediterranean Sea have embryos in the ovicells through all or most of the year; for instance, *Chorizopora brongniarti* (Aud.) and *Puellina gattyae* (Dusk) have embryos at least 9 months a year. Gautier also observed that the breeding season was longer in deeper-living colonies. Geographically there seem to be large variations. For example, Gautier noticed no embryos in *Escharella immersa* (Flem.) in the Mediterranean, although Ryland observed embryos almost throughout the year in *E. immersa* from around the Isle of Man (Ryland, 1967).

Many species in temperate areas such as Sweden and Norway have embryos during summer and autumn, for example, *Cryptosula* and *Crisia* (Ryland, 1963). The reproductive activity seems to follow the annual variation in sea temperature at the surface, with a peak in early autumn (Fig. 36a). Deeper forms can also have the same fluctuation (Fig. 36b) which indicates that temperature cannot be the only factor. There is probably a correlation between phytoplankton production and breeding activity which can explain the autumn peak.

Although most temperate species have a breeding optimum in early autumn there are many exceptions. *Alcyonidium polyoum,* for example, lives mainly on *Fucus serratus* in shallow water and has a reproductive period from January to

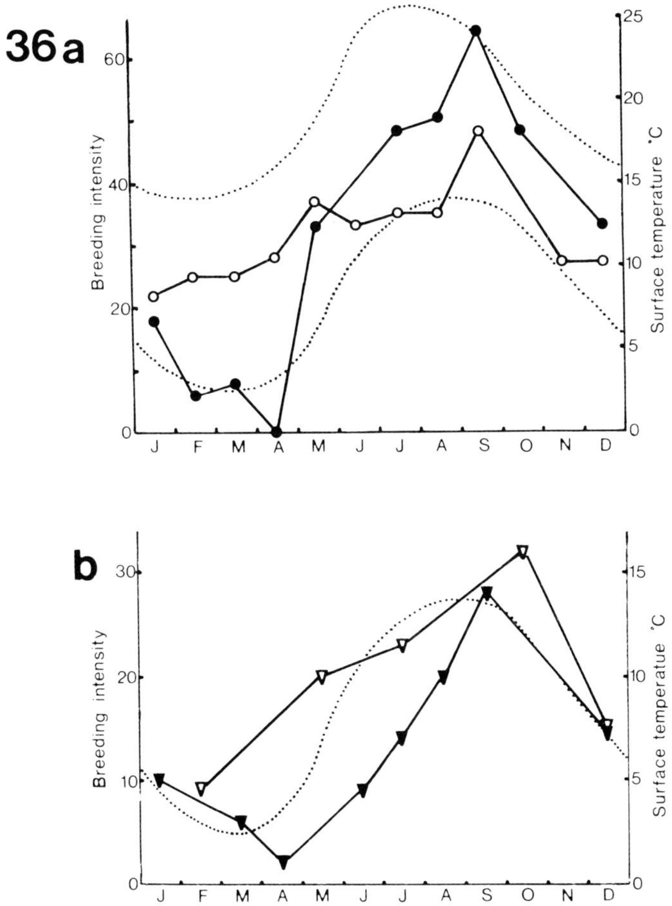

Fig. 36. Seasonal pattern of reproductive activity. Redrawn from Ryland (1967). (a) ○——○, abundance of developing embryos of bryozoans in western Mediterranean (Gautier, 1962); ●——●, abundance of developing embryos of bryozoans in Bergen, Norway (Ryland, 1963); upper dotted line, sea surface temperature in Bay of Naples according to Hapgood (1959); lower dotted line, sea surface temperature at Espegrend, near Bergen, according to Tambs-Lyche (1957). (b) △——△, embryo abundance in deep (75 m) fjord bryozoans near Bergen (Ryland, 1963); ▲——▲, embryo abundance in shallow fjord bryozoans near Bergen (Ryland, 1963); dotted line, sea surface temperature at Espegrend, near Bergen, according to Tambs-Lyche (1957); temperature below 75 m is almost constantly 7–8°C, according to Burdon-Jones and Tambs-Lyche (1960).

March on the Swedish west coast with a maximum during February, when the sea surface temperature is the lowest of the year. Matricon (1963) has investigated reproduction in *A. polyoum* living on *Fucus serratus* at Roscoff, France. These colonies had a maximum fertility in September. The cause of that diversity is unknown and the reasons could be many, for instance, a geographical difference or two breeding periods a year, as in *Bicellariella* and *Bugula* (Ryland, 1970).

According to Ryland (1970), forms with short-lived colonies have a brief, but intensive period of reproduction, considering a given colony, but because of successive overlapping generations, the yearly breeding season of the species is extensive. Eggleston (1972) has observed that species living in shallow water have a very short period of reproduction. In some forms there are special conditions which influence reproduction; for example, those bryozoans living on animals that molt annually have their reproductive period correlated with the molting period of the host [e.g., *Triticella koreni* (Ström, 1969; Eggleston, 1971)]. *Triticella* has an annual life cycle and dies after the breeding period, a situation rather common among other bryozoans. Many species (e.g., *Flustra foliacea*) are perennial but have annual breeding.

Generally, it is very difficult to quantify breeding seasons and breeding activity because of numerous interacting variables, such as latitude, depth, temperature, substrate, and phytoplankton production, which can modulate reproduction. The role of these different factors varies from species to species; there probably is also intraspecific variation from region to region.

ACKNOWLEDGMENTS

I wish to express my cordial thanks to Mrs. Kerstin Ahlfors for redrawing the text figures and to Mr. Nigel Rollison for translating the manuscript.

REFERENCES

Banta, W. C. (1968). *Mimosella cookae,* new species (Bryozoa, Ctenostomata) with a review of the family Mimosellidae. *Bull. S. Calif. Acad. Sci.* **67**, No. 4, 245–254.

Barrois, J. (1877). Recherches sur l'embryologie des Bryozoaires. *Trav. Stn. Zool. Wimereux* **1**, 1–305.

Bassler, R. S. (1953). "Treatise on Invertebrate Paleontology," Part G. Univ. of Kansas Press, Lawrence.

Bobin, G., and Prenant, M. (1954). Sur un Bryozoaire perforant, *Terebripora comma* (Soule) trouvé en Mediterranée. *Arch. Zool. Exp. Gén.* **91**, 130–144.

Borg, F. (1926). Studies on Recent cyclostomatous Bryozoa. *Zool. Bidr. Uppsala* **10,** 181–507.

Borg, F. (1933). A revision of the recent Heteroporidae (Bryozoa). *Zool. Bidr. Uppsala* **14**, 253–394.

Braem, F. (1896). Die geschlechtliche Entwicklung von *Paludicella ehrenbergii. Zool. Anz.* **19**, 54–57.

Braem, F. (1897). Die geschlechtliche Entwicklung von *Plumatella fungosa. Zoologica (Stuttgart)* **10**, 1–96.

Braem, F. (1908). Die geschlechtliche Entwicklung von *Fredericella sultana. Zoologica (Stuttgart)* **20**, 1–38.

Braem, F. (1940). *Victorella sibogae* Harmer. *Z. Morphol. Oekol. Tiere* **36**, 267–278.

Braem, F. (1951). Über Victorella und einige ihrer nächsten Verwandten sowie über die Bryozoenfauna des Ryck bei Greifswald. *Zoologica (Stuttgart.)* **102**, 1–59.

Brien, P. (1953). Etude sur les Phylactolémates (évolution de la zoécie, bourgeonnement d'accroisement, bourgeonnement statoblastique, embryogénèse, l'ontogénèse multiple). *Ann. Soc. R. Zool. Belg.* **84**, 301–444.

Brien, P. (1960). Classe des Bryozoaires. *In* "Traité de Zoologie" (P.-P. Grassé, ed.), Vol. 5, Part 2, pp. 1053–1379. Masson, Paris.

Calvet, L. (1900). "Contribution à l'Histoire Naturelle des Bryozoaires Ectoproctes Marins." Coulet et fille, Montpellier.

Corrêa, D. D. (1948). A embryologia de *Bugula flabellata* (J. V. Thomson) (Bryozoa Ectoprocta). *Univ. Sao Paulo, Fac. Filos, Cienc. Ret., Bol. Zool.* **13**, 7–71.

Davenport, C. B. (1890). *Cristatella:* The origin and development of the individual in the colony. *Bull. Mus. Comp. Zool.* **20**, Part 4, 101–151.

Eggleston, D. (1971). Synchronization between moulting in *Calocaris macandreae* (Decapoda) and reproduction in its epibiont *Triticella koreni* (Polyzoa Ectoprocta). *J. Mar. Biol. Assoc. U.K.* **51**, No. 2, 404–410.

Eggleston, D. (1972). Patterns of reproduction in the marine Ectoprocta of the Isle of Man. *J. Nat. Hist.* **6**, 31–38.

Franzén, Å. (1956). On spermiogenesis, morphology of the spermatozoon and biology of fertilization among invertebrates. *Zool. Bidr. Uppsala* **31**, 355–483.

Gautier, Y. V. (1962). Recherches écologiques sur les Bryozoaires Chilostomes en Mediterranée Occidentale. *Recl. Trav. Stn. Mar. Endoume* **38**, 1–434.

Gerwerzhagen, A. (1913). Untersuchungen an Bryozoen, (vorläufige Mitteilung). *Sitzungsber. Heidelb. Akad. Wiss., Math.–Naturwiss. Kl. B* **9**, 1–16.

Harmer, S. F. (1890). On the origin of the embryos in the ovicells of cyclostomatous Polyzoa. *Proc. Cambridge Philos. Soc.* **7**, 48.

Harmer, S. F. (1893). On the occurrence of embryonic fission in cyclostomatous Polyzoa. *Q. J. Microsc. Sci. [N.S.]* **34**, 199–241.

Harmer, S. F. (1915). The Polyzoa of the Siboga Expedition. Part I. *Siboga Exped., Monogr.* **28a**, 1–180.

Harmer, S. F. (1926). The Polyzoa of the Siboga Expedition. Part II. Cheilostomata Anasca. *Siboga Exped., Monogr.* **28b**, 1–331.

Hastings, A. B. (1941). The British species of *Scruparia* (Polyzoa). *Ann. Mag. Nat. Hist.* [11] **7**, 465–472.

Hastings, A. B. (1943). Polyzoa. Part I. *Discovery Rep.* **22**, 301–510.

Hyman, L. H. (1959). "The Invertebrates," Vol. 5. McGraw-Hill, New York.

Jägersten, G. (1972). "Evolution of the Metazoan Life-Cycle, Comprehensive Theory." Academic Press, New York.

Jebram, D. (1973). Zooid individuality and brooding organs (Bryozoa). *Z. Morphol. Tiere* **75**, 255–258.

Joliet, L. (1877). Contribution à l'histoire naturelle des Bryozoaires des côtes de France. *Arch. Zool. Exp. Gén.* **6,** 193–304.

Kraepelin, K. (1892). Die deutschen Süsswasserbryozoen II: Entwicklungsgeschichtlicher Teil. *Abh. Wiss. Ver. Hamburg* **12,** 1–67.

Levinsen, G. M. R. (1909). "Morphological and Systematic Studies on the Cheilostomatous Bryozoa." Natl. Forfotterers Forlag, Copenhagen.

Marcus, E. (1926). Beobachtungen und Versuche an lebenden Meeresbryozoen. *Zool. Jahrb., Abht. Sys* **52,** 1–102.

Marcus, E. (1934). Uber *Lophopus crystallinus* (Pall.). *Zool. Jahrb., Abt. Anat. Ontog. Tiere* **58,** 501–606.

Marcus, E. (1937a). Bryozoarios marinhos Brasileiros. I. *Univ. Sao Paulo, Fac. Filos., Cienc. Let., Bol. Zool.* **1,** 1–224.

Marcus, E. (1937b). Bryozoen von St. Helena. *Vidensk. Medd. Dansk Naturhist. Foren.* **1,** 183–252.

Marcus, E. (1938). Bryozoarios marinhos Brasileiros. II. *Univ. Sao Paulo, Fac. Filos., Cienc. Let., Bol. Zool.* **2,** 1–196.

Marcus, E. (1941). Sôbre Bryozoa de Brazil. *Univ. Sao Paulo, Fac. Filos., Cienc. Let., Bol. Zool.* **25,** 3–208.

Matricon, J. (1963). Dégénérescence du polypide femelle et formation d'une poche incubatrice chez *Alcyonidium polyoum* (Hassal) (Bryozoaire Cténostome). *Arch. Zool. Exp. Gén.* **102,** 79–93.

Mawatari, S. (1952). On *Watersipora cucullata* (Busk). II. *Misc. Rep. Res. Inst. Nat. Resour. (Tokyo)* **28,** 17–27.

O'Donoghue, C. (1924). The Bryozoa (Polyzoa) collected by the SS "Pickle." *Union S. Afr., Fish. Mar. Biol. Surv., Rep.* **3,** 15–65.

Osburn, R. C. (1912). The Bryozoa of the Woods Hole region. *Bull. U.S. Bur. Fish.* **30,** 203–266.

Pace, R. M. (1906). On the early stages in the development of *Flustrella hispida* (Fabricius), and on the existence of a "yolk nucleus" in the egg of this form. *Q. J. Microsc. Sci.* [N.S.] **50,** 435–478.

Paltschikova-Ostroumova, M. W. (1925). Kurze Bemerkung über der Ovidukt bei Bryozoen. *Zool. Anz.* **65,** 100–102.

Prouho, H. (1890). Recherches sur la larve de *Flustrella hispida:* Structure et métamorphose. *Arch. Zool. Exp. Gén.* [2] **8,** 409–459.

Prouho, H. (1892). Contribution à l'histoire des Bryozoaires. *Arch Zool. Exp. Gén.* [2] **10,** 557–656.

Robertson, A. (1903). Embryology and embryonic fission in the genus *Crisia. Univ. Calif., Berkeley, Publ. Zool.* **1,** 115–156.

Ryland, J. S. (1963). Systematic and biological studies on Polyzoa (Bryozoa) from western Norway. *Sarsia* **14,** 1–60.

Ryland, J. S. (1967). Polyzoa. *Oceanogr. Mar. Biol.* **5,** 343–369.

Ryland, J. S. (1970). "Bryozoans." Hutchinson University Library, London.

Seeliger, O. (1906). Über die Larven und Verwandtschaftsbeziehungen der Bryozoen. *Z. Wiss. Zool.* **84,** 1–78.

Silén, L. (1942). Origin and development of the Cheilo-ctenostomatous stem of Bryozoa. *Zool. Bidr. Uppsala* **22,** 1–59.

Silén, L. (1944). The anatomy of *Labiostomella gisleni* Silén (Bryozoa Protocheilostomata). *K. Sven. Vetenskaps. Akad. Handl.* [3] **21,** No. 6, 1–111.

Silén, L. (1945). The main features of the development of the ovum, embryo and ooecium in the ooeciferous Bryozoa Gymnolaemata. *Ark. Zool.* **35A,** No. 17, 1–34.

Silén, L. (1947). On two new groups of Bryozoa living in shells of Mollusks. *Ark. Zool.* **38B**, No. 1, 1–7.

Silén, L. (1948). On the anatomy and biology of Penetrantiidae and Immergentiidae (Bryozoa). *Ark. Zool.* **40A,** No. 4, 1–48.

Silén, L. (1966). On the fertilization problem in the gymnolaematous Bryozoa. *Ophelia* **3,** 113–140.

Silén, L. (1972). Fertilization in the Bryozoa. *Ophelia* **10,** 27–34.

Soule, J. D. (1950a). A new species of *Terebripora* from the Pacific (Bryozoa Ctenostomata). *J. Wash. Acad. Sci.* **40,** No. 11, 378–381.

Soule, J. D. (1950b). Penetrantiidae and Immergentiidae from the Pacific (Bryozoa Ctenostomata). *Trans. Am. Microsc. Soc.* **69,** 359–367.

Stach, L. W. (1938). Observations on *Carbasea indivisa* (Busk) (Bryozoa). *Proc. Zool. Soc. London, Ser. B* **108,** 389–399.

Ström, R. (1969). Sexual reproduction in a stoloniferous bryozoan, *Triticella koreni* (G.O. Sars). *Zool. Bidr. Uppsala* **38,** 113–127.

Waters, A. W. (1907). *Tubucellaria:* Its species and ovicells. *J. Linn. Soc. Zool.* **30,** 126–132.

Waters, A. W. (1912). A structure in *Adeonella (Laminopora) contorta* (Michelin) and some other Bryozoa, together with remarks on the Adeonidae. *Ann. Mag. Nat. Hist.* [8] **9,** 489–500.

Waters, A. W. (1913). The marine fauna of British East Africa and Zanzibar from collections made by Cyril Crossland. Bryozoa-Cheilostomata. *Proc. Zool. Soc. London* Parts 3–4, pp. 458–537.

Waters, A. W. (1914). The marine fauna of British East Africa and Zanzibar from collections made by Cyril Crossland. Bryozoa-Cyclostomata, Ctenostomata and Endoprocta. *Proc. Zool. Soc. London* Parts 3–4, pp. 831–858.

Woollacott, R., and Zimmer, R. (1971). Attachment and metamorphosis of the cheilo-ctenostome bryozoan *Bugula neritina* (Linné). *J. Morphol.* **134,** 351–382.

Woollacott, R., and Zimmer, R. (1972). Origin and structure of the brood chamber in *Bugula neritina* (Bryozoa). *Mar. Biol.* **16,** 165–170.

Zirpolo, G. (1933). *Zoobothryon verticillatum* (Della Chiaje). *Mem. Accad. Nuovi Lincei* **17,** 109–441.

3

Structure and Classification of Gymnolaemate Larvae

RUSSEL L. ZIMMER and ROBERT M. WOOLLACOTT

I. INTRODUCTION

The larvae of numerous gymnolaemates (Ctenostomata and Cheilostomata) have been illustrated during the past century, especially during the late 1800's and early 1900's. An excellent bibliography to this literature has been compiled by Nielsen (1971). The pelagic and planktotrophic cyphonautes larvae of *Membranipora* and *Electra* are among the most commonly recognized of any marine larva because of their regular presence and often incredible abundance in the plankton. The brooded larvae of other genera have only a brief dispersal phase and are rarely taken from the plankton. However, as their release is light triggered, these larvae may be readily obtained by maintaining mature colonies in dishes overnight and collecting the released larvae the following morning.

In the following section the morphological variations in gymnolaemate larvae are assessed. We have approached this comparative survey through a systemic analysis. This review provides a framework on which a morphological classification of larval types can then be based.

II. LARVAL STRUCTURE

Terms for the structure and orientation of bryozoan larvae are multiple and not always obvious in their meaning. A median sagittal section of a hypothetical larva is represented and labeled in Fig. 1 to indicate the major terms employed in this chapter.

Several of the terms should perhaps be justified. The designations of oral and aboral are widely used, despite the fact that few bryozoan larvae have mouths. The oral and aboral fields are derived from the vegetal and animal regions of the embryo, respectively. The two fields are separated by a ring of heavily ciliated cells, the corona. This regionation of the larva is suggested as early as the 32-cell stage of cleavage. The mesendodermal cells are proliferated from near the vegetal pole, but an open blastopore is usually absent. If the embryo is provided with a

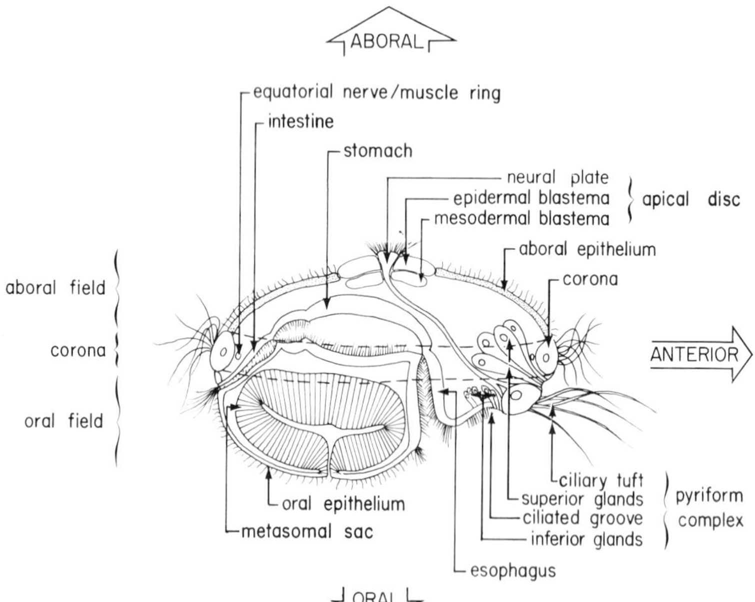

Fig. 1. Diagram of a hypothetical gymnolaemate larva represented in midsagittal section with general larval components illustrated.

larval mouth, this originates near the vegetal pole also. The terms dorsal and ventral are used for the aboral and oral fields by some scholars but this can be misleading. The use of the terms anterior (for the region bearing the pyriform complex) and posterior is a matter of convenience in a sense, for the larva usually swims with the aboral pole forward or at an angle from this toward the anterior end.

The majority of larvae of gymnolaemates are without a satisfactory name; only the term cyphonautes is well defined and widely used. For brooded larvae the term buguliform larvae has been applied at times, but this has the obvious disadvantage of implying a genus. Maturo (1971) suggested the term coronate larva for the noncyphonautes larvae of gymnolaemates, in reference to their distinctive band of ciliated cells, the corona. Although a corona is also present in shelled larvae and in the larvae of cyclostomes, we believe "coronate larva" is a useful term. We have utilized it to indicate all gymnolaemate larvae except those which are cyphonautes (shelled, planktotrophic) or shelled, lecithotrophic ones. As the larvae of cyclostomes are apparently of only one type, the term cyclostome larva is diagnostic.

A. Aboral Field

The aboral field is typically convex, but it ranges in extent from most of the external surface in shelled forms to a small fraction of that surface in larvae with extensive coronas. The central region of the aboral field is consistently differentiated as an apical disc. The remaining aboral epithelium may be essentially uniform, but usually is specialized into several tissues.

1. Apical Disc

The apical disc (also referred to as the calotte, polar calotte, ciliated disc, retractile disc, disciform organ, or apical complex) is a composite structure centered around the former animal pole. In cyphonautes larvae and similar forms, the term apical knob would be more descriptive of the small rounded projection that forms only a fraction of the aboral field (Fig. 2). In coronate larvae, the disc is slightly concave, flattened, or slightly convex, and often has a diameter more than half that of the larva (Fig. 12). In species with extensive coronas, such as *Cellepora plumosa* (Fig. 13), *Bugula neritina* (Figs. 14 and 15), and *Bowerbankia pustulosa* (Fig. 16), the apical disc occupies virtually the entire free aboral surface.

Typically, the disc has a small, unpigmented, central area in the form of a circle or crescent and a peripheral zone that frequently is decorated in a radial pattern; bands of this pattern may be alternately pigmented and unpigmented (Fig. 5). The disc may bear regionalized or overall ciliation and for many species of coronate larvae is indicated to be bordered by a single ring of long immotile "stereocilia."

The central zone constitutes a neural plate (apical plate, sensory plate, sensory disc, apical sense organ, etc.). This designation is based on morphological evidence and the topological equivalence of this region with the larval "brains" of other invertebrate larvae. The tall columnar cells of this plate may be ciliated (*Electra pilosa,* Kupelwieser, 1905; *Alcyonidium polyoum,* Hondt, 1973a) or provided only with apical microvilli (*Bugula neritina,* Woollacott and Zimmer, 1971); see Figs. 2, 3, and 4, respectively.

Basally the plate cells join a fibrillar cord reported to be both neural and muscular (Figs. 1, 3, and 4). In *Alcyonidium polyoum* synaptic vesicles have been observed in the basal extremities of neural plate cells, and synapselike approximation of the cells of the plate and those of the neuromuscular cord have been figured (Hondt, 1973a). No specific sensory or coordinative function has been demonstrated for the neural plate, but the neuromuscular cord and

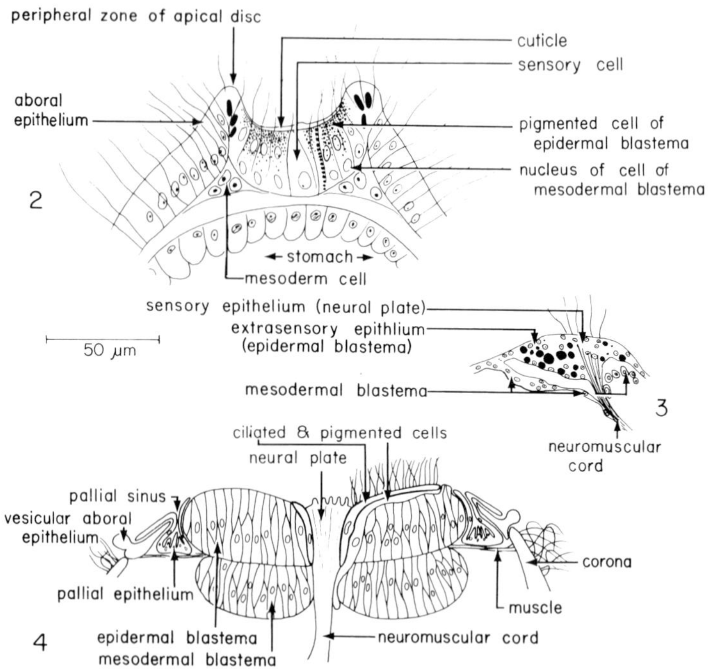

Figs. 2, 3, and 4. Midsagittal section through the apical discs of three gymnolaemate larvae illustrating the wide variation in development of the blastema. Drawn to same scale.

Fig. 2. *Electra pilosa.* From Kupelwieser (1905).

Fig. 3. *Alcyonidium polyoum.* From Seeliger (1906).

Fig. 4. *Bugula neritina.* From Woollacott and Zimmer (1971).

other larval nerves do unite the plate with the effector and other sensory systems of the larva, namely, the corona, the larval eyespots (if present), and the pyriform complex. Normally, this entire neural complex of the larva is histolyzed shortly after settlement of the larva (e.g., Calvet, 1900). Hondt (1975) indicated that the apical plate forms part of the tentacle sheath in the ancestrular polypide of *Alcyonidium polyoum* and the entire sheath in *Bowerbankia imbricata.* In any event, it seems certain that the larval and ancestrular nervous systems are totally independent.

Excluding putative sensory cells that may exist on its surface in certain species, the peripheral zone of the apical disc is nonsensory and composed totally or in large part of columnar epithelial cells. The knoblike discs of *Electra pilosa* (Fig. 2), as well as the massive one of *Bugula neritina,* have only columnar cells, but in some coronate larvae the central or the cortical regions of the peripheral zone may be composed of cuboidal rather than columnar cells. At least the columnar cells (and probably also the cuboidal ones where they exist) collectively constitute an epidermal blastema (epidermal thickening, external thickening, or extrasensory epithelium).

Underlying all or part of the epidermal blastema is a layer of cells, the mesodermal blastema (mesodermal thickening or internal thickening) (Figs. 1–4). Although properly an internal structure, the mesodermal blastema is introduced here because of its intimate association with and cytological similarity to the epidermal blastema from which it probably originated and with which it shares an allied fate. In earlier works, the two blastemas were often represented as forming a single stratified epithelium with nuclei arranged in two layers [e.g., Kupelwieser, 1905, for *Electra pilosa* (Fig. 2); Vigelius, 1886, for *Bugula calathus*].

Calvet (1900) appears to be the first to have appreciated fully the significance of the two blastemas and he figured these structures for a large number of species. He recognized that these elements were variably developed in different species and correctly deduced that together they were responsible for the formation of the epidermal and mesodermal components of the polypide of the ancestrula.

It is significant that blastemas have been found, when looked for, in all but two studies. In an ultrastructural analysis of the larva of *Alcyonidium polyoum,* Hondt (1973a) specifically indicated that he could find neither the epidermal nor mesodermal anlagen. However, in none of his figures does he show the cells bordering the apical sensory cells, so it is difficult to evaluate his findings. In earlier studies on the same species, Harmer (1887), Seeliger (1906), and Zschiesche (1909) (Fig. 3) illustrated what almost certainly are the blastemas; Hondt did not comment on the discrepancies between his observations and those of earlier workers. In a subsequent paper (Hondt, 1975), the same author reported the apical disc of *Bowerbankia imbricata* to consist of but one cell type (presumably neural); in contrast the apical disc of *Bowerbankia pustulosa* (Fig. 15) has the usual composite structure (Calvet, 1900).

The ultrastructure of the blastemas has been examined only for *Bugula neritina* (Woollacott and Zimmer, 1971), a species Calvet (1900) recognized to have exceptionally well-developed blastemas. In *B. neritina,* the cells of both blastemas (Fig. 4) are narrow and columnar; distinctive features of the cytoplasm are the droplets of densely staining reserve material, numerous microtubules paralleling the lateral cell surfaces, and, above all, a wealth of free ribosomes, indicative of the blastemal nature of these cells.

The epidermal blastemal cells in *Bugula neritina* possess microvilli but no cilia and reach the surface only in about 50 narrow bands radiating over the surface of the apical disc. The remainder of the surface of the peripheral zone is occupied by a corresponding number of thin, straplike ciliated cells that extend radially from the central cap to the margin of the disc. Mesodermal pigment cells underlie and are visible through the delicate ciliated cells (Fig. 4) as the dark radiating spokes of a sunburst pattern (Fig. 5). The significance of these radial cells is not clear in this or other species, but Barrois (1886) suggested such elements were part of the neurosensory complex.

2. *Aboral Epithelium*

The epidermal tissue between the oral margin of the apical disc and the aboral margin of the corona may be composed of uniformly unspecialized cuboidal or columnar cells (e.g., *Bulbella,* Braem, 1951) (Fig. 11), but usually has regional specializations.

In shelled larvae, most of the aboral epithelium is squamous and involved in production of the larval shells (Fig. 7). However, in the cyphonautes of *Electra pilosa,* the tissues surrounding the apical disc and lying just above the corona are vesicular and those cells between the anterior and posterior margins of the paired valves are ciliated (Kupelwieser, 1905).

The most common differentiation in coronate larvae is for part of the aboral epithelium to be depressed in a circular furrow or sinus at the margin of the apical disc. This pallial sinus may be only a superficial and somewhat open groove (Fig. 12) [e.g., *Alcyonidium polyoum* (as *A. mytili*), Barrois, 1877], may be lined by highly specialized cells that are virtually sealed off from the exterior by the intimate approximation of the lips of the furrow (Figs. 4 and 15) (e.g., *Bugula neritina,* Calvet, 1900; Woollacott and Zimmer, 1971), or may be extremely deep reaching nearly to the oral field except in the region of the pyriform complex (Fig. 16) (e.g., *Bowerbankia pustulosa,* Calvet, 1900).

The ultrastructure of the cells lining the pallial sinus has been examined in both *Alcyonidium polyoum* and *Bugula neritina.* In the former, the cells are columnar with clavate microvilli at their free surfaces. The abundance of densely granular endoplasmic reticulum and the concentration of strongly PAS-positive

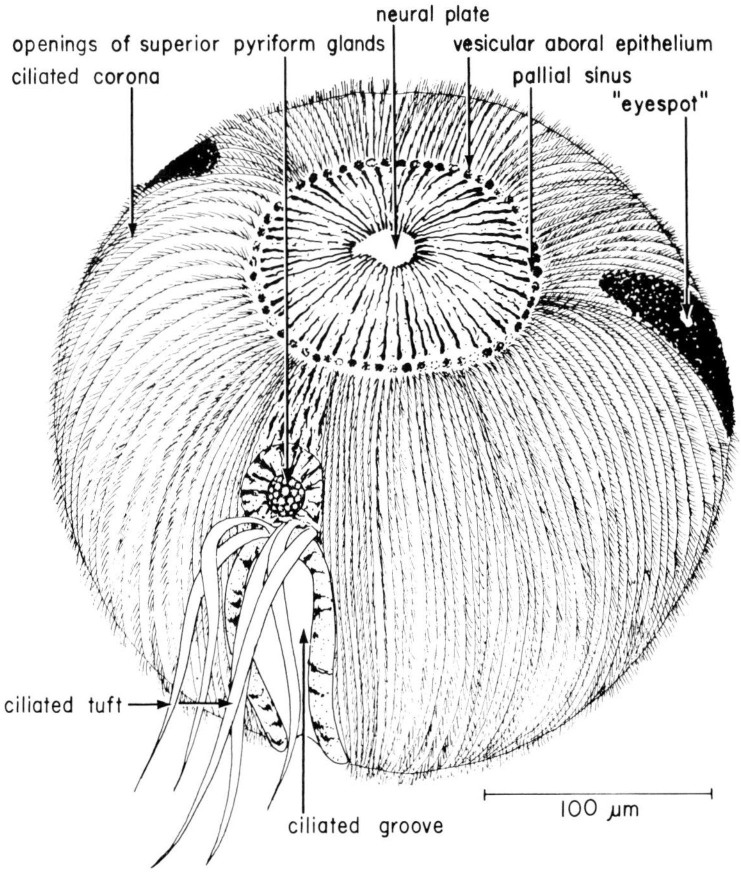

Fig. 5. *Bugula neritina.* Oblique anterior-aboral view of larva. From Woollacott and Zimmer (1971).

"granulations" at the apex of these cells indicate the secretory nature of this tissue; Hondt (1973a) noted the similarity between these cells and those of the metasomal sac.

The pallial sinus tissue in *Bugula neritina* is strikingly different from that in *Alcyonidium.* Virtually cut off from the surface (Fig. 4), the *Bugula* cells appear to form an entangled rete rather than an epithelium. The cells contain prominent Golgi complexes and numerous microtubules, but the paucity of other cytoplasmic organelles and inclusions in the extensive ground cytoplasm is striking (Woollacott and Zimmer, 1971).

The dichotomy between the cytology of the pallial sinus epithelia in these two species may seem incongruous. In fact, it is easily explicable as the pallial sinus epithelium in *Alcyonidium* will contribute a portion of the body wall epidermis with its thick cuticle, but in *Bugula neritina,* it will eventually form the delicate squamous epithelium of the tentacle sheath.

When a pallial sinus is formed, it may involve essentially all or, more commonly, only part of the aboral epithelium. The junctions of the pallial epithelium with the apical disc and with the remaining aboral epithelium are made by distinctive cells at least in *Bugula neritina* (Woollacott and Zimmer, 1971) and in *Alcyonidium polyoum* and *Flustrellidra hispida* (Hondt, 1975). At metamorphosis these cells either contribute to the tentacle sheath or are lysed.

The aboral epithelium is frequently specialized at its junction with the corona into glandular cells (Fig. 15). Protrusions of the apical surfaces of these cells are filled with vesicular globules and collectively form a ring, the vesicular collarette, immediately aboral to the corona.

In general, the aboral epithelium, excepting that which is differentiated into a vesicular glandular tissue, is retained through metamorphosis as a component of the body wall epidermis of the ancestrular cystid or as the tentacle sheath of the ancestrular polypide (or perhaps contributes to both). The vesicular gland cells of both cyphonautes and coronate larvae undergo histolysis at metamorphosis.

B. Corona

The corona constitutes the locomotory organ in all gymnolaemate larvae. It characteristically originates from a supraequatorial ring of 8 cells in the 32-cell stage during embryogenesis. Corrêa (1948) indicated these cells have the same cell lineage as trochoblasts of annelidan and molluscan trochophores, but this is patently impossible if one compares the cleavage sequences. The coronal cells provide a convenient landmark in developmental studies, for they undergo relatively few divisions and consequently are larger than other cells and soon become ciliated. The number of cells in the definitive corona is 32 in most species, but is more than 300 in *Bugula neritina.* Invagination of the oral face in cyphonautes and its flattening in certain other larvae results in a displacement of the corona from near the equator to the lower margin of the larva. The corona remains as a complete and uniform ring in all species except those with true cyphonautes larvae; in these, the coronal ring is very tenuous at each side of the inhalant aperture (see below and Fig. 7).

As indicated in Section III, the corona may form only a narrow band or may constitute most of the larval surface. Even in the latter case, the corona is formed of only a single row of cells.

The degree to which the free surface of these cells is occupied by cilia is variable. Only a narrow band of cilia is found on the coronal cells of *Alcyonidium polyoum* (Hondt, 1973a) (Fig. 12), but the elongated coronal cells of

Bugula neritina are virtually completely ciliated at their apical surface (Figs. 14 and 15).

In gymnolaemate larvae, metachronal waves pass around the corona in a clockwise manner when the larvae are viewed from the aboral pole. In forms that are symmetrical around the oral–aboral axis, the larvae usually spin rapidly in a clockwise direction when observed from the aboral pole and swim with that pole foremost. Larvae that are elongated along their anterior–posterior axis usually swim without rotation in the direction of the aboral pole or at an angle between that pole and the anterior end of the larva.

The fine structure of the corona has been described for both *Bugula neritina* (Woollacott and Zimmer, 1971) and *Alcyonidium polyoum* (Hondt, 1973a), but is not detailed here as the corona is another transitory larval tissue destined for destruction at metamorphosis.

C. Oral Field

The oral field of gymnolaemate larvae may be extensive or reduced and is centered on the former vegetal pole of the embryo. This field is basically convex in coronate larvae, but is deeply invaginated in cyphonautes and is flattened in shelled lecithotrophic and certain coronate larvae.

Invariably, to our knowledge, in both shelled and coronate larvae an anteriorly positioned pyriform complex and a posteriorly or centrally positioned invagination, the metasomal sac (see Section II, C, 2), are developed in the sagittal midline of the oral field. It should be noted that in coronate larvae with highly developed coronas that "invade" the oral field, the pyriform complex appears to lie within the coronal region rather than constituting part of the oral field (Fig. 5). In these larvae, however, coronal cells flank the lateral margins of the complex but do not interrupt its continuity with the oral field in the median line. Except for vesicular gland cells which may border the corona, the remaining surface of the oral field is covered by a simple, ciliated squamous epithelium in most species. In cyphonautes, shelled lecithotrophic, and some coronate larvae a mouth is present between the pyriform complex and the opening to the metasomal sac (Figs. 7, 8, and 11). In true cyphonautes, which are planktotrophic, the oral field surrounding the mouth is deeply recessed as the vestibule and an anus is formed posterior to the metasomal sac.

1. Pyriform Complex

The pyriform complex or organ is a conspicuous feature of all gymnolaemate larvae (Fig. 1). In living larvae, the characteristic flicking motion of the ciliary tuft or vibratile plume attracts one's attention. This tuft is composed of several bundles (five in *Callopora dumerili,* Silén, 1944b) of multiple cilia (Calvet, 1900;

Woollacott and Zimmer, 1971). It beats only in the sagittal plane and at the completion of its effective stroke it is directed orally and momentarily passes through a shallow groove, the ciliated cleft or groove, that lies in the midline below it.

The ciliated groove is deepest near the base of the ciliary tuft and is lined by ciliated epidermal cells. In optical sections, the cilia, at the margin of the groove, appear immotile and are arranged in one or more paired bands.

Intimately associated with the ciliary tuft and ciliated groove are clusters of flask-shaped unicellular glands. Opening to the surface just above the base of the ciliated tuft are cells of the superior glandular "field." The inferior glandular "field" consists of a median group of cells that open between the base of the ciliary tuft and the beginning of the ciliated groove and a pair of lateral clusters of gland cells that open on the margin of the groove. Cytologically, these inferior and superior gland cells appear identical, and collectively they occupy a considerable portion of the interior of the larva.

At their basal ends, the ciliated cells of the complex are joined with the neural plate by the neuromuscular cord. Gymnolaemate larvae appress the ciliary tuft and ciliated groove against the substrate prior to metamorphosis, as if testing for a suitable site for attachment. The action of the ciliary tuft could remove portions of the surface film for "analysis" either by the tuft itself or the cells of the ciliary groove into which it enters on completion of its effective stroke. Most authors agree that substrate selection is accomplished by the pyriform complex, but this hypothesis, although obvious, lacks any experimental confirmation.

Jägersten (1972) has suggested that the complex originally served as a creeping organ at a time when the ancestral adult was benthic but vagile rather than attached. Locomotion does not appear to be the primary function of the complex now, but the gliding movements of the larvae during the thigmotactic phase appear to be mediated by the ciliation of the groove. It has also been suggested that the pyriform complex could aid in steering the cyphonautes during its pelagic phases (Atkins, 1955b).

A possible feeding role of the pyriform complex has also been suggested. For the anenteric coronate larva of *Callopora,* Silén (1944b) conjectured that the plume could sweep particles from the water and transport them to the ciliated cleft for use in nutrition. As coronate larvae live for only a few hours in the plankton, it is questionable whether this putative function would be of any utility to such larvae. Kupelwieser (1905) indicated that the ciliation of the pyriform complex directs a food-bearing current into the inhalant chamber of the vestibule in the planktotrophic cyphonautes of *Electra.* Neither Atkins (1955b) nor Strathmann (1973) in their analyses of feeding in cyphonautes observed such a current, however.

The function or functions of the superior and inferior gland clusters are also enigmatic. Atkins (1955a) did note that cyphonautes larvae after pressing the pyriform complex against the substrate would often be "tethered" to that surface by a delicate thread reportedly secreted by the complex. The gland cells might also function to adhere the cilia of the tuft into groups or to make sticky the surface of these cilia so they could more effectively collect particles. Even if the complex served all these functions, the secretory reserves would seem to exceed the need for mucus by far. In *Bugula neritina* after metamorphosis, the degenerating gland cells still contain an abundance of secretory granules.

2. *Metasomal Sac*

The metasomal sac (adhesive sac, internal sac) is an invagination of the oral epithelium and is invariably present in the larvae of gymnolaemates. The dimensions of the sac range from small in cyphonautes (Fig. 7) (e.g., Kupelwieser, 1905) and coronate larvae of the stoloniferans *Bowerbankia* (Fig. 16) and *Amathia* (Barrois, 1887; Calvet, 1900) to massive, occupying one-half or more of the larval interior in the coronate larvae of cellularioids (Fig. 15) (e.g., Calvet, 1900).

Calvet (1900) in his superb comparative study and Kupelwieser (1905) for *Electra pilosa* have distinguished two regions of the metasomal sac (Fig. 7). Calvet and Kupelwieser noted that cells of the metasomal sac near its opening were filled with large, densely staining secretory vesicles. At the onset of metamorphosis as the sac is everted the cells release their contents against the substrate. With further evagination the deeper columnar cells of the metasomal sac contact the cement and are anchored by it to the substrate. The precise fate of the cement-producing cells is not specified, but the deeper cells form the attachment disc and, in some cases, other parts of the ancestrular epidermis.

Woollacott and Zimmer (1971) concluded that the metasomal sac of *Bugula neritina* (Fig. 15) is differentiated into three regions called the neck, wall, and roof. The epidermis forming the neck region is conspicuous because of the large, darkly staining inclusions found throughout its cells. The secretion of this material at the onset of metamorphosis was suggested to provide the calyx which encases the transforming larva. This delicate extracellular envelope has not been observed in other species and its function is not clear. This neck region topologically and cytologically corresponds to the cement-producing zone noted by Kupelwieser and Calvet. For *Bugula neritina,* Woollacott and Zimmer (1971) reported that the adhesive material is directly produced by cells of the attachment disc, but subsequent observations indicate that secretory products of the neck region may also be involved. The neck region is lysed during metamorphosis with the corona and other transitory larval tissues.

The majority of the metasomal sac in *Bugula neritina* is formed by the wall region, the cells of which have a highly polarized regionation of organelles. The apices of these cells are filled with pale-staining vesicles. Continuing basally, one would find next a zone of mitochondria, small densely staining inclusions, and Golgi bodies presumably involved in packaging of secretory material. A nuclear zone separates this last region from the basal section of the cell in which rough endoplasmic reticulum is compacted. The wall region of the metasomal sac in *Bugula neritina* is folded upon itself, thus virtually eliminating the sac's lumen. This region is destined to form the entire epidermis of the free surface of the ancestrular cystid after metamorphosis.

In *Bugula neritina* there is an innermost region, the roof, which is composed of very elongate cells, the apices of which are filled with numerous, large secretory vesicles or droplets containing a pale-staining, PAS-negative material believed to assist in anchoring the larva at metamorphosis. This region forms the rather small attachment disc of the erect ancestrula. Hondt (1975) reported that the metasomal sac of *Bowerbankia imbricata* does not correspond to that of other larvae, but is constituted of two epidermal tissues normally part of the exposed oral field (see Section II, C, 5).

3. Mouth and Anus

As mentioned earlier, the presence of a mouth and anus or a mouth alone or the absence of both is related to the degree of development of the larval gut. The digestive system will be considered with other internal structures, but as the entrance to and exit from it are associated with the oral field they are noted here.

4. The Vestibule

In the planktotrophic cyphonautes, the entrance to the esophagus (or pharynx) is recessed deeply into the interior of the larva as a conical invagination of the oral field, the vestibule (mantle cavity). The vestibule (Figs. 6 and 7) has become highly specialized as a food-collecting device that delivers phytoplankters to the digestive system. A beautiful analysis of the morphology and function of the vestibule was achieved by Atkins (1955b) and was confirmed by Strathmann (1973). In these studies it was demonstrated that the vestibule is functionally divided into an inhalant and exhalant chamber by ciliated lateral ridges. These ridges, as the tentacles of the adult, bear lateral, laterofrontal, and frontal cilia that create the feeding current and separate food items from it. The corona of the planktotrophic cyphonautes is considerably less developed where it borders the inhalant aperture than elsewhere; in all other larvae, none of which feed or have a vestibule, the corona is a uniform band of cilia.

The vestibule then is a remarkable adaptive elaboration of the oral epithelium. It will be lysed with other purely larval tissues after metamorphosis.

5. The Epithelium of the Exposed Oral Field

The free surface of the oral field is clothed by so-called oral epithelium. At its junction with the corona, this epithelium may be produced into a glandular tissue forming an oral vesicular collarette (Fig. 15). In *Bugula*, with the exception of the gland cells just noted, the oral epithelium consists of squamous to cuboidal ciliated cells.

In *Alcyonidium polyoum*, Hondt (1973a) indicated a band of oral field cells immediately beneath the corona to be of special significance; he referred to them as "M" cells. They possess a poorly developed granular endoplasmic reticulum around the large basal nucleus which has a conspicuous nucleolus. Small mitochondria and an occasional Golgi body are found more apically. The cytoplasm, "*dense et très colorable au citrate de plomb,*" contains vacuoles and granules. Hondt indicated that the cells have an undifferentiated appearance and asserted they are homologous with the cells of epidermal blastemas of other larvae, for he claimed they form the epidermal components of the polypide anlage during metamorphosis (Hondt, 1973a).

Hondt (1975) reported that subcoronal cells with the same cytology and fate as those of *A. polyoum* also occur in *Bowerbankia imbricata*. In this stoloniferan ctenostome, however, these undifferentiated cells have invaginated at the oral pole and surround a second epidermal tissue that has also invaginated. The tissue of the metasomal sac which normally occupies the position taken by these two cell layers is reported to be totally absent in *B. imbricata*.

Presumably undifferentiated epidermal cells flanking the pyriform complex were reported in the ascophoran *Schizoporella unicornis* and the stoloniferan *Amathia lendigera* (as *Serialaria lendigera*) to form mesoderm of the ancestrular polypide (Barrois, 1879, 1886).

With the possible exception of these undifferentiated cells in *Alcyonidium, Bowerbankia, Schizoporella,* and *Amathia* (see Chapter 4), the general oral epithelium of gymnolaemate larvae is resorbed at metamorphosis and plays no role in the formation of the ancestrula.

D. Internal Tissues

With few exceptions, the interior of gymnolaemate larvae is filled solidly with cells: blastocoelic spaces are usually obliterated, coelomic cavities apparently never occur, and a hollow gut is rarely developed. The internal elements include endodermal derivatives, nerves, muscles, the mesodermal blastema, and other mesodermal cells.

1. The Gut and Gut Rudiments

A complete and functional digestive tract is found only in true cyphonautes, but a complete gut is also reported for the lecithotrophic larvae of *Tendra zostericola* (Repiachoff, 1875). The nonfeeding larvae of *Alcyonidium* (Prouho, 1892), *Nolella* (Prouho, 1892), *Victorella, Sundanella,* and *Bulbella* (Braem, 1951), and *Triticella koreni* (Ström, 1969) have evanescent digestive systems of varying degrees of completeness. In the vast majority of coronate larvae the cells internalized at gastrulation do not form even traces of a tubular gut.

The functional gut of the cyphonautes of *Electra pilosa* appears in a number of the detailed drawings by Kupelwieser (1905) (Fig. 7), but the most thorough description of the larval gut is provided by Atkins (1955b) for *Membranipora membranacea* larvae. The digestive tract of this latter cyphonautes consists of a rather simple ciliated tube that is differentiated into an esophagus, stomach, and intestine. Only the esophagus is provided with extrinsic muscles, a layer of circular fibers which form a sphincter at its entrance. A band of secretory cells is found on the floor of the stomach and intestine. Cells of the lateral walls of the stomach appear to function in intracellular digestion.

Whether fully developed and functional or incomplete, the larval digestive system is transitory. In cyphonautes, its histolysis is accomplished shortly after metamorphosis (e.g., Kupelwieser, 1905). In forms with nonfunctional guts, degenerative changes may begin even during embryogenesis. For example, in *Alcyonidium polyoum* a pharynx and sacciform stomach are present in the early embryo; the former is degraded during metamorphosis but the latter is very short-lived and has usually degenerated by the larval stage (Barrois, 1877; Hondt, 1973a, 1974).

The majority of coronate larvae have no traces of a tubular gut, although endodermal cells presumably are produced at gastrulation. Within the interior of these larvae, cells filled with large yolk deposits and lipid droplets are abundant; such parenchymal cells probably are endodermal derivatives. In the cyphonautes lipid reserves are also concentrated in parenchymal cells during its planktotrophic existence. During elaboration of the ancestrula, these cells and transitory larval tissues accumulate adjacent to the polypidal rudiment in a mass that gradually is reduced as the polypide becomes functional and the ancestrula produces its buds. The depleted cells are stated to form a residual brown body attached to the cecum of the ancestrular polypide (Calvet, 1900, for *Bugula simplex* as *B. sabatieri*).

2. *Nerves and Muscles*

As previously mentioned, the apical sensory cap and the pyriform complex are united by a neuromuscular cord located in the interior of the larva. Presumably, at least the neural components of that cord would be of ectodermal origin. The source of the muscular components could be either ectodermal, ectomesodermal, or entomesodermal. Prouho (1892) indicated the cord to originate from a pair of precociously segregated entomesodermal cells in *Membranipora membranacea.* This observation has not been repeated to the best of our knowledge.

In addition to this median nerve there is a conspicuous neuromuscular ring immediately beneath the corona (e.g., Calvet, 1900; Woollacott and Zimmer,

1972; Hondt, 1973a). Detailed descriptions of the nervous system are available for various larvae (e.g., Kupelwieser, 1905; Seeliger, 1906) and certain aspects of the ultrastructure have been considered by Hondt (1973a) and Woollacott and Zimmer (1971, 1972).

Many coronate larvae possess putative photoreceptors that usually appear as orange or red pigment cup ocelli. These eyespots (Fig. 5) most commonly occur near the equator in association with the corona but they may be found alternatively on the oral and aboral fields. Their occurrence is not universal within or restricted to larvae that have a positively phototactic phase and their number, distribution, or presence may be highly variable even within a genus (Ryland, 1960). The fine structure and organization of the eyespots of *Bugula neritina* have been reported by Woollacott and Zimmer (1972). This information is briefly reviewed by Ryland in Chapter 12 of this volume.

All gymnolaemate larvae possess a variety of striated and nonstriated muscles. Presumably they are derived from entomesodermal elements, but this is by no means certain; in larvae of other phyla, the larval musculature is frequently of ectomesodermal origin. Our knowledge of bryozoan early development is insufficient to resolve this question at the present time.

The full larval musculature has been described in detail only for the cyphonautes of *Electra pilosa* (Kupelwieser, 1905) and the shelled larva of *Flustrellidra hispida* (as *Flustrella hispida,* Prouho, 1890). Certain muscles have been described for other species but it is difficult to establish homologies. For example, in *Alcyonidium polyoum,* the eversion of the metasomal sac is accomplished indirectly by wall muscles that extend from above to below the corona and which on contraction shorten the oral–aboral axis, but in *Bugula neritina,* it is an equatorial muscle beneath the corona that increases the internal body pressure to evert the sac (Figs. 14 and 15). For details on the muscle system, the reader is referred to the original literature: Prouho (1890) for *Flustrellidra hispida* (as *Flustrella hispida*); Kupelwieser (1905) for *Electra pilosa;* and Hondt (1973b) for *Alcyonidium polyoum* and a brief comparative analysis.

The larval neuromuscular system including its epidermal elements are replaced by adult components at metamorphosis (Calvet, 1900) with the apparent exception of certain muscles in the larva of *Flustrellidra hispida* which are retained as the parietals of the ancestrula (Prouho, 1890).

3. *The Mesodermal Blastema*

The mesodermal blastema has already been considered when discussing the epidermal blastema located on the aboral field (see Section II, A, 1). As indicated there, both morphological and cytological reasons exist for considering this tissue to be derived from ectodermal tissue, specifically the epidermal blastema.

4. *Other Mesodermal Components*

In addition to the above mesodermal components, amoeboid mesenchymal cells may occur throughout the larval interior. At metamorphosis they differentiate as phagocytes involved in the degradation of larval transitory tissues and as cells of the funicular and muscular systems of the ancestrula (Calvet, 1900). Other mesodermal cells may be found as an epithelial sheet lining the basal ends of parts or all of the larval ectodermal derivatives. Presumably such tissue would form the somatic peritoneum of the ancestrular cystid after metamorphosis. A variation on this theme is found in *Bugula neritina* (Woollacott and Zimmer, 1971). The pigmented mesodermal cells that form the somatic mesothelium of the ancestrula are not arranged in an epithelium in the larva but are largely intercalated between the basal ends of adjacent coronal cells (Fig. 15).

III. LARVAL TYPES

There is considerable variation in the external appearance of gymnolaemate larvae, but their organization is based on a common plan. The relative sizes of individual components of this basic morphology account for most of the variability, but certain features such as larval shells and functional digestive tracts are restricted to only a small number of species. Only a few of the possible combinations of variability have been exploited, permitting one to categorize larvae on the basis of their gross morphology. Within these main larval types, differences in dimensions, number of coronal cells, pigmentation, and the number and distribution of eyespots (when present) may permit the identification of a given larva to species, but are not useful above this level.

There have been several previous attempts to categorize gymnolaemate larvae on the basis of their development and morphology (e.g., Barrois, 1877; Korschelt and Heider, 1899; Marcus, 1940). In the classification suggested below, we have attempted to retain the strong points of previous systems and to include all larvae that have been adequately described or illustrated. The proposed classification may not be adequate to accommodate the full larval variation that exists in gymnolaemates, for our knowledge of larvae is still limited to but a fraction of the known adult species.

We should like to stress that the assemblages of species represented in all but a few of the following larval categories may not be natural ones. The larvae have been grouped on the basis of their gross morphology without other considerations. On the basis of a single criterion—presence or absence of larval shells—we have initially distinguished gymnolaemate larvae as either shelled or coronate. Two and five subdivisions, respectively, have been recognized within these two major categories.

A. Shelled Larvae

Gymnolaemate larvae provided with shells constitute two distinctive types, one planktotrophic and the other lecithotrophic. Members of the first group are so familiar as cyphonautes larvae that we have used that name rather than "planktotrophic shelled larvae" to contrast with "lecithotrophic shelled larvae." The term "cyphonautes-like" has been used for the lecithotrophic forms, but we believe such usage implies a closer relationship to the planktotrophic ones than exists. Both categories of shelled larvae have a narrow corona that is asymmetrically positioned because the oral field is flattened or invaginated. Additionally, they have very small apical discs and lack pallial sinuses. This suite of characters is also found in Type O coronate larvae.

1. Cyphonautes Larvae (Shelled, Planktotrophic Larvae)

Cyphonautes larvae (Figs. 6 and 7) are obligatorily planktotrophic and have the larval body strongly compressed bilaterally. The lateral surfaces of the aboral epithelium produce the paired, flattened, chitinous shells or valves that are joined by adductor muscles. The shells are triangular, hyaline, and smooth in the more familiar species (e.g., *Electra pilosa* and *Membranipora membranacea*), but range to elliptical, opaque, particle-encrusted ones in *Electra crustulenta.* The oral field is deeply invaginated as a conical vestibule with the mouth at its apex. Except in very young larvae the corona does not form a continuous uniform ring as in all other forms but is interrupted—at least in part—at the level of the inhalent aperture of the vestibule into "preoral" and "postoral bands." The larval gut is complete and is functional throughout the long pelagic phase. The small metasomal sac is situated between the mouth and anus. Species with true cyphonautes are oviparous. The small yolk-poor eggs are shed directly into the seawater via intertentacular organs except in a few species with simple external brooding for a brief period. Cyphonautes larvae are characteristic of the genera *Conopeum, Electra,* and *Membranipora** (all anascan cheilostomes of the superfamily Malacostegoidea, see Nielsen, 1971) and are reported for the closely related *Tendra repiachowi* (e.g., Ostroumoff, 1886a,b, as *Membranipora repiachowi*) and *Pyripora catenularia* (Abbott, 1973). Larvae of this type have also been described for the ctenostomes *Alcyonidium albidum* (Carnosa, Halcyonelloidea) (Prouho, 1892), *Farella repens*† (Stolonifera, Walkerioidea) (Marcus, 1926), and *Hypophorella expansa* (Stolonifera, Walkerioidea) (Prouho, 1892).

2. Shelled, Lecithotrophic Larvae

In addition to true cyphonautes, shelled larvae are known only for the species *Flustrellidra hispida* (Fig. 8) (as *Flustrella hispida,* e.g., Pace, 1906), *Pherusella*

*However, see Cook (1962) re. *M. seurati*

†The early development of this form indicates the larva is a true cyphonautes, but the later, presumably shelled stages are not known.

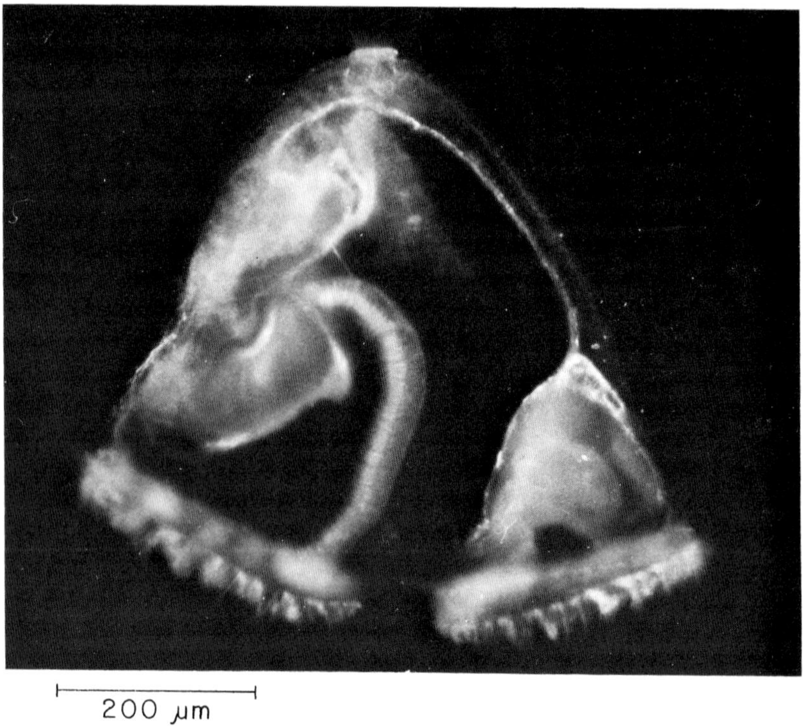

Fig. 6. *Membranipora* sp. Dark-field light photomicrograph of right-lateral view of cyphonautes larva. See Fig. 7 for identification of larval structures.

brevituba (Soule, 1953), and *P. tubulosa* (as *Pherusa tubulosa,* Prouho, 1892). Larvae of these three halcyonelloid, carnose ctenostomes are only slightly compressed bilaterally if at all and have short oral–aboral and long anterior–posterior axes. In consequence of the larval shape, the shells are more or less rectangular and curved along their shorter dimension. The metasomal sac is extensive. The larvae are provided with a gut as far as is known, but this is incomplete posteriorly and presumably without function. Differentiation of the shelled larvae from large, yolky eggs is completed within the parental body. The pelagic phase is assumed to be brief.

It is possible that three of the species described below as Type O coronate larvae have close affinities with this group.

B. Coronate Larvae

Coronate larvae include all those gymnolaemate larvae lacking shells. No planktotrophic coronate larvae are known; nutrients for the complete develop-

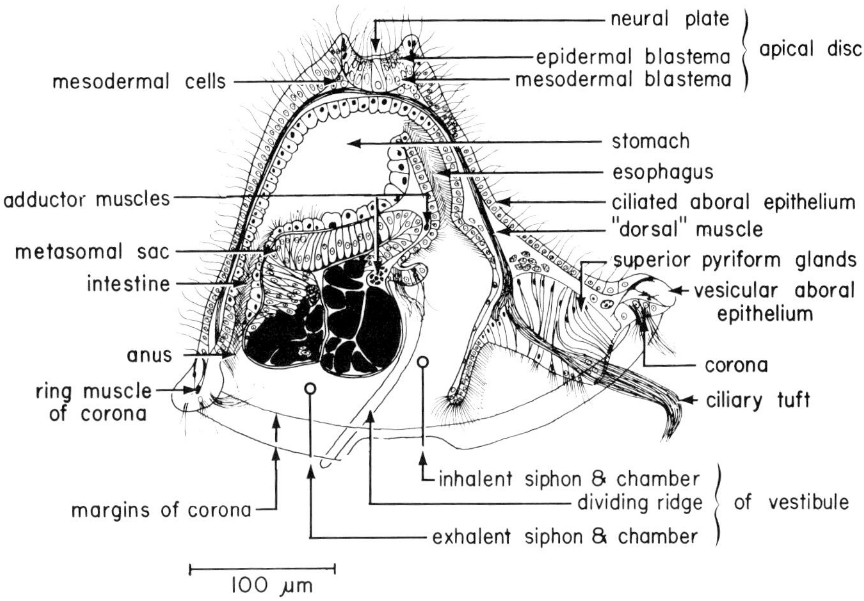

Fig. 7. *Electra pilosa.* Right-lateral view of midsagittal section of cyphonautes larva. Because the section is median, the two lateral valves do not appear; the aboral epithelium between the valves is ciliated. The extent of the corona and the position of the ridges dividing the vestibule are shown although they lie outside the median plane. From Kupelwieser (1905).

ment are provided within the egg itself or by the parent during embryogenesis. The endoderm rarely shows even a partial differentiation into a tubular digestive system. The size of the metasomal sac varies greatly between types of coronate larvae. With the exception of species placed in Type O, coronate larvae always have a large apical disc and a part of the aboral epithelium is always invaginated as a pallial groove that ranges from a shallow depression to an extremely deep fissure. The extent and placement of the corona are diagnostic of three types of coronate larvae. The remaining two types share a fourth pattern of corona development, but are distinguished by differences in sizes of their pallial sinuses (and metasomal sacs).

1. Type O Coronate Larvae [Coronate Larvae with Narrow Coronas That Are Displaced Orally (O) due to Flattening or Invagination of the Oral Field]

This small group of five larvae is almost certainly polyphyletic and its members may be more closely allied with shelled larvae than with coronate ones.

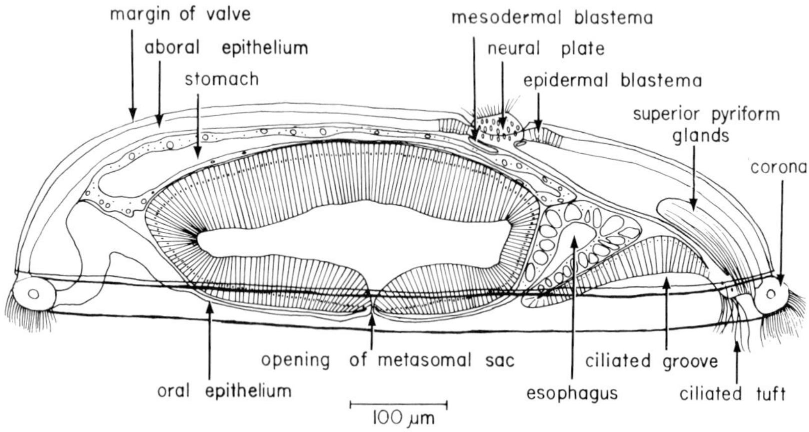

Fig. 8. *Flustrellidra hispida.* Right-lateral view of midsagittal section through larva. The two heavy lines indicate the extent of the corona. Modified from Pace (1905) and Prouho (1890).

In congruence with shelled forms, but in contrast with all other coronate larvae, Type O coronate larvae have a flattened or invaginated oral field so that the narrow corona is at the basal (oral) margin of the larval body, possess a small, knoblike apical disc, and lack a pallial groove.

The larvae of the ctenostomes *Triticella koreni* (Stolonifera, Walkerioidea) and *Alcyonidium duplex* (Carnosa, Halcyonelloidea) (Fig. 9) have the triangular profile of a cyphonautes, but are not strongly compressed laterally and, of

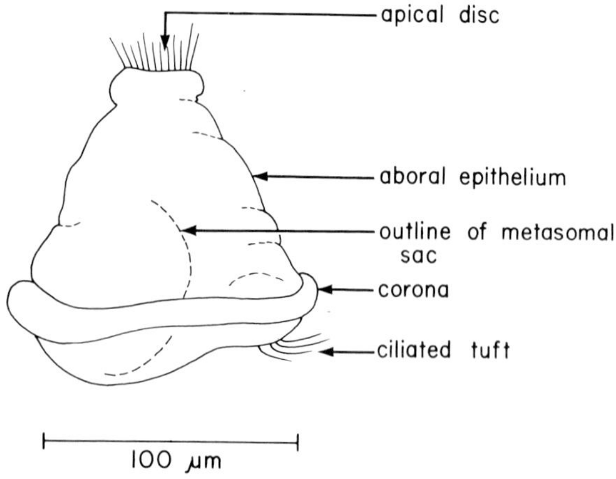

Fig. 9. *Alcyonidium duplex.* Right-lateral view of larva. From Prouho (1892).

course, lack shells. In both species only a few embryos are produced at one time and each is brooded within its own egg envelope attached externally to the tentacle sheath of the parent. Both larvae have relatively small metasomal sacs.

In *T. koreni,* the embryo spends only about 24 hours in the egg envelope before release into the plankton where both the metasomal sac and pyriform complex develop. After a week the larvae appear fully differentiated, but they may survive several additional months. Metamorphosis has not been observed. *Triticella koreni* larvae are the only bryozoan larvae other than cyphonautes in which there is development of the vestibule. However, there is no evidence that they feed while in the plankton for, although there are also rudiments of an esophagus and stomach, a mouth is lacking (Ström, 1969). The morphological similarities of this larva to a cyphonautes—especially the development of a ciliated vestibule—are impressive. In a phylogenetically based classification, this species would probably be appended to the shelled, planktotrophic group.

Despite the gross topographical similarities of *A. duplex* larvae with cyphonautes, the oral field is flattened (not invaginated) and there is no trace of a digestive system. The larva metamorphoses shortly after hatching from its egg envelope. Prouho (1892), who provided the only information on this larva, considered it to have affinities with cyphonautes. (It may be noteworthy that its congener *A. albidum* does possess a typical cyphonautes; however, other species of *Alcyonidium* have coronate larvae of Type E.)

The larva of *Scruparia* sp. (Cheilostomata, Anasca, Scruparioidea) (Fig. 10) is known from a single illustration of a living larva (as *Eucratia chelata,* Barrois, 1887); however, on this limited basis it appears identical with that of *Flustrellidra hispida* (shelled lecithotrophic larva) but lacks shells. Unfortunately, no other scruparioid larva has been examined; as this small superfamily contains the genus *Labiostomella* which Silén (1942, 1944a) considers the most primitive "cheilo-ctenostome," further work on this group would be of great interest.

The larva of *Thalamoporella evalinae* (Cheilostomata, Anasca, Coelostegoidea) has been illustrated only in surface view (Marcus, 1938), but one can see that the oral field is flattened, the apical disc is small, and the aboral field is extensive and apparently not produced into a pallial sinus (ps). Marcus (1938) stated that larvae of *Thalamoporella evalinae* are similar only to those of *Scruparia* sp. described above. In addition to possessing similar larvae, there is another curious coincidence in that individuals of these species brood multiple embryos in a two-valved ovicell. Such multiple brooding within an ovicell is otherwise unknown to our knowledge and two-valved ovicells are reported only in these two genera and in *Alysidium parasiticum* (Levinsen, 1909), another species of the superfamily Coelostegoidea, but one for which there is no information on development. In any event, the placement of these genera in two widely separated superfamilies does not correlate with the available information on their reproduction. The larva of one other coelostegoid

Discoporella umbellata has been described (Maturo, 1971) but apparently belongs to Type AEO/ps coronate larvae described below.

The larva of *Bulbella abscondita* (Fig. 11) (Ctenostomata, Carnosa, Paludicelloidea) (Braem, 1951) bears a strong superficial resemblance to the shelled lecithotrophic larva of *Flustrellidra hispida* in development of the gut and conformation of the apical disc, aboral epithelium, corona, and oral field; the shell-less larva of *Bulbella* is not bilaterally compressed or elongated as is the larva of *Flustrellidra* and its metasomal sac is much smaller. In *B. abscondita,* as in *Alcyonidium duplex,* the development is lecithotrophic and occurs within the egg envelope which adheres to the external body wall of the parent. The larvae of *Bulbella* are easily distinguished from the Type E coronate larvae of *Victorella muelleri,* the only other member of the superfamily Paludicelloidea for which developmental information is available (Braem, 1951), but both are brooded in an identical manner and share an otherwise unique pattern of metamorphosis.

2. *Type E Coronate Larvae [Coronate Larvae with Narrow, Equatorially (E) Positioned Coronas]*

(Coronate larvae of Types O and E both have narrow coronas but they have been separated, nominally, only on the basis of position of the corona. This is a character of minor significance, but a comparison of the synopses of the two larval types reveals important morphological distinctions.)

In Type E coronate larvae, the oral–aboral axis is usually relatively short and there may be lengthening along the anterior–posterior axis. The apical disc reaches its greatest diameter in this group. A pallial sinus is present as a shallow furrow. The oral hemisphere usually is flattened anteriorly in the region of the

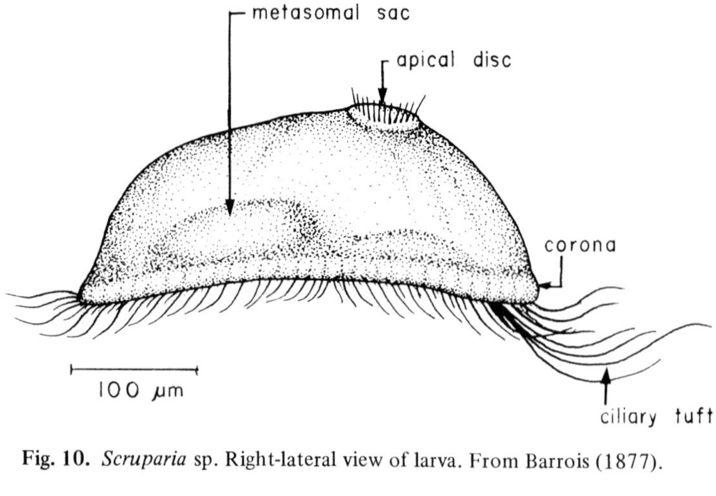

Fig. 10. *Scruparia* sp. Right-lateral view of larva. From Barrois (1877).

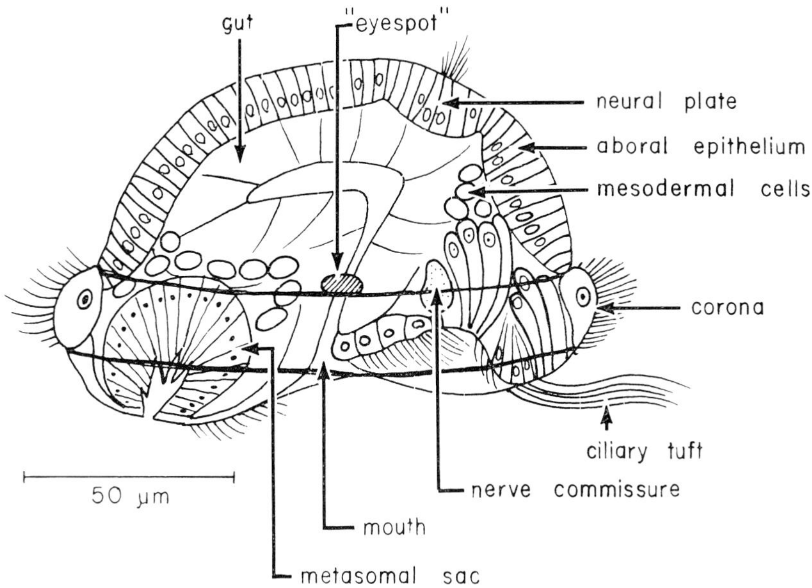

Fig. 11. *Bulbella abscondita.* Right-lateral view of midsagittal section through larva. The two heavy lines indicate the extent of the corona. From Braem (1951).

ciliated groove and bulges posteriorly due to the metasomal sac. In most species there is no development of a digestive system. A complete larval gut is reported for *Tendra zostericola* (Repiachoff, 1875), however, and transitory larval guts are found in other species; for example, in *Alcyonidium polyoum* (Fig. 12), a pharynx and stomach appear during early embryogenesis, but the latter has degenerated before the larva is fully differentiated.

Coronate larvae with narrow equatorial coronas are found among carnose ctenostomes in the halcyonelloids *Alcyonidium polyoum* and *A. variegatum* (Prouho, 1892) and the paludicelloid *Victorella muelleri* (Braem, 1951), in malacostegoid anascan cheilostomes such as *Membraniporella* nitida* (as *Membranipora nitida,* Barrois, 1877) and *Tendra zostericola* (as *Membranipora zostericola,* Ostroumoff, 1886a,b), and in ascophoran cheilostomes including *Smittina pappilifera* (Stach, 1938), *Watersipora cucullata* (Wiseley, 1958), and *Parasmittina nitida* (Humphries, 1975). (See also the following larval type concerning the difficulty of categorizing certain ascophoran larvae.)

*This genus is placed in the suborder Anasca, superfamily Malacostegoidea by Ryland (1970), but has been assigned to the suborder Cribrimorpha by others (e.g., Nielsen, 1971).

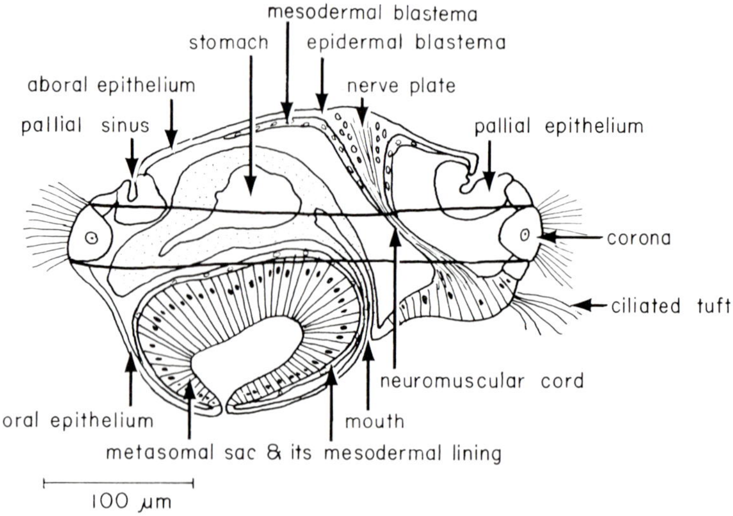

Fig. 12. *Alcyonidium polyoum.* Right-lateral view of midsagittal section through larva. The two heavy lines indicate the extent of the corona. Modified from Harmer (1887) and Seeliger (1906).

3. *Type AE Coronate Larvae [Coronate Larvae with Expanded Coronas That Are Aboral and Equatorial (AE) in Position]*

A number of ascophorans have larvae that are similar to the preceding group with the exception that the coronal cells are considerably higher and appear to have grown preferentially toward the aboral pole (Fig. 13). Correlated with the heightening of the coronal cells, their ciliation is no longer localized, but found over the entire cell surface. The apical disc is of only modest dimension, but occupies virtually all of the remaining aboral field. The pallial furrow is no longer a broadly opened trough, but has its walls appressed, so that in surface view it appears only as an annular cleft. The oral field is usually flattened or convex anteriorly but bulges posteriorly. The larvae have a short oral–aboral dimension and often are slightly elongated. A larval gut is not developed.

Coronate larvae with aborally expanded coronas are found in true ascophorans such as *Cellepora pumicosa* (Fig. 13) (Barrois, 1877), *Catenicella cantei* and *Savignyella lafonti* (Marcus, 1939), as well as in the gymnocystid *Escharoides coccineus* (as *Discopora coccineus,* Barrois, 1877).

It is difficult to ascertain whether a number of ascophoran larvae figured by Barrois (1877) including *Porella concinna* (as *P. loevis*) and an unidentified species of *Escharina* (or *Lepralia*) are of this or the preceding type.

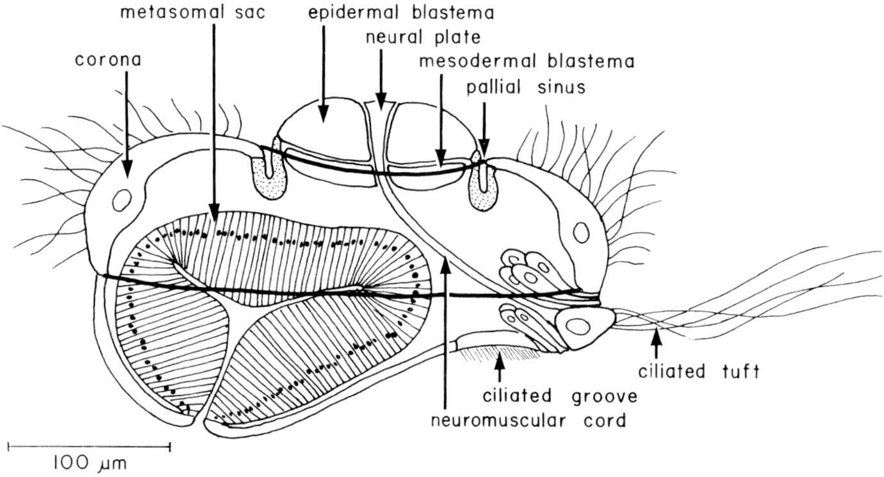

Fig. 13. *Cellepora plumosa.* Right-lateral view of midsagittal section through larva. The two heavy lines indicate the extent of the corona. Modified from Barrois (1877); a possible conformation of the internal structures has been indicated.

4. Type AEO/ps Coronate Larvae [Coronate Larvae with Expanded Coronas That Are Aboral, Equatorial, and Oral (AEO) in Position and with Small Pallial Sinuses (ps)]

In larvae of this type (Figs. 5, 14, and 15), expansion of the corona both orally and aborally has limited the polar fields to small circles. The apical disc is of medium dimensions and is bordered by a sharply invaginated but shallow pallial sinus. The epidermal and mesodermal blastemas are very prominent, reaching their greatest development in this larval type. The subpallial aboral epithelium is a very minor ring. The oral–aboral and anterior–posterior dimensions are about equal and the coronal cells are nearly as long as the oral–aboral height. The invasion of the corona into the oral field is clearly indicated by the fact that the pyriform complex is embraced by the coronal field rather than being at its oral margin. With reduction of the oral field, the metasomal sac which is quite large opens at or near the oral pole rather than more posteriorly. The digestive system is completely lacking.

Larvae of this type are characteristic of the cellularioid cheilostomes [various species of *Scrupocellaria* and *Bugula* (Figs. 5, 14, and 15), for example, Calvet, 1900]. If the term buguliform larvae is to be used, it would be appropriate, therefore, to restrict it to this type. The larvae is also found in *Cellaria salicornia* (as *C. fistulosa,* Calvet, 1900) the only pseudostegoid studied, the malacostegoid *Cupuladria doma,* the coelostegoid *Discoporella umbellata depressa* (Maturo, 1971), certain true ascophorans such as *Celleporella hyalina* (Ryland,

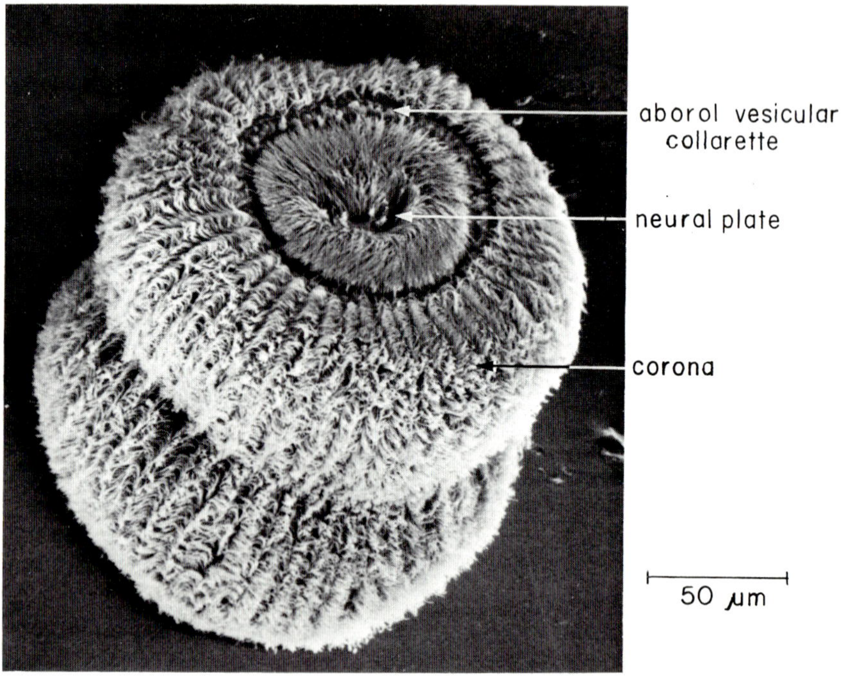

Fig. 14. *Bugula neritina.* Scanning electron micrograph of the larva. A majority of the surface is formed by the orally-aborally elongate coronal cells the dense cilia of which are arrested in metachronal waves. The equatorial neuromuscular ring of this specimen is partially contracted giving the larva its characteristic "dumbbell" appearance. Photomicrograph courtesy of Dr. John A. Long.

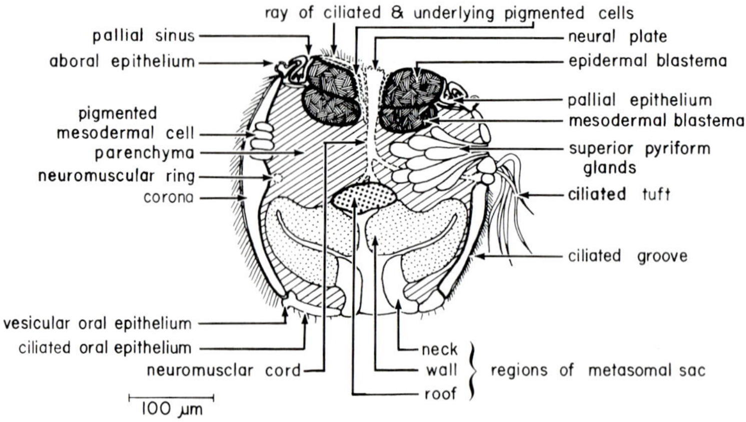

Fig. 15. *Bugula neritina.* Right-lateral view of midsagittal section through larva. From Woollacott and Zimmer (1971).

1962), *Chorizopora brongniarti, Schizobrachiella sanguinea* (as *Schizoporella sanguinea*) (Calvet, 1900), *Schizoporella carvalhoi* (Marcus, 1939), *Stylopoma duboisii* (Cook, 1973), the gymnocystidean *Metrarabdotos cookae* (Cook, 1973), and apparently the walkerioid stoloniferan ctenostome *Walkeria uva* (Barrois, 1877).

5. *Type AEO/PS Coronata [Coronate Larvae with Expanded Coronas That Are Aboral, Equatorial, and Oral in Position (AEO) and with Exceptionally Developed Pallial Sinuses (PS)]*

Larvae (Fig. 16) of this final category are elongated in the oral–aboral axis. The apical disc is relatively small and its blastemas are modest, especially in comparison with those of the previous type. An exceptionally developed pallial sinus extends nearly to the oral pole except in the anterior midline where it is blocked by the glands of the pyriform complex. The small metasomal sac opens at the center of the very reduced oral field. There is no trace of a larval gut.

This larval type is unique to, and perhaps characteristic of, the vesicularioid stoloniferan ctenostomes [e.g., *Amathia lendigera* and *Bowerbankia pustulosa* (Fig. 16), Calvet, 1900]. In *Zoobotryon verticillatum*, the only other vesiculari-

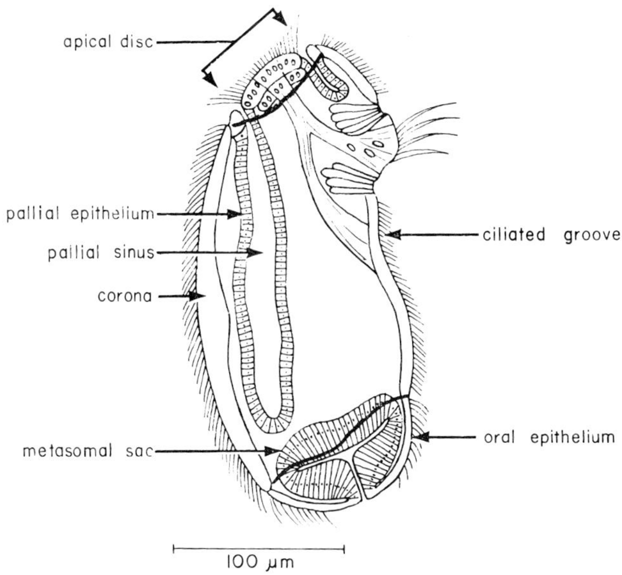

Fig. 16. *Bowerbankia pustulosa.* Right-lateral view of midsagittal section through larva. The heavy lines indicate the extent of the corona. Modified from Barrois (1877) and Calvet (1900).

TABLE I
Systematic List of Gymnolaemate Species Cited in This Chapter with the Designated Categories of Their Larvae

Species		Larval category
Order Ctenostomata		
Suborder Stolonifera		
Superfamily Walkerioidea		
Farella repens		Cyphonautes
Hypophorella expansa		Cyphonautes
Triticella koreni		Type O
Walkeria uva		Type AEO/ps
Superfamily Vesicularioidea		
Amathia lendigera		Type AEO/PS
Bowerbankia pustulosa		Type AEO/PS
Zoobotryon verticillatum		Type AEO/PS or Type AEO/ps?
Suborder Carnosa		
Superfamily Paludicelloidea		
Bulbella abscondita		Type O
Victorella muelleri		Type E
Superfamily Halcyonelloidea		
Alcyonidium albida		Cyphonautes
Alcyonidium duplex		Type O
Flustrellidra hispida		Shelled Lecithotrophic
Pherusella brevituba		Shelled Lecithotrophic
Pherusella tubulosa		Shelled Lecithotrophic
Alcyonidium polyoum		Type E
Alcyonidium variegatum		Type E
Order Cheilostomata		
Suborder Anasca		
Superfamily Inovicelloidea		Undocumented for any species
Superfamily Scruparioidea		
Scruparia sp.		Type O
Superfamily Malacostegoidea		
Conopeum sp.	in all species examined	Cyphonautes
Electra sp.	in all species examined	Cyphonautes
Membranipora sp.	in all species examined	Cyphonautes
Pyripora catenularia		Cyphonautes
Tendra repiachowi		Cyphonautes
Tendra zostericola		Type E
Membraniporella nitida		Type E
Cupuladria doma		Type AEO/ps
Superfamily Cellularioidea		
Bugula spp.	in all species examined	Type AEO/ps
Scrupocellaria spp.	in all species examined	Type AEO/ps
Superfamily Coelostegoidea		
Thalamoporella evelinae		Type O
Discoporella umbellata		Type AEO/ps
Superfamily Pseudostegoidea		
Cellaria salicornia		Type AEO/ps

TABLE I *continued*

Species	Larval category
Suborder Cribrimorpha	Undocumented for any species
Suborder Gymnocystidea	
Escharoides coccineus	Type AE
Metrarabdotos cookae	Type AEO/ps
Suborder Ascophora	
Parasmittina nitida	Type E
Smittina pappilifera	Type E
Watersipora cucullata	Type E
Escharina sp.	Type E or Type AE?
Porella concinna	Type E or Type AE?
Cellepora pumicosa	Type AE
Catenicella cantei	Type AE
Savighyella lafonti	Type AE
Celleporella hyalina	Type AEO/ps
Chorizopora brongniarti	Type AEO/ps
Schizobrachiella sanguinea	Type AEO/ps
Schizoporella carvalhoi	Type AEO/ps
Stylopoma duboisii	Type AEO/ps

oid studied (Zirpolo, 1933), the larva either belongs to this final type or the preceding one, Type AEO/ps; information on the extent of the pallial sinus is not sufficient to make the distinction.

IV. CONCLUSIONS

Canu and Bassler suggested that bryozoan taxonomists should place heavy emphasis on larval characters. They stated: "The great obstacle in bryozoology is the lack of knowledge of the larval system" (Canu and Bassler, 1920, p. 53) and "the family is characterized by its larval system" (Canu and Bassler, 1920, p. 71). These authors did not list which characters they considered of diagnostic value and those traits that are mentioned seem of doubtful utility. For example, although a large volume of information on larvae of the family Bugulidae was available to them, the only larval character mentioned for this family is "The larva is higher than wide" (Canu and Bassler, 1929, p. 187)! As is evident from the foregoing sections on larval structure and types, their basic premise that each family has a characteristic larva is not tenable. Data on development are essential, however, if we are to begin to unravel the complex radiation that has occurred within the Gymnolaemata.

Our analysis of the structure of bryozoan larvae emphasizes that these larvae share a common suite of morphological features. Although the larvae frequently appear superficially dissimilar, these differences largely can be accounted for by changes in the relative proportions of their basic components. The different combinations of allometric changes provide a basis for categorizing the larvae of bryozoans.

The classification of larvae that we have presented in this chapter does not correspond well with that of the adult gymnolaemates. The cause of these discrepancies is apparently that the adult and larval stages of the life cycle have been subjected to different selection pressures which resulted in each undergoing an independent radiation (see Chapter 15).

It appears to us that the occurrence of a given larval type in only one of the major adult taxa (but not necessarily to all members of that taxon) provides some degree of assurance that the larval category is a valid one. For example, the restriction of types O, AE, and AEO/PS coronate larvae to halcyonelloid carnose ctenostomes, ascophorans, and vesicularioid stolonate ctenostomes, respectively, suggests these larval groups are "natural" ones. Conversely, the occurrence of multiple larval types within a given adult taxon (for example, an exceptional variability in larvae is found within the genus *Alcyonidium*) may suggest that the adult taxon is polyphyletic.

A review of Table I provides other examples of disparities between larval and adult systematics, but most graphically depicts the fragmentary state of our knowledge. For example, of the four suborders of cheilostomes, we have no information on larvae of the Cribrimorpha and but two observations on those of the Gymnocystidea. Within the Anasca only the superfamily Cellularioidea seems adequately sampled for us to conclude that we probably know the range of larval morphologies to be expected; observations on three of the four other superfamilies are absent or limited to a single report. Although the larvae are known for a number of ascophorans it is far too early to make conclusions concerning distribution patterns of larval types in that suborder. The situation in ctenostomes is no better. A clear correlation of larval and adult categories exists only for the vesicularioid stolonates. Of the other three suborders, information is most fragmentary for the paludicelloid Carnosa which many scholars consider the most primitive group of gymnolaemates (e.g., Ryland, 1970). Unfortunately, the distinctive genus *Monobryozoon* and the families Immergentiidae, Penetrantiidae, and Terebriporidae remain unexplored.

We should like to stress that we are not proposing a scheme for the derivation of one larval type from another. For example, it would seem extremely simple to derive coronate larval Type AE from Type E, but we do not imply that this happened. However, we are convinced that this analysis of larval structure and a reevaluation of the life cycle of bryozoans (see Chapter 4) force one to accept that the larvae of early bryozoans were provided with a functional larval gut and presumably were planktotrophic. Acceptance of that point does not require one to conclude that these ancestral larvae were cyphonautes-like or even shelled.

However, as cyphonautes are the only existing gymnolaemate larvae that are obligatorily planktotrophic, one must accord them some special consideration. The fact that apparently identical cyphonautes are found in ctenostomes and in cheilostomes may indicate that they are indeed close to the ancestral form.

ACKNOWLEDGMENTS

We gratefully acknowledge the extensive resources made available by the libraries of the Allan Hancock Foundation (U.S.C.) and the Museum of Comparative Zoology (H.U.). Mrs. Elaine Jahn cheerfully and expertly assisted with many aspects of the preparation of this chapter. The scanning electron micrograph of the larva of *Bugula neritina* (Fig. 14) was kindly taken by Dr. John A. Long (School of Medicine, University of California, San Francisco). Our work was facilitated by grants from the National Science Foundation and by support from the Harvard Graduate Society and the William F. Milton Fund (H.U.).

REFERENCES

Abbott, M. B. (1973). Seasonal diversity and density in bryozoan populations of Block Island Sound (New York, U.S.A.). *In* "Living and Fossil Bryozoa" (G. Larwood, ed.), pp. 37–51. Academic Press, New York.

Atkins, D. (1955a). The cyphonautes larvae of the Plymouth area and the metamorphosis of *Membranipora membranacea* (L.). *J. Mar. Biol. Assoc. U.K.* **34**, 441–449.

Atkins, D. (1955b). The ciliary feeding mechanism of the cyphonautes larva (Polyzoa Ectoprocta). *J. Mar. Biol. Assoc. U.K.* **34**, 451–466.

Barrois, J. (1877). Recherches sur l'embryologie des Bryozoaires (=Mémoire sur l'embryologie des Bryozoaires). *Trav. Stn. Zool. Wimereux* **1**, 1–305.

Barrois, J. (1879). Mémoire sur la métamorphose des bryozoaires. *Ann. Sci. Nat., Zool.* [6] **9**, 1–67.

Barrois, J. (1886). Mémoire sur la métamorphose de quelques Bryozoaires. *Bibl. Ecole Haut. Etud., Sect. Sci. Nat.* **32**(5) 1–94.

Braem, F. (1951). Über *Victorella* und einige ihrer nächsten Verwandten sowie die Bryozoenfauna des Ryck bei Greifswald. *Zoologica (Stuttgart)* **102**, 1–59.

Calvet, L. (1900). Contribution à l'histoire naturelle des Bryozoaires Ectoproctes marins. *Trav. Inst. Zool. Univ. Montpellier [N.S.]* 8, 1–488.

Canu, F., and Bassler, R. S. (1920). North American early Tertiary Bryozoa. *U.S. Natl. Mus., Bull.* **106**, 1–880.

Canu, F., and Bassler, R. S. (1929). Bryozoa of the Phillipine Region. *U.S. Natl. Mus., Bull.* **100** (9), 1–685.

Cook, P. L. (1962). The early larval development of *Membranipora seurati* (Canu) and *Electra crustulenta* (Pallas), Polyzoa. *Cah. Biol. Mar.* **3**, 57–60.

Cook, P. L. (1973). Settlement and early colony development in some Cheilostomata. *In* "Living and Fossil Bryozoa" (G. Larwood, ed.), pp. 65–71. Academic Press, New York.

Corrêa, D. D. (1948). A embryologia de *Bugula flabellata* (J. V. Thompson) (Bryozoa Ectoprocta). *Univ. Sao Paulo, Fac. Filos., Cienc. Let., Bol. Zool.* **13**, 7–71.

Harmer, S. F. (1887). Sur l'embryogénie des Bryozoaires Ectoproctes. *Arch. Zool. Exp. Gén.* [2], **5**, 443–458.

Hondt, J. L. d' (1973a). Etude anatomique, histologique et cytologique de la larve d'*Alcyonidium polyoum* (Hassall, 1841), Bryozoaire Ctenostome. *Arch. Zool. Exp. Gén.* **114**, 537–602.

Hondt, J. L. d' (1973b). La musculature de la larve d'*Alcyonidium polyoum* (Hassall), Bryozoaire Ctenostome. Comparaison avec la musculature larvaire d'autres Bryozoaires Eurystomes. *C.R. Hebd. Séances Acad. Sci.* **276,** 1449–1452.

Hondt, J. L. d' (1974). La métamorphose larvaire et la formation du "cystide" chez *Alcyonidium polyoum* (Hassall, 1841), Bryozoaire Ctenostome. *Arch. Zool. Exp. Gén.* **115,** 577–605.

Hondt, J. L. d' (1975). Etude anatomique et cytologique comparée de quelques larves de Bryozoaires Ctenostomes: *In* "Bryozoa 1974" (S. Pouyet, ed.), Docum. Lab. Geol. Fac. Sci. Lyon, H.S., Vol. 3, No. 1, pp. 125–134. Univ. Claude Bernard, Lyon.

Humphries, E. M. (1975). A new approach to resolving the question of speciation in smittinid bryozoans (Bryozoa, Cheilostomata). *In* "Bryozoa 1974" (S. Pouyet, ed.), Docum. Lab. Geol. Fac. Sci. Lyon, H.S., Vol. 3, No. 1, pp. 19–35. Univ. Claude Bernard, Lyon.

Jägersten, G. (1972). "Evolution of the Metazoan Life Cycle." Academic Press, New York.

Korschelt, E., and Heider, K. (1899). "Textbook of the Embryology of Invertebrates," Vol. II. MacMillan, New York.

Kupelwieser, H. (1905). Untersuchungen über den feineren Bau and die Metamorphose des Cyphonautes. *Zoologica (Stuttgart)* **47,** 1–50.

Levinsen, G. M. R. (1909). "Morphological and Systematic Studies on the Cheilostomatous Bryozoa." Nat. Forfatt. Forlag, Copenhagen.

Marcus, E. (1926). Beobachtungen und Versuche an lebenden Meeresbryozoen. *Zool. Jahrb. Abt. Syst. Oekol. Geogr. Tiere* **52,** 1–102.

Marcus, E. (1938). Bryozoarios marinhos brasileiros. II. *Univ. Sao Paulo, Fac. Filos., Cienc. Let. Bol., Zool.* **2,** 1–196.

Marcus, E. (1939). Bryozoarios marinhos brasileiros. III. *Univ. Sao Paulo, Fac. Filos., Cienc. Let. Bol., Zool.* **3,** 111–354.

Marcus, E. (1940). Mosdyr (Bryozóa eller Polyzóa). *Dan. Fauna* **46,** 1–401.

Maturo, F. S. (1971). Larval behavior and metamorphosis in the Cupuladriidae (Bryozoa, Anasca). *Proc. 2nd Int. Conf., Inter. Bryozool. Assoc., 1971.*

Nielsen, C. (1971). Entoproct life-cycles and the entoproct/ectoproct relationship. *Ophelia* **9,** 209–341.

Ostroumoff, A. A. (1886). Contribution à l'étude zoologique et morphologique des Bryozoaires du Golfe de Sébastopol. III. Données sur l'histoire du développment. *Arch. Slaves Biol.* **2,** 184–190, 329–355.

Pace, R. M. (1906). On the early stages in the development of *Flustrella hispida* (Fabricius), and on the existence of a "yolk nucleus" in the egg of this form. *Q. J. Microsc. Sci. [N.S.]* **50,** 435–478.

Prouho, H. (1890). Recherches sur la larve de *Flustrella hispida* (Gray) structure et métamorphose. *Arch. Zool. Exp. Gén.* [2] **8,** 409–459.

Prouho, H. (1892). Contribution à l'histoire des Bryozoaires. *Arch. Zool. Exp. Gén.* [2] **10,** 557–656.

Repiachoff, W. (1875). Zur Entwickelungsgeschichte der *Tendra zostericola. Z. Wiss. Zool.* **25,** 129–142.

Ryland, J. S. (1960). Experiments on the influence of light on the behaviour of polyzoan larvae. *J. Exp. Biol.* **37,** 783–800.

Ryland, J. S. (1962). Biology and identification of intertidal Polyzoa. *Field Studies* **1**(4), 33–51.

Ryland, J. S. (1970). "Bryozoans". Hutchinson University Library, London.

Seeliger, O. (1906). Über die Larven und Vervandtschaftsbeziehungen der Bryozoen. *Z. Wiss. Zool.* **84,** 1–78.

Silén, L. (1942). Origin and development of the cheilo-ctenostomatous stem of Bryozoa. *Zool. Bidr. Uppsala* **22,** 1–59.

Silén, L. (1944a). The anatomy of *Labiostomella gisleni* Silén (Bryozoa Protocheilostomata). *K. Sven. Vetenskapskad. Handl.* [3] **21,** 1–111.

Silén, L. (1944b). The main features of the development of the ovum, embryo and ooecium in the ooeciferous Bryozoa Gymnolaemata. *Ark. Zool.* **35A,** 1–34.

Soule, J. D. (1953). Post-larval zoarial development in Carnosa (Bryozoa Ctenostomata). *Bull. S. Calif. Acad. Sci.* **52,** 88–92.

Stach, L. W. (1938). Observations on *Carbasea indivisa* Busk (Bryozoa) *Proc. Zool. Soc. London, Ser. B* **108,** 389–399.

Strathmann, R. (1973). Function of lateral cilia in suspension feeding of lophophorates (Brachiopoda, Phoronida, Ectoprocta). *Mar. Biol.* **23,** 129–136.

Ström, R. (1969). Sexual reproduction in a stoloniferous bryozoan, *Triticella koreni* (G.O. Sars). *Zool. Bidr. Uppsala* **38,** 113–128.

Vigelius, W. J. (1886). Zur Ontogenie der marinen Bryozoen. *Mitt. Zool. Stn. Neapel* **6,** 499–541.

Wisely, B. (1958). The settling and some experimental reactions of a bryozoan larva, *Watersipora cucullata* (Busk). *Aust. J. Mar. Freshwater Res.* **9,** 362–371.

Woollacott, R. M., and Zimmer, R. L. (1971). Attachment and metamorphosis of the cheilo-ctenostome bryozoan *Bugula neritina* (Linné). *J. Morphol.* **134,** 351–382.

Woollacott, R. M., and Zimmer, R. L. (1972). Fine structure of a potential photoreceptor organ in the larva of *Bugula neritina* (Bryozoa). *Z. Zellforsch. Mikrosk. Anat.* **123,** 458–469.

Zirpolo, G. (1933). *Zoobothryon verticillatum* (Dell Chiaje). *Mem. Accad. Nuovi Lincei* [2] **17,** 109–442.

Zschiesche, A. (1909). Untersuchungen über die Metamorphose von *Alcyonidium mytili. Zool. Jahrb., Abt. Anat. Ontog. Tiere* **28,** 1–72.

4

Metamorphosis, Ancestrulae, and Coloniality in Bryozoan Life Cycles

RUSSEL L. ZIMMER and ROBERT M. WOOLLACOTT

I. INTRODUCTION

During the last century many workers have contributed to our understanding of the structure of specific bryozoan larvae* and have reported a variety of ways

*See Chapter 3 for the classification of larval types used in this chapter.

in which their metamorphosis is accomplished. Yet when one attempts to correlate the organization of the bryozoan larva with that of the resulting ancestrula and to relate the life histories of bryozoans to those of other marine invertebrates, the result is often an unsatisfying and fragmentary synthesis.

A part of this confusion may result from the facts that the larvae are usually small and opaque and that the events of metamorphosis involve a number of tissues as well as a series of rapid morphogenetic movements. Since a majority of studies have been based only on observations of living material, one can appreciate why many of these reports are either superficial or contradictory or both.

Of the three existing classes, the Phylactolaemata usually is considered to possess the greatest number of primitive features as adults, but its development has been severely modified due to their existence in a freshwater habitat. Of the remaining two classes, the development of gymnolaemates seems less altered, and this class contains the only bryozoan larvae provided with a functional gut. Even in these planktotrophic forms, however, metamorphosis is highly derived. Attempts to interpret the ontogeny of gymnolaemates (or other bryozoans) invoking types of developmental patterns known for other phyla have not been successful (however, see Nielsen, 1971). We suggest that adaptations usually considered as unique to adults must also be taken into account to elucidate the bryozoan life cycle. The profound influence of coloniality and the ability to replace the internal parts or polypides in adult bryozoans is commonly recognized. The impact of these two factors on the early stages of the life cycle has not been appreciated, however. It is our hypothesis that the appearance of these two "adult" events has been accelerated into early development and that an assessment of their contribution to the total life cycle will provide a useful key to ordering and understanding the diverse and sometimes enigmatic patterns of bryozoan reproduction.

Although the unnecessary proliferation in bryozoan terminology has been justly cited, we will use the specialized terms zooecium, cystid, and polypide for their economy of words. The term polypide refers to the movable parts of the individual or zooid and includes the lophophore and gut with their coelomic linings, as well as the nerve ganglion, and certain muscles. Cystid refers to the cellular parts of the body wall, namely, the epidermis and the somatic coelomothelium. Finally, the term zooecium refers to the exoskeleton secreted by the cystid. A typical feeding functional zooid has all three components, of course, but the totipotence of the cystid to produce the other parts and the periodic degeneration of the polypide into a brown body and its subsequent replacement from the cystid wall justifies the retention of these specialized terms—the cystid is more than a simple body wall. The term metamorphosis is used here to include all the events from settling or attachment of the larva through the differentiation of the ancestrula.

With respect to a functional zooid, the lophophore is extended from the frontal face at or near the distal end. The opposite end and face are proximal and basal, respectively. We have avoided use of the terms anterior and posterior for distal and proximal, respectively, and of ventral and dorsal for frontal and basal, respectively. Considering the position of the mouth, one might think the anterior end of the larva and the distal end of the ancestrula would correspond; in fact, however, the opposite is true.

II. GYMNOLAEMATA

A. General Considerations

The metamorphosis of modern gymnolaemates probably differs considerably from that of their predecessors. We believe that the larvae of gymnolaemates were originally planktotrophic and provided with a functional gut and food-gathering apparatus. Considering the metamorphic transformations characteristic of other phyla, one would expect planktotrophic larvae of ancestral and of modern gymnolaemates to retain the food-gathering and processing machinery of the larva and utilize it, at least in part, after settlement. That this does not occur in the modern species is well known, but what happened in primitive gymnolaemates can only be hypothesized.

The metamorphosis of all modern gymnolaemates involves two functionally and temporally distinctive phases, one typified by morphogenetic movements and the other by histogenic differentiation.

The first phase is a rapid one, usually requiring less than 20 minutes. It consists of a more or less complicated sequence of morphogenetic movements effecting the eversion (and attachment) of the metasomal sac and complex segregation of the larval structures according to their eventual fates. In general, various components of the larva may be functional only during larval life, only during adult life, or during both phases, but bryozoans seem to possess only tissues that are strictly larval or strictly adult. During the metamorphosis of a gymnolaemate larva, the tissues of the first category, which are largely epidermal but include the digestive tract or other endodermal derivatives, are cut off into the interior of the individual. With the internalization of the transitory larval cells, the external body wall, including the epidermis and its underlying mesodermal layer, constitutes virtually all of the adult tissues and may be considered a cystid. Near the center of its free face, this cystid is provided with a special blastogenic zone which, in essence, is the apical disc of the larva. The invagination of this two-layered blastogenic region as a vesicle into the interior completes the first phase of metamorphosis. The resultant disc-shaped or hemispherical body is termed the preancestrula until completion of the histogenic phase of metamorphosis.

The second phase of the metamorphosis requires a period of one to six days, depending on the species (Hyman, 1959) and the temperature at which transformation takes place. The critical feature of this phase is the progressive histogenic differentiation of the ancestrular polypide from the two-layered vesicle noted above. There is also the regressive breakdown of the transitory larval tissues, and the cystid of the preancestrula assumes the definitive shape of the cystid of the ancestrula; this secretes the zooecium of this first individual. In consequence of these changes, the deceptively simple preancestrula is transformed into an ancestrula, the first functional zooid of the colony.

In the following section, these distinctive phases of metamorphosis are elucidated in detail through description of the transformations for selected species. Subsequently an analysis of metamorphosis in gymnolaemates is presented.

B. Patterns of Metamorphosis

1. Phase 1 of Metamorphosis–Transformation of the Larva to the Preancestrula

a. The General Pattern. The first phase of metamorphosis involves three characteristic steps, but there may be subsequent elaborations. In a number of species only the basic steps are employed in production of the preancestrula and the epidermis of the cystid after involution of the transitory larval tissues is about equally derived from the aboral field and the metasomal sac. Reasons for considering this pattern to be a basic one include the facts that (a) it involves only the three basic morphogenetic movements, (b) it occurs in larvae that are not highly specialized, (c) it is also the pattern found in cyclostomes, and (d) it serves as an excellent model from which the variations can be derived.

We shall describe the early metamorphosis of *Alcyonidium polyoum* as reported by Zschiesche (1909), *Flustrellidra hispida* (as *Flustrella hispida*) as documented by Prouho (1890), and continue with that of *Electra pilosa* as studied by Kupelwieser (1905). Subsequently, we shall describe variations on this first phase of metamorphosis and finally will summarize and evaluate the observations.

i. *Alcyonidium polyoum* (Type E coronate larva). *Alcyonidium* is one of the most extensively studied gymnolaemates in terms of larval structure and metamorphosis. That its larva is relatively primitive is indicated by the fact that it has rudiments of the larval digestive system. However, its metamorphosis is not complicated by the presence of larval valves.

After an exploratory examination of the available substratum, the larva attaches temporarily by means of secretions from the pyriform complex. Permanent settlement is soon made by eversion and attachment of the metasomal sac. The everted sac is initially conical with the base flattened against the substratum and the truncated apex joining with the oral field of the larva (Fig. 1). A progressive flattening of the walls of the cone against the substrate shortens the cone and widens the diameter of the annular junction between the sac and the larva proper. With these changes all those tissues between this junction of the metasomal sac with the oral field and the junction of the aboral epithelium with the corona begin involution into the "larval" interior. A doughnut-shaped cavity surrounded by the involuted tissue is thus produced. The involuted tissues constitute transitory larval epidermal structures and these soon dehisce from their former connections with the aboral epithelium and metasomal sac. The now free oral edge of the aboral epithelium and the free aboral edge of the metasomal sac fuse with each other, sealing the transitory larval tissues internally and reestablishing externally a continuous epidermal sac. With its internal mesodermal sheath, this sac constitutes a cystid of the preancestrula. At about the same time the apical disc region is invaginated as a vesicle.

The mechanisms of these morphogenetic movements are not clear from Zschiesche's analysis, but Hondt (1972) indicated that there are paired longitudinal muscles joining the coronal tissue and the aboral epithelium that could assist in initiation of the involution of the transitory larval tissues. Subsequently (1974), he stated that eversion of the metasomal sac is accomplished by a pair of smooth muscles that run from the sub- to the supracoronal epithelium.

ii. *Flustrellidra hispida* (shelled lecithotrophic larva). The metamorphosis of the shelled larvae of *Flustrellidra hispida* has been concisely and convincingly described by Prouho (1890). The everted metasomal sac is attached to the

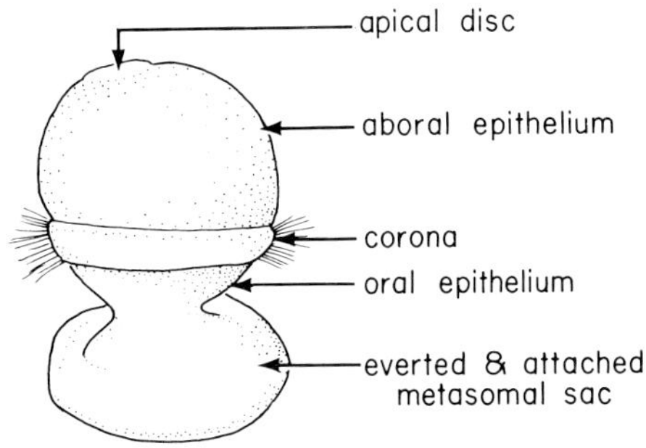

Fig. 1. *Alcyonidium mytili.* Individual after attachment by the everted metasomal sac, but before involution of the corona and other transitory tissues has started. From Zschiesche (1909).

substratum, presumably by the "granular substance" produced in the neck region of the sac as this material is no longer visible after metamorphosis. The mechanism of eversion is not clearly stated, but Prouho's notation that the shells open after relaxation of the adductor muscle may suggest that the adductor was indirectly responsible for the eversion.

The oral field is retracted inward by paired anterior and posterior lateral muscles, drawing the corona with it. Subsequently, by its contraction as a drawstring, the ring muscle located at the topological aboral margin of the corona (Fig. 2) completes the involution of the transitory larval tissues (Figs. 3 and 5). The fusion of the edge of the adhesive sac with that of the aboral epidermis transforms the metamorphosing individual into a closed sac (Figs. 3–5).

Invagination of the apical disc region follows, but this differs from that process in other species. Prouho (1890) indicated that the sensory portions and the slightly separated blastemal portions are not invaginated simultaneously, but that the more central sensory plaque invaginates as a vesicle first and then is surrounded by the invaginating blastemal regions (Fig. 4). The mechanism(s) of these invaginations are not stated.

iii. *Electra pilosa* (cyphonautes larva). The metamorphosis of *Electra pilosa* has been elegantly described by Kupelwieser (1905). Eversion of the metasomal sac is accomplished by the coordinated contraction of the entire larval musculature. This contraction causes a rounding of the larval body, rupturing the adductor muscles which held the valves together. The valves separate from each other except at points near the oral and the aboral ends of their anterior margins. The valves, pivoting on these two points, will eventually form the right and left halves of a flattened "roof" over the preancestrula, their anterior margins overlapping in the midline (Fig. 6). The everted metasomal sac progressively flattens against the substratum forming the attachment disc. Concomitantly, smooth muscle fibers joining the sac with each of the valves draw the valves down over the metamorphosing mass, compressing it into an elliptical disc (Figs. 6 and 7). The body which had been bilaterally compressed during larval life is now flattened in the oral–aboral axis.

Figs. 2, 3, and 4. Larva and metamorphosis of *Flustrellidra hispida.* All after Prouho (1890) as *Flustrella hispida.*

Fig. 2. Cross section at level of apical disc of free-swimming larva to illustrate relationship of structures before metamorphosis.

Fig. 3. Cross section at level of apical disc of metamorphosing individual after attachment by metasomal sac and involution of larval transitory tissues but before apical disc is fully retracted and histogenic events have commenced.

Fig. 4. Cross section at level of apical disc of slightly older preancestrula; apical disc is now retracted and histolysis of larval transitory tissues has begun.

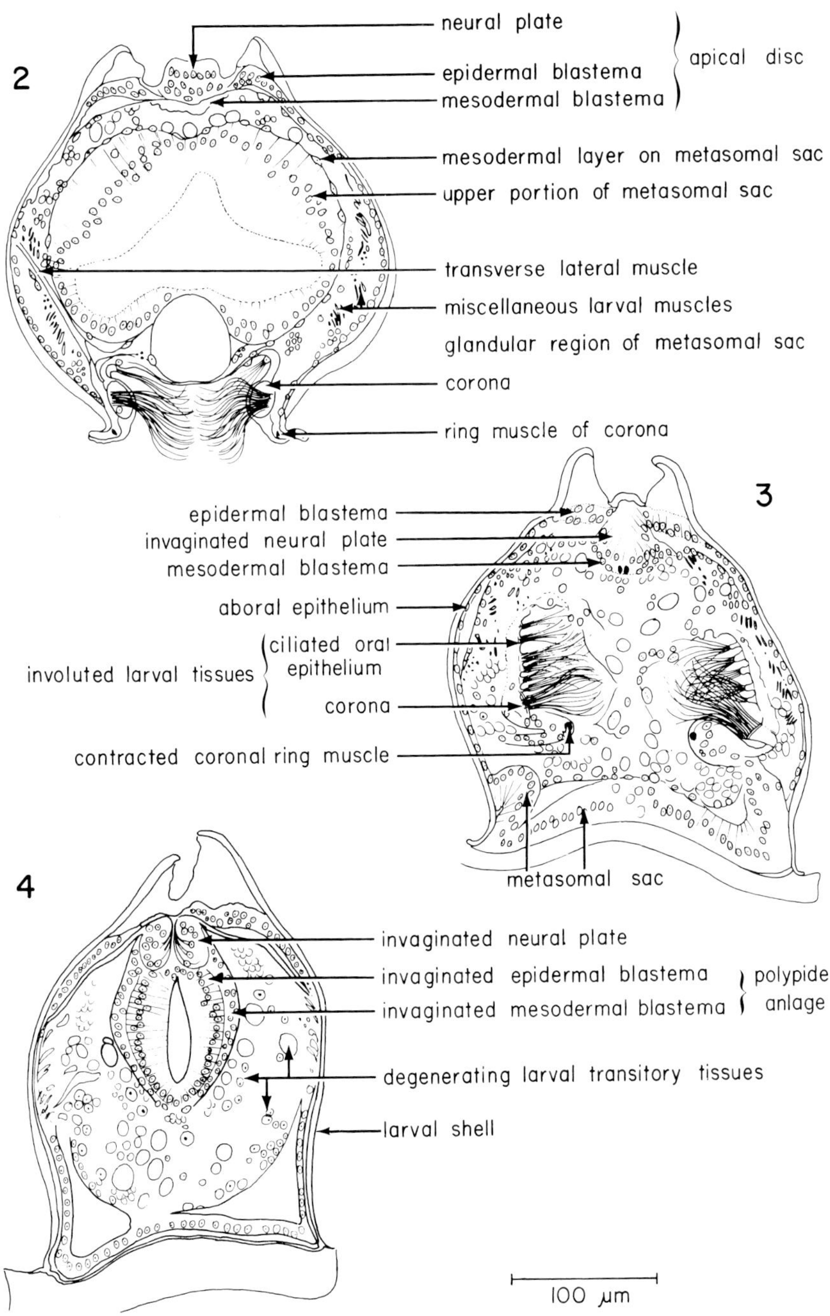
2
neural plate
epidermal blastema
mesodermal blastema
apical disc
mesodermal layer on metasomal sac
upper portion of metasomal sac
transverse lateral muscle
miscellaneous larval muscles
glandular region of metasomal sac
corona
ring muscle of corona
3
epidermal blastema
invaginated neural plate
mesodermal blastema
aboral epithelium
ciliated oral epithelium
involuted larval tissues
corona
contracted coronal ring muscle
metasomal sac
4
invaginated neural plate
invaginated epidermal blastema
invaginated mesodermal blastema
polypide anlage
degenerating larval transitory tissues
larval shell
100 μm

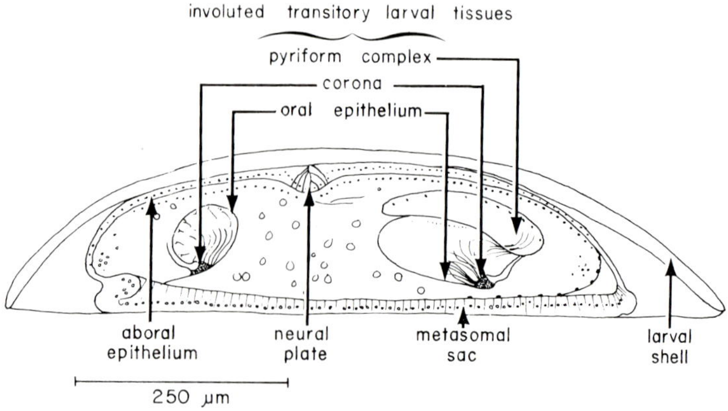

Fig. 5. *Flustrellidra hispida.* Right-lateral view of midsagittal section of metamorphosing individual at same stage illustrated in Fig. 3. Note the involuted larval transitory tissue. After Prouho (1890) as *Flustrella hispida.*

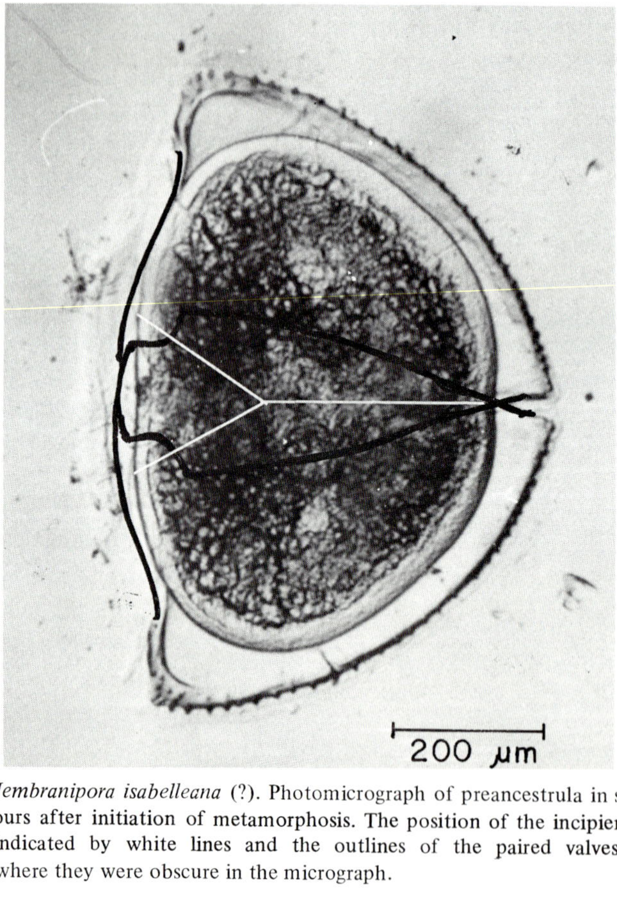

Fig. 6. *Membranipora isabelleana* (?). Photomicrograph of preancestrula in surface view about 20 hours after initiation of metamorphosis. The position of the incipient Y-shaped septum is indicated by white lines and the outlines of the paired valves have been emphasized where they were obscure in the micrograph.

Around the entire margin of the disc, the corona, the aboral vesicular tissue adjacent to the corona, and all parts of the oral field except the attached metasomal sac are withdrawn centripetally into the interior of the incipient preancestrula forming an annular groove (Fig. 7). A possible mechanism for this involution was suggested by Kupelwieser (1905) and seems plausible if one analyzes the preceding events: the eversion of the invaginated metasomal sac not only reduces the internal volume of the metamorphosing individual but also increases its effective surface area. Consequently, the cell mass when it is flattened into a disc, has too much surface area for the given volume and an internal "*Vakuum*" results. The attachment of the internal sac to the substratum and stiffening of the aboral epithelium by the valves would leave only the intervening tissue susceptible to this negative pressure. Suction would seem to be a reasonable mechanism for involution of the larval transitory tissues in other species as well, if the aboral epithelium is resistant to deformation because of its own rigidity or some other support.

At the upper and lower lips of the groove the involuted tissues join the oral margin of the aboral epithelium and the aboral margin of the metasomal sac, respectively. The two lips will fuse all around the equator of the flattened disc, cutting off internally the transitory epidermal tissues and the digestive tract of the larva. In consequence, the external body wall of the disciform preancestrula consists essentially of an extensive attachment disc (the basal body wall) derived from the metasomal sac and the opposing (frontal) surface derived from the aboral epithelium that secreted the larval shells (the apical disc now forms a minor part of this surface). The erect lateral, proximal, and distal walls are

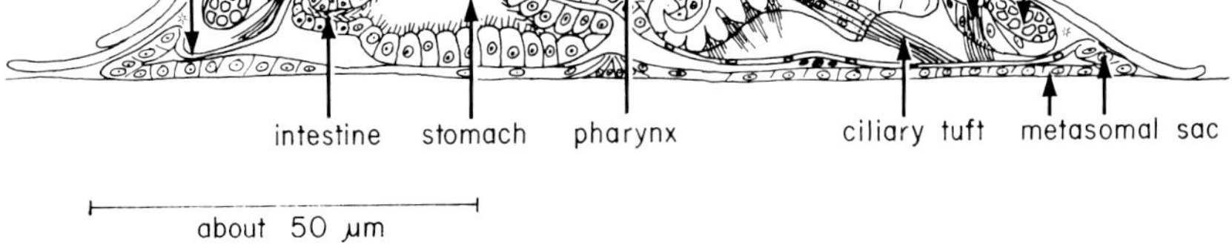

Fig. 7. *Electra pilosa.* Right-lateral view of midsagittal section of early preancestrula near completion of phase of rapid morphogenetic movements but before isolation of the larval transitory tissues internally and fusion of the aboral epithelium with metasomal sac tissue (at asterisks). Modified from Kupelwieser (1905).

probably derived from the metasomal sac also, but their partial or complete origin from the aboral epithelium cannot be precluded.

The early phase of metamorphosis is completed by the invagination of the small apical disc into the interior of the preancestrula. The mechanism of this movement is not apparent.

iv. Other species. This idealized pattern of metamorphosis has also been reported for *Bugula simplex* (as *B. sabatieri*) and other *Bugula* species (all Type AEO/ps coronate larvae) (Calvet, 1900). As explained in the subsequent section, we believe that Calvet may have been in error.

Considering larval morphology, we would expect that this pattern of formation of the preancestrula will be found to be characteristic of cyphonautes, lecithotrophic shelled, and Types O and E coronate larvae, but this is highly conjectural considering the paucity of detailed information on metamorphosis. Exceptions to this generalization are known for Type O and E coronate larvae as noted below in Section II, B, 1, b, i, (d).

b. Variations from the General Pattern. Numerous variations from this basic pattern have been reported in the literature, but many of these appear to be the result of misinterpretation or error and were often based only on the study of living material or whole mounts not sections. However, there is convincing evidence (a) that the origin of the cystid epidermis does not always involve the more-or-less equal participation of the metasomal sac and aboral epithelium and (b) that attachment by the metasomal sac is not always a permanent one.

i. Variation in origin of the epidermis of the cystid. (a) *Bugula neritina* (Type AEO/PS coronate larvae). The metamophosis of this cosmopolitan, fouling bryozoan has been examined by Calvet (1900), Lynch (1947), Mawatari (1946a, 1951), and Woollacott and Zimmer (1971). The first two scholars indicated that the pallial sinus epithelium and metasomal sac contribute to the epidermis of the cystid, but the latter three reported that the metasomal sac alone is involved. Woollacott and Zimmer (1971) documented their study with photomicrographs, providing irrefutable evidence that the pallial epithelium is invaginated with the apical disc, leaving only the metasomal sac to sheathe the preancestrula. It seems inescapable but to conclude that Calvet (1900) and Lynch (1947) were in error concerning the fate of the pallial epithelium; hypothetically, geographic differences could occur within this familiar, cosmopolitan species, but it is doubtful that members of disjunct populations would differ drastically in their metamorphoses.

In the following summary based on Woollacott and Zimmer (1971), certain details of the metamorphosis of *B. neritina* unrelated to formation of the cystid epidermis are included to supplement the information already provided on the processes and mechanisms of the transformation of the larva to the preancestrula.

The metamorphosis of *B. neritina* is somewhat more complex than those of *Alcyonidium* (Type E coronate), *Flustrellidra* (lecithotrophic shelled), and *Electra* (cyphonautes). As should be expected, the initial step of metamorphosis is eversion and attachment of the metasomal sac, but in this case the eversion is achieved by the contraction of the equatorial neuromuscular ring. When first attached the larva is tilted forward as seen in Fig. 8. Soon however, the larva reorients itself so that the oral-aboral axis is perpendicular to the substrate (Figs. 9 and 11). In this species, only the roof region of its metasomal sac is involved in formation of the attachment disc. The wall region remains reflected upon itself as it was when invaginated (Fig. 10) and forms a thick double pad at the margin of the attachment disc (Figs. 11–13). As the metasomal sac is everted, the cells of the neck region are believed to release the contents of the numerous vesicles contained in their apical cytoplasm. This material is believed to anchor the attachment disc as well as to form a thin transparent sheath, the calyx (Figs. 9, 12–14), over the entire surface of the metamorphosing individual.

In a fashion directly comparable to that observed in the previously described examples, the corona and certain immediately adjacent tissues (the oral epithelium, the pyriform complex, the neck region of the metasomal sac, and the vesicular and oral ciliated epithelia) move centripetally into the interior of the incipient preancestrula (Figs. 12 and 13). This process begins near the oral margin of the corona and occurs after the eversion of the metasomal sac and reorientation of the incipient preancestrula are completed. Thus the involution of the transitory larval structures does not seem to be a consequence of negative pressure in this species, but the mechanism(s) of this movement is (are) unknown.

As the corona and its associated tissues are enfolded, the pallial epithelium is pulled out of its tightly packed configuration within the pallial groove and is drawn orally by the receding transitory tissues (Fig. 12). At the completion of these movements, the pallial epithelium has been stretched over the surface just vacated by the corona and its oral margin is contiguous with the aboral margin of the wall region of the metasomal sac (Fig. 13). Fusion of these margins accompanies the loss of the original connections of the pallial epithelium and metasomal sac with the internalized transitory tissues.

At this stage, the configurations of the preancestrulae of *Alcyonidium, Flustrellidra, Electra,* and *Bugula* are topologically identical (compare Figs. 3, 7, and 13). However, the nonattached surface of the preancestrulae consists of a columnar secretory epithelium in *Alcyonidium,* the cuboidal valve-secreting mantle epithelium in *Electra* and *Flustrellidra,* and the delicate squamous pallial epithelium in *Bugula;* these tissues are all homologous and each represents a specialization of the general epithelium of the aboral field.

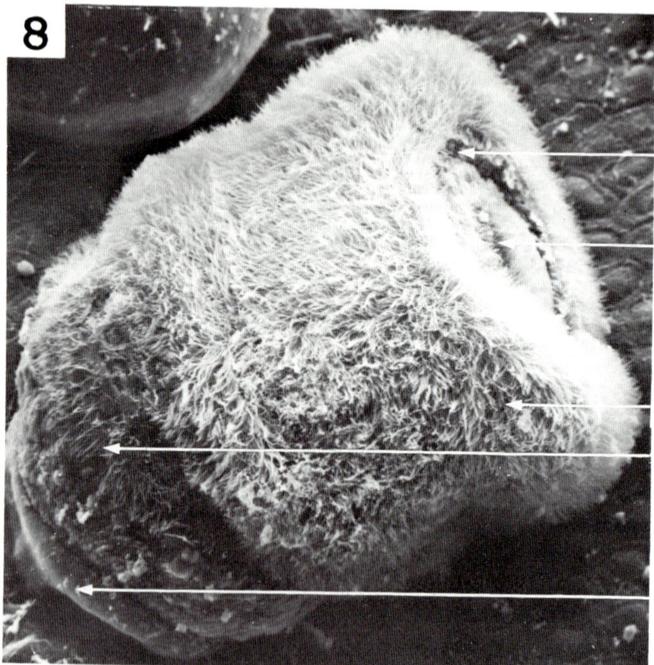
8
pallial sinus
apical disc
corona
oral epithelium
everted metasomal sac

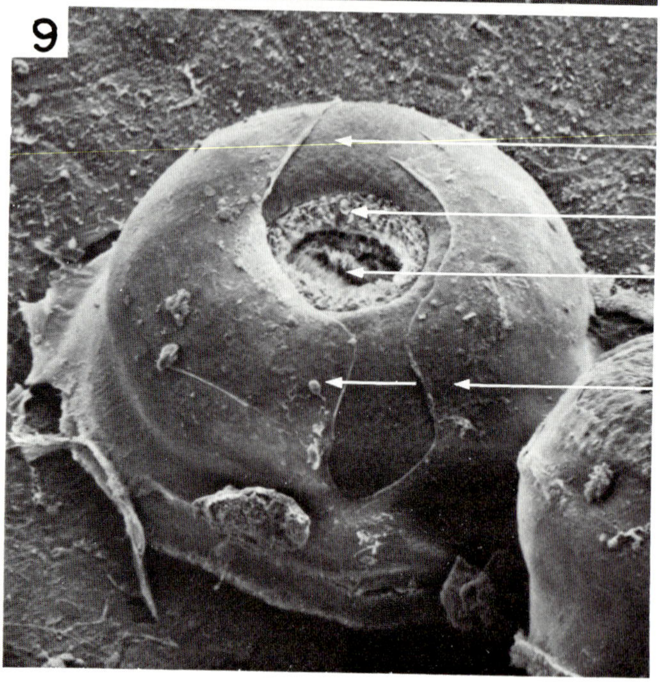
9
epidermis from metasomal sac
contracted pallial epithelium
partially retracted apical disc
calyx
50 μm

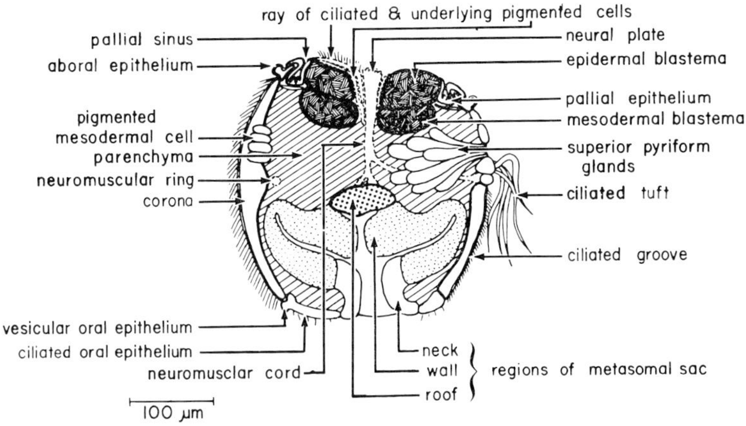

Fig. 10. *Bugula neritina.* Right-lateral view of midsagittal section of larva. From Woollacott and Zimmer (1971).

Immediately, however, in *Bugula*, the pallial epithelium begins a progressive contraction until it is reduced to a narrow ring around the apical disc (Figs. 9 and 14). Microfilaments apparently effect this contraction in much the same fashion as they effect contraction of the caudal epidermis in tunicates (e.g., Cloney, 1966). With this contraction the wall region is pulled out of its reflected position and drawn aborally. Thus the surface that was once covered largely by the corona (Figs. 8, 10, and 11), is transitorily sheathed by the pallial epithelium (Fig. 13), and finally is permanently covered by the wall region of the metasomal sac (Figs. 9 and 14). At the center of the dome-shaped free surface there remains the apical disc surrounded by a narrow annulus of contracted pallial epithelium (Figs. 9 and 14). These elements are drawn internally, presumably by the tetanic contraction of muscle strands that unite the sensory complex at the center of the apical disc with the roof region of the metasomal sac. The wall

---

Figs. 8 and 9. *Bugula neritina.*

Fig. 8. Scanning electron micrograph of individual fixed just after attachment of metasomal sac but before involution of larval transitory tissues began. Viewed from right side. Anterior region with pyriform complex is tilted forward against substratum. Compare with Fig. 11.

Fig. 9. Scanning electron micrograph of incipient preancestrula fixed just before invagination of apical disc completes phase of rapid morphogenetic movements. In this specimen the calyx is incomplete, permitting visualization of the apical region of the specimen. Compare with Fig. 14. Scanning electron micrographs courtesy of Dr. John A. Long.

11

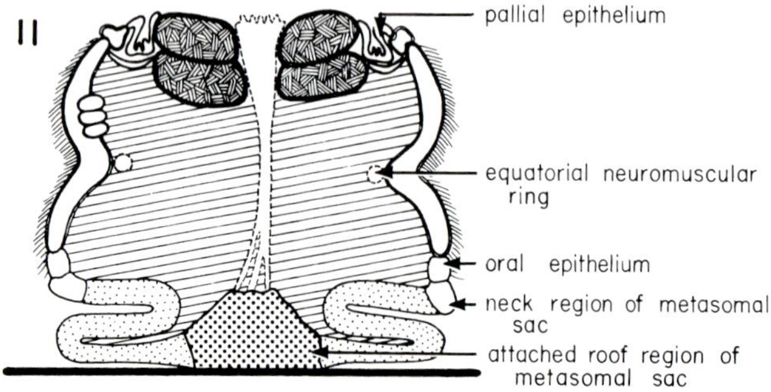
pallial epithelium
equatorial neuromuscular ring
oral epithelium
neck region of metasomal sac
attached roof region of metasomal sac

12

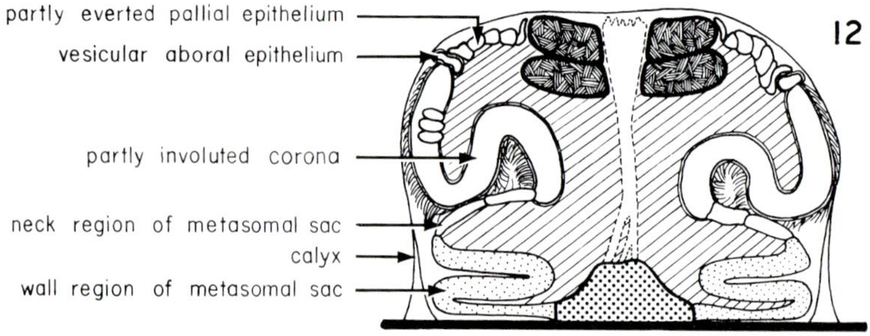
partly everted pallial epithelium
vesicular aboral epithelium
partly involuted corona
neck region of metasomal sac
calyx
wall region of metasomal sac

13

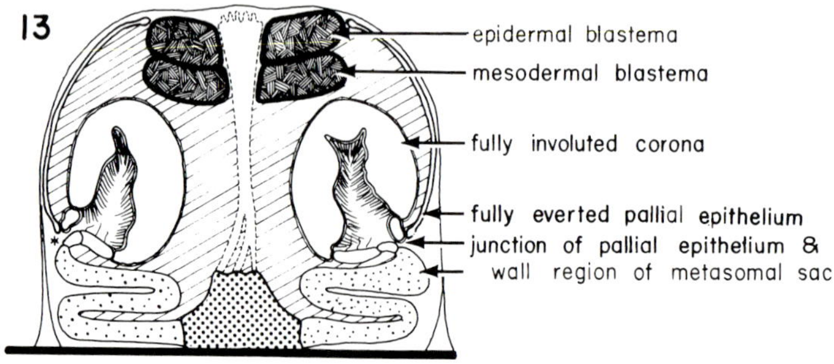
epidermal blastema
mesodermal blastema
fully involuted corona
fully everted pallial epithelium
junction of pallial epithelium & wall region of metasomal sac
*

14

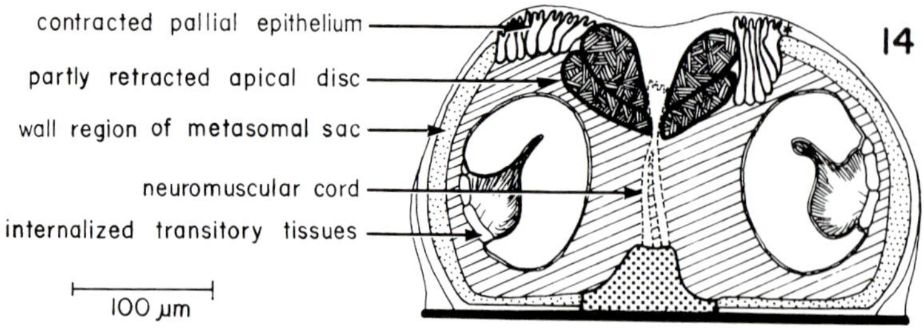
contracted pallial epithelium
partly retracted apical disc
wall region of metasomal sac
neuromuscular cord
internalized transitory tissues
100 μm
*

portion of the metasomal sac closes over the apical pole of the preancestrula as if constricted by a drawstring.

Thus in *Bugula neritina* the entire epidermis of the preancestrular cystid is derived from the roof and wall regions of the metasomal sac, and the pallial epithelium (the only remaining component of the aboral epithelium) has been invaginated with the apical disc and does not contribute to the cystid epidermis, but rather will form the tentacle sheath later.

It is appropriate to return momentarily to the question of whether or not Calvet (1900) and Lynch (1947), two careful scholars, may have erred concerning formation of the cystidal epidermis: although the displacement of the thick epithelium of the metasomal sac over the free surface as it replaces the delicate pallial epithelium is conspicuous *in vivo,* the event is not easily interpreted without sections of a sequence of stages fixed during the few seconds when the pallial epithelium contracts. As Lynch (1947) observed only living specimens and Calvet (1900) apparently prepared few sections (only of *B. simplex*?), it is probable they overlooked or misinterpreted the critical stages of contraction of the pallial epithelium and, in consequence, erroneously concluded that this tissue contributes to the epidermis of the cystid.

(b) *Other species possessing Type AEO/ps coronate larvae.* As indicated above, the preancestrular epidermis originates exclusively from the metasomal sac in *B. neritina.* The same origin has been reported for other cellularioids including *B. californica* (Mawatari, 1946b) and unidentified species of *Bugula* and *Scrupocellaria* (Braem, 1951). In still other cellularioids, a portion of this epidermis is reportedly derived from the aboral (pallial) epithelium according to Barrois (1886, *B. flabellata*), Calvet [1900, *B. simplex* (as *B. sabatieri*), *B. stolonifera* and the pseudostegoid *Cellaria salicornia* (as *C. fistulosa*)], and Vigelius (1886, *B. calathus*), or from the corona (!) by Grave (1930, *B. simplex*).

Figs. 11, 12, 13, and 14. *Bugula neritina:* Diagrams of median longitudinal sections at four stages in completion of first phase of metamorphosis involving rapid morphogenetic movements. All figures modified from Woollacott and Zimmer (1971).

Fig. 11. Individual after eversion and attachment of the metasomal sac has been accomplished; note partially contracted equatorial neuromuscular ring that was involved in eversion of metasomal sac. Compare with Fig. 8.

Fig. 12. Corona and other transitory larval tissues (vesicular aboral epithelium, oral epithelium, and neck region of metasomal sac) are partially involuted.

Fig. 13. Involution of the transitory tissues is complete but they retain their surface connections. In living specimens the pallial epithelium and wall region of the metasomal sac would next lose their connections with the involuted transitory tissues and fuse with each other (at asterisk).

Fig. 14. Contraction of the pallial epithelium (shown only partially accomplished on left side of figure) has drawn wall region of metasomal sac over most of incipient preancestrula. Partial contraction of the neuromuscular cord has partly retracted the apical disc.

The latter observation is certainly based on error or is a misprint, and we suspect that reports of involvement of the pallial epithelium resulted from the oversight or misinterpretation of contraction of that tissue as Calvet and Lynch apparently did in *B. neritina.*

It is obvious that additional clarification of the metamorphosis of Type AEO/ps coronate larvae is essential, but we believe it is probable that species with such larvae have preancestrulae and ancestrulae in which the epidermis is exclusively derived from the metasomal sac and in which the delicate pallial epithelium is retracted internally with the invagination of the voluminous apical disc.

(c) *Vesicularioid stoloniferan ctenostomes.* The morphology of cornate larvae of Type AEO/PS, which apparently are limited to the vesicularioid stolonate ctenostomes, dictates that their metamorphosis must be quite different from that of the externally similar larvae of Type AEO/ps. Rather than having a voluminous metasomal sac and a small pallial epithelium, AEO/PS coronate larvae have a small metasomal sac but a greatly hypertrophied pallial epithelium. One would deduce that the preancestrula must possess a small attachment disc derived from the metasomal sac and have a large free surface provided by the pallial epithelium. This is what Ostroumoff (1886a,b) has documented for *Bowerbankia stationis* (as *Vesicularia stationis*). Barrois (1886), however, came to a significantly different conclusion: for *Amathia lendigera* (as *Serialaria lendigera*) he reported the entire cystid epidermis, including the attachment disc, to originate from the pallial epithelium. It is difficult to attribute confidence to Barrois' assertions, considering that he provided only highly diagrammatic drawings to support his observations and reported not one but a whole suite of singular peculiarities in this metamorphosis; he claimed both the metasomal sac and apical disc to be without function, the corona to make the initial attachment on settling, and the polypide to have a unique origin involving mesodermal and ectodermal masses that are from the pyriform organ or adjacent tissue and from the epidermis near the apical disc, respectively.

Hondt (1975) also believed that the pallial epithelium forms all the epidermis of the ancestrular cystid in *Bowerbankia imbricata.* The mechanism of attachment is not indicated, but he claimed the metasomal sac of this species does not correspond to that of other gymnolaemates, consisting entirely of two epidermal tissues that normally are at the larval surface. One of these invaginated tissues is stated to form the epidermis of the ancestrular polypide and the other degenerates.

Eiben (1976) reported the metasomal sac to form a small attachment disc in *Bowerbankia gracilis,* but did not indicate the origin of the rest of the ancestrular body wall.

It is apparent that additional work is essential to clarify the pattern(s) of metamorphosis of this larval type.

(d) *Bulbella abscondita (Type O coronate larva) and Victorella muelleri (Type E coronate larva).* The origin of the entire cystidal epidermis (including the attachment disc!) from the aboral epithelium has been convincingly described and figured for the brackish water paludicelloid ctenostomes *Bulbella abscondita* and *Victorella muelleri* [as *Paludicella (Tanganella) muelleri*] (Braem, 1951). (Figs. 15–17). Braem observed that the small metasomal sacs of their larvae function only transitorily at settling. The initial attachment to the substratum (Fig. 16) is lost during the simultaneous inrolling with the corona of the *entire* oral face, including the metasomal sac (Fig. 17). The permanent attachment disc is formed by the oral margin of the aboral or pallial epithelium which thus is the unique source of the epidermis of the ancestrula. Conceivably, Braem could have been misled if he did not study and figure the median sections of his series, but his publications indicate he was a most careful and thorough worker. Until a reexamination of these species can be made it is appropriate to accept his observations.

Nielsen [(1971) and Chapter 16] attached considerable importance to this metamorphosis as the secondary attachment of these ctenostomes is strongly reminiscent of the attachment of the entoproct *Pedicillina.*

ii. Loss of the original attachment made by the metasomal sac. (a) As indicated above, *Bulbella abscondita* and *Victorella muelleri* are initially attached by the everted metasomal sac, but a reattachment is made via the aboral epithelium (Braem, 1951). The aboral epithelium was also indicated to form the permanent attachment in *Bowerbankia lendigera* after a transitory attachment by two lobes of the corona (Barrois, 1886), but the reliability of this report is questionable.

(b) *Zoobotryon verticillatum (Type AEO/PS coronate larva?)* A most unusual reattachment has been documented in the vesicularioid stoloniferan *Zoobotryon verticillatum* (Zirpolo, 1933; Bullivant, 1968). During the first phase of metamorphosis, the metasomal sac is everted and makes the normal attachment. After completion of the preancestrula, however, a stolon several hundred microns in length is produced. At its tip the stolon expands as an adhesive disc that is applied against the substratum. Subsequently, the initial attachment is loosened and the preancestrula is elevated above the substratum by the now erect stolon; there it completes its normal differentiation into a functional ancestrula (Fig. 18). Whether the stolon is proliferated from tissue that originated as the metasomal sac or as aboral (pallial) epithelium is not known, but probably of little consequence. It is important to note that as a consequence of the reattachement, and the proliferations of asexual buds from the stolon just below the ancestrula, the ancestrula occupies a distal, not proximal, position in the colony and what appears to be an ancestrula devoid of a polypide at the base of the colony is only the expanded tip of the primary stolon. [It is of interest to note at this time that the normal zooids of *Zoobotryon*—as in certain other stoloniferans—do not produce a replacement polypide, but are dehisced after

15

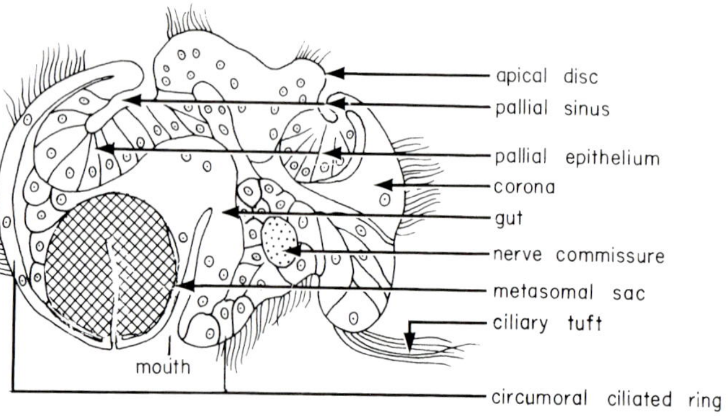

16

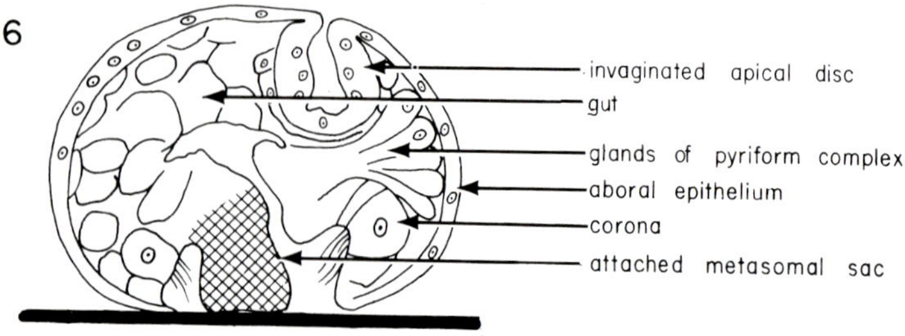

17

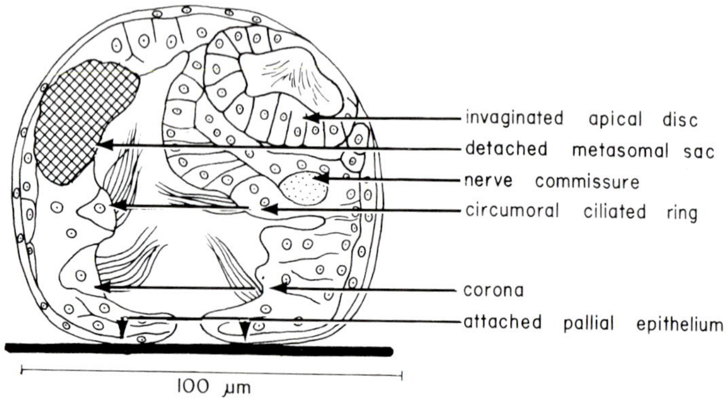

Figs. 15, 16, and 17. Right-side views of midsagittal sections of larval and early metamorphic stages of paludicelloid carnose ctenostomes. All after Braem (1951).

Fig. 15. *Victorella muelleri.* Larva. Note unique circumoral ciliated ring.

Fig. 16. *Bulbella abscondita.* Individual after eversion and attachment of metasomal sac. The transitory tissues have partly involuted, carrying the aboral epithelium toward the oral pole.

Fig. 17. *Victorella muelleri.* The everted and once-attached metasomal sac of the incipient preancestrula has become detached and withdrawn internally with the usual transitory larval tissues. The pallial epithelium has almost completed its formation of the entire epidermis and now attaches the individual to the substratum.

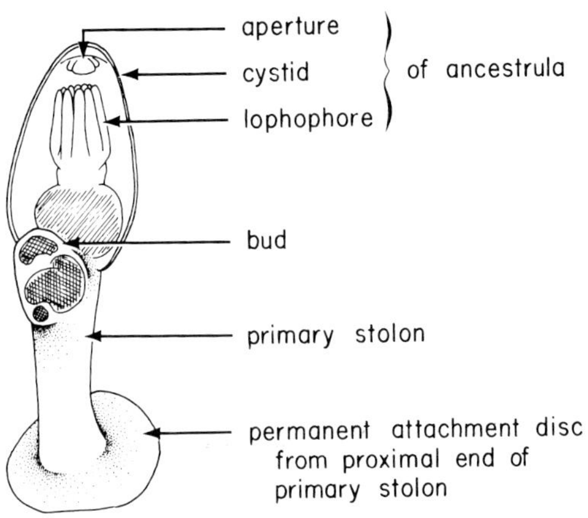

Fig. 18. *Zoobotryon verticillatum.* Fully differentiated ancestrula after proliferation of the primary stolon, loss of the original attachment made at settling, and reattachment by the tip of the primary stolon. Modified from photomicrograph from Bullivant (1968).

degeneration of the original gut and lophophore (Ries and Schölzel, 1934; see also Hyman, 1959, and Ström, 1969). Although the exact fate of the ancestrular cystid is not known in *Zoobotryon,* it also is probably dehisced after its brief functional phase, leaving the colony totally devoid of its ancestrula zooid.]

(c) *Interstitial and vagile species.* Finally, it is appropriate to mention that several forms have become adapted for life in sand or mud and therefore lack the usual attachment to a solid substratum. In the unique ctenostome genus *Monobryozoon* the "colony" consists of a single autozooid provided with stolons that anchor it within its interstitial environment (see Ott, 1972). Members of the genus *Kinetoskias* are aberrant cellularioids each consisting of a tuft of branches supported on a long stalk constituted of interwoven kenozooids (see Silén, 1950). A rootlike mass of rhizoids at the base of the stalk secures the colony within muddy deposits. Unfortunately nothing is known of either the larvae or the metamorphosis of these two genera, but it is obvious that attachment via the metasomal sac is transitory at best. Within the Anasca, several genera produce discoidal or cone-shaped lunulitiform colonies that are semivagile on the surface of sand, thanks to their vibracula. Maturo (1971) observed the larvae of *Cupuladria doma* and *Discoporella umbellata depressa* to settle on single granitic sand grains using the metasomal sac. Although such a substratum would anchor the minute ancestrula, it would not secure even a small colony. Thus, regardless of whether the attachment made at metamorphosis is actually lost or the colony simply carried the sand grain with it, the colony in effect is unattached.

2. *Phase 2 at Metamorphosis–Transformation of the Preancestrula to the Ancestrula*

At the completion of the first phase of metamorphosis the preancestrula consists essentially of (1) the cystidlike body wall, which contains and is still connected with (2) the invaginated apical disc, and also encloses (3) a mass of internal cells and tissues. A progressive differentiation of these three components during the slow second phase of metamorphosis results in the translation of the preancestrula into the functional ancestrula.

Normally each preancestrula differentiates into a unique individual–the ancestrula–and, near the completion of its histogenesis, the ancestrula proliferates asexually one or more autozooidal buds or, in stolonate ctenostomes, a proliferative stolon from which autozooids will be elaborated. However, this basic pattern is not inviolable: the preancestrula may differentiate as a compound rather than a simple ancestrula, produce precocious buds, or do both. Although technically these events probably are features involving differentiation of the cystid, they are considered after the normal pattern has been described in Sections II,B,2,d and II,B,2,e.

a. Differentiation of the Preancestrular Cystid to the Cystid and Zooecium of the Ancestrula. As indicated earlier, the body wall of the preancestrula is usually a cystid provided with both an epidermis and a lining mesodermal sheet. The mesothelium is not always differentiated at the completion of the first phase of metamorphosis, however. In *Bugula neritina,* for example, the fenestrated somatic peritoneum is fabricated by the amoeboid wandering of pigmented mesodermal cells. During larval life, these were largely compacted between adjacent coronal cells (Fig. 10) and are entrained with the transitory larval structures as those are involuted during the first phase of metamorphosis. The mesodermal cells emigrate centrifugally from the regressing corona to line the basal ends of the former metasomal sac tissues during the histogenic phase of metamorphosis (Woollacott and Zimmer, 1971).

The initial step in elaboration of the exoskeleton is proliferation of a flexible cuticle from the epidermis of the preancestrula. This primary component of the zooecium is produced between the attachment disc and the substratum as well as over the free surface. In true cyphonautes (with the possible exception of *Hypophorella expansa,* see Ehlers, 1876) and in *Flustrellidra hispida* (lecithotrophic shelled larva) the larval valves are also underlain by this cuticle and will soon fall off. This pattern is not true of *Pherusella brivituba* (lecithotrophic shelled larva) [and presumably *Hypophorella expansa* (cyphonautes)] for the larval valves remain attached to the ancestrula even in advanced colonies (Zimmer, personal observations; Ehlers, 1876). The originally delicate cuticle may be later thickened in ctenostomes, and is partially underlain by sclerites of calcium carbonate in cheilostomes.

The disclike, hemispherical, or hemiovoid shape of the preancestrula may also be the eventual form of the ancestrula (e.g., some carnose ctenostomes), but usually it is modified: the definitive ancestrulae of encrusting cheilostomes take many shapes ranging in outline from circles to triangles to rectangles, pentagons, or hexagons. In stolonate ctenostomes and nonencrusting cheilostomes (e.g., *Bugula*) the preancestrula elongates into an upright cylinder during the histogenic phase of metamorphosis (Fig. 19). The erect ancestrula is usually stabilized by the proliferation of a stolon (and rhizoids) in the stoloniferan ctenostomes (Fig. 20) or may be buttressed by rhizoids in erect cheilostomes (Fig. 19). Such rhizoids usually originate near the proximal end of the ancestrula [e.g., *Bugula neritina* (Mawatari, 1951)], but in *Scrupocellaria* two distally produced rhizoids form a mechanically stable tripod with the ancestrula itself (Fig. 21) [e.g.,

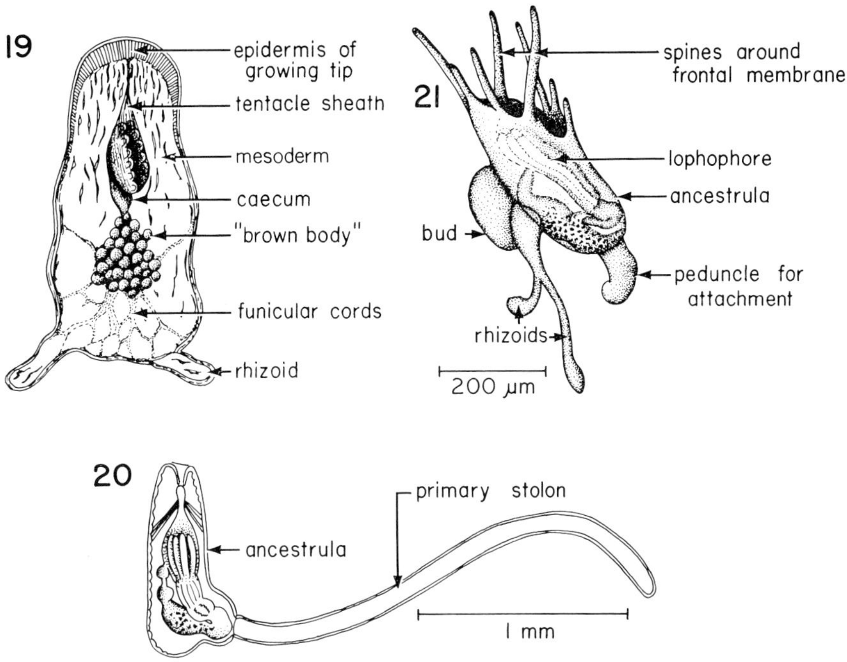

Figs. 19, 20, and 21. Variations in ancestrular morphology.

Fig. 19. *Bugula calathus.* Elongating incipient ancestrula with the cecum of its partially differentiated polypide embedded in the brown body-like remains of the nutrient mass. Basal rhizoids support the erect ancestrula. From Calvet (1900).

Fig. 20. *Amathia lendigera.* The erect ancestrula is stabilized by the primary stolon which will proliferate all future buds.

Fig. 21. *Scrupocellaria scruposa.* The completed ancestrula is supported by rhizoids originating near its distal end. Figures 20 and 21 from Barrois (1877).

Scrupocellaria scruposa (Barrois, 1877)]. The ancestrulae are often smaller than normal individuals, but may be larger than or equal to them.

In addition to size and shape differences, ancestrulae and normal individuals of the colony often vary considerably in their morphology. Such variations are modest in ctenostomes usually—for example, the ancestrular frontal of *Alcyonidium hirsutum* is smooth, lacking the small tubercles which characterize the species—but may be striking in cheilostomes. Perhaps the most extreme variation is found in many ascophorans which possess "anascan" ancestrulae with much or all of the frontal uncalcified (e.g., *Schizoporella,* Waters, 1924; *Fenestrulina,* personal observations). When the ancestrular frontal is fully calcified, the ancestrulae of ascophorans do possess asci, but these may not be characteristic of the species: in *Haplopoma* the ancestrula ascus communicates with the exterior via the narrow sinus of the aperture, but in normal zooids an ascopore opens on the frontal surface independently of the aperture (Ryland, 1970).

Ancestrulae with membranous frontals often have erect marginal spines developed around the uncalcified region (Fig. 22). Normal zooids of species with

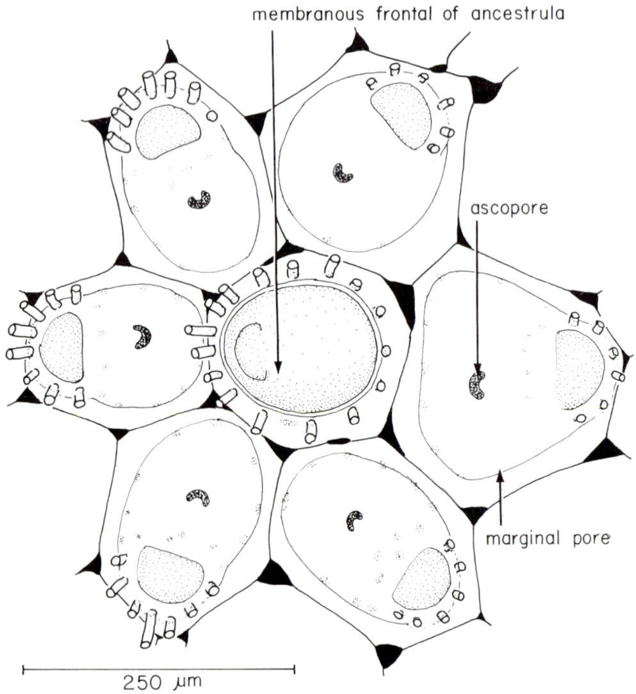

Fig. 22. *Fenestrulina malusi.* Surface view of tata ancestrula and the periancestrular zooids.

such "tata" ancestrulae usually have fewer marginal spines than the ancestrula or may totally lack them.

The occurrence of and degree of calcification of the frontal surface of the ancestrula are not correlated with the extent of calcification in normal zooids, as noted above, nor does it seem related to the systematic position of the species. For example, within the family Smittinidae, different species have full, partial, or no calcification of the frontal (see Soule and Soule, 1972).

Not only does the degree of calcification vary between ancestrulae and normal zooids, but even the nature of calcification is different. Lutaud (1953) demonstrated that calcification in the ancestrulae of the gymnocystid *Escharoides* differs from that of normal zooids not only in details of direction of spread of calcification from centers of deposit, but more significantly in that only a single layer is deposited on that part of the frontal which is calcified rather than having a secondary "pleurocyst" in addition to this "holocyst" as in normal individuals. In association with this lack of secondary calcification, ancestrulae generally lack tremopores and aureolae.

The generalization that secondary calcification does not occur in ancestrulae and that they consequently lack tremopores and areolae or marginal pores is refuted by Cook's (1973) study of *Metrarabdotos*; the ancestrula has all the complications of normal zooids including marginal pores, a tuberculate region on the frontal shield, and a secondary orifice.

These differences in calcification, spination, and ascopore formation may have phylogenetic significance (e.g., Hyman, 1959), but this is highly speculative.

b. Differentiation of the Polypide from Its Anlage. A detailed consideration of the differentiation of the polypide rudiment in the ancestrula was last given by Zschiesche in 1909. Prior to this analysis of metamorphosis in *Alcyonidium polyoum* (Type E coronate larva) (as *A. mytili*), the two most detailed studies were provided by Prouho (1890) for *Flustrellidra hispida* (lecithotrophic shelled larva) (as *Flustrella hispida*) and Calvet (1900) for *Bugula simplex* (Type AEO/ps coronate larva) (as *Bugula sabatieri*). The following description of "normal" polypide differentiation from the epidermal and mesodermal blastemas is composited from these three studies, but includes some observations of Kupelwieser (1905) for *Electra pilosa* (cyphonautes) and Woollacott and Zimmer (1971) for *Bugula neritina* (Type AEO/ps coronate larva).

It should be noted that the origin of the polypide anlage from tissues other than the blastemas and the occurrence of ancestrulae lacking polypides have been reported. These possible variants are considered at the end of this subsection, but it is appropriate to clarify at this point why the blastemas are thought to be the fundamental (if not the only) source of the polypide: (a) The five species mentioned above have larvae representative of true cyphonautes (probably the least modified existing gymnolaemata larvae), other shelled larvae, and both simple (Type E) and advanced (Type AEO/ps) coronate larvae; (b) within these few representatives, the blastemas range from their simplest expression (in *Electra,* Fig. 7) to their most complex one (in *Bugula,* Fig. 10); and (c) on a more subjective note, the five studies are all well documented and illustrated—collectively they present a cohesive and consistent pattern of derivation of the polypide from the two blastemas.

i. The normal pattern. As previously noted, the apical disc, consisting of a central neural plate and a peripheral epidermal blastema, together with the underlying mesodermal blastema are invaginated as a unit of the end of the first phase of metamorphosis.* A narrow annulus of aboral (or pallial) epithelium is also carried in as a cone that, at its apex, is continuous with the superficial epidermis of the cystid and, at its base, joins the margin of the epidermal blastema. The apical disc is deformed on invagination into a "cup" that is two layered throughout most of its extent. This "cup" with the invaginated aboral epithelium constitutes a more or less spherical vesicle joined by a narrow stalk with the center of the free surface of the preancestrula.

This narrow connection with the epithelium of the cystid is usually reported to be transitory. Prouho (1890) and Kupelwieser (1905) related that, after severing its surface connection, the aboral epithelium persists as a thin roof over the thicker blastemal portions of the vesicle; Calvet (1900), however, indicated the nonblastemal cells degenerate and are replaced by amoebocytes. In either case, a closed vesicle formed predominantly of the two blastemas results. In *Bugula neritina,* Woollacott and Zimmer (1971) observed that the epithelial cone (specifically derived from the pallial epithelium) retains its connection at the surface and will directly form the tentacle sheath; the pore left by invagination of the apical disc will persist as the aperture, shifting distally as histogenesis of the ancestrula proceeds.

Normally the neural plate of the larva forms a central plug in the base of this vesicle. Observation on the fate of these cells are limited, but Calvet (1900) for *Bugula simplex* stated they degenerate after dehiscence from the apical disc; the pore left through both blastemal layers by the loss of the nerve cells is repaired by mesenchymal cells. In *Flustrellidra hispida,* the neural elements invaginate separately from the blastemas and degenerate after they are surrounded by the blastemal vesicle (Fig. 4) [Prouho (1890)]. Hondt (1974, 1975) reported that cells of this plate form part or all of the tentacle sheath in the ancestrulae of *Alcyonidium polyoum* and *Bowerbankia imbricata,* respectively.

*An exception to this generalization occurs in *Flustrellidra hispida:* Prouho (1890) observed the neural cells [which are separated from the epidermal blastema by a ring of undifferentiated cuboidal cells in the larva (Fig. 2)] invaginate first as a separate vesicle and later are surrounded by the invagination of the blastemal complex (Fig. 4).

The two-layered vesicle, now devoid of neural elements, represents the primordium of the ancestrular polypide. Soon a constriction in a plane nearly parallel to the future frontal surface of the ancestrula, partially divides the vesicle into two unequal sections (Fig. 23). The larger of these is already or soon becomes regionated into a thinner portion that will produce the tentacle sheath and a more massive primordium of the pharynx and the base of the lophophore (Fig. 24). The curved smaller section is more basally positioned and will differentiate as the stomach, cecum, and intestine according to most studies [Prouho (1890) claimed the stomach originates from a separate aggregation of cells in *Alcyonidium polyoum*]. The two sections remain joined at the site where the anus opens into the tentacle sheath. According to most reports, the pharynx and stomach are usually separated at this stage, but Calvet (1900) reported both ends of the primordium of the stomach and intestine remain joined with the primordium of the lophophore, pharynx, and tentacle sheath.

Tentacle buds appear on the lophophoral base which is usually elliptical at this stage. Initially, the tentacles are arranged in an interrupted ring (or horseshoe) (Fig. 19) since the anal-most tentacles develop only later or are more widely spaced than other adjacent pairs. In the median line between this tentacle gap the cerebral ganglion of the future polypide will invaginate as an epidermal vesicle near the junction of the lophophoral base and the pharynx (Fig. 25). The pharynx soon develops its triangular profile and characteristic cytology, but is still blind at its lower end (except in *Bugula simplex*).

The tentacle buds are initially oriented perpendicular to the future frontal surface and a thickening immediately above them on that surface is said to be the site at which the tentacle sheath will reestablish contact with the body wall (e.g., Prouho, 1890). With elongation of the tentacles, the polypide rudiment which was oriented normal to the frontal surface rotates through 90° and the site of the future aperture migrates to near the distal end of the incipient ancestrula. The tentacles of the now radially symmetrical lophophore are directed distally and are aligned with the long axis of the cystid (Figs. 25 and 26).

It seems natural to assume that the distal end of the ancestrula would correspond with the anterior end of the larva, and Prouho (1890) stated this is the case in *Flustrellidra*; apparently, the opposite is usually true. That the proximal end of the ancestrula corresponds with the anterior end of the larva is evident in the figures of Robertson (1908), O'Donoghue (1926), and Atkins (1955) for species of *Membranipora* and *Electra* and has been confirmed in *Schizoporella unicornis* by Barrois (1879) and *Bugula neritina* by Lynch (1947) (see also Ryland and Stebbings, 1971; Jebram, 1973). The significance of this orientation is obscure, but it may be noted that in gymnolaemate zooids, the anus is distal to the mouth rather than proximal to it as in most animals.

The stomach, cecum, and intestine complete their differentiation from their rudiment and subsequently the stomach and pharynx will join (Figs. 25 and

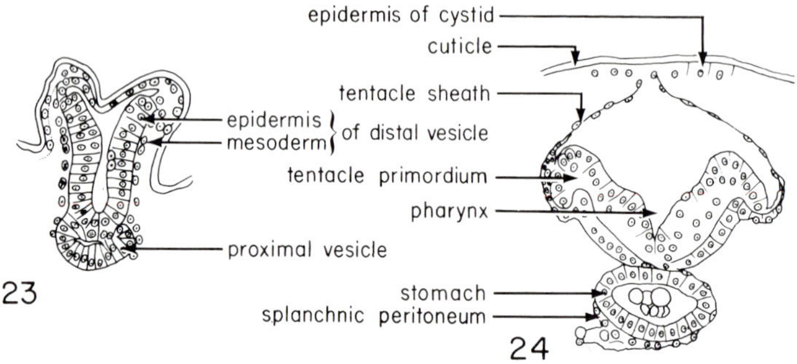

23 24

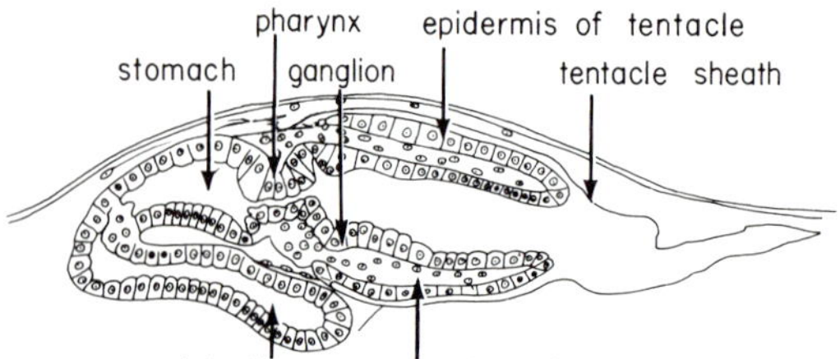

25

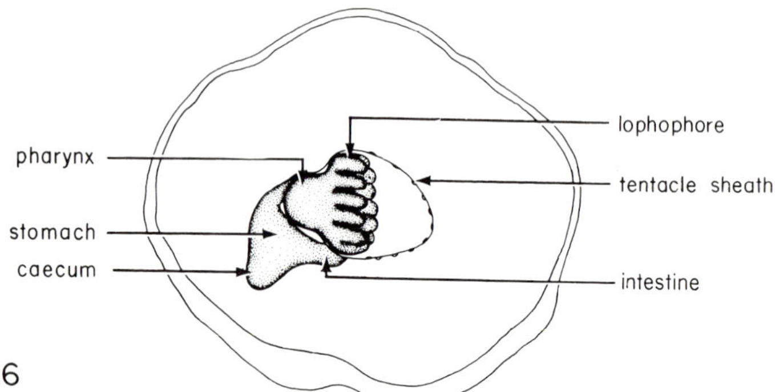

26

Figs. 23, 24, 25, and 26. *Alcyonidium polyoum.* Stages in histogenesis of the ancestrular polypide. All from Zschiesche (1909).

Fig. 23. Constriction of the original two-layered blastemal vesicle has produced a larger distal vesicle and a smaller proximal one. Longitudinal section.

Fig. 24. Differentiation of the larger vesicle into tentacle sheath, lophophore, and pharynx is in progress. Frontal section.

Fig. 25. Elaboration of the ancestrular polypide is nearly complete. Note origin of cerebral ganglion. Junction of intestine with tentacle sheath not shown in this plane. Near median sagittal section.

Fig. 26. Surface view of stage comparable to that illustrated in Fig. 24.

26). The cecum (Prouho, 1890; Calvet, 1900) or stomach proper (Woollacott and Zimmer, 1971) is embedded within the nutritive mass (see below) during the morphogenesis of the polypide (Fig. 19). As its growth is correlated with a diminution of the nutritive mass, the polypide apparently receives nutrients from the latter, perhaps through the wall of the gut. We know of no convincing documentation that such nutrition actually occurs, the form(s) in which nutrients are transferred, or the mechanism(s) of nutrient uptake by the polypide. Humphries (1977) reported that pieces of the nutritive mass are pinched off into the stomach during the differentiation of the ancestrula of *Parasmittina nitida*, but her observations were made on living specimens.

When the polypide is complete, the nutrient mass will be reduced to a small, dark spherule that Calvet (1900) equated with a brown body (Fig. 19). We do not disagree with this interpretation for one would expect such a residual body to persist after intensive phagocytic activity. However, it does not correspond precisely to the brown bodies remaining after polypide degeneration as different tissues were histolysed. With disappearance of the nutrient mass, the presence of funicular cords associated with the gut and occupying the former site of the mass are evident. Calvet (1900) indicated that the funicular system is derived from leukocytic mesenchymal cells associated with this mass.

With the exception that it is initially associated with a nutritive mass rather than a funicular complex as in autozooids, the polypides of an ancestrula and of regenerating zooids and asexual buds are identical in their morphogenesis. This fact is significant for it indicates that in all these cases, the formation of the polypide is an asexual process.

Information on the elaboration of the major muscles of the ancestrula is very limited but apparently they develop *de novo* from larval mesenchyme with one exception. The parietal elements associated with protrusion of the lophophore are reported to differentiate *in situ* from mesenchyme retained from the larva (Calvet, 1900), but in *Flustrellidra hispida* they constitute persistent larval muscles (Prouho, 1890). The lophophore retractors originate from a thickening in the mesodermal layer of the original vesicle that was situated near the future proximal end of the ancestrula (Prouho, 1890), and the occlusar muscles of the operculum from an equivalent but distal thickening of that layer (Prouho, 1890).

Although we have not so specified, the epidermal and mesodermal blastemas differentiate in parallel during optimal elaboration of the polypide. The epidermal blastema produces the epithelia of the lophophore and the gut, and the mesodermal blastema the mesocoel (including the ring canal and tentacular extensions) of the lophophore and the splanchnic mesothelium of the gut. As indicated earlier, the somatic peritoneum is usually established earlier as part of the cystidal body wall. Consequently, the mesodermal lining of the main visceral cavity or metacoel of gymnolaemates has a composite origin and the cavity itself forms neither as schizocoel nor archenteric pouch, but is technically blastocoelic in derivation. Woollacott and Zimmer (1971) have commented on the uniqueness of this formation.

With the conclusion of this extensive differentiation the polypide is protrusible and the ancestrula can become a functional zooid.

In general, normal ancestrular polypide differentiation is insufficiently known to assess whether the slight variations noted have any significance other than at the species level. Soule (1954), however, studied polypide differentiation in asexually produced buds of a variety of ctenostomes and found two basic patterns in the temporal appearance of the retractor, parietal, and apertural muscles. The two sequences observed are correlated with the divisions Carnosa and Stolonifera, supporting the diphyletic status of the ctenostomes. Thus, the careful analysis of polypide differentiation might well reveal additional patterns of variation that are of systematic significance.

ii. Variations in polypide formation. *Polypide formation from tissues other than the epidermal and mesodermal blastemas.* Although the origin of the polypide rudiment from tissues other than the epidermal and mesodermal blastemas has been reported, especially in the early literature, we doubt that such variations actually exist. In many cases, possible origins of components of the polypide anlagen were "suggested" rather than observed. In most cases the greatest uncertainty related to the origin of the mesoderm of the polypide, but occasionally an unusual origin of the epidermal tissues has been reported.

Hondt (1972, 1973, 1974) studied the cytology and metamorphosis of larvae of *Alcyonidium polyoum,* using plastic-embedded materials with light and electron microscopy, and claimed a unique origin of the polypide rudiment. After the usual attachment of the larva by the everted metasomal sac, the pallial epithelium is evaginated and moves downward to join the metasomal sac, completing the cystid wall as the intervening cells (including a supracoronal zone of aboral epithelium, the corona, a band of distinctive infracoronal cells, and the usual transitory components of the oral field) are involuted. The internalized cells form a vesicle in the form of a flattened spheroid rather than a torus as in other species. The corona, pyriform complex, and supracoronal epithelium are soon lysed, but the infracoronal or "M" cells (as Hondt refers to them) come to line the entire vesicle as the transitory cells degenerate. Simultaneously, this vesicle is sheathed by mesenchyme cells, largely derived from mesodermal elements that encased the neuromuscular cord in the larva. Now two-layered, the vesicle constitutes the primordium of the polypide according to Hondt.

The apical disc which normally forms the polypidal rudiment, has invaginated as in other forms (it apparently invaginates before rather than after the involution of the corona and its associated cells in this species), but the vesicle formed is stated to give rise only to the tentacle sheath.

As the metamorphosis of *Alcyonidium polyoum* and the apparently conspecific form *Alcyonidium mytili* has been previously described to occur in the usual fashion (see especially Zschiesche, 1909), it is most enigmatic that Hondt refers to these works only in passing and makes no attempt to explain the differences between his conclusions and previous observations on the same and different species.

In the summary of his 1974 paper, Hondt does indicate that he has observed the typical formation of the polypide from the invaginated apical disc in *Flustrella hispida, Microporella ciliata,* and *Carbasea papyracea.* Relative to the evaluation of Hondt's report, this statement is partially reassuring, but he could have provided direct confirmation of his unusual interpretation of metamorphosis in *Alcyonidium* by providing photomicrographs of median sections of the critical stages rather than only highly diagrammatic figures. Hondt (1975) also reported undifferentiated infracoronal cells form the epidermis of the ancestrular polypide in *Bowerbankia imbricata* (Type AEO/PS coronate larvae). In this stoloniferan ctenostome, however, these cells are not superficial as in *A. polyoum* but have invaginated at the oral pole, occupying, with transitory epidermal tissue, the position normally taken by the metasomal sac. Hondt did not specify how the larva attaches in the complete absence of the sac, but did state that the monotypic cells of the apical disc are the primordium of the tentacle sheath and that the extensive pallial sinus epithelium forms the entire epidermis of the ancestrular cystid. [See, however, reports on the metamorphosis of *Bowerbankia pustulosa* (Ostroumoff, 1886a,b) and of *B. gracilis* (Eiben, 1976).] It may also be noted that Barrois (1879, 1886) claimed aggregations of columnar cells of the larval epidermis on each side of the pyriform complex in the larva provide the mesoderm of the polypide in various gymnolaemates (e.g., *Schizoporella unicornis* and *Amathia lendigera*). These cells (which Barrois called "m" cells!) would correspond in position with parts of the ring of "M" cells reported by Hondt, but their function would be quite different.

In summary, Hondt (1972, 1974, and 1976) has suggested a unique and aberrant origin for the polypide but has provided no incontrovertible evidence in the form of photographic documentation. He has neither acknowledged the existence of previous studies on the same species (which relate fundamentally different conclusions) nor attempted to correlate his observations with previous studies other than to claim he has observed typical formation of the polypide in several species. The exceptional nature of Hondt's claims requires that he justify them more carefully.

A final variation to be considered relative to the ancestrular polypide is the report that it may be totally absent in stoloniferan ctenostomes. Hyman (1959, p. 362) stated: "Calvet made a general statement that the stoloniferous ancestrula does not develop a polypide and this was confirmed by Silén (1942a) for

Walkeria and *Vesicularia* (our Figs. 27 and 28, p. 122) and also appears to be the case in *Mimosella* (Harmer, 1915). However, an ancestrula of *Amathia lendigera* (Fig. 20) with a fully developed polypide was illustrated by Barrois (1877).

Polypides have also been figured in the ancestrulae of *Zoobotryon verticillatum* (Fig. 18) (Zirpolo, 1933; Bullivant, 1967), *Walkeria uva* (as *Vesicularia uva,* Ostroumoff, 1886a,b), and *Bowerbankia stationis* (as *Vesicularia stationis,* Ostroumoff, 1886a,b). In fact, in no stoloniferan in which the metamorphosis has been followed is a polypide-deficient ancestrula produced. The observations of Silén (1942a) on *Walkeria uva* and *Vesicularia spinosa,* of Silén (1944) on *Labiostomella gisleni,* and, according to Hyman (1959), by Harmer (1915) on *Mimosella gracilis* were all based on identification of the ancestrula in the established colony. From examination of very young colonies of *Triticella koreni,* Ström (1969) concluded that the ancestrula of this stoloniferan does give rise directly to a zooid, as in other bryozoans, as well as to the stolon. However, his reference to the small adnate vesicle from which this "primary zooid" and stolon arise as the "ancestrula" is confusing. It is evident from his Fig. 3F (in which the "primary zooid" is functional before the stolon has produced any buds), that the entire complex, not just the "ancestrula" is derived from the larval cell mass. Probably, considering its position, this small vesicle constitutes the remains of the preancestrula after proliferation of the stolon and "primary zooid" (see the following paragraph on *Zoobotryon,* however). Significantly, Ström noted that "the primary zooid disappears in most cases during an early phase of following growth of the colony" (1969, p. 123); subsequently, it would appear that the ancestrula was indeed "polypideless."

The curious reattachment of the developing colony of *Zoobotryon* by the expanded tip of the primary stolon was reported earlier (see Bullivant, 1968). It was noted then that this expanded tip could easily be misinterpreted as an ancestrula that failed to develop a polypide. It seems quite plausible that the metamorphosis of *Zoobotryon* or that of *Triticella* may provide an explanation for the minute polypideless "ancestrulae" reported by Calvet (1900), Harmer (1915), and Silén (1942a and 1944). However, it is also feasible that the nutrient store of the preancestrula could be used to produce a proliferative stolon that would bud at least one functional zooid rather than produce a polypide in the ancestrular cystid. The definitive resolution of this puzzle will require the observation of metamorphic stages or perhaps very young colonies of each species in question.

c. Degradation of the Transitory Larval Tissues. In addition to the rudiment of the polypide, the preancestrular cystid is filled with transitory larval tissues, including (a) epidermal tissues involuted during the first phase of metamorphosis and (b) mesodermal and endodermal elements derived from the few cells that were internalized at gastrulation. The endodermal cells are sometimes differentiated as a tubular gut, but more commonly consist of nutrient-laden parenchymal elements. The mesodermal components include differentiated larval nerves and muscles and unspecialized mesenchymal cells that will differentiate as short-lived phagocytes as well as the funicular and muscular systems of the polypide; the latter two components have been considered above under differentiation of the polypide.

The fact that the larval transitory tissues are "degraded" carries a negative connotation, but in fact an opposite interpretation is appropriate. No longer capable of any useful direct function after the attachment of the larva, these tissues provide a significant source of nutrients for the growth of the incipient ancestrula. At least the epidermal components (and probably the internal ones as well) could have been cast off from the transforming larva to undergo autolysis or ingestion by other organisms. Their "recycling" at metamorphosis reflects an enormous economy of energy by the bryozoan.

Degradation of the transitory tissues begins within minutes of the onset of metamorphosis or even earlier. Rudimentarily developed digestive tracts may degenerate during embryogenesis (e.g., in *Alcyonidium polyoum,* Harmer, 1887). In addition, necrotic changes in other transitory tissues are already apparent in the larva according to several workers (e.g., Calvet, 1900). The completion of the breakdown process requires virtually all of the histogenic phase or longer.

As indicated above, the cells destined for recycling aggregate in a nutritive mass associated with the basal region of the polypide rudiment initially and later with the stomach region that differentiates there. Reduction of this nutritive mass is accomplished by autolysis and phagocytosis. At the completion of the histogenic phase, the nutrient mass is but a small, pigmented, opaque mass and the ancestrula contains a spacious body cavity (Fig. 19). Both Vigelius (1886) and Calvet (1900) referred to the remnant as a brown body such as is found after polypide degeneration in normal zooids. The parallel is not precise, however, for different tissues are involved in the two cases; moreover, one would expect a comparable dark indigestible product from any extensive histolysis. Nevertheless, a possible correspondence cannot be ignored.

The degenerative changes occurring in each of the various transitory tissues have been documented by Prouho (1890), Calvet (1900), Kupelwieser (1905), Seeliger (1906), Zschiesche (1909), Hondt (1974), and others. The process is similar, even in particulars, to that observed during normal polypide replacement, for details of which the reader is referred to Chapter 10 in this volume.

d. Precocious Budding of the Preancestrula–Composite Ancestrulae: Precocious bud formation from the preancestrula during the metamorphosis occurs in a variety of gymnolaemates, but is best known for ascophorans. Several examples will provide an introduction to the concept of composite ancestrulae in which the preancestrula forms not only an ancestrular zooid, but simultaneously produces at least the cystids for one or more additional individuals.

27

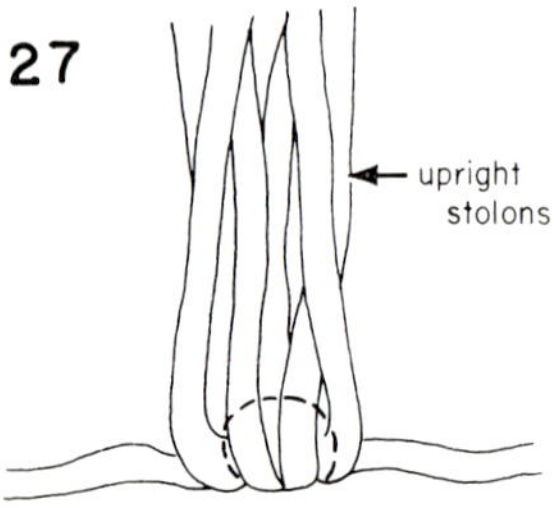

28

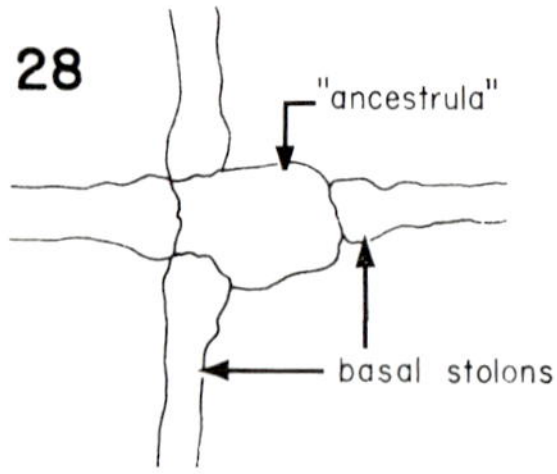

29

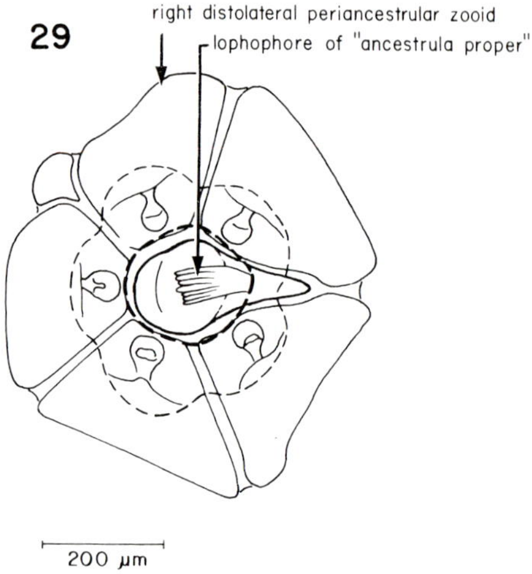

Fig. 27. *Vesicularia spinosa.* Surface view of portion of zoarium with main stalk formed of upright stolons, horizontal stolons, and assumed ancestrula (dotted line) which is devoid of a polypide. From Silén (1942a).

Fig. 28. *Walkeria uva.* Surface view of portion of zoarium with assumed ancestrula (which is devoid of a polypide) and four stolons. From Silén (1942a).

Fig. 29. *Smittina papillifera.* Composite ancestrula in surface view. The five periancestrular individuals have rudiments of their polypides and their cystids have enlarged somewhat from their original size (the thin dotted line indicates their extent in the preancestrula $1\frac{3}{4}$ hours after metamorphosis began). The ancestrula proper" has formed its polypide precociously; the outline of its cystid $1\frac{3}{4}$ hours after initiation of metamorphosis is indicated by the heavy dashed line. From Stach (1938).

Stach (1938) has described in detail the metamorphosis of the ascophoran *Smittina papillifera.* The preancestrula is initially hemispherical, but becomes tri- and then pentalobed in outline within a few minutes of the attachment of the larva. Septa divide the lobed structure into a central chamber that is destined to form the ancestrula proper and five periancestrular units that represent the cystids of a single distal, paired distolateral, and paired proximolateral "buds" (Fig. 29). The central unit normally produces its polypide first, but differentiation of the internal parts of the "buds" soon follows (subsequently the ancestrula proper and the proximolateral buds collectively contribute to a proximally positioned bud that completes the periancestrular ring). Although the fate of the apical disc was not followed—presumably it formed only the rudiment of the polypide of the ancestrula proper—we concur with Stach that only the central individual corresponds to the typical ancestrula and the five periancestrular zooids represent precocious buds. Stach related this precocious budding to the large nutrient content of the larva, but this does not seem a sufficient explanation: for example, the larvae of *Bugula neritina* are packed with yolk-laden cells and exceed larvae of *Smittina* in dimension but produce simple ancestrulae.

The same pattern with five periancestrular zooids has been deduced for various ascophorans by Waters (1925a,b; 1926a,b). For *Smittina collifera,* Mawatari (1946c) observed that the preancestrula forms three lobes rather than five, but the proximolateral buds form shortly later. Another variation from this general theme occurs in *Metrarabdotos unguiculatum cookae* (Cook, 1973). A similar segmentation of the preancestrula cystid occurs, but the pair of proximolateral buds are considerably smaller than the other four subdivisions, and are destined to form kenozooids. The distal and distolateral daughter zooids differentiate slightly later than does the ancestrula proper and with the paired heterozooids encircle the ancestrula. Only the daughter zooids of this composite ancestrula produce other zooids and the resulting colony is fan-shaped rather than circular. The significance of this precocious formation of buds from the preancestrula is obscure, but it might facilitate a more rapid and symmetrical growth of the colony.

e. Compound Ancestrulae (with Ancestrular Twins and Triplets) and Compound–Composite Ancestrulae. The occurrence of compound ancestrulae is best known in the genus *Membranipora* (but not the closely related genera *Electra* and *Conopeum*). Members of this genus have cyphonautes larvae which produce at metamorphosis a disc-shaped preancestrula that is soon divided by a Y-shaped septum into a pair of large lateral cystids, each destined to form its own polypide, and a small median chamber, the first bud which will differentiate later (Fig. 28). The simultaneously produced and synchronously developing individuals represent twin ancestrulae proper—hence the ancestrula is compound—but the existence of a precocious bud also makes it composite.

In *Membranipora isabelleana* at least, the pattern of precocious budding is more complex than just indicated for the genus. Shortly after the initial partitioning of the preancestrula four or five diminutive chambers are cut off in an arc from the distal and lateral margins of each of the two larger cystids, which now contain polypidal rudiments of the ancestrulae proper. Of these eight to ten units, the pair adjacent to the distomedial bud will progressively differentiate as distomedial autozooids and the more lateral members will initially develop as spines representing heterozooids that flank the ancestrulae proper (Fig. 30), but later may enlarge as autozooids (Zimmer, personal observations). In this species, then, the preancestrular mass completely provides the ancestrular twins and additionally contributes primordia for the distomedial zooid, the paired distolateral zooids, and six to eight periancestrular spinozooids.

Compound ancestrulae producing triplets rather than twin ancestrulae proper are known for five species of *Stylopoma, Cupuladria, Hippopodina,* and *Dis-*

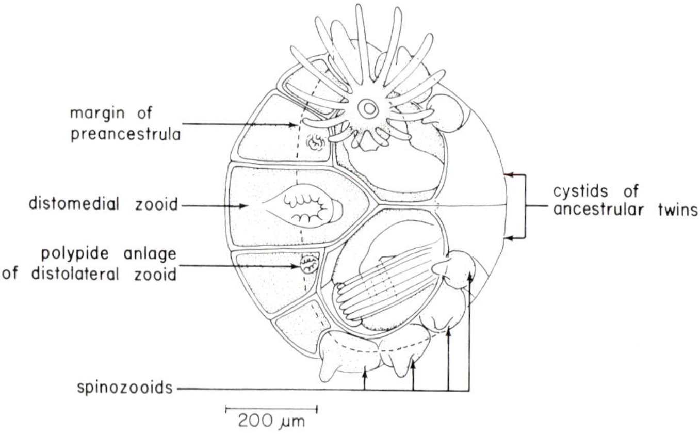

Fig. 30. *Membranipora isabelleana* (?). Young zoarium developing from the compound composite ancestrula. All individuals–the ancestrular twins, distomedial zooid, paired distolateral zooids and spinozooids–originated from the preancestrula.

coporella. In the simplest of these examples (Cook, 1973), the preancestrula of the ascophoran *Stylopoma duboisi* is initially a simple hemisphere, but soon becomes divided by three symmetrically positioned septa. A central mass of pigmented cells is equally shared by the three sectors. The exact nature of this pigmented mass is unknown, but it more likely represents the nutrient mass rather than the blastemal complex. Polypides differentiate simultaneously within the three chambers, and shortly later six buds arise synchronously, two growing between each of the three notches formed by adjacent arms of the compound ancestrula.

A slight variation on this theme was reported by Maturo and Middleton (1972) for *Stylopoma informata.* The preancestrula of this species also undergoes an initial tripartitioning, but before polypide formation is initiated, two parallel septa subdivide each of the three initial sectors into a pair of lateral chambers and a larger central one. Within the same cultures, about one-half of the preancestrulae form polypides in the three larger chambers first and the six smaller chambers subsequently, but in the other half, polypides form in all nine chambers simultaneously. It seems evident that this species also has a triple ancestrula proper, but that budding is quite precocious producing a compound, composite ancestrula.

In still another example, the unattached lunulitiform anascans *Cupuladria doma* and *Discoporella umbellata depressa* initially produce hemispherical preancestrulae, but each soon becomes trilobed. After development of partitions, each of the lobes develops simultaneously into a typical zooid, an ancestrula proper, and a distally positioned vibraculum, a heterozooid (Maturo, 1971).

Håkansson (1973) has proposed that the subdivision of the preancestrula in *Cupuladria* is additionally complicated by the partitioning off of a "narrow extrazooidal hypostegal coelom" around the entire margin (see also Tavener-Smith, 1973). In any event, the ancestrula is both compound and composite, but in this case the "buds" are heterozooids rather than autozooids.

Eitan (1972) also reported triplet ancestrulae in *Hippopodina feegeensis.* The significance of twinning in *Membranipora* and tripling in *Cupuladria, Discoporella,* and *Stylopoma* is obscure. It may not be mere coincidence that colonies of *Membranipora* grow as a more or less uniform circle around the ancestrula, but those of *Conopeum* and *Electra* are characteristically of a highly irregular stellate form. Triplet formation in the lunulitiform species could be adaptive, ensuring the symmetrical growth of the vagile, cone-, or disc-shaped colonies. Silén (1942b) related that colony growth in a few bryozoans, including several species of *Cupuladria,* follows spiral rather than radial lines from the ancestrula. He concluded that the "value" of such spiral growth is the symmetrical form of the resulting colony (however, see Håkansson, 1973).

C. Analysis and Interpretation

There is general agreement that, within the phylum Bryozoa, early life history is least modified in the class Gymnolaemata. Therefore, it is logical to initiate an evaluation of bryozoan metamorphosis with that group. It cannot be overemphasized, however, that the ontogeny of modern gymnolaemates is highly derived. As should be evident from the preceding sections, the development and metamorphosis of gymnolaemates is not uniform, but the variations that do occur are simply permutations "on a common theme." They provide few clues to the sequence of evolutionary changes that has resulted in the distinctive pattern of gymnolaemate ontogeny.

Before proceeding further, it is appropriate to state that a resolution to the age-old question of the relative primitiveness of plankotrophic cyphonautes and anenteric coronate larvae is not of critical importance for an analysis of the transformation of modern gymnolaemates, for metamorphosis is fundamentally the same regardless of the development of the gut. An answer to the conundrum is important, however, if one is to speculate on evolution of the life cycle of gymnolaemates or the developmental interrelationships between the three classes of the Bryozoa. Therefore, we reiterate that we consider bryozoan development initially involved a plankototrophic phase, despite the fact that a large majority of modern gymnolaemates produce anenteric larvae that are either fully lecithotrophic or are nourished by the parent during embryogenesis. In the case of echinoderms, Fell (1967) has presented arguments that larve with lecithotrophic development may be more basic than planktotrophic ones. Strathmann (e.g., 1975), however, has critically reexamined Fell's hypothesis, which was summarily dismissed by Hyman (1959), and rejected it; utilizing the available literature and the results of extensive personal research on feeding behavior in larval echinoderms he has reaffirmed the nearly catholic opinion that indirect, planktotrophic development is primitive. Jägersten (1972) and Strathmann (1975) both emphasized the generally accepted concept that an evolutionary loss is irreversible. Thus a lecithotrophic larva, lacking a plankton collecting and digesting system, has reached a cul-de-sac nutritionally (unless there is precocious development of the adult nutritive system), but a planktotrophic larva has the flexibility to remain planktotrophic or to assume a lecithotrophic development. Strathmann (1975) also noted that it is simpler to explain existing patterns of development in echinoderms if we assume that planktotrophy can be lost but not regained. Certainly this argument is persuasive in bryozoans, for partial to complete guts present transitorily during early embryogenesis of certain lecithotrophic larval species are most difficult to interpret unless one assumes that they represent stages in the progressive loss of a once functional system.

In analyzing the ontogenetic history of modern gymnolaemates it is obvious that a number of alterations on the ancestral pattern must have occurred. Such modifications can be specified, but the causative factors can only be speculated on. Almost certainly these effectors of evolutionary change have interacted extensively and therefore are difficult to isolate. In the following analysis we have attempted to consider the observed changes in order of the profoundness of their modification of ontogeny. The first of these changes is apparently unique to bryozoans, but others are commonly observed in other animal groups, both colonial and/or noncolonial. We have frequently invoked the process of "acceleration" in which the genetic program for a specific structure (and hence function) is transcribed earlier than is normal or necessary. Such precocious development would have serious disadvantages to the organism if it did not have some preadaptive value. Jägersten (1972) has called attention to the fact that acceleration is a plausible process to explain otherwise enigmatic aspects in the ontogenetic cycles of many animals. He has coined the descriptive and useful term "adultation" to specify the accelerated differentiation of strictly adult structures during preadult phases of the life cycle.

Significant deviations of the ontogeny of modern gymnolaemates from a simple life cycle include the following:

1. The ancestrular polypide is formed *de novo* during metamorphosis by budding from the cystid.

In the cystid–polypide duality of bryozoans, a pluripotent "body wall" is capable of producing a complete and functional individual by the asexual proliferation of the gut and eversible lophophore (or budding of an entire new individual). This unusual potency makes possible a life history in which the immediate product of sexual reproduction need only be a cystid, not a total zooid. Regardless of whether or not their larvae are provided with a functional digestive system, modern bryozoans have exploited this capability—metamorphosis of the larva results directly (but transitorily) in a cystid only. Of necessity, therefore, polypide replacement, originally only a postmetamorphic event, is now a critical step in formation of the ancestrula. The potential of an individual to replace its polypide is a well-recognized phenomenon of the adult bryozoan, but its significance in formation of the ancestrula has been largely ignored.

It may be mentioned that a comparable periodic replacement of the internal parts is found in another colonial group, the tunicates (limited to a few species of ascidians), and in the solitary holothuroids. In neither of these forms has the rejuvenation process been accelerated into the early stages of the life history, however.

Seeliger (1906) " explained" the degradation of the larval digestive tract in true cyphonautes on the basis that the larva lacks an excretory system; hence its digestive tract becomes filled with waste material and must be replaced. However, the absence of excretory systems in other planktotrophic larvae—including echinoderms and specifically holothuroids—is not correlated with a degeneration and replacement of the larval digestive tract at metamorphosis.

The significance of polypide replacement remains obscure even in adult bryozoans although it has been studied extensively (see Chapter 10). We suggest that the precocious occurrence of this phenomenon during metamorphosis of gymnolaemates is not so much correlated with an absolute physiological need (e.g., excretion), but rather with the facilitation of adaptation to sessile life. The only existing planktotrophic bryozoan larvae, the cyphonautes, are provided with both a complete gut and a "filter" system to concentrate and collect plankton. The former probably could be used directly after metamorphosis, but the cone-shaped vestibule perhaps was inadequate to meet the nutritional needs of the sessile ancestrula. Strathmann (1975) correlated length of the ciliary band involved in feeding with size of the organism and found cyphonautes to be inexplicably ill-equipped. Replacement of the entire "polypide" via cystidal budding may have provided a convenient alternative solution to replacement of only the food-gathering system. Adoption of this mechanism is additionally a most advantageous preadaptation to the development of anenteric larvae in life cycles in which embryogenesis is supported by nutrients from the parent which are either stored in the egg or shunted to the embryo via a placenta. There would be no need to retain even rudiments of the larval digestive system as an alternative primordium for the adult system is available.

At this time, it may be appropriate to mention that although we know of no other organisms in which the ontogenetic history has an exact parallel, other colonial forms have employed related options made possible by their asexual capacities. In such forms, zygote-produced individuals may never function as feeding zooids and may undergo immense morphological simplification; invariably one or more asexual buds are produced precociously and are nourished to a state of functional independence. These in turn are responsible for proliferation of further individuals. Such life histories in which the product of sexual reproduction has only a blastogenic function are found in a diverse group of invertebrates including ascidians, thaliaceans, entoprocts, and hydrozoans.

Considering the widespread occurrence in metazoans of an ontogenetic sequence in which the founder of the colony has only a limited asexual function, it may be somewhat surprising that bryozoans produce a fully differentiated ancestrula.

The immense asexual potency of the gymnolaemate cystid would seem to obviate the necessity to "replace" the ancestrular polypide: the cystid wall itself has full competency to produce the primordia of other individuals by budding. The fact that a polypide is normally differentiated in the ancestrula may be correlated with a need to procure additional nutrients for the complete differentiation of its buds. The differentiation of an ancestrular polypide in forms such as *Smittina papillifera* (Stach, 1938) would, however, seem unnecessary, for apparently there are abundant nutritive reserves. In *Stylopoma informata* (Maturo and Middleton, 1972) the preancestrula directly produces nine functional zooids, in some cases. There are some indications that stoloniferan ctenostomes have ancestrulae that fail to develop their own polypide and proliferate only generative stolons, but these would be rare exceptions if true (see Section II, B, 2, b).

2. The development of a fully constituted planktotrophic larva has been abandoned in most species in favor of a simplified anenteric larva.

With the exception of the few species producing true cyphonautes larvae, modern gymnolaemates have resorted to a reproductive strategy in which there is extensive brood protection of a very few embryos. The embryos either arise from large, nutrient-rich eggs or are nourished by the parent via a placenta. Characteristically this shift involves the partial to complete failure of the entoderm to differentiate as a tubular gut. Such simplification of the larva also occurs in other groups in correlation with the assumption of a "more direct" life cycle, but the causative factors may not be the same. As indicated above, we conjecture that the unique derivation of the ancestrular polypide in bryozoans has facilitated the evolutionary adoption of anenteric larvae in this group.

3. Certain morphological structures strictly associated with the sessile adult phase are preformed during embryogenesis.

In all gymnolaemates, the process of acceleration is evidenced by the development of the metasomal sac during early embryogenesis. This structure serves for the attachment of the larva at the onset of metamorphosis and later (with possible rare exception) forms part or all of the preancestrular and ancestrular body wall. Clearly the metasomal sac is associated with the sessile not the larval phase of the life cycle. Its precocious development does facilitate metamorphosis but the structure originally has no direct value to the larva. Secondarily, the sac has been utilized as the embryonic component of the placentalike system employed in *Bugula neritina* (Woollacott and Zimmer, 1975) and in certain other bryozoans for extraembryonic nutrition.

Another example of strictly adult structures found in the larva are the blastemal components. This complex would be totally enigmatic if one tried to interpret it in terms of larval biology, but is readily explained in terms of the differentiation of the ancestrula. The great variation in development of the epidermal and mesodermal blastemas may have some evolutionary significance for they are minute in true cyphonautes and reach their greatest development in cellularioids and some ascophorans.

4. There is no significant postmetamorphic growth of the individual produced from the larva.

The fact that the definitive ancestrula of gymnolaemates is close in size to the larva which produced it is highly unusual, for in most animal groups there is a very rapid postmetamorphic growth phase. Obviously this feature is related to the coloniality of bryozoans and more specifically to the minute size of their individual zooids.

5. There may be the precocious anticipation of colony formation involving production of composite, compound, and composite–compound "ancestrulae."

This feature is restricted to relatively few species and seems to be without general significance. When present it may confer some selective advantage by facilitating the symmetrical and perhaps more rapid growth of the colony.

III. CYCLOSTOMATA

A. Development and Metamorphosis

As is well known, the development of cyclostomes takes place in special reproductive zooecia, the gonozooids. This female individual normally produces but a single egg, but after its fertilization the zygote produces multiple embryos by fragmentation of the early cleavage stage (Borg, 1926; Robertson, 1903). This polyembryonic proliferation of large numbers of genetically identical young provides an unusual opportunity for studies on phenotypic variation in a metazoan, but has never been utilized to our knowledge.

An introduction to the literature as well as numerous figures and photographs of cyclostome larvae, their structure, and their metamorphosis have been provided by Nielsen (1970). The following summary of cyclostome development is based on Nielsen's 1970 study and personal communication with him.

The larvae of cyclostomes have been simplified to an extreme. Externally they appear as small (about 100 μm) spherules, completely ciliated externally. They are usually slightly longer than wide and are often flattened perceptibly on one side or part of one side. Inconspicuous pores at the poles persist from the invaginations of the aboral and oral fields. The invaginated aboral field is formed of a simple cuboidal to columnar epithelium that has secreted a cuticle; the invaginated oral field consists only of the metasomal sac (Fig. 31). The entire outer surface of the larva is provided by the heavily ciliated corona. This is composed of a large number of cells arranged in a cylindrical epithelial sheet rather than as a single ring of cells as in gymnolaemates. Internally a layer of flattened mesodermal cells is situated beneath the central part of the aboral epithelium. Between this mesodermal plaque and the invaginated metasomal sac are highly vacuolated cells of uncertain significance. There are no recognizable traces of a digestive system, pyriform complex, neural plate, neuromuscular cord, or other differentiated neural and muscular components in either the embryonic or larval stages.

The metamorphosis of cyclostomes usually begins within a few minutes of their release from the gonozooid. Lacking a pyriform complex or other distinguishable sense organs, the cyclostome larva would seem to have limited abilities in selecting an appropriate site for attachment. However, Nielsen (1970) did observe that larvae glide with the flattened face pressed against the substratum for brief periods prior to settlement, suggesting a testing phase as seen in gymnolaemates.

The initial step of metamorphosis, as in gymnolaemate larvae, is eversion of the metasomal sac. In cyclostomes this is protruded as a cone which rapidly conforms to the contour of the substratum and anchors the larva. Concurrently, the aboral field with its covering cuticle is also everted and soon the corona begins to roll into the interior. After involution of the locomotory corona is accomplished, the margins of the metasomal sac and the aboral epidermis are contiguous. They fuse with each other and lose their connections with the corona which is now cut off internally in a ring-shaped cavity. Thus, the metasomal sac and the aboral epithelium form the entire epidermis of this hemispherical stage. Internally, the corona and vacuolated cells degenerate. The columnar cells near the center of the aboral epithelium invaginate as a vesicle, pushing the underlying mesodermal layer deeper within the preancestrula or primary disc (Fig. 32).

The metamorphosis to this stage is, in principle, identical with that of gymnolaemates. It would be logical to assume that the invaginated vesicle and its associated mesoderm in cyclostomes represent the polypidal rudiment. Earlier observations (Barrois, 1886) in fact suggest this is true, but these studies were based exclusively on surface observations and optical sections. In his recent and thorough study, Nielsen (1970) described a different origin for the polypide, one involving only the layer of flattened mesodermal cells. These are reported to multiply, fold together into a sac-shaped vesicle, and become segregated into inner and outer epithelia. The inner layer is reported to form the epidermis of the lophophore and gut and the outer, the mesoderm of the polypide and the membranous sac. The invaginated epithelium from the aboral pole is reported to form only the tentacle sheath. His observations thus support the early report of Ostroumoff (1887), who also claimed that the "mesodermal" layer migrated away from the aboral epithelium, became bilaminar, and differentiated as the polypide.

As the polypide initiates its development, the primary disc becomes calcified and proliferates a tubular cylinder from its free surface (Fig. 33). The distal parts of the ancestrular polypide will extend into this broad cylinder and eventually the tentacles will be protruded and retracted through the open aperture of the

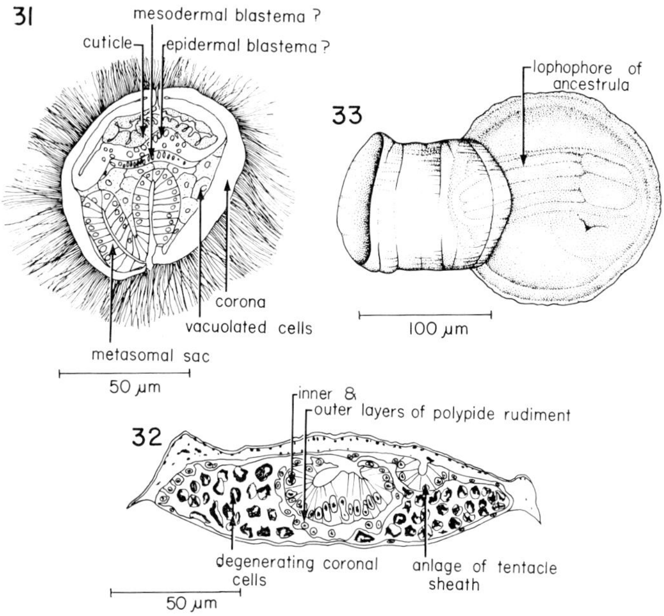

Figs. 31, 32, and 33. *Berenicea patina.* Various developmental stages. All from Nielsen (1970).

Fig 31. Near median longitudinal section of free-swimming larva (redrawn from photomicrograph).

Fig. 32. Median section through a young primary disc (or preancestrula) after completion of the sequence of morphogenetic movements and initiation of the histolysis of the involuted corona (redrawn from photomicrograph).

Fig. 33. Surface view of completed ancestrula with functional polypide (shown retracted).

tube when the ancestrula becomes functional. That region of the tube nearest the anal side of the polypide will be the site of proliferation of the two primary buds of the incipient colony. The primary zooids formed will produce secondary buds in turn and this process will continue, resulting in a fan-shaped colony with progressive generations of zooids arranged in arcs in encrusting forms.

B. Analysis of Cyclostome Development and a Comparison with Gymnolaemates

As previously indicated, the larvae of cyclostomes can be derived by extreme reduction from those of gymnolaemates. In essence, of the strictly larval tissues, only the locomotory corona remains. The oral and aboral fields are deeply invaginated as the metasomal sac and aboral epithelium, respectively. The latter is uniform in composition and has secreted a closely adherent cuticle. Internally there is the equivalent of a mesodermal blastema and some enigmatic vacuolated cells.

Somewhat surprisingly the metamorphosis of cyclostomes is remarkably similar to that in Gymnolaemata. In both, the transformation involves an early portion characterized by rapid morphogenetic movements and a slow second phase of histogenic differentiation. The times required for these two phases in cyclostomes are comparable to those required in gymnolaemates–about 15 minutes (for one species) and 5–7 days (for five species), respectively (Nielsen, 1970).

Dissecting the metamorphosis more finely, the parallel is continued. The rapid initial phase in cyclostomes consists of precisely the same three steps considered fundamental to gymnolaemates: (1) eversion and attachment of the metasomal sac; (2) involution of the larval transitory tissues (and an unfolding of the deeply invaginated aboral face); and (3) invagination of the central portion of the aboral face. At the completion of this phase the primary disc of cyclostomes and the preancestrula of typical gymnolaemates are both cystids with the wall comprised of about equal contributions from the metasomal sac and aboral epithelium.

Similarly the histogenic phases of gymnolaemates and cyclostomes have as their primary step the differentiation of the polypide, but there is also the lysis of the transitory larval tissues and the development of the cystid and zooecium of the functional ancestrula from the cystid of the preancestrula or primary disc.

The existence of an aborally secreted cuticle in the larvae of cyclostomes and certain gymnolaemates suggests that larvae of their common ancestor also possessed this feature even though its expression is different in the two modern groups. As the larval valves of true cyphonautes are shed shortly after metamorphosis and replaced by a permanent cuticle, their presence would seem to be a disadvantage to the larva. Therefore, Jägersten (1972) concluded that the valves must represent an "ancient adult character" without importance in modern forms. However, as the valves of *Pherusella brevituba* (Zimmer, personal observations) and the aboral cuticle of cyclostomes (Nielsen, 1970) persist into colonial life, it is probable that the presence of the larval cuticle is a precocious anticipation of the adult exoskeleton. Additionally, as noted in Section II,B,1, an aboral cuticle could be of importance at metamorphosis by facilitating the internalization rather than the dehiscence of the larval transitory tissues. These larval tissues are a considerable source of energy and their retention could more than offset the disadvantage conferred by the increased density of the larva possessing a cuticle. In most gymnolaemates, alternative mechanisms of internalizing the transitory larval tissues may have obviated the selective advantage of a larval exoskeleton.

The origin of the ancestrular polypide from larval components is the only area of apparent discord preventing a one-to-one matching of the stages of metamorphosis in cyclostomes and gymnolaemates. Cyclostomes are provided with what appear to be the equivalents of the epidermal and mesodermal blastemas that form this polypide in gymnolaemates, at least in gross morphology, position, and early movements at metamorphosis. However, Nielsen, on the basis of extensive comparative observations, concluded that "the whole polypide . . . appears to develop from the inner layer" (1970, p. 255). Admittedly, the epidermal component in cyclostomes is provided with a cuticle produced during embryogenesis which should make its identification reasonably straightforward. However, with the proliferation of cells, the epithelium could have separated away from the cuticle. Neither Neilsen (1970) nor Ostroumoff (1887) provided a clear explanation of how the single mesodermal layer became doubled. Certain of Nielsen's photographs, especially those for *Berenicea patina,* are such that it would be easy to interpret them in terms of gymnolaemate development. A reassessment of the origin of the ancestrular polypide seems appropriate considering these facts.

It is difficult to obtain irrefutable evidence for any suggested pattern of metamorphosis, considering the problems that result from extensive morphogenetic movements, compression, and cell proliferation. We believe, however, that knowledge of the detailed cytology of the central aboral epithelium and the underlying mesodermal layer could be most instructive, even if based only on the larva. Unfortunately, no ultrastructural studies are available.

It would also be of interest to know which of the three epidermal regions of the embryo—the incipient aboral epithelium, incipient corona, or incipient metasomal sac—is (are) involved in nutrient uptake during embryogenesis. Although extraembryonic nutrition in cyclostomes has never been documented to our knowledge, published figures (Borg, 1926) of sections through a functional gonozooid indicate there is considerable growth of the secondary or tertiary embryos after they are cut off from the primary or secondary embryos, respectively. The metasomal sac in cyclostomes invaginates precociously (e.g., Borg, 1926), suggesting that it is not the tissue involved in extraembryonic nutrition as it is in certain gymnolaemates (Woollacott and Zimmer, 1975).

One might also question whether only mesoderm could proliferate structures normally derived from all three so-called primary germ layers as would be required by Ostroumoff's (1887) and Nielsen's (1970) interpretation. This should not be disturbing, for differentiation of the ancestrula and normal bud formation if gymnolaemates clearly demonstrate that the sanctity of primary germ layers is not absolute. Further, the exceptional "inventiveness" of budding in tunicates dictates that one should be cautious in making deductions concerning budding (e.g., Brien, 1972).

IV. PHYLACTOLAEMATA

A. Development and Metamorphosis

The development of phylactolaemates is also highly modified, but the pattern is uniform within the group as it was in cyclostomes. One could conclude that their ontogeny is simply a direct development, but certain details of their early embryology and a comparison of their life history with that of other bryozoans suggest another alternative as indicated in the analysis and interpretation of this section.

Many features of phylactolaemate development (especially its early phases) remain obscure. For example, the precise mode by which the zygote becomes enclosed within the "embryo sac" of the parent and the relationship of this invagination of the body wall to other brood chambers found in the phylum have not been fully explained.

Early cleavage produces an irregular mass of blastomeres showing no evidence of the biradial arrangement that characterizes gymnolaemates.

Further division produces an elongated, one-layered sac presumably equivalent to a blastula. Proliferation of an internal layer results in a hollow, two-layered "cystid." The internal cells, at least functionally, are the equivalent of mesoderm, not entoderm. Their proliferation from the thickened distal end of the elongate "blastula" stage was reported for *Plumatella* by Braem (1897), but the same author observed the cells to be produced from the opposite pole in *Fredericella* (Braem, 1908).

Braem (1897), Davenport (1891), and Kraepelin (1892), but no subsequent workers, reported the internalization of short-lived endodermal cells at early cleavage. Braem (1908) and Brien (1953), both studying *Plumatella,* observed a few large transitory cells attached to the distal pole of the two-layered cystid (Fig. 34). Braem (1908) interpreted the cells as entodermal and the embryo as an exogastrula. Brien (1953) denied this, but later (1960) suggested the cells might have a nutritive function. Marcus (1934) also reported external entoderm cells for *Fredericella.* From these conflicting reports on entoderm, it seems safe to conclude that this "germ layer," if present at all, has no ultimate role in the development.

Nourished by the parent through a placental attachment (Fig. 34), the two-layered cystid develops a budding zone at one pole and one to four buds, depending on the species, differentiate into functional polypides while still within the parental embryo sac. There may be additional buds which remain as simple invaginations of the cystid wall. Before its release, the elongated "larva" develops a ringlike fold of the body wall near its equator and this grows upward over the budding end until only a small pore remains. This fold, the mantle, develops ciliation and the "larva" soon breaks free from the parent (Fig. 35). It will swim about briefly with the polypide and mantle-bearing end trailing and eventually settles on the "anterior" pole. The presence of long cilia and putative nervous tissue at the "anterior" end (Marcus, 1926) suggests that this region may be sensory and function in site selection as well as in attachment.

After settlement, the mantle is reflected down toward the attachment area, exposing the polypide-bearing region. Thus as soon as attachment is accomplished, the polypides can be extruded and the colony grows by asexual proliferation of new polypides. The mantle is variously reported either to be resorbed gradually at this external site (Braem, 1890) or to be invaginated into the interior before lysis (Fig. 36) (Brien, 1953).

B. Analysis of the Developmental History of Phylactolaemates

The curious "larva" and its metamorphosis in phylactolaemates could be interpreted as components in a pattern of direct development, but in fact they appear to be the result of extreme extension of some of the changes observed in gymnolaemates.

There is a remarkable compression and simplification of the initial phases of embryogenesis. Blastomeres of the early cleavages are in no precise pattern. There is at best only a hint at the process of gastrulation and the two-layered "embryo" that is produced is, in fact, an anenteric cystidlike larva. The only typical larval feature is the "mantle" which clearly corresponds with the corona of other bryozoans.

The stages of metamorphosis seen in gymnolaemates and in cyclostomes are greatly simplified and in large part accomplished precociously in the period before release from the parent.

The specific events of typical bryozoan metamorphosis as observed in both gymnolaemates and cyclostomes include early morphogenetic movements of (1) eversion and attachment of the metasomal sac, (2) involution of the transitory

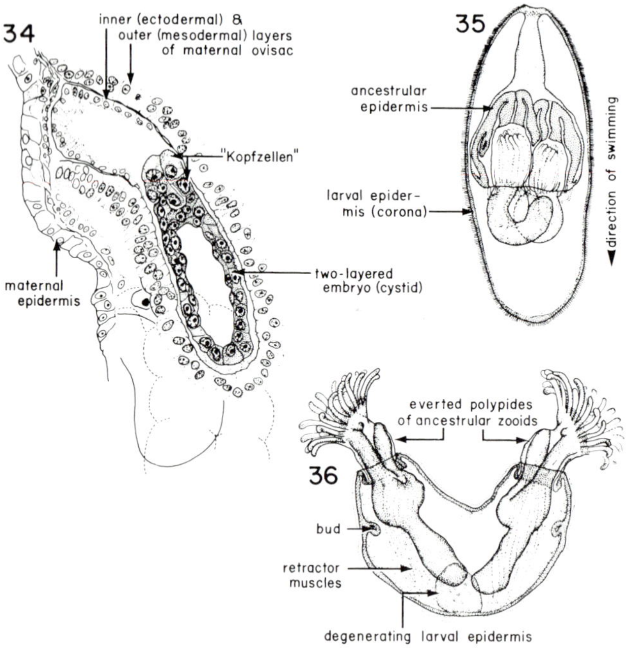

Figs. 34, 35, and 36. *Plumatella fungosa.* Various developmental stages. All from Brien (1953).

Fig. 34. Median longitudinal section of two-layered embryo within the maternal embryo sac. Note the "*Kopfzellen*" at distal end of the embryo.

Fig. 35. Surface view of a free swimming "larva." In effect this is a fully differentiated but unattached ancestrula.

Fig. 36. Young colony or ancestrula in surface view. At the attached end the ciliated epithelium (corona) of the "larva" has been invaginated and is undergoing histolysis.

larval tissues, and (3) invagination of the polypide rudiment. Histogenic differentiation involves (1) simultaneous elaboration of the ancestrular polypide, (2) differentiation of the cystid and zooecium, and (3) histolysis of the strictly larval structures.

Of these six events, not all are fully represented in phylactolaemates. The equivalent of the metasomal sac is remarkably simple and is never invaginated during embryogenesis, but does serve in attachment at the end of the short dispersion phase. Involution of the transitory larval tissues and their subsequent histolysis occur rather normally; however, these tissues are minimal and they do not contribute nutrients for the differentiation of the ancestrular polypides as in gymnolaemates. Additionally, the blastemal region at the aboral pole is not invaginated in the same sense. Buds are proliferated internally and the entire region is overgrown by the mantle fold prior to release. In further contrast, the differentiation of the blastemal tissue into functional polypides occurs before release from the parent in phylactolaemates rather than constituting a major part of the histogenic phase of metamorphosis as seen in gymnolaemates. Finally the "larval" cystid has essentially completed its differentiation as the body wall and exoskeleton of the ancestrula prior to the free-swimming phase.

V. SUMMARY

1. A surprisingly precise interpretation of metamorphosis was achieved by Calvet in 1900, but the accuracy of his observations has been seldom appreciated and was clouded by subsequent less critical work. It is hardly to our credit that one of the most elegant cytological and histological studies on a bryozoan larva and its metamorphosis was accomplished in 1905 by Kupelwieser or that so little new knowledge has been gained in the intervening seven decades.

2. There exist distinct morphological similarities between the larvae of gymnolaemates, cyclostomes, and phylactolaemates. The least modified larvae are the cyphonautes of gymnolaemates and the most altered are the "larvae" of phylactolaemates.

3. Metamorphosis of gymnolaemates is a twofold process involving first a rapid phase of complicated morphogenetic movements and then a longer period of histogenic differentiation. Events of this transformation occur in the following sequence:

   a. Morphogenetic phase
      (1) Eversion of the metasomal sac and anchoring of the larva
      (2) Rolling into the interior of the larval transitory tissues and union of the metasomal sac and pallial sinus epithelia to form the external surface of the transforming larva
      (3) Invagination of the apical disc region as the polypide rudiment

   b. Histogenic phase
      (1) Differentiation of the ancestrular polypide
      (2) Completion of the ancestrular cystid and zooecium
      (3) Histolysis of the larval transitory tissues, providing nutrients for the two above steps

4. The larvae of cyclostomes are remarkably simple and can be considered an epidermal sac that is differentiated as a metasomal sac, corona, and a mesodermally-lined apical disc which have precise homologies in gymnolaemate larvae. There is no suggestion of even a transient gut. The metamorphosis of cyclostomes is easily related to and consists of the same morphogenetic and histogenic events listed above for gymnolaemates. However, there are indications that the polypides of these two classes may have different origins.

5. The intimate similarities between the metamorphoses of gymnolaemates and cyclostomes either indicate that this pattern of transformation evolved in a common ancestor or was reached independently by convergent evolution. Considering the number and specificity of the similarities the latter possibility seems remote and as the gymnolaemates and cyclostomes diverged at least by the Ordovician, it seems probable that the basic features of their metamorphosis were established more than 500 million years ago and have remained fundamentally unchanged.

6. The life cycle of phylactolaemates previously has been considered as strikingly different from that of gymnolaemates, but we suggest that considerable similarities exist in their ontogenies. Phylactolaemate development is considered here as indirect because:

a. During embryogenesis there is the possible formation of a transitory entoderm
b. During the short-lived dispersal phase the "larva" possesses transitory structures for locomotion and attachment
c. At transformation a series of morphogenetic movements takes place which is similar to that observed in the metamorphosis of other bryozoans

However, in phylactolaemates there has been a drastic suppression of larval structures and a pronounced acceleration in the appearance of adult features. Not only the formation of rudiments of the first zooids, but also their precocious differentiation into functional zooids and the proliferation of secondary buds occur before the "larva" is released from the parent.

7. If our analysis is correct, the ontogenies of all three classes of the Bryozoa are variations on an essentially comparable theme. The dominant elements of this pattern are the progressive reduction of the "individual" normally produced by the zygote to a cystidal sac and the concurrent acceleration of the adult process of polypide replacement into the time of metamorphosis. Additionally, in some cases, there is a precocious anticipation of the colonial mode of life at metamorphosis in that a colony of several individuals rather than a simple ancestrula is produced from the larva.

ACKNOWLEDGMENTS

Access to the literature reviewed in this chapter was made available by the libraries of the Allan Hancock Foundation and the Museum of Comparative Zoology. Dr. John A. Long (U.C.S.F. Medical School) kindly donated the scanning electron micrographs and Mrs. Elaine Jahn provided expert assistance with the preparation of this manuscript. We are especially grateful to Dr. Donald P. Abbott (Hopkins Marine Station) for several invaluable discussions. Drs. Kristian Fauchald (University of Southern California), Robert Fernald (Friday Harbor Laboratories), Claus Nielsen (University of Copenhagen), and David Pawson

(U.S. National Museum of Natural History) read early drafts of this chapter and offered valuable suggestions for its improvement. Our work on this chapter was facilitated by grants from the National Science Foundation and by support from the Harvard Graduate Society and the William F. Milton Fund (H.U.).

REFERENCES

Atkins, D. (1955). The cyphonautes larvae of the Plymouth area and the metamorphosis of *Membranipora membranacea* (L.). *J. Mar. Biol. Assoc. U.K.* **34**, 441–449.

Barrios, J. (1877). Recherches sur l'embryologie des Bryozoaires (=Mémoire sur l'embryologie des Bryozoaires). *Trav. Stn. Zool. Wimereux* **1**, 1–305.

Barrois, J. (1879). Mémoire sur la métamorphose des bryozoaires. *Ann. Sci. Nat., Zool.* [6] **9**, 1–67.

Barrois, J. (1886). Mémoire sur la métamorphose de quelques Bryozoaires. *Bibl. Ecole Haut. Etud., Sect. Sci. Nat.* **32**(5), 1–94.

Borg, F. (1926). Studies on recent cyclostomatous Bryozoa. *Zool. Bidr. Uppsala* **10**, 181–507.

Braem, F. (1890). Untersuchungen über die Bryozoen des süssen Wassers. *Zoologica (Stuttgart)* **2**(6), 1–134.

Braem, F. (1897). Die geschlechtliche Entwicklung von *Plumatella fungosa. Zoologica (Stuttgart)* **10**(23), 1–96.

Braem, F. (1908). Die geschlechtliche Entwicklung von *Fredericella sultana* nebst Beobachtungen über die weitere Lebensgeschichte der Kolonien. *Zoologica (Stuttgart)* **20**(5), 1–38.

Braem, F. (1951). Uber *Victorella* und einige ihrer nächsten Verwandten sowie die Bryozoenfauna des Ryck bei Greifswald. *Zoologica (Stuttgart)* **102**, 1–59.

Brien, P. (1953). Etude sur les Phylactolaemates. *Ann. Soc. R. Zool. Belg.* **84**, 301–444.

Brien, P. (1960). Classe des Bryozoaires. *In* "Trâité de Zoologie," Vol. 5, fascicule 2, pp. 1053–1355. Masson, Paris.

Brien, P. (1972). La reproduction asexuée. Un aspect des études faites à la station biologique de Roscoff, de 1924 à 1969. *Cah. Biol. Mar.* **13**, 659–679.

Bullivant, J. S. (1968). Attachment and growth of the stoloniferous ctenostome bryozoan, *Zoobotryon verticillatum. Bull. South. Calif. Acad. Sci.* **67**, 199–202.

Calvet, L. (1900). Contribution à l'histoire naturelle des Bryozoaires Ectoproctes marins. *Trav. Inst. Zool. Univ. Montpellier [N.S.]* **8**, 1–488.

Cloney, R. A. (1966). Cytoplasmic filaments and cell movements: Epidermal cells during ascidian metamorphosis. *J. Ultrastruct. Res.* **14**, 300–328.

Cook, P. L. (1973). Settlement and early colony development in some Cheilostomata. *In* "Living and Fossil Bryozoa" (G. Larwood, ed.), pp. 65–71. Academic Press, New York.

Davenport, C. B. (1891). Observations on budding in *Paludicella* and some other Bryozoa. *Bull. Mus. Comp. Zool.* **20**, 101–152.

Ehlers, E. (1876). *Hypophorella expansa.* Ein Beitrag zur Kenntniss der minierenden Bryozoen. *Abh. K. Ges. Wiss. Göttingen Phys., Kl.* **21**, 1–156.

Eiben, A. (1976). Einfluss von Benetzungsspannung und Ionen auf die Substratbesiedlung und das Einsetzen der Metamorphose bei Bryozoenlarven *(Bowerbankia gracilis). Mar. Biol.* **37**, 249–254.

Eitan, G. (1972). Types of metamorphosis and early astogeny in *Hippopodina feegeensis* (Busk) (Bryozoa: –Ascophora). *J. Exp. Mar. Biol. Ecol.* **8**, 27–30.

Fell, H. B. (1967). Echinoderm ontogeny. *In* "Treatise on Invertebrate Paleontology" (R.C. Moore, ed.), Vol. **S** (Echinodermata), pp. 60–85. Univ. of Kansas Press, Lawrence, Kansas.

Grave, B. H. (1930). The natural history of *Bugula flabellata* at Woods Hole, Massachusetts, including the behavior and attachment of the larva. *J. Morphol.* **49**, 355–379.

Håkansson, E. (1973). Mode of growth of the Cupuladriidae (Bryozoa, Cheilostomata). *In* "Living and Fossil Bryozoa" (G. Larwood, ed.), pp. 287–298. Academic Press, New York.

Harmer, S. F. (1887). Sur l'embryogénie des bryozoaires ectoproctes. *Arch Zool. exp. gén.* [2] Ser. **5**, 443–458.

Harmer, S. F. (1915). The Polyzoa of the Siboga expedition. Part I. Entoprocta, Ctenostomata and Cyclostomata. *Siboga Exped.* **28a**, 1–180.

Hondt, J. L. d' (1972). Métamorphose de la larve d'*Alcyonidium polyoum* (Hassall), Bryozoaire Ctenostome. *C.R. Hebd. Acad. Sci. Paris* **275**, 767–770.

Hondt, J. L. d' (1973). Etude anatomique, histologique et cytologique de la larve d'*Alcyonidium polyoum* (Hassall, 1841), Bryozoaire Ctenostome. *Arch. Zool. Exp. Gén.* **114**, 537–602.

Hondt, J. L. d' (1974). La métamorphose larvaire et la formation du "cystide" chez *Alcyonidium polyoum* (Hassall, 1841), Bryozoaire Ctenostome. *Arch. Zool. Exp. Gén.* **115**, 577–605.

Hondt, J. L. d' (1975). Etude anatomique et cytologique comparée de quelques larves de Bryozaires Ctenostomes. *In* "Bryozoa 1974" (S. Pouyet, ed.). Docum. Lab. Geol. Fac. Sci. Lyon, H.S. **3**(1), pp. 125–134. Univ. Claude Bernard, Lyon.

Humphries, E. M. (1977). Larval behavior and post-larval development in *Parasmittina nitida* morphotype B (Bryozoa, Cheilostomata). *Am. Zool.* **17**, 5–20.

Hyman, L. H. (1959). "The Invertebrates," Vol. 5. McGraw-Hill, New York.

Jägersten, G. (1972). "Evolution of the Metazoan Life Cycle." Academic Press, New York.

Jebram, D. (1973). The importance of different growth directions in the Phylactolaemata and Gymnolaemata for reconstructing bryozoan phylogenies. *In* "Living and Fossil Bryozoa" (G. Larwood, ed.), pp. 565–576. Academic Press, New York.

Kraepelin, K. (1892). Die deutschen Süsswasser-Bryozoen. Eine Monographie. II. Entwickelungsgeschichtlicher Teil. *Abh. Naturw. Hamburg* **12**, 1–67.

Kupelwieser, H. (1905). Untersuchungen über den feineren Bau und die Metamorphose des Cyphonautes. *Zoologica (Stuttgart)* **47**, 1–50.

Lutaud, G. (1953). Progression de la calcification au cours de la metamorphose de la larve chez *Escharoides coccinea* Abildgaard, Bryozoaire Chilostome. *Arch. Zool. Exp. Gén.* **91**(1), 36–50.

Lynch, F. L. (1947). The behavior and metamorphosis of the larva of *Bugula neritina* (Linnaeus): Experimental modification of the length of the free-swimming period and the responses of the larvae to light and gravity. *Biol. Bull. (Woods Hole, Mass.)* **92**, 115–150.

Marcus, E. (1926). Beobachtungen und Versuche an lebenden Süsswasserbryozoen. *Zool. Jahrb., Abt. Syst. Oekol. Geogr. Tiere* **52**, 279–350.

Maturo, F. S. (1971). Larval behavior and metamorphosis in the Cupuladriidae (Bryozoa, Anasca). Presented at the 2nd International Conference of the International Bryozoological Association (unpublished).

Maturo, F. S., and Middleton, M. A. (1972). Alternative patterns of ancestrula development in the bryozoan *Stylopoma informata* (lonsdale) (Gymnolaemata, Cheilostomata, Schizoporellidae). *Assoc. Southeast. Biol. Bull.* **19**, 84.

Mawatari, S. (1946a). On the hatch-out, swimming and attachment of the larvae of *Bugula neritina* Linnaeus. *Shigen Kagaku Kenkyūjo Tanhē*, **9**, 1–15.

Mawatari, S. (1946b). On the larva of the hemifusa bryozoan and its metamorphosis. *Misc. Rep. Res. Inst. Nat. Resour. (Tokyo)* **9**, 21–28.

Mawatari, S. (1946c). Metamorphosis and formation of the young colony in *Smittina collifera* Robertson. *Misc. Rep. Res. Inst. Nat. Resour. (Tokyo)* **10**, 31–35.

Mawatari, S. (1951). The natural history of a common fouling bryozoan, *Bugula neritina* (Linnaeus). *Misc. Rep. Res. Inst. Nat. Resour. (Tokyo)* **19–21**, 47–54.

Nielsen, C. (1970). On metamorphosis and ancestrula formation in cyclostomatous bryozoans. *Ophelia* **7**, 217–256.

Nielsen, C. (1971). Entoproct life-cycles and the entoproct/ectoproct relationship. *Ophelia* **9**, 209–341.

O'Donoghue, C. H. (1926). Observations on the early development of *Membranipora villosa* Hincks. *Contrib. Canad. Biol. Fish. [N.S.]* **3**(8), 1–17.

Ostroumoff, A. A. (1886). Contribution a l'étude zoologique et morphologique des Bryozoaires du Golfe de Sébastopol. III. Données sur l'histoire du développement. *Arch. Slaves Biol.* **2**, 184–190 and 329–355.

Ostroumoff, A. A. (1887). Zur Entwicklungsgeschichte der cyclostomen Seebryozoen. *Mitt. Zool. Stn. Neapel* **7**, 177–189.

Ott, J. S. (1972). *Monobryozoon bulbosum* n. sp., a new solitary interstitial bryozoan from the West Atlantic coast. *Cah. Biol. Mar.* **13**, 421–428.

Prouho, H. (1890). Recherches sur la larve de *Flustrella hispida* (Gray) structure et métamorphose. *Arch. Zool. Exp. Gén.* [2] **8**, 409–459.

Ries, E., and Schölzel, G. (1934). Beiträge zu Problemen der vitalen Färbung. I. Cytologische und cytophysiologische Ergebnisse von Vitalfärbungen an *Zoobotryon* (Bryozoa) und *Gastroblasta* (Hydromedusae). *Zellforsch. Mikrosk. Anat.* **20**, 523–564.

Robertson, A. (1903). Embryology and embryonic fission in the genus *Crisia. Univ. Calif., Berkeley, Publ. Zool.* **1**, 115–156.

Robertson, A. (1908). The incrusting chilostomatous Bryozoa of the west coast of North America. *Univ. Calif., Berkeley, Publ. Zool.* **4**, 253–344.

Ryland, J. S. (1970). "Bryozoans." Hutchinson University Library, London.

Ryland, J. S. (1974). Behaviour, settlement and metamorphosis of bryozoan larvae: a review. *Thalassia Jugoslavia* **10**, 239–262.

Ryland, J. S., and Stebbings, A. R. D. (1971). Settlement and oriented growth in epiphytic and epizoic bryozoans. *In:* "Fourth Europ. Mar. Biol. Symp." (D. J. Crisp, ed.), pp. 105–123. Cambridge Univ. Press, Cambridge.

Seeliger, O. (1906). Über die Larven und Vervandtschaftsbeziehungen der Bryozoen. *Z. Wiss. Zool.* **84**, 1–78.

Silén, L. (1942a). Origin and development of the cheilo-ctenostomatous stem of Bryozoa. *Zool. Bidr. Uppsala* **22**, 1–59.

Silén, L. (1942b). On spiral growth of the zoaria of certain Bryozoa. *Ark. Zool.* **34A**, 1–22.

Silén, L. (1944). The anatomy of *Labiostomella gisleni* Silén (Bryozoa, Protocheilostomata). *K. Sven. Vetenskapskad. Handl.* [3] **21**(5), 1–111.

Silén, L. (1950). On the mobility of entire zoids in bryozoans. *Acta Zool. (Stockholm)* **31**, 349–386.

Soule, D. F., and Soule, J. D. (1972). Ancestrulae and body wall morphogenesis of some

Hawaiian and Eastern Pacific Smittinidae (Bryozoa). *Trans. Amer. Microsc. Soc.* **91,** 251–260.

Soule, J. D. (1954). Post-larval development in relation to the classification of the Bryozoa Ctenostomata. *Bull. South. Calif. Acad. Sci.* **53,** 13–34.

Stach, L. W. (1938). Observations on *Carbasea indivisa* Busk (Bryozoa). *Proc. Zool. Soc. London, Ser. B.* **108,** 389–399.

Strathmann, R. (1975). Larval feeding in echinoderms. *Am. Zool.* **15,** 717–730.

Tavener-Smith, R. (1973). Some aspects of skeletal organization in Bryozoa. *In* "Living and Fossil Bryozoa" (G. Larwood, ed.), pp. 349–359. Academic Press, New York.

Vigelius, W. J. (1886). Zur Ontogenie der marinen Bryozoen. *Mitt. Zool. Stn. Neapel* **6,** 499–541.

Waters, A. W. (1924). The ancestrula of *Membranipora pilosa* L., and of other cheilostomatous Bryozoa. *Ann. Mag. Nat. Hist. Ser.* [9] **14,** 594–612.

Waters, A. W. 1925a). Ancestrulae of cheilostomatous Bryozoa. Part II. *Ann. Mag. Nat. Hist. Ser.* [9] **15,** 341–352.

Waters, A. W. (1925b). Ancestrulae of cheilostomatous Bryozoa. Part III. *Ann. Mag. Nat. Hist. Ser.* [9] **16,** 529–545.

Waters, A. W. (1926a). Ancestrulae and frontal of cheilostomatous Bryozoa. Part IV. *Ann. Mag. Nat. Hist. Ser.* [9] **17,** 425–439.

Waters, A. W. (1926b). Ancestrulae of cheilostomatous Bryozoa. Part V. *Ann. Mag. Nat. Hist. Ser.* [9] **18,** 424–433.

Woollacott, R. M. and Zimmer, R. L. (1971). Attachment and metamorphosis of the cheilo-ctenostome bryozoan *Bugula neritina* (Linné). *J. Morphol.* **134,** 351–382.

Woollacott, R. M., and Zimmer, R. L. (1975). A simplified placenta-like system for the transport of extraembryonic nutrients during embryogenesis of *Bugula neritina* (Bryozoa). *J. Morphol.* **147,** 355–378.

Zirpolo, G. (1933). *Zoobothryon verticillatum* (Dell Chiaje). *Mem. Accad. Nuovi Lincei* [2] **17,** 109–442.

Zschiesche, A. (1909). Untersuchungen über die Metamorphose von *Alcyonidium mytili. Zool. Jahr., Abt. Anat. Ontog. Tiere* **28,** 1–72.

5

Ultrastructure, Mineralogy, and Development of Bryozoan Skeletons

PHILIP A. SANDBERG

I. INTRODUCTION

Calcareous skeletons are produced by modern bryozoans belonging to the order Cheilostomata (class Gymnolaemata) and the order Cyclostomata (class Stenolaemata). Cheilostomes, which may have very weakly calcified (less than 25% dry weight $CaCO_3$) to very heavily calcified skeletons (Schopf and Manheim, 1967), occur in rocks as old as Jurassic (Pohowsky, 1973). Cyclostome skeletons are well calcified and occur in rocks as old as Paleozoic, where they occur with skeletons of other, extinct stenolaemates (orders Trepostomata, Cryptostomata, and Cystoporata).

Fossil members of the order Ctenostomata (class Gymnolaemata) occur as natural molds (body fossils) bored into various shell substrates (Pohowsky, 1974, 1976). Traces of $CaCO_3$ have been reported from modern ctenostome bryozoans (Schopf and Manheim, 1967; Ryland, 1970), but not as organized skeletons. Ctenostomes and weakly calcified cheilostomes may rarely be preserved as molds formed by overgrowth by encrusting organisms (Voigt, 1966). A few examples of inferred soft parts have been reported from fossil stenolaemates (McKinney, 1969; Boardman, 1971, 1973; Utgaard, 1973). Obviously these molds and possible soft-part relics are rare, and routine study of fossil bryozoans (other than ctenostomes and phylactolaemates) is dependent on skeletal material. The morphological features of bryozoan skeletons which form the basis of classifications differ between the tubular stenolaemate bryozoans and the box-like gymnolaemate cheilostomes. In most stenolaemate bryozoans the exterior frontal surface is virtually filled by the orifice and, of external features, only the colonial morphology and the type and distribution of heterozooecia are available to differentiate taxa. These latter characters are insufficient and, because of a fair degree of external homeomorphy among even quite distantly related stenolaemates, taxonomic separation in that group (particularly Paleozoic stenolaemates) has relied heavily on characters visible in variously oriented petrographic thin sections. The necessity for using thin sections is amplified by the common occurrence of ancient stenolaemates in crystalline limestones or indurated shales.

In the study of skeletal structure, light microscopic examination of thin sections provides information on the optical orientation of skeletal crystals, general texture (laminated, granular, etc.) and distribution of major skeletal units, prominent skeletal accretionary increments and discontinuities (ultrastructural unit boundaries, cuticles, etc.). In order to determine the morphology of individual crystals, the details of arrangement of crystals, and the nature of crystal-organic matrix relationships it is necessary to use scanning electron microscopy (SEM). It has been found that structure seen in the light microscope may often be a second-order expression of a finer structure visible only by means of SEM. An example of this is the quite different number of laminations in the lamellar skeletal units of *Metrarabdotos tenue* determined by means of light microscopy (Cheetham *et al.*, 1969) and by SEM (Sandberg, 1971).

There is somewhat of a paradox with respect to the study of ultrastructure in bryozoan skeletons. Fine structural detail, although available in the commonly used thin sections, has been only a minor part of most studies of stenolaemates. Although petrographic thin sections (in contrast to histological microtome sections of decalcified material) were only very rarely used in most studies on cheilostomes, skeletal structure has been the prime subject of a number of studies on that group (Schneider, 1957, 1963; Bronstein and Prenant, 1936; Bobin and Prenant, 1968; Sandberg, 1971, 1973, 1975; Tavener-Smith and Williams, 1970, 1972; the latter also includes stenolaemate ultrastructure).

Frontal surfaces of cheilostome zooecia are occupied by large membranes or by calcified walls with varying amounts of morphological detail including avicularia, ovicells, several types of frontal pores (Banta, 1973), and secondary calcification. As a result of these abundant, readily available frontal wall features, cheilostome classification has not made significant use of either internal detail (from sections) or fine structure of the calcified walls. Frontal wall layer types (olocyst, tremocyst, pleurocyst) proposed by Canu and Bassler (1920) were differentiated only on the basis of pore distribution and hyaline or opaque-chalky appearance, not on fine structure. Those skeletal wall categories have been questioned (Banta, 1970) and shown to be inhomogeneous with respect to ultrastructural and even mineralogical makeup (Sandberg, 1971).

For the reasons discussed above, ultrastructural details in cheilostome bryozoans have been little studied until very recently. Mineralogical and ultrastructural data are being accumulated at an increasing rate, but as yet are insufficient to relate meaningfully to the total morphologically based classification scheme for cheilostomes. Virtually all of my own scanning electron microscope (SEM) study of skeletal ultrastructure has been on cheilostomes; the data included herein on stenolaemates come mostly from the literature or from personal communications by colleagues working with stenolaemates. The orientation of this chapter is accordingly biased toward cheilostomes. This bias in part reflects available data, because little SEM study of stenolaemate ultrastructure has been published.

Earlier studies of fine skeletal structure have been undertaken in diverse invertebrate groups for various purposes including: (1) to determine both the applicability of skeletal structural features as taxonomic characters (such as Raup's studies of echinoid crystal orientation, 1960, 1962) and the level of taxonomic significance of ultrastructurally based groupings; (2) to investigate the ontogenetic development of skeletons and the nature of biomineralization processes (Wilbur and Watabe, 1967; Wilbur and Simkiss, 1967; Goreau, 1959), at least as reflected by their calcified end products; or, (3) to develop probable phylogenetic lineages in skeletal development (Williams, 1968).

All of these clearly cannot be covered in this chapter on bryozoan skeletons. Given the limitations of scope and available data, this chapter attempts briefly to characterize known skeletal ultrastructures, their mineralogic and taxonomic distribution, as well as the relationships of skeletal ultrastructures, organic matrix, accretionary laminations, and cuticles to inferred modes of wall development.

Note that the term "ultrastructure," as used in this chapter, refers to the skeleton and not to cytological features.

II. INTERPRETATION AND APPLICATION OF SKELETAL ULTRASTRUCTURE

The understanding of structure of organisms at any given time in history has been closely tied to the limits of the observational techniques then available. Taxonomic arrangements, which are and have been primarily based on such structural data, reflect the limitations of the current mode of observation. For any very small organism this simply means that it will not be encountered or clearly understood until resolution of available observational methods reaches its level. Coupled with this observational limit, and largely dependent on it, is a limit of understanding of the biology of the organism, a problem particularly true for small organisms, or those studied first primarily as fossil or dead modern shells. This was certainly the case for the foraminifers. In the fifth century B.C., Herodotus wrote about, but misinterpreted, the large nummulites which comprise much of the limestones of the Pyramids. Early workers with smaller foraminifers in the eighteenth and nineteenth centuries regarded them as small cephalopods or worm tubes (Loeblich and Tappan, 1964). With development of modern microscopes, the fine structures of foraminifera have become differentiable, and more satisfactory taxonomic systems have developed.

For bryozoans the problems of understanding are not so clear-cut. Bryozoans have the added complication of aggregate colonial skeletons which are often large enough to distinguish and even characterize with the naked eye or with a simple hand lens. Because the colonial morphology was large enough to "recognize clearly" and zooecial detail was small enough to be commonly unnoticed, there developed taxonomic systems based (with false security) on gross morphological characters. Earlier genera often grouped, on the basis of colonial form, species which are now placed in numerous genera in different families.

Bryozoan colonial morphologies are dependent on mode of wall formation, budding pattern, zooecial shape, arrangement of calcified and uncalcified wall portions (as in radicle-jointed cellariiform colonies). Some of those features may be quite similar among cheilostomes which are otherwise distinct. Therefore, considerable homeomorphy exists in colonial morphology among what we now recognize (with higher resolution) as quite diverse cheilostome taxa. Light microscope and, more recently, electron microscope data on zooecial characters have shown how generally unsatisfactory a taxonomy based on colonial form really is.

Light microscope study of zooecial detail has resulted in a greatly refined, more natural classification as compared to that based on naked-eye study of colonial morphology. An uncritical extension of that improvement might project that a similar major taxonomic refinement would result from study of skeletal ultrastructure. Indeed, much earlier work on ultrastructure may have been approached with a reductionist philosophy that if we could just see small enough we could decipher all taxonomic relationships. Such a philosophy is predicated on the notion that, just as organisms each possess a distinctive morphology which separates them from others of similar rank in the hierarchy, so also ultrastructural details of skeletal parts are taxonomically correlated, but at a finer level of distinction because of the higher resolution.

In fact, the latter does not seem to be generally the case. Ultrastructural types are closely controlled by organic matrix types and distribution, as well as by mineralogy of the skeletal carbonate. When a similar biochemical environment of calcification occurs in members of different taxa, similar ultrastructural types may be produced in the skeletons. Clearly those biochemical details of skeletal genesis are rather poorly understood, so we can say little about equivalency of distantly related organisms with respect to precise environments of biomineralization. Nevertheless, we can say that quite similar ultrastructures occur in skeletons of taxa separable even at the phylum level. At the other end of the scale, distinctly different ultrastructural types have been found to characterize species of the same genus. Hansen (1972) found different ultrastructures in tests of two species of the foraminiferal genus *Turrilina.* Sandberg (in preparation) has found that species of *Schizoporella* may have quite different frontal wall structure. *Schizoporella floridana* has a thin lamellar calcite primary frontal, overlain by a spherulitic (transverse fibrous) aragonite superficial layer (Sandberg, 1971, Plate 4, Figs. 2 and 3); *S. unicornis* has a calcite primary frontal composed of a crudely lamellar layer and a thick spherulitic calcite layer. The overlying spherulitic aragonite layer of *S. unicornis* appears considerably thinner than that of *S. floridana.* Preliminary comparison of my SEM data on *Umbonula ovicellata* with SEM illustrations of *U. littoralis* presented by Tavener-Smith and Williams (1970) suggest a marked difference in skeletal ultrastructure between those two species.

These examples of interphylum similarities and interspecific differences do not mean that skeletal ultrastructure is taxonomically useless. Certain ultrastructures, such as the monocrystalline calcite plates of echinoderms and the crossed-lamellar ultrastructure of molluscan shells, are unique to those taxa. Those structures are quite unlike what might be expected in a coprecipitated inorganic carbonate. The particular organisms are exerting a high degree of biological interference (Sandberg, 1975) with calcification, and thus produce unique, distinctly biogenic structures. However, in some other groups, biological interference is less, and the resulting skeletal ultrastructure more closely approximates inorganic carbonates (ooids, cements), and also becomes more similar among diverse taxa. A good example of this is the similarity of fibrous aragonite in cheilostome bryozoans, scleractinian corals, some codiacean algae, sclerosponges, submarine cements, and some ooids (Sandberg, 1975).

Even when a high degree of biological interference exists, individual ultrastructural types may show great similarity between different taxa. An example of this is the equivalence of the screw-dislocation lamellae in skeletons of some cheilostome bryozoans (Plate I, Figs. 2 and 4); stenolaemates (Tavener-Smith and Williams, 1972, Plate 20, Fig. 105); and inarticulate brachipods (Williams, 1971, Plate 5, Fig. 3). These similarities in ultrastructure can complicate the determination of source organisms for small fragments of carbonate shells in petrographic studies (such as the work of Hay *et al.,* 1970), but even there the dimensional properties of the fragments should eliminate much of the uncertainty.

Within any single taxon (phylum, class, or lower) ultrastructure can be useful for taxonomic characterization, primarily in the form of the ontogenetic successions of ultrastructural types through the shell thickness (Williams, 1968; Taylor *et al.,* 1969; Kobayashi, 1971). That is, although similar or even identical individual skeletal ultrastructures may occur in other closely or distantly related taxa, the succession of ultrastructures developed by members of any given taxon is often distinctive and may even be unique.

III. STENOLAEMATE BRYOZOA

A. Skeletal Mineralogy

Modern stenolaemate bryozoans (cyclostomes) all possess calcite skeletons. Magnesium content of the calcite, although less than in cheilostomes, appears to be quite variable (Schopf and Manheim, 1967; Poluzzi and Sartori, 1973, 1976) ranging from about 2 to 10 mole% $MgCO_3$. Similar magnesium content variations in skeletons of ancient stenolaemates and the resulting variations in diagenetic alterations (Folk, 1973, 1974; Sandberg, 1975) could account for much of the observed textural variation in fossil stenolaemate bryozoans. Although some strongly altered examples are known (e.g., *Nicholsonella* spp.), most fossil stenolaemates studied at the microscopic level have been found to preserve well-ordered skeletal structure. The high degree of preservation suggests original calcite skeletons with a generally low $MgCO_3$ content.

B. Ultrastructural Types

Despite the common use of thin sections for identification of fossil stenolaemates, ultrastructures in stenolaemate bryozoan skeletons have only very recently become the subject of detailed study. Early descriptions of fine structure referred to lamellar or granular skeletal units visible at the light microscope level. More recent studies, often utilizing the SEM, have reported variations or new types such as fibrous, granular-prismatic, nonlaminated, "yellow calcite," or hyaline, and have provided illustrations of ultrastructure (e.g., Blake, 1973; Blake and Towe, 1971; Boardman and Cheetham, 1969; Boardman and

Towe, 1966; Brood, 1972; Cutler, 1973; Gautier, 1973; Nye, 1968; Söderqvist, 1968; Tavener-Smith, 1968, 1969; Tavener-Smith and Williams, 1972; Utgaard, 1973).

As has so often been found in modern ultrastructural studies, skeletal similarities exist among species of major taxonomic groups, even among those of different phyla. Thus, the fibrous layer of stenolaemate bryozoans, especially cyclostomes (Tavener-Smith and Williams, 1972, Figs. 113, 116, and 117; Brood, 1973, Plate 2, Fig. 3), shows strong resemblance to the foliated layer of pelecypods, especially oysters (Taylor *et al.*, 1969, Plate 6). Similar structures also occur among cheilostome bryozoans (Plate I, Fig. 3).

Information on growth surfaces of various ultrastructural types is most readily available from modern cyclostomes or the many Cenozoic or even Mesozoic cyclostomes which are easily cleaned of loose sedimentary matrix. In contrast, zooecia of most Paleozoic stenolaemates contain calcite cement, and the zoaria are frequently embedded in crystalline limestones or dense shales. Because the younger cyclostomes commonly occur in loose, removable sediments, and because those bryozoans possess a greater diversity of frontal surface characters than other stenolaemates, their classification has made less use of internal detail.

C. Skeletal Growth Modes

For this and subsequent discussion in this chapter, it is necessary to define "exterior" and "interior" walls. Exterior walls (Silén, 1944, p. 436) have been characterized as walls "developed by intussusception from the walls of the primary zooid" or walls "that separate the colony from its environment" (Boardman and Cheetham, 1973, pp. 163, 138). Calcified exterior walls, which are "derived by reinforcement of the ectocyst" (Banta, 1970, p. 39) are thus intimately associated with cuticle on their outer surfaces and thicken away from that cuticle. Interior walls (Silén, 1944, p. 436) have been characterized as walls "developed by apposition from the interior face of an exterior wall or from another interior wall" (Banta, 1970, p. 39). Boardman and Cheetham (1973) have suggested that exterior walls are those which expand the total colonial coelomic volume as they grow, whereas interior walls partition preexisting coelomic space.

Interior skeletal walls will commonly thicken on both sides forming the walls referred to (particularly in stenolaemates) as "compound" (e.g., Boardman, 1971). Interior walls in stenolaemate skeletons may sometimes thicken on one side only (as in the case of diaphragms), in which case they are called "simple," as are exterior walls, which also calcify on one side only. Neither type of interior wall is associated with cuticle. The instances where cuticle is incorporated into the upper edges of the transverse interior walls in cheilostome skeletons are

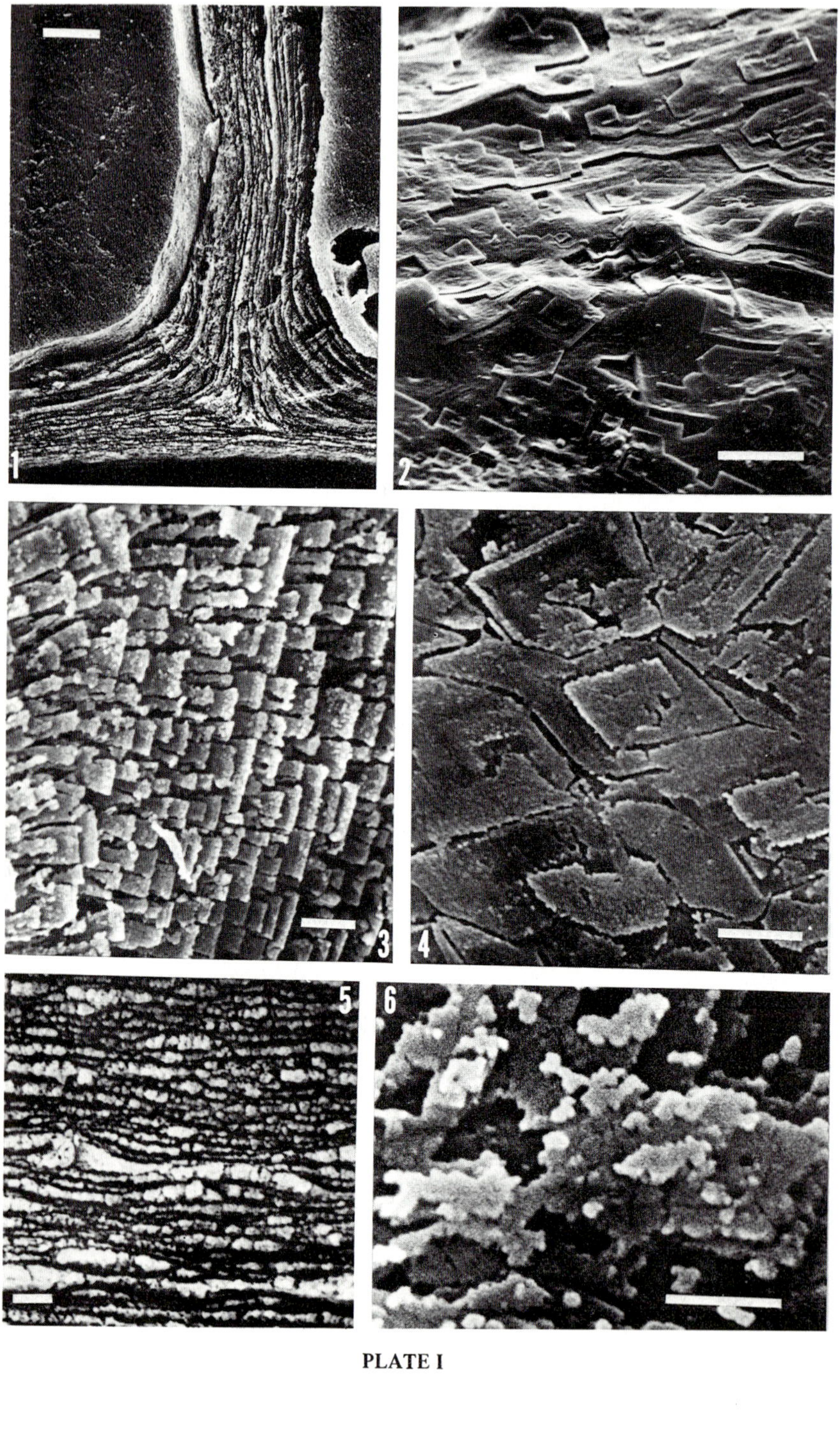

PLATE I

probably related to a coelomic expansion by a frontally directed growth vector. In the more extreme case of frontal budding (Banta, 1972) this frontal wall expansion by intussusception is clear.

Interpretations of skeletal development in extinct stenolaemates are made on the basis of distribution and orientation of skeletal units, accretionary surfaces, inferred interior and exterior walls, and inferred cuticle locations. The basis for the models utilized (Boardman, 1971; Boardman and Cheetham, 1969, 1973) is the work of Borg (1926) on modern cyclostomes. Borg (1926, p. 196) described a wall structure which he calls a "double wall." In double-walled forms the wall (from the outside in) was composed of cuticle, epithelium, mesodermal layer, coelomic cavity (hypostegal), mesodermal layer, epithelium, calcified wall, epithelium, mesodermal layer, body cavity (coelom). In what have come to be known as single-walled forms, the wall (from the outside in) is composed of cuticle, calcified wall, epithelium, mesodermal layer, body cavity (coelom) (Borg, 1926, p. 191). In double-walled forms, in contrast to single-walled forms, the cuticle is a cover around the colony as a whole but is not in direct contact with the calcified walls except on the single-walled basal portion of the ancestrular region and its extension (see Boardman, 1971, Fig. 2; Utgaard, 1973, Figs. 1–6; Boardman and Cheetham, 1973, Figs. 8 and 9). An added complication,

---

PLATE I

Fig. 1. *Schizobrachiella sanguinea* (Norman) (BMNH 1911.10.1.1367) Etched longitudinal section of a basal–distal wall junction (calcite). Note continuation of lamellae out of basal walls into both sides of distal wall, as well as presence of underlying basal platform (Sandberg, 1973). Bar scale represents 10 μm.

Fig. 2. *Labioporella calypsonis* Cook (BMNH 1964.7.2.31) Unetched frontal surface of cryptocyst (calcite). Note numerous screw-dislocations. Bar scale represents 5 μm.

Fig. 3. *Tessaradoma gracilis* (Sars) (BMNH 1911.10.1.841) Etched exterior frontal surface. Note elongated, superimposed laths of the foliated structure (calcite). Bar scale represents 1 μm.

Fig. 4. *Metrarabdotos tenue* (Busk) (USNM 209434). Interior basal surface of gonoecium. Note marginally accreting rhombic crystals and few screw-dislocations. Bar scale represents 2 μm.

Fig. 5. *Petralia undata* MacGillivray (BMNH 1883.10.15.27-31) Etched longitudinal section of basal wall (calcite). Note the subcrystal units of the flattened crystals of the irregular lamellae. Bar scale represents 1 μm.

Fig. 6. *Metrarabdotos tenue* (Busk) (USNM 209434) Exterior surface of frontal wall superficial (calcite) layer. Note numerous levels of simultaneous accretion of calcite. Bar scale represents 1μm.

which can occur in either mode of growth, is the interior simple walls (such as diaphragms) which, although they lack cuticle, are still calcified from one side only. As discussed in Section IV and elsewhere (Sandberg, 1976), planar spherulitic ultrastructure is usable to differentiate interior and exterior walls (with allowances for one-sided, simple interior walls). These ultrastructural data can aid in statements of inferred growth modes of extinct stenolaemates. In stenolaemates, as in cheilostomes, the location and direction of accretion of secondary thickening layers show the sites of wall growth. If thickening occurs on both sides of a wall (with no intervening discontinuity where a cuticle might have been) then that wall is a compound interior wall. Because the planar spherulitic ultrastructure layer deposited directly against cuticle (Sandberg, 1971, 1976) is very thin and not really distinguishable in section, the wall-thickening vectors just discussed can be useful in determining interior or exterior wall structure. Also useful in this respect is the distribution of skeletal units, such as the common granular "primary" and lamellar "secondary" units. However, in several respects the designations "primary" and "secondary" are questionable. The "primary" layer, which can occur by itself at the growing edge (Hinds, 1975, p. 883, Fig. 6), may continue to be deposited as rodlike extensions throughout "secondary" deposition time. This ordinal naming of skeletal units is even less desirable with cheilostome skeletons, in which four or more ultrastructural types may occur and in which repetition of successions may develop (Sandberg, in preparation).

IV. CHEILOSTOME BRYOZOA

A. Skeletal Mineralogy

Cheilostome bryozoan skeletons are constructed of calcite, aragonite, or both (Lowenstam, 1954; Schopf and Manheim, 1967; Rucker, 1967, 1968; Rucker and Carver, 1969; Sandberg, 1971, 1976; Poluzzi and Sartori, 1973, 1976). Magnesium concentration in calcites is more in skeletons of living cheilostomes (5–12 mole% $MgCO_3$, mean about 8 mole% $MgCO_3$, Poluzzi and Sartori, 1976) than in those of living cyclostomes. Aragonitic cheilostome skeletons or skeletal parts have low Mg content, but Sr/Ca levels at or near that of seawater (near 1/100, Dodd, 1967; Schopf and Manheim, 1967).

In bimineralic skeletons the two $CaCO_3$ polymorphs are segregated into discrete regions. The most common arrangement is a calcite initial skeleton with superficial aragonite layers on the frontal surface only (Cheetham *et al.*, 1969; Sandberg *et al.*, 1969). In forms in which the basal walls beyond the ancestrula are interior walls (without basal cuticle), aragonite may also occur or even only occur as superficial exterior basal thickenings (Greeley, 1969; E. Håkansson, unpublished).

Aragonite may less commonly occur on basal interior surfaces, usually as a thick wedge near the distal end of the basal wall, at the junction with the transverse wall (Plate V, Fig. 1). In the fossil record, occurrence of aragonitic basal exterior thickenings is commonly reflected, in diagenetically altered fossil specimens, by basally open calcite partial zooecia produced by solution loss of the more soluble aragonite (Greeley, 1969, Plates 33 and 34). Calcite replacement of aragonite, as well as recrystallization of calcite accompanying loss of magnesium, is discussed elsewhere (Sandberg, 1975).

B. Ultrastructural Types

Lamellar and spherulitic ultrastructures predominate in cheilostome bryozoan skeletons, and ultrastructural types are somewhat mineralogically controlled. Kobayashi (1969, 1971) found ultrastructures in pelecypods to be mineralogically dependent, with much greater variety for aragonite (only two of the eleven ultrastructures observed were totally calcite). In cheilostome bryozoans the situation is reversed. Sandberg (1971) found aragonite to occur in cheilostome skeletons only as elongate fibers or needles arranged perpendicular to the walls or in interfering spherulitic arrays (Plate IV, Figs. 4 and 6; Plate V, Fig. 1; see also Sandberg, 1971, Plate 3, Figs. 1–8 and Plate 4, Figs. 1–3). Recent work has shown that aragonite between closely spaced organic sheets may develop a more blocky form (Plate II, Fig. 4; Sandberg, 1976, Plate 2, Figs. 3 and 6).

Calcite in cheilostome skeletons occurs in a variety of crystal morphologies and arrangements, primarily as variants of lamellar, spherulitic, or dendritic growth. Sandberg (1971) differentiated parallel and transverse ultrastructures. Parallel ultrastructures included: parallel fibrous (referred to in this chapter as "planar spherulitic"), lamellar, massive, and a crudely lamellar type called crystal stacks. Transverse calcitic structures included: flattened fans of calcite laths and a structure called "cell-mosaic." The latter name was proposed by Levinsen (1909, p. 4) in allusion to the similarity between the surface appearance of that layer and the pattern of cells in cheilostome epithelial layers. The "cell mosaic" wall structure illustrated by Levinsen (1909, Plate 19, Fig. 10c) and Sandberg (1971, Plate 1, Figs. 5 and 6) is now recognized as a system of interfering spherulitic arrays of calcite crystals, separated by complex sutured boundaries (Plate V, Fig. 2). The possible one-to-one relationship between individual cells of the secretory epithelium and individual suture-bounded units of the "cell-mosaic" structure, suggested by Sandberg (1971, p. 140), is unlikely. Comparable interlocking, crenulate boundaries of the spherulitic crystal arrays occur in ooids from Great Salt Lake, Utah, and are a function of the compete-

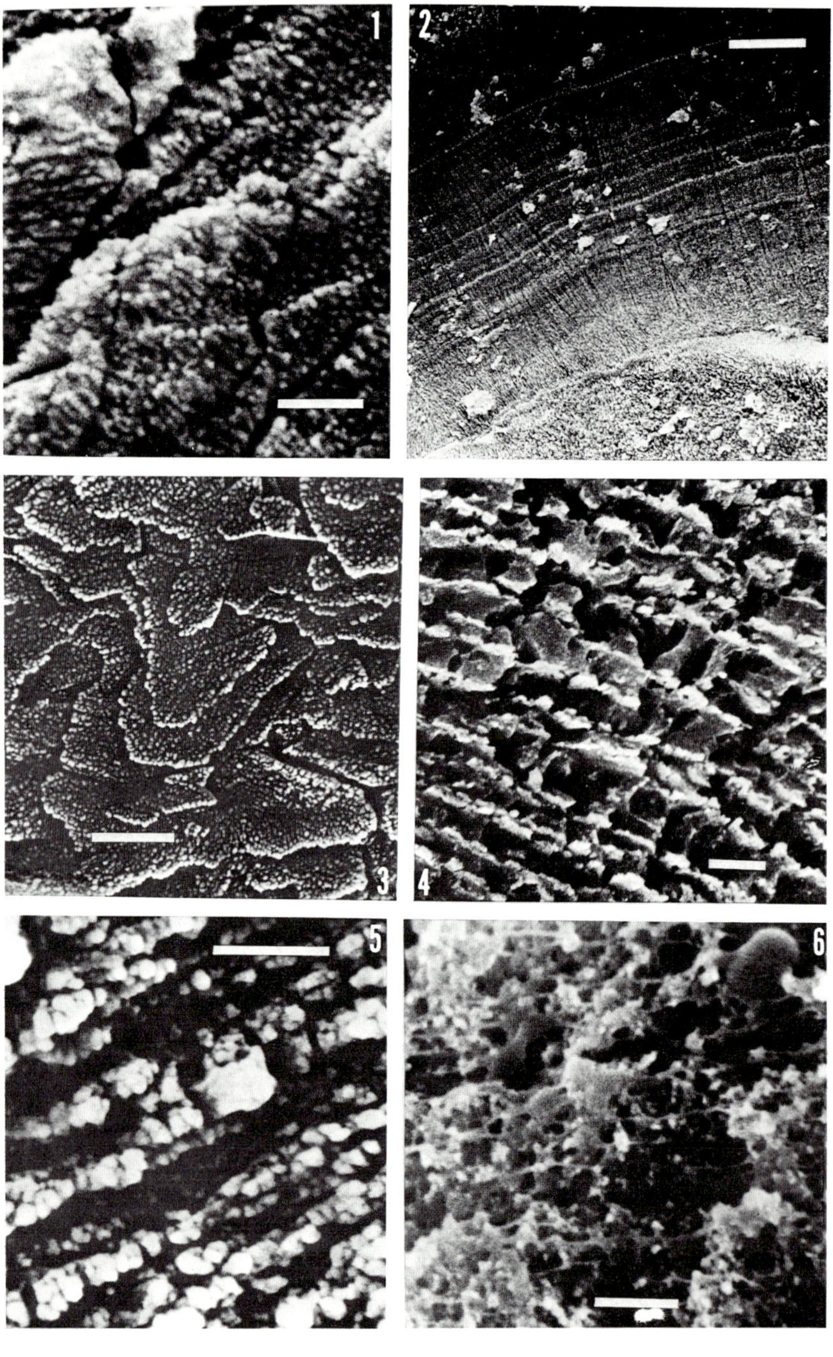

PLATE II

tive interference of the adjacent conical arrays. The ends of those spherulitic clusters are highly variable in size, and the resemblance of the resulting irregular pattern to the epithelial cell-mosaic of Levinsen (1909) must be fortuitous.

In cheilostomes, lamellar skeletal units (Plate IV, Fig. 5; Plate VI, Fig. 5) have been found to be separated by proteinaceous sheets visible in transmission electron micrographs of decalcified specimens (Tavener-Smith and Williams, 1972). The organic interlamellar sheets are evidenced in etched SEM specimens as raised, thin ridges (Plate II, Figs. 4 and 6; Plate VI, Fig. 6). Histological details of skeletal genesis in cheilostomes are poorly known, and it is not known whether a compartmentalization effect by the organic interlamellar sheets results in the tabular crystallites, as suggested for pelecypods (Bevelander and Nakahara, 1969; Wada, 1972). Massive zooecial lining layers and transverse ultrastructures constructed of spherulitic calcite ("cell mosaic" of Sandberg, 1971) and spherulitic aragonite ("transverse fibrous" of Sandberg, 1971) may show very numerous, very closely spaced accretionary increments (Plate V, Fig. 2). Except in very heavily etched specimens one does not usually see any organic sheets marking former locations of the growth surface (Plate II, Fig. 6). In some spherulitic aragonite toward the outer surface, however, organic sheets occur (Plate V, Fig. 3; Sandberg, 1976, Plate 2, Figs. 3 and 6). The structure is

---

PLATE II

Fig. 1. *Pentapora foliacea* (BMNH 1911.10.1.1561) Etched tangential fracture section through lateral wall (calcite). Bar scale represents 0.5 μm.

Fig. 2. *Posterula sarsi* (Smitt) (BMNH 1911.10.1.1360A-B) Interior surface of ovicell endoecium (calcite). Note planar spherulitic ultrastructure with accretionary banding. Bar scale represents 10 μm.

Fig. 3. *Labioporella calypsonis* Cook (BMHN 1964.7.2.31) Etched tangential fracture section of interior of frontal wall (calcite). Note lineations of subcrystal elements. Bar scale represents 2 μm.

Fig. 4. *Adeonella gibbera* (USNM 209435) Etched longitudinal section of zooecial lining near basal–distal corner. Note the blocky aragonite crystals and the numerous organic sheets. Bar scale represents 1 μm.

Fig. 5. *Hippopetraliella magna* (BMNH 1896.2.28.1) Etched longitudinal section of basal wall (calcite). Note subcrystal elements of the lamellar crystals. Bar scale represents 10 μm.

Fig. 6. *Melicerita obliqua* (BMNH 1967.2.8.119) Heavily etched transverse section of lateral wall (calcite) initial skeleton (not zooecial lining). Note numerous organic sheets parallel with wall surfaces (aligned with top and bottom of figure) and abundant intervening organic matrix. Bar scale represents 1 μm.

somewhat similar to that found in the cephalopod *Nautilus* by Mutvei (1972, Plates 1 and 2, Figs. 1–3), in which initially organic-poor prismatic aragonite (like transverse fibrous) is, toward the inner shell surface, cut by numerous organic sheets into the "inner nacreous layer." The aragonite crystal orientations of the nacreous and prismatic layers are identical and only the presence of organic sheets serves to differentiate the nacreous layer. In cheilostome skeletons, when the organic sheets acquire the close spacing characteristic of *Nautilus,* the crystal morphology may alter to the broader, blocky shapes discussed earlier (see also Sandberg, 1975, Plate 4, Fig. 4) As yet little work has been done on the optic properties of cheilostome bryozoan skeleton crystals. C-axis orientation is often perpendicular to the tabular crystals of lamellar layers and parallel to the long axis of elongate crystals of spherulitic aragonite. Towe and Cifelli (1967, pp. 744–745) in their study of foraminiferal wall structure discussed the problem of defining what constitutes a crystal (Plate 6, Fig. 5). As they suggested, what one perceives as a single crystal depends greatly on "the nature and resolving power of the instruments at his disposal." Treatment of the specimen prior to observation will also greatly affect interpretation of what is a crystal. In several cheilostome skeletal ultrastructures, what appear in unetched specimens to be single crystals (Plate I, Fig. 2) are seen after etching to be composed of parallel arrays of numerous smaller crystal units of presumably uniform optic orientation (Plate I, Fig. 5; Plate II, Figs. 1, 3, and 5).

In an earlier publication (Sandberg, 1973) the term "double-walled," based on Borg's (1926) cyclostome model, was used for describing the growth mode in skeletons of some cheilostome bryozoans. In fact, many of these "double-walled" cheilostomes, unlike the cyclostome skeletons on which Borg's concept was defined, do not have all interior walls above the protecial wall of the ancestrula (Boardman and Cheetham, 1973, p. 138). Because of the complexity and great diversity of modes of skeletal development among cheilostomes the use of Borg's term has been discontinued. Instead, it is recognized that, as discussed below, cheilostome skeletons may be constructed mainly of exterior walls, mainly of interior walls, or of variable intermediate combinations. The descriptive term "interior-walled" has been used to characterize cheilostomes in which the bulk of the zoarial skeleton (and perhaps all but the ancestrular basal wall) is composed of interior walls.

The "standard" pattern of cheilostome skeletal development may be characterized as predominantly exterior-walled growth with cuticulate exterior walls laterally, basally, and generally frontally; noncuticulate, interior walls occur distally (the transverse wall) and in the cryptocysts of cryptocystideans (Silén, 1944; Banta, 1969; Ryland, 1970). It has been recognized (Boardman and Cheetham, 1973, p. 163) that interior skeletal walls could be limited to just the pore plates separating zooecia in linear series (as in *Pyripora, Wilbertopora,* and other flask-shaped, uniserial cheilostomes; Cheetham, 1954; Thomas and Larwood, 1956, 1960).

Recent histological (Banta, 1968, 1969, 1970, 1972; Håkansson, 1973) and skeletal ultrastructure (Sandberg, 1973) studies have shown that many cheilostomes depart from this growth mode, and a considerable number have skeletons built almost totally of interior walls (Text-Fig. 1F; Plate III, Figs. 2 and 6). This latter group includes a variety of discoidal or linguiform types (cupuladriids, mamilloporids, conescharellinids: Håkansson, 1973; Sandberg, 1973) as well as some cellariids (Sandberg, 1973) and *Euthyrisella* (see Harmer, 1902).

In between these extremes are a great variety of modes of skeletal genesis, separable primarily on positional relationships of cuticle, hydrostatic mechanism (membranous frontal or ascus), coelomic spaces, secretory epithelia, initial calcified wall, and secondary wall thickenings in the frontal (Text-Fig. 1, p. 160). The simplest growth mode (Membranidea: like Text-Fig. 1A but without cryptocyst) has been omitted. The relationship of cuticle and calcified wall, direction of wall thickening, and the presence or absence of hypostegal coelomic spaces are shown in Text-Fig. 1. Where cuticle is shown against skeleton (in simple exterior walls) planar spherulitic ultrastructure is to be expected.

These growth modes are not intended to represent all possibilities for cheilostome skeletal growth. The interior-walled ascophoran, only relatively recently recognized, is paralleled by an unillustrated interior-walled anascan (e.g., *Cellaria*). As detailed study of cheilostome skeletal structure progresses, additional growth modes may be discovered. The number of reasonable combinations is, however, limited by topological relationships and the necessity for eliminating or at least minimizing interference of skeletal parts with the functioning of the hydrostatic mechanism. Because these modes of skeletal genesis strongly influence distribution of ultrastructural units and are thus largely decipherable on the basis of skeletal data it is useful to discuss them here in some detail.

C. Modes of Wall Development in Cheilostome Bryozoa

Observations and models of wall development in cheilostomes have been primarily concerned with the formation of the frontal wall. This emphasis on frontal walls is partly due to a widely accepted view of consistent occurrence of exterior walls on the lateral and basal surfaces of zooecia and interior walls distally, between successive zooecia. Thus, walls of obvious structural diversity (frontals) have been more studied that those other walls of assumed homogeneity. Nevertheless, because calcified frontal walls are complex and closely correlated with the structure and functioning of the hydrostatic mechanism, it is not unwarranted to give them greater attention.

In this discussion of growth modes and classifications based on frontal wall

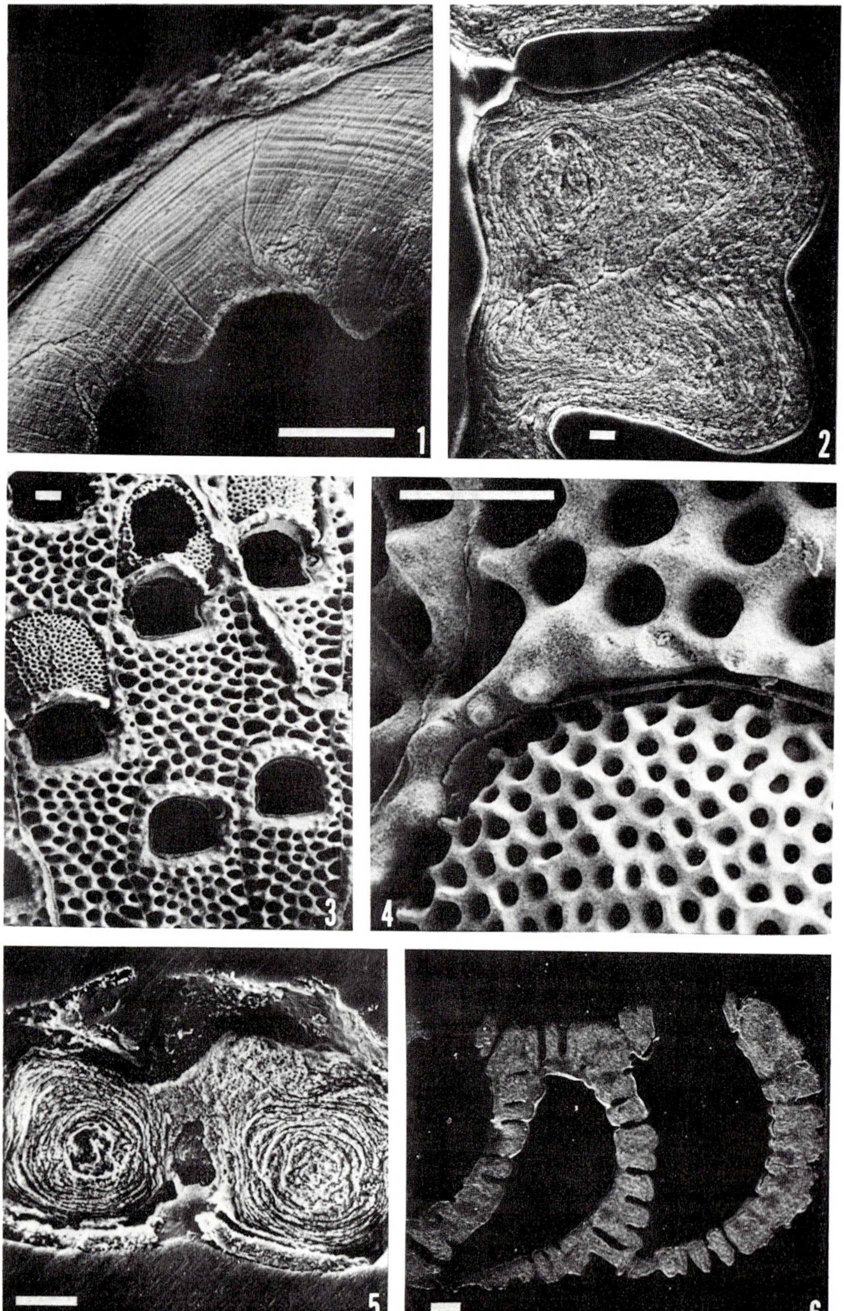

PLATE III

structure there are a number of terms which should be defined:

Cryptocyst—A calcified interior frontal wall between the polypide and frontal membrane in anascans, and between the ascus and the hypostegal coelom in ascophorans (modified from Banta, 1970).

Gymnocyst—A calcified exterior frontal wall which is a continuation of the calcified lateral exterior walls calcifying against the original frontal membrane and not involving epifrontal extensions by intussusception. The gymnocyst is often separated from the lateral walls by a flexure, but the boundary may be indistinct.

Pericyst—"A calcified frontal wall formed by fusion of spines" (from Banta, 1970). This condition is observable in cribrimorphs, but is an inference for umbonuloid ascophorans.

Ascus—"An ectocyst-lined cavity beneath the frontal wall of some cheilostomes; communicates with the outside through either an ascopore or a cameral pore (the opening into the ascus, covered by the poster of the operculum)" (combines the definitions of ascus and cameral pore of Banta, 1970).

Epistegal Space—"The space above the frontal membrane in forms with umbonuloid development. It is analogous in function with an ascus, but develop-

---

PLATE III

Fig. 1. *Posterula sarsi* (Smitt) (BMNH 1911.10.1.1360A-B) Exterior surface of ectoecium of ovicell (calcite) with marginal overgrowth by superficial frontal thickening of distal zooecium. Note planar spherulitic ultrastructure and accretionary bands. Bar scale represents 10 μm.

Fig. 2. *Veleroa* ? sp. (USNM 209436) Etched longitudinal section of several zooecia (calcite). Bar scale represents 100 μm.

Fig. 3. *Hippopetraliella magna* (BMNH 1896.2.28.1) Frontal view of several zooecia, including three with ovicells. The frontal walls are calcite overlain by superficial aragonite. Bar scale represents 100 μm.

Fig. 4. *Hippopetraliella magna* (BMNH 1896.2.28.1) Detail of distal edge of an ovicell from the same colony. Note that the ectoecium is calcified only up to the general frontal surface, and that the endoecium shows superficial thickening on upper (inner) surface. Bar scale represents 100 μm.

Fig. 5. *Watersipora subovoidea* (BMNH 1970.6.1.32) Etched transverse section of a frontal wall (calcite). Note the accretion on both upper and lower surfaces of frontal in this cryptocystidean ascophoran. Bar scale represents 10 μm.

Fig. 6. *Veleroa* ? sp. (USNM 209436) Detail of a distal wall in Fig. 2. Note accretionary laminations which show wall thickening primarily on distal side of wall. Bar scale represents 10 μm.

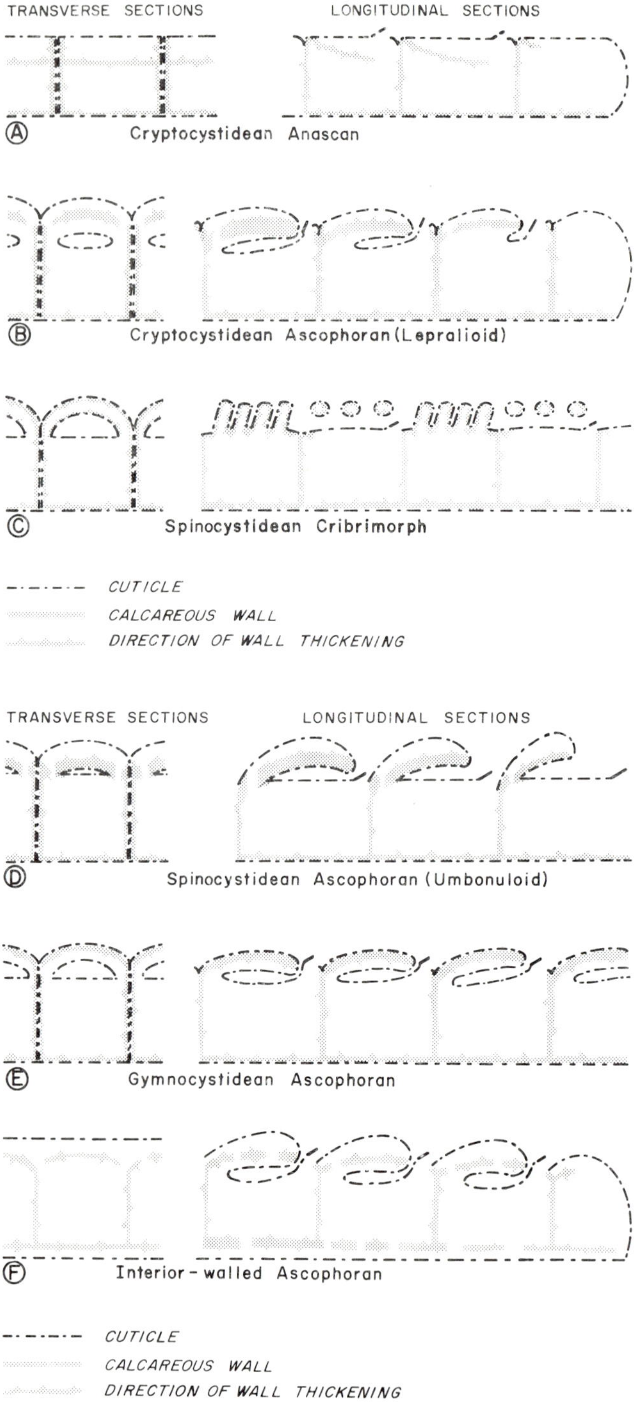
TRANSVERSE SECTIONS
LONGITUDINAL SECTIONS
Ⓐ Cryptocystidean Anascan
Ⓑ Cryptocystidean Ascophoran (Lepralioid)
Ⓒ Spinocystidean Cribrimorph
CUTICLE
CALCAREOUS WALL
DIRECTION OF WALL THICKENING
TRANSVERSE SECTIONS
LONGITUDINAL SECTIONS
Ⓓ Spinocystidean Ascophoran (Umbonuloid)
Ⓔ Gymnocystidean Ascophoran
Ⓕ Interior-walled Ascophoran
CUTICLE
CALCAREOUS WALL
DIRECTION OF WALL THICKENING

mentally similar in its relationships to a cribrimorph epistege" (from Cook, 1973).

Epistege—"A cavity between the frontal membrane and the pericyst of cribrimorphs; communicates with the exterior through intercostal lacunae" (from Banta, 1970).

Hypostegal coelom—A coelomic space separated from the main coelom by a calcified wall. A hypostegal coelom may occur basally (as in *Cupuladria*), but is more commonly developed frontally. Between the hypostegal coelom and the main coelom in anascans there is the cryptocyst; in ascophorans there is a calcified wall (cryptocyst or pericyst) and a compensation device (ascus or epistegal space). In the former group, communication between the two coelomic spaces is usually around the distal end of the cryptocyst. In the latter group communication is through marginal areolar pores.

A number of schemes of classification have been developed on the basis of frontal wall features and on the absence or presence and mode of development

---

Text-Fig. 1. Generalized diagrammatic compilations of the major growth modes in cheilostome bryozoans. Membranidea omitted. See discussion, p. 157.

(A) **Cryptocystidean Anascan**—The cryptocyst may thicken on both top and bottom surfaces, but commonly develops strong asymmetry with later thicker and perhaps ultrastructurally distinctive calcification on the upper surface (Sandberg, 1971, Plate 2, Fig. 12). A narrow proximal gymnocyst is shown in the longitudinal section, and planar spherulitic ultrastructure would be expected there. In this and all other modes, interior walls (here distal walls only) thicken on both sides unless or until interfered with (see B).

(B) **Cryptocystidean Ascophoran** (*Lepralioid*)—The frontal wall commonly shows evidence of thickening on both upper and lower surfaces. Secondarily, the calcification may be halted on the lower side as the developing ascus reaches that surface. Superficial thickening on the upper frontal surface may be considerable and is often quite distinctive in ultrastructure and even mineralogy.

(C) **Spinocystidean Cribrimorph**—The preexisting functional frontal membrane is overarched by calcified spines, each of which thickens inward toward its lumen.

(D) **Spinocystidean Ascophoran** (*Umbonuloid*)—The calcified frontal wall (with planar spherulitic ultrastructure on its lower surface) overarches a preexisting, functional frontal membrane and thickens on its upper surface, often with distinctive ultrastructure or mineralogy.

(E) **Gymnocystidean Ascophoran**—The structure and development of the ascus in this group is not clearly understood but appears to be by invagination if forms with true gymnocystal frontals are considered. Evidence in skeletal structure indicates wall thickening inward and downward, but this is apparently mostly completed during early ontogeny (before ascus formation).

(F) **Interior-walled Ascophoran**—Of all the growth modes this is the least clearly understood. It appears to characterize a rather great variety of cheilostomes, including comparable anascans (Sandberg, 1973), in which wall thickening and absence of cuticles indicate presence of hypostegal coelomic spaces surrounding the skeleton not only frontally, but also basally (in colonies which are not bilamellar or cylindrically branching). This is obviously a preliminary simplification, because forms with both cuticle incorporated into the skeleton and basal skeletal thickening have been encountered.

of a compensation device. The cheilostomes have commonly been subdivided into the suborders Anasca, Cribrimorpha, and Ascophora, with various anascan super-families or families based on the nature of gymnocyst and cryptocyst development (Bassler, 1953; Ryland, 1970).

Harmer (1902) recognized, among the Ascophora, calcified frontal walls which developed by two modes: umbonuloid frontals formed by overarching of a membranous frontal which had an operculum and parietal muscles; lepralioid frontals appeared "at first sight to result from the direct calcification of a *Membranipora*-like membranous wall" (Harmer, 1902, p. 14) with later development of the ascus below the wall. Subsequent histological and fine structural study of lepralioid taxa mentioned by Harmer (1902) and other structurally similar forms has shown the frontal wall to be a cryptocyst, rather than a gymnocyst as Harmer's original comment would imply (Banta, 1970, 1972; Cook, 1973; Boardman and Cheetham, 1969).

Harmer (1957) separated the Ascophora into the Ascophora Vera and the Ascophora Imperfecta. In the latter group the frontal membrane formed first and persisted as the "floor of the ascus" when overarched by the developing calcareous frontal wall. This is equivalent to the development of umbonuloid frontal walls of Harmer (1902), Cook (1973), and others. Ultrastructural characteristics of the frontal walls of members of some of the families (e.g., Umbonulidae, Exechonellidae, Arachnopusiidae, Metrarabdotosidae) included in the Ascophora Imperfecta by Harmer (1957) and Banta (1970, p. 52) show that they have umbonuloid frontal wall development. The implication is that umbonuloids and "imperfect ascophorans" may be synonymous, but other families placed in the Ascophora Imperfecta must be studied.

Silén (1942) subdivided the Cheilostomata into six "sections" based largely on the nature of the frontal wall. Most taxa fall into one of the four major "sections": (1) Membranidea (membranous frontal); (2) Cryptocystidea (calcified wall below frontal membrane, ascophorans *excluded* by Silén, 1942, p. 53); (3) Gymnocystidea (calcified wall developed above a membranous frontal as an evagination, calcified on its upper surface; the lower surface of the evagination is uncalcified and forms the roof of the ascus), and (4) Spinocystidea (calcified wall developed by fusion of spines overarching a membranous frontal). In the latter group, Silén's orientation of the calcified and uncalcified layers of the fused-spine fold was contrary to that found in acknowledged spinocystideans (see Text-Fig. 1D, and discussion below).

Confusion as to what constitutes a gymnocyst has caused problems in ascophoran classification. As illustrated by Silén (1942) the calcified frontal walls of neither the Spinocystidea nor the Gymnocystidea would be gymnocysts (see definitions above). The exterior wall surface on the basal side of an umbonuloid frontal wall has been called a gymnocyst (Tavener-Smith and

Williams, 1970) or an extension of a gymnocyst (Cook, 1973). However, all that a gymnocyst and an umbonuloid wall have in common is that they are both calcified exterior walls in the frontal region of the zooecium. If the definition of a gymnocyst is not limited (as defined above) to calcification on the inner side of the original frontal membrane, then any exterior wall calcification in the frontal region of the zooecium could be called a gymnocyst. A true gymnocystidean frontal is simply a calcification inside the original membranous frontal, as a continuation of the calcified lateral exterior walls (Text-Fig. 1E). The umbonuloid polyspinocystidean frontal (Text-Fig. 1D) originates first marginally as a cryptocyst (Sandberg, in preparation), then emerges (by intussusception) through the frontal membrane as an epifrontal fold (Tavener-Smith and Williams, 1970) in which initial exterior calcification is on the lower (basal) side. This exterior wall surface is a totally new structure, that is not a direct continuation of the calcified exterior lateral walls or any proximal gymnocyst. It is seemingly homologous with the lower or medial exterior surface of a series of fused spines.

Silén's interpretation (1942, pp. 43–45) of development of the calcified frontal and the compensation device in both spinocystideans and gymnocystideans appears to have been greatly influenced by his nonacceptance of origin of an ascus by invagination or cell proliferation from the proximal lip of the primary orifice and by the assumption that the compensation device of spinocystideans is, in fact, an ascus (a complete, membranous sac) and not simply an epistege (limited frontally by calcified spines separated by lacunae) or an epistegal space (limited frontally by a solid, spine-derived calcified sheet, as in umbonuloids). His arguments on the homology of ascus floor and the original membranous frontal relied heavily on examples of cribrimorphs in which a spinous frontal shield overarches the original membranous frontal which then "becomes" the ascus floor. The ascus roof was generated in his hypothetical model (Silén, 1942, Figs. 51 and 52, pp. 43, 44, and 47–49) by the supposedly uncalcified lower surface of that spinous epifrontal shield. It is difficult to reconcile the separation of the calcified spines on the upper surface of that epifrontal fold with the supposed continuity of the membranous "ascus roof" on the lower surface of that fold. Because the "floor of the ascus" in spinocystideans is the original frontal membrane, it functions perfectly well as a compensation device without having to introduce some hypothetical ascus roof. Indeed, in young umbonuloid zooids near colony margins, the membranous frontal functions in polypide protrusion well before completion of the calcified epifrontal shield (Cook, 1973, p. 249). Furthermore, Silén's inferred reconstruction does not correspond to known cribrimorph or umbonuloid spinocystideans, in which the lower surface of the shield is always calcified (see Plates I–VI and Text-Fig. 1D; Banta, 1970, Figs. 22 and 23; Boardman and Cheetham, 1973, Fig. 18). In such forms only the upper surface may remain uncalcified, forming the hypostegal coelom of umbonuloids or early ontogenetic stages of some cribrimorphs (Sandberg, in preparation).

Silén made the assumption that his gymnocystideans develop in a manner similar to his spinocystideans, that is, the original membranous frontal was supposedly overarched by a fold calcified on its upper surface only, and presented a hypothetical model for formation of the gymnocystidean ascus and calcified frontal (1942, pp. 49–52, Fig. 51). In his discussion of the mode of calcified frontal development in gymnocystideans, Silén (1942, p. 50) used the example of *Mucronella coccinea* (Abildgaard) as described by Hincks (1880, p. CV). That same species was, however, included by Harmer (1902, p. 332) among a group of species in which umbonuloid frontal development had been observed. My material of that species from the British Museum (Natural History) (BMNH 1911.10.1.1031) shows an umbonuloid frontal surrounded by marginal areolae.

Silén suggested (1942, pp. 51 and 52) that the calcified frontal of the ancestrula of *Microporella ciliata* must be a gymnocyst rather than a spinous frontal shield because it lies within the ring of frontally projecting spines (see Hincks, 1880). However, Harmer (1902, pp. 325 and 326), observed that wall region to be steeply sloping downward and interpreted it as a cryptocyst. Although there may be exceptions, it appears generally true that frontal spines arise at or very near the inner margin of the gymnocyst (sometimes as a result of fluting of the gymnocystal edge, Harmer, 1902, p. 292), and that calcification within that ring of spines, i.e., toward the center of the frontal, produces a cryptocyst.

The unusual development of the frontal in *Pseudolepralia ellisinae* (Silén, 1941, 1942) is very difficult to interpret on the basis of the available illustrations. From what has been observed of the ultrastructural development of a great variety of cheilostomes, the developmental pattern inferred for *Pseudolepralia* is unique.

The implications of these observations is that Silén's original Section Gymnocystidea did not include forms in which the calcified frontal wall was a true gymnocyst (as defined earlier in this paper). The examples given included umbonuloids, cribrimorphs, and the unusual *Pseudolepralia*. Ascophorans in which true gymnocystal frontals have been determined on the basis of ultrastructure (Sandberg, 1976) and on the absence of hypostega and marginal areolae (Banta, 1973) are calcified against the inner surface of the original frontal membrane. In such forms, that membrane can no longer function as a hydrostatic device and an ascus must have developed by invagination from the proximal lip of the orifice. That mode of ascus formation was not accepted by Silén (1942), and that is probably why such developmental types were excluded by him from the Gymnocystidea.

Gymnocystideans appear, in contrast to Silén's prediction (1942, pp. 57 and 58) to be a rare structural type among ascophoran cheilostomes. Excluding the spinocystideans originally included in the group by Silén, the only cheilostome which may fit the original concept is the enigmatic *Pseudolepralia ellisinae* Silén. Modifying the concept of the Gymnocystidea to forms with a true gymnocystal frontal (as discussed by Banta, 1973) results in only a very slight increase in the number of genera (*Hippothoa, Eurystomella, Figularia,* the latter perhaps a cribrimorph?). Until the details of the calcified frontal development in *Pseudolepralia* can be determined, it is difficult to say how that genus relates to the other three genera with true gymnocystal calcified frontals. In any case, gymnocystideans are a clearly minor developmental group among cheilostomes. There are certain structural–topological disadvantages of gymnocystidean morphology, relative to cryptocystideans and spinocystideans that may well have resulted in selection against gymnocystidean development. The upper frontal surface of the calcified gymnocystidean frontal wall is covered and protected only by a cuticle, whereas in cryptocystideans and umbonuloid spinocystideans there is an overlying hypostegal coelom intervening. Thickening of the calcified frontal wall is downward and inward in gymnocystideans and restricted by development of the ascus. Subsequent frontal thickening or repair, and lateral fusion of calcified frontal walls of adjacent zooecia (as found in the umbonuloid *Metrarabdotos,* see Cheetham *et al.,* 1969), would not be expected in gymnocystideans. However, the unusual development of what appears to be a superficial frontal layer in the gymnocystidean *Hippothoa* (see Sandberg, 1971, Plate 2, Figs. 10 and 11) is still to be explained.

The Gymnocystidea of Ryland (1970, pp. 40 and 159) correspond, in terms of the taxa included and of location of initial calcification in the frontal wall, to the ascophoran spinocystideans of Banta (1970) (= umbonuloids of Harmer, 1902). Boardman and Cheetham (1969, p. 222) suggested the possible occurrence in some umbonuloids of calcification on the frontal rather than basal side of the epifrontal fold. Such calcification would produce a gymnocystidean-like form similar to Silén's interpretation of *Pseudolepralia,* quite distinct structurally and functionally from what is now recognized as an umbonuloid. Accepting that umbonuloids [as exemplified by the taxa included by Harmer (1902) and Banta (1970)] are inferred polyspinocystideans with initially basal calcification in the epifrontal fold and subsequent frontal thickening, they should be clearly separated from gymnocystideans in which calcification relationships in the frontal are exactly reversed and do not involve an epifrontal fold.

Banta (1970) suggested that much of the difficulty with classification of ascophoran cheilostomes arises from confusion of ascus origin and frontal wall origin. Observations of early ontogenetic development (Cook, 1973), as well as the ultrastructural details (especially the planar spherulitic layer) of frontal walls (see Section IV, D), show that, at least for umbonuloids, origins of the frontal and the "ascus" are linked. The frontal calcified wall in umbonuloids develops as an evagination over a preexisting, functional, cuticulate frontal membrane. Therefore, the formation of the "ascus" or rather the epistegal space, is inseparable from formation of the umbonuloid calcified frontal wall (Text-Fig. 1D). Ascus formation in gymnocystideans (Text-Fig. 1E) is not well understood, but may well be similarly linked to calcified frontal wall development.

The epistegal space of umbonuloids was considered by Banta (1970) and others as synonymous with an ascus. However, Banta (1970) differentiated an epistege in cribrimorphs. Given the postulated origin of umbonuloids from cribrilinid cribrimorphs (Text-Figs. 1C and D) (now additionally supported by findings of Sandberg, in preparation) one could postulate an early "umbonuloid" with frontal thickening but with minute lacunae. The floor of the compensation device in both the cribrimorph epistege and the umbonuloid epistegal space is homologous with the frontal membrane of Membranidea and quite ontogenetically different from the floor of the ascus of Cryptocystidea and presumably of Gymnocystidea. It would seem preferable, therefore, to consider the epistegal space in umbonuloids as only a functional analog of the ascus, as suggested by Cook (1973).

In the Cryptocystidea, as exemplified by *Watersipora* and *Schizoporella* (Banta, 1970, 1972) and *Adeonella* (Cook, 1973), ascus development is distinctly separate from and subsequent to frontal wall formation (Text-Fig. 1B). In other Ascophora Vera, which include interior-walled forms (Text-Fig. 1F) (Håkansson, 1973; Sandberg, 1973), ascus–frontal wall development relationships are mostly unknown.

This section on wall development in cheilostomes has pointed out some of the problems of morphogenetically-based systems of classification. Given the orientation of this chapter as a whole the obvious next question is: "Are there ways in which ultrastructural data can be used to resolve some of these problems?" The next section will attempt to demonstrate that the answer to that question is a definite, encouraging "Yes."

D. Ultrastructural Clues to Wall Genesis

In the study of skeletal wall genesis in cheilostomes or other bryozoans it is critical to know whether any given wall is exterior or interior. In modern material, doubled intercalary cuticles (Plate V, Fig. 3) between two adjacent calcified walls (sometimes called "duplex" walls, but actually two mirror-image simple exterior walls) serve as excellent indicators of the exterior wall origin. In fact, the presence of cuticle and the resulting lack of cohesion between walls allows another means of testing for exterior–interior wall relationships, that is the tendency of colonial skeletons which are fractured, crushed, or treated with

NaOCl or H_2O_2 to separate into linear series of zooids if intercalary cuticles are present in lateral walls (or in basal walls of bilamellar colonies) (Banta, 1968, p. 499; Banta, 1972, p. 69; Ryland, 1970, p. 87). Absence of such separation, as between zooids in a linear series or, more significantly, in lateral walls (*Myriapora:* Sandberg, 1973, Figs. 2, 5, and 17; *Cellaria:* Sandberg, 1971, Plate 2, Fig. 8), is a strong indication of interior-wall development. A rather large number of cheilostomes show a lack of separation between adjacent linear series, suggesting a lack of exterior lateral walls. One must be careful to distinguish forms in which lateral walls are initially divided by intercalary cuticle, but in which later frontal thickening occurs with a breakdown in lateral cuticles and a bridging of secondary calcification between laterally adjacent zooecial rows (e.g., *Metrarabdotos:* Boardman and Cheetham, 1973, Fig. 18). Breakage tests, even after NaOCl treatment, should be interpreted with caution and preferably verified with SEM examination of specimens.

In fossil material, even when rather strongly recrystallized, the former locations of intercalary cuticles may still be evident, that is, those cuticles have apparently influenced the pattern of recrystallization (Sandberg, 1975). It should be emphasized that it is the presence of calcified walls to either side of the cuticle and the consequent "interruption" of the skeleton by the doubled intercalary cuticle which allows the recognition of those walls as exterior. For exterior walls that are not directly adjoined by another exterior wall, determination of exterior wall origin is correspondingly more difficult because of the absence of those intercalary cuticle clues.

Therefore, one of the major questions investigated in some recent ultrastructural studies (Sandberg, 1976) was: "Are there ultrastructural clues usable for recognition of exterior skeletal walls which can be applied to fossil or dead modern bryozoans?" In order to answer this, a number of skeletal walls known to be exterior walls on histological or morphological grounds were studied by means of SEM to determine the nature of the ultrastructure present. In almost all of these cases the initial calcification against cuticle was found to be a thin layer of acicular calcite or aragonite crystals arranged in strips or wedges of two-dimensional or planar spherulitic structure with accretionary banding (Plate II, Fig. 2; Plate III, Fig. 1; Plate IV, Fig. 3) (Sandberg, 1976). This structure was earlier called parallel fibrous ultrastructure (Sandberg, 1971), but the term "planar spherulitic" is used in this chapter. Levinsen (1909), Bronstein and Prenant (1936), and Schneider (1957, 1963) reported oriented bands or strips of crystalline skeletal material in walls (mainly basal) of various cheilostomes. Because that structure was observed in transmitted light it was difficult to determine if it was internal, external, or throughout the wall thickness. By means of the SEM it has been possible to demonstrate that those bands of generally longitudinally oriented acicular crystals form a thin, initial, exterior skeletal layer on the cuticle.

The association between "planar spherulitic" ultrastructure and histologically or morphologically determined cuticulate exterior skeletal walls is so strong that one might be tempted to use the name "exterior wall ultrastructure." However, such genetic terms are prejudicial and best avoided in favor of descriptive structural terms like "planar spherulitic." This is particularly true since it appears that outer surfaces of some exterior skeletal walls may lack this structure, e.g., the presumably secondarily calcified basal window in *Fenestrulina malusii* (Sandberg, 1971, Plate 3, Fig. 11). More significantly, similar planar spherulitic ultrastructure has been found on the undersides of coral dissepiments and tabulae (Barnes, 1970; Sorauf, 1971, 1974; Wells, 1969) which form by one-sided growth with thickening on the top. Planar spherulitic ultrastructure certainly does occur on exterior walls in stenolaemate bryozoans (Söderqvist, 1968; Tavener-Smith and Williams, 1972; Brood, 1973) and, in analogy to the coral dissepiments, could occur on undersides of simple interior walls in stenolaemates as well (Sandberg, 1976).

I have found that planar spherulitic ultrastructure is less clearly developed in aragonitic walls. Furthermore, in some exterior skeletal walls such as the non-areolate frontal wall of *Hippothoa hyalina* only a very crude approximation of planar spherulitic ultrastructure occurs. With these various qualifications and limitations, planar spherulitic ultrastructure appears to be an excellent criterion for recognition of exterior calcified walls in bryozoans. The near identity of that ultrastructure between cheilostomes and cyclostomes makes it reasonable to predict that exterior walls in skeletons of other stenolaemates would also possess similar ultrastructure.

Two general characteristics of planar spherulitic ultrastructure are its restriction to a very thin exterior layer and the presence of faint to pronounced transverse accretionary lines marking successive positions of the advancing front of calcification (Sandberg, 1971, p. 137–138, Plate 2, Figs. 1, 2, and 4; Schneider, 1957, p. 253, Figs. 1 and 5). The oriented, intermittently deposited crystals are laid down along a linear front on the surface of the cuticle, which is being extended by intusseception. The growth vector for both the cuticle and the planar spherulitic layer is distal, parallel with the wall. Subsequent thickening of the wall does not occur along a linear front, but rather over the broad inner surface of the wall. Therefore, the growth vector has shifted to be essentially perpendicular to the wall, and accretionary units are parallel to the deposition face, the inner wall surface. Successions of several distinctive ultrastructures are common above that initial planar spherulitic layer.

The planar spherulitic ultrastructure of exterior skeletal walls does not always occur in bandlike arrangements of stripes. In the ancestrula, which is preformed and then calcified, Schneider (1957, Figs. 2 and 6) found that calcification occurs as scattered two-dimensional radial spherulites which grow out and intersect one another. The "band-pattern" only occurred in later zooids as the ancestrular cuticle was expanded by intusseception. Schneider (1957, pp. 251 and 252) observed that the calcite bands of later zooids arose by competitive interference of the most distal spherulites of the ancestrula (compare discussions on spherulitic growth by Barnes, 1970; Bryan and Hill, 1941).

Nonlinear arrangements of planar spherulitic ultrastructure occur in some anascan cheilostomes which are weakly calcified. Levinsen (1909, Plate 19, Fig. 8a) discussed and illustrated what he called a plate-mosaic in *Flustra securifrons,* and Sandberg showed a similar structure (1976, Plate 1, Figs. 6 and 9) from *Megapora ringens* and *Onychocella angulosa.* In *Flustra,* the slight calcification probably does not occur in close association with the distally advancing front of cuticle intussusception, but somewhat after formation of the zooid. In the case of the basal window of *Fenestrulina malusii* (Sandberg, 1971, Plate 3, Figs. 11 and 12) the ultrastructure against cuticule is not planar spherulitic. It appears there that an area of wall commonly uncalcified has subsequently been calcified (perhaps because of separation from an algal substrate) well after zooid formation. Surrounding portions of the basal walls in zooecia in that same colony show good planar spherulitic ultrastructure.

The implications of these various observations are as follows:

1. Planar spherulitic ultrastructure, whether as elongate bands or as radial planar spherulites, is a good indicator of exterior, cuticulate walls.

2. Because of possible distinctiveness of ancestrular calcification and the weak or irregular development of parallel fibrous ultrastructure on some known exterior walls, caution should be exercised in interpreting the absence of that ultrastructure as an indication that the wall in question is not exterior. It should be noted, however, that internal consistency exists. That is, in all forms studied where planar spherulitic ultrastructure occurred on any exterior wall it occurred on all other exterior walls which calcified at the time of initial wall development (note the distinctive, later ultrastructure of secondarily calcified basal walls in *Fenestrulina malusii* discussed above).

Given these characteristics of planar spherulitic ultrastructure, how can it be used (in conjunction with ultrastructural type and accretion direction of subsequent wall-thickening layers) to determine mode of skeletal genesis for any given cheilostome?

As pointed out by Cook (1973) and Banta (1970), recognition of umbonuloid or lepralioid origin of calcified frontal walls is difficult without dissection or sectioning of freshly fixed material or observation of preserved growing edges with membranous walls intact or of wall growth in living colonies. Two ultrastructural clues are usable for separation of these groups. Lepralioid frontal walls (cryptocysts) generally thicken not only frontally but also basally, at least until formation of the ascus shuts down potential calcification on the lower frontal surface. Therefore, such walls in section will commonly show accretion on both sides, sometimes concentrically (Plate III, Fig. 5; Plate V, Fig. 5; Plate VI, Fig. 4)

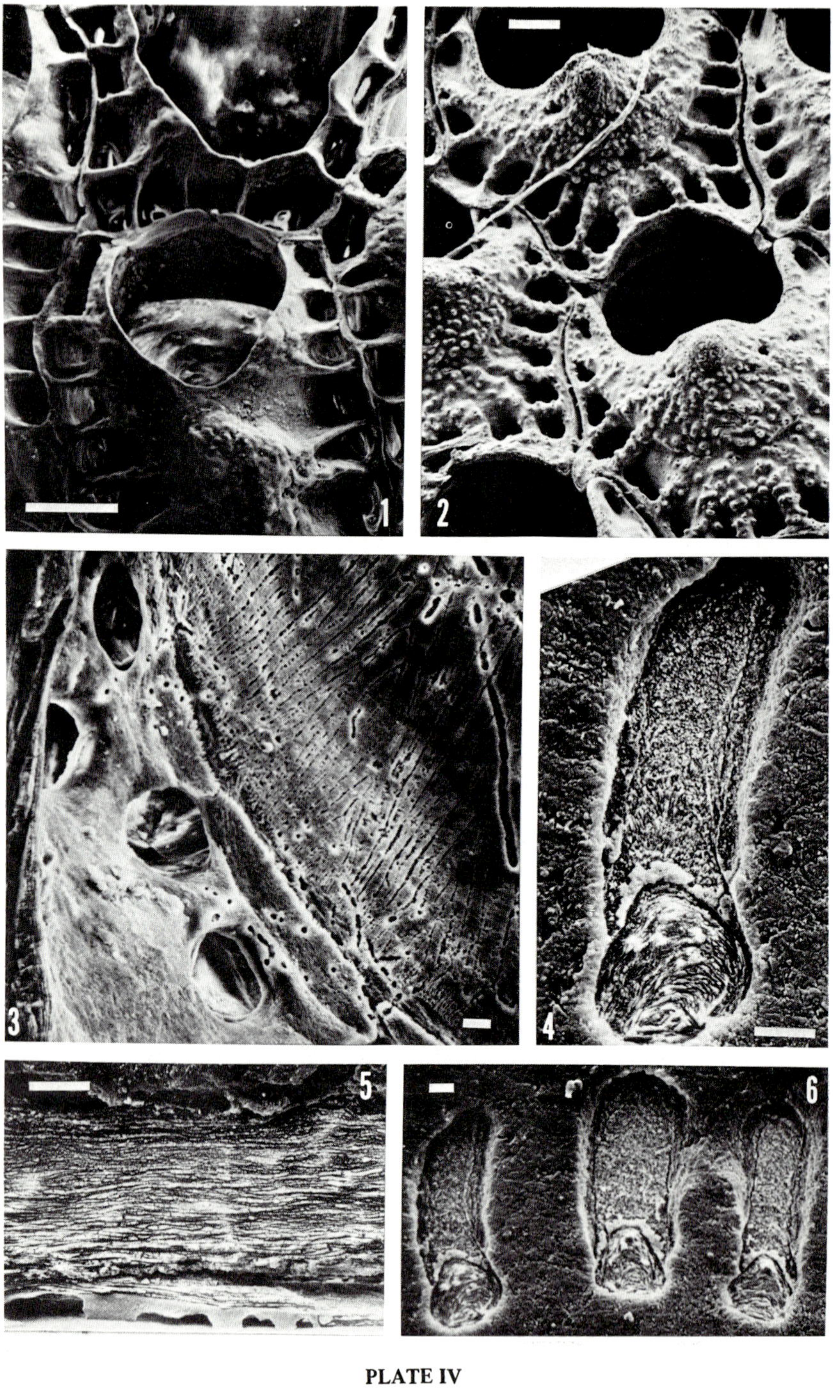

PLATE IV

(Sandberg, 1971, Plate 3, Fig. 10). Umbonuloid frontals, calcifying as simple exterior walls away from cuticle, add skeletal material on one side only and will lack such concentric accretion. Furthermore, the lower surface of the frontal wall in umbonuloids will show parallel spherulitic ultrastructure (Plate IV, Fig. 3). Gymnocystidean frontal walls will be characterized by a similar succession of planar spherulitic ultrastructure on an exterior surface followed by thickening on only one side, however, the succession will be inverted, exterior side upward (Text-Figs. 1D and E).

Interior-walled cheilostomes will lack planar spherulitic ultrastructure, and, in interior skeletal walls, thickening (as in lepralioid frontals) can occur on both sides although not necessarily symmetrically (Plate III, Fig. 2 and 6). Furthermore, the accretionary laminations of interior-walled forms may be quite distinctive.

In most simple exterior walls in skeletons of cheilostomes, the accretionary layers are nearly parallel to the wall surface, suggesting that the wall originated as a thinly calcified surface and later thickened over much or all of its inner surface.

The interpretation should be qualified with the suggestion by Boardman and Towe (1966) of edgewise growth of numerous lamellae simultaneously in steno-

---

PLATE IV

Fig. 1. *Posterula sarsi* (Smitt) (BMNH 1911.10.1.1360A-B) Frontal exterior (calcite) view. Cuticle in upper transverse wall separates off distal quarter of orifice in this umbonuloid ascophoran. Bar scale represents 100 μm.

Fig. 2. *Umbonula verrucosa* (Esper) (BMNH 1963.3.6.8) The cormidial orifices of this species are each contributed to by skeletal parts (calcite) of four zooecia. Bar scale represents 100 μm.

Fig. 3. *Escharoides contorta* (BMNH 1970.1.4.14) Interior surface of calcite frontal wall (distal toward top of figure) with marginal areolae and, toward upper right, the underside of the frontal wall is an exterior surface with planar spherulitic ultrastructure. Bar scale represents 100 μm.

Fig. 4. *Hippopetraliella magna* (BMNH 1896.2.28.1) Detail of Fig. 6. Note the roughly concentrically laminated calcite of the initial frontal and the considerable thickening by superficial spherulitic aragonite on the upper surface only. Bar scale represents 10 μm.

Fig. 5. *Petralia undata* MacGillivray (BMNH 1883.10.15,27-31) Etched longitudinal section of basal wall (calcite). Note the irregular lenticular nature of the lamellar structure. Bar scale represents 10 μm.

Fig. 6. *Hippopetraliella magna* (BMNH 1896.2.28.1) Etched transverse section of several bars of meshlike frontal wall structure. Bar scale represents 10 μm.

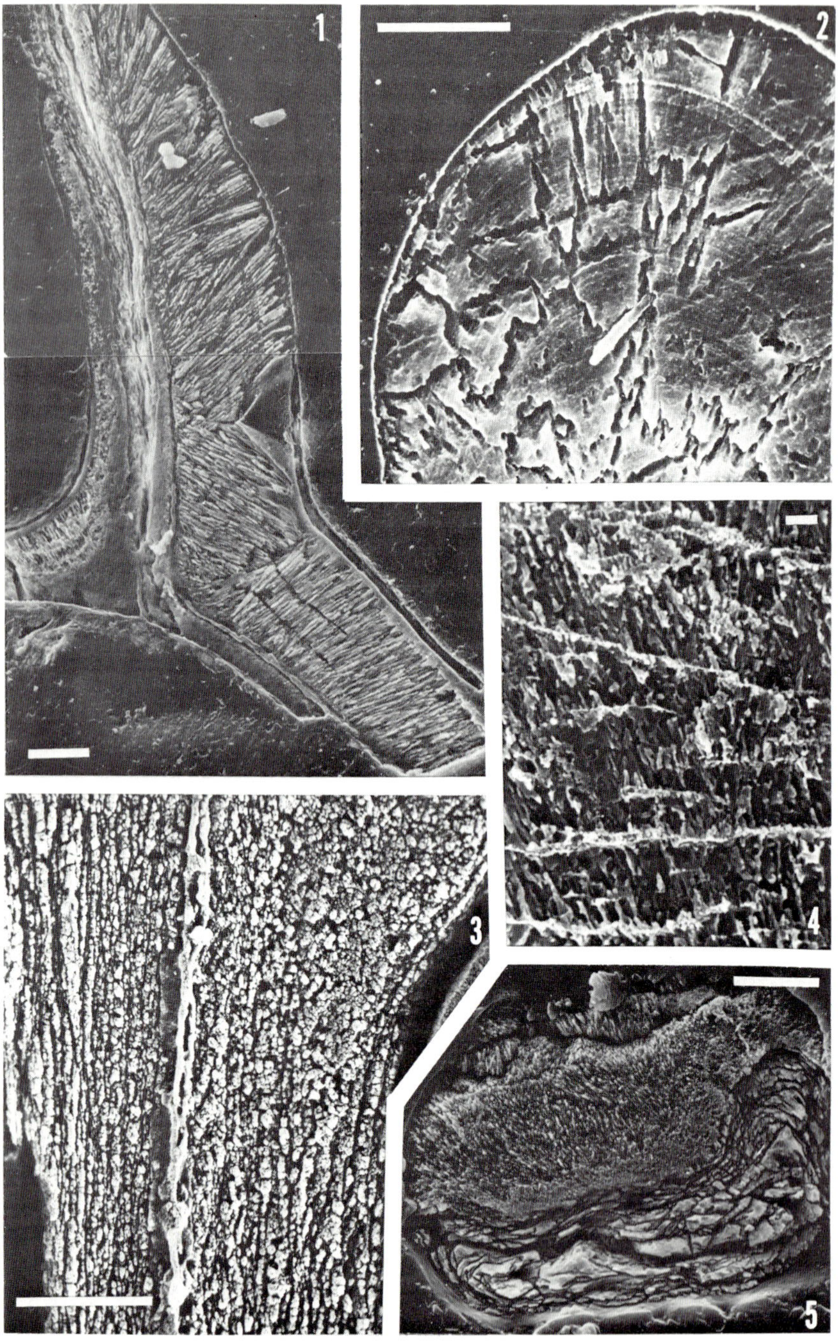

PLATE V

laemates. SEM observations I have made on presumably unbroken distal edges of skeletons of some cheilostome colonies suggest that numerous, superimposed, distinct lamellae of basal walls may be growing simultaneously along their leading edges, producing a stair-step pattern similar to that observed in some pelecypod nacre, but with a much closer spacing of the steps. Even on broad growth surfaces, simultaneous addition of skeletal material on several laterally expanding superimposed layers is not uncommon in cheilostomes (Plate I, Fig. 6).

In contrast to cheilostome skeletons with the "standard" growth mode (mainly exterior walls), skeletons of interior-walled forms may often show accretionary increments at high angles to the wall surfaces and convex distally (Sandberg, 1973, Figs. 3–5, and 12), indicating distal advance of a calcification front which may be nearly as broad as the final skeletal wall thickness.

Boardman and Cheetham (1973) stressed that varying degrees of colonial integration may occur in diverse bryozoans. In recent detailed studies of cheilostome morphology it has become apparent that the sort of clearly defined individual zooecial limits which characterize uniserial forms (e.g., *Pyripora*) do not readily transfer to cheilostomes in general. In the "standard" morphological pattern for multiserial cheilostomes (discussed earlier in this paper) the distal

---

PLATE V

Fig. 1. *Petraliella* sp. (USNM 209433) Etched longitudinal section of basal–distal wall junction (distal is toward the left). The thin calcite initial skeleton is overlain by a thick zooecial lining of spherulitic aragonite. Bar scale represents 10 μm.

Fig. 2. *Flabellopora arculifera* Canu and Bassler (USNM 209437) Etched longitudinal section of zooecial lining (calcite) near orifice. Note the very numerous, closely spaced accretionary bands and the irregular boundaries of the obliquely sectioned spherulitic calcite arrays. Bar scale represents 10 μm.

Fig. 3. *Exechonella antillea* (BMNH 1972.3.3.37) Etched longitudinal section of aragonite frontal wall (frontal direction is up). Note the numerous though somewhat irregularly spaced, discontinuous organic sheets running transverse to the spherulitic aragonite. Bar scale represents 1 μm.

Fig. 4. *Porella remotorostrata* (BMNH 1962.6.6.1) Etched longitudinal section of the basal walls (calcite) of two zooecia of a bilamellar colony. A doubled intercalary cuticle (raised white structure running vertically in center of photo) separates the two basal walls. Bar scale represents 10 μm.

Fig. 5. *Petraliella* sp. (USNM 209433) Etched transverse section of part of frontal wall. Note the concentrically lamellar calcite initial frontal overlain on the upper surface only by spherulitic superficial aragonite. Bar scale represents 10 μm.

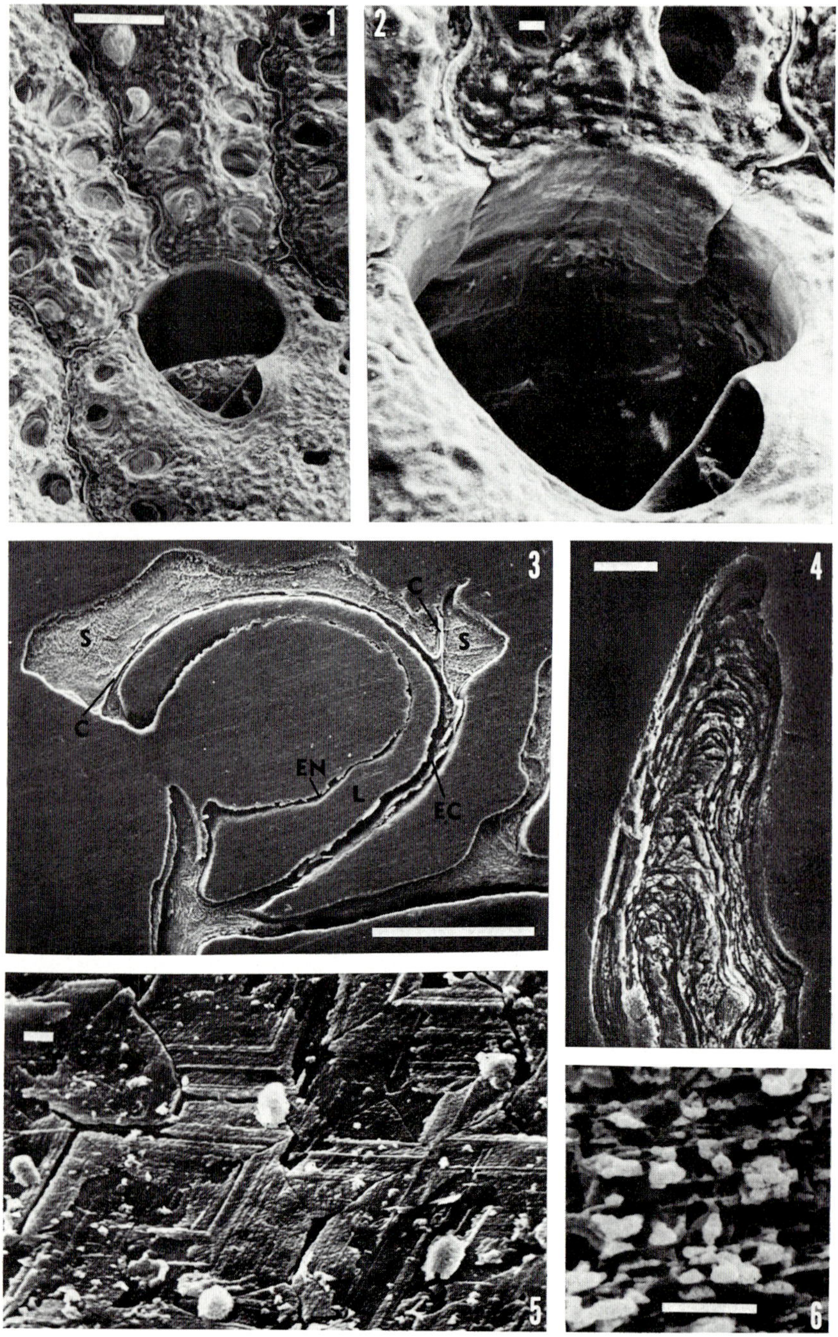

PLATE VI

calcified walls (Plate I, Fig. 1) in linear series are products of two zooids. Definition of limits of successive zooecia in those walls is at best difficult and rather arbitrary. In interior-walled forms, which lack intercalary cuticles, this indistinguishability and arbitrariness of individual boundaries in skeletal parts extends to lateral and basal walls as well.

Whereas one is more willing to accept the varying degree of integration in basal and vertical walls of zooids, there is a natural tendency to regard the frontal of a zooecium as clearly individual. It is therefore striking that ovicelled "individuals" of various cheilostomes have been found to be cormidial complexes (Beklemishev, 1970) involving two or more individuals (Woollacott and Zimmer, 1972; Cheetham and Cook, in press). Interpretation of ovicell growth is aided by the use of planar spherulitic ultrastructure distribution in conjunction with sectioning and dissection. Although it is quite difficult to tell in frontal view (even with the SEM), it is commonly evident in sections that the lumen

---

PLATE VI

Fig. 1. *Posterula sarsi* (Smitt) (BMNH 1911.10.1.1360A-B) Frontal view of orifice in part constructed by the proximal ends of two distally budded individuals (calcite walls). Bar scale represents 100 μm.

Fig. 2. *Posterula sarsi* (Smitt) (BMNH 1911.10.1.1360A-B) Detailed view of orifice in Fig. 1. Note the distinct zooecial boundaries (doubled intercalary cuticles present but not visible here) dividing the distal margin of the orifice. Planar spherulitic ultrastructure is only faintly evident at this magnification. Bar scale represents 10 μm.

Fig. 3. *Pentapora foliacea* (BMNH 1911.10.1.1561) Etched longitudinal section of an ovicel. The lumen (L) is surrounded by the calcified endoecium (EN) and ectoecium (EC). The latter (EC) is overlain by superficial frontal calcification (S), apparently from several zooecia, as indicated by the distribution of cuticles (C) between the superficial calcification and the ectoecium. The connection of the ovicell lumen (L) to the distal zooecium is clearly shown in other zooecia of this colony. Bar scale represents 100 μm.

Fig. 4. *Labioporella calypsonis* Cook (BMNH 1964.7.2.31) Etched longitudinal section of cryptocyst distal margin (distal direction up). Note the irregular but clearly concentric accretions on both upper (toward right) and lower surfaces of cryptocyst (calcite). Bar scale represents 10 μm.

Fig. 5. *Tremogasterina robusta* (BMNH 1966.2.24.1) Interior surface of a basal wall. Note spaced seeding of calcite nuclei and marginal accretion and impingement of rhombic calcite crystals. Bar scale represents 1 μm.

Fig. 6. *Adeona* sp.. (BMNH 1892.1.28.91) Etched longitudinal section of areolar pore lining (aragonite). Note abundant organic sheets and blocky aragonite crystals. Bar scale represents 1 μm.

(Plate VI, Fig. 3) of the flattened, expanded spine or spines forming the ovicell connects with the distal individual, not the fertile zooid.

Presence or absence of planar spherulitic ultrastructure on inner and/or outer surfaces (endoecium and ectoecium, respectively) of ovicells (Plate II, Fig. 2; Plate III, Figs. 3 and 4) shows that there is a great diversity of modes of ovicell development in cheilostomes. Ectoecium or endoecium may variably be calcified or uncalcified throughout some or most of their extent. When the ectoecium is uncalcified above the general frontal surface (Plate III, Figs. 3 and 4), thickening on the upper surface of the calcified endoecium (which has an exterior-wall lower surface) can produce a direct analog of an umbonuloid frontal wall. Such "umbonuloid" ovicells do occur in cribrimorphs, the inferred ancestral forms of umbonuloid ascophorans (Harmer, 1902), but they also occur in cryptocystideans (Plate III, Figs. 1, 3, and 4).

Perhaps as striking as the cormidial nature of ovicelled zooecia is the finding that, for many umbonuloids, the secondary zooecial orifice may be cormidial. In umbonuloids the calcareous frontal wall develops in an epifrontal fold from the lateral and proximal zooecial margins (Harmer, 1902; Tavener-Smith and Williams, 1970; Cook, 1973). The cuticle of the underlying membranous frontal is caught into the upper edge of the transverse wall distally. However, the operculum is relatively close to the distal end of the membranous frontal, and very little calcification of any gymnocystal wall proximally toward the operculum occurs. Subsequent thickening of the frontal walls produces a secondary orifice which is clearly constructed by two or even three or four individuals (Plate IV, Figs. 1 and 2; Plate VI, Figs. 1 and 2). The arrangement of planar spherulitic ultrastructure along the distal lip of the orifice shows that surface to be an exterior calcified wall and requires a quite different reconstruction of tissue–skeleton relationships (Text-Fig. 1D) than that previously given (Cheetham, 1968, Fig. 3, Cheetham *et al.*, 1969, Fig. 1B; Boardman and Cheetham, 1969, Fig. 4).

Investigations of skeletal ultrastructure of bryozoans have only begun to utilize the potential of the SEM. Nevertheless, usefulness of planar spherulitic ultrastructure, location and type of subsequent skeletal thickening, accretionary laminations, and cuticle have been clearly demonstrated. Less understood is the significance of specific ultrastructural types or successions of ultrastructural types. The combined study (in section and on growth or fracture surfaces) of the nature and positional relationships of cuticles and calcareous ultrastructure has already supplied abundant information on skeletal genesis in bryozoans. Continued investigations of this type will hopefully direct more attention to organic matrix type and distribution, optic orientation of skeletal $CaCO_3$ elements, and the loss or retention of fine texture and major element composition of the skeleton during fossilization. Such investigations could yield additional valuable data for both ontogenetic and phylogenetic studies of bryozoan skeletal genesis.

ACKNOWLEDGMENTS

This study constitutes part of the work undertaken during tenure of a NATO Postdoctoral Fellowship in Science (1972–1973) at the British Museum (Natural History), London, and part of a study supported by a grant from the National Science Foundation (GB27346).

Numerous discussions throughout the year with Pat Cook, Senior Scientific Officer (Bryozoa Section, Zoology), were most helpful in developing this paper. The technical assistance of P. J. Chimonides of that section during the year is also gratefully acknowledged. Aspects of this study have benefitted from the comments and criticisms of Drs. Richard S. Boardman and Alan H. Cheetham (U.S. National Museum), Dr. John Utgaard (Southern Illinois University, Carbondale), and Dr. William C. Banta (American University, Washington, D.C.).

Modern bryozoan material for this study was loaned from the collections of the U.S. National Museum and the British Museum (Natural History), and illustrated material has been redeposited in those collections.

Scanning electron micrographs were taken on Cambridge Stereoscan instruments in the Center for Electron Microscopy, University of Illinois [microscope acquired with funds from the National Science Foundation (GA-1237), the U.S. Public Health Service (FR-07030), and the University of Illinois Research Board]; the British Museum (Natural History) (BMNH); the Geological Institute, London; the Department of Geology, Imperial College, London; and the Geological Institute, University of Copenhagen.

Lowell Bostrom gave valuable assistance in preparing the plates.

REFERENCES

Banta, W. C. (1968). The body wall of cheilostome Bryozoa. I. The ectocyst of *Watersipora nigra* (Canu & Bassler). *J. Morphol.* **125,** 497–506.

Banta, W. C. (1969). The body wall of cheilostome Bryozoa. II. Interzooidal communication organs. *J. Morphol.* **129,** 149–170.

Banta, W. C. (1970). The body wall of cheilostome Bryozoa. III. The frontal wall of *Watersipora arcuata* Banta, with a revision of the Cryptocystidea. *J. Morphol.* **131,** 37–56.

Banta, W. C. (1972). The body wall of cheilostome Bryozoa. V. Frontal budding in *Schizoporella unicornis floridana* Osburn. *Mar. Biol.* **14,** 63–71.

Banta, W. C. (1973). The significance of areolae in cheilostome Bryozoa. *In* "Living and Fossil Bryozoa" (G. P. Larwood, ed.), pp. 209–219. Academic Press, New York.

Barnes, D. J. (1970). Coral skeletons: An explanation of their growth and structure. *Science* **170,** 1305–1308.

Bassler, R. S. (1953). *In* "Treatise on Invertebrate Paleontology" (R. C. Moore ed.), Vol. G (Bryozoa), pp. 1–253. Geological Society of America, Lawrence, Kansas.

Beklemishev, V. N. (1970). "Principles of Comparative Anatomy of Invertebrates," Vol. I. Univ. of Chicago Press, Chicago, Illinois.

Bevelander, G., and Nakahara, H. (1969). An electron microscope study of the formation of the nacreous layer in the shell of certain bivalve molluscs. *Calcif. Tissue Res.* **3,** 84–92.

Blake, D. B. (1973). Coloniality and polymorphism in Bryozoa of the families Rhabdomesidae and Hyphasmoporidae (Order Rhabdomesonata). *In* "Animal Colonies" (R. S.

Boardman, A. H. Cheetham, and W. A. Oliver, Jr., eds.), pp. 361–376. Dowden, Hutchinson, & Ross, Inc., Stroudsburg, Pennsylvania.

Blake, D. B., and Towe, K. M. (1971). Acanthopore ultrastructure in the Paleozoic bryozoan *Idioclema insigne* Girty. *J. Paleontol.* **45**, 913–917.

Boardman, R. S. (1971). Mode of growth and functional morphology of autozooids in some Recent and Paleozoic tubular Bryozoa. *Smithson. Contrib. Paleontol.* **8**, 1–51.

Boardman, R. S. (1973). Body walls and attachment organs in some recent cyclostomes and Paleozoic trepostomes. *In* "Living and Fossil Bryozoa" (G. P. Larwood, ed.), pp. 231–246. Academic Press, New York.

Boardman, R. S., and Cheetham, A. H. (1969). Skeletal growth, intracolony variation, and evolution in Bryozoa: A review. *J. Paleontol.* **43**, 205–233.

Boardman, R. S., and Cheetham, A. H. (1973). Degree of colony dominance in stenolaemate and gymnolaemate Bryozoa. *In* "Animal Colonies" (R. S. Boardman, A. H. Cheetham, and W. A. Oliver, Jr., eds.), pp. 121–220. Dowden, Hutchinson, & Ross, Inc., Stroudsburg, Pennsylvania.

Boardman, R. S., and Towe, K. M. (1966). Crystal growth and lamellar development in some Recent cyclostome Bryozoa. *Geol. Soc. Am. Program Annu. Meet.* Abstract, p. 20.

Bobin, G., and Prenant, M. (1968). Sur le calcaire des parois autozoéciales d'*Electra verticillata* (ell. et Sol.), Bryozoaire Chilostome, Anasca. *Arch. Zool. Exp. Gén.* **109**, 157–191.

Borg, F. (1926). Studies on Recent cyclostomatous Bryozoa. *Zool. Bidr. Uppsala* **10**, 1–507.

Bronstein, G., and Prenant, M. (1936). Texture cristalline du calcaire dans la paroi basale chez les Bryozoaires Chilostomes. *Bull. Soc. Zool. Fr.* **61**, 411–421.

Brood, K. (1972). Cyclostomatous Bryozoa from the Upper Cretaceous and Danian in Scandinavia. *Stockholm Contrib. Geol.* **26**, 1–464.

Brood, K. (1973). Paleozoic Cyclostomata (a preliminary report). *In* "Living and Fossil Bryozoa" (G. P. Larwood, ed.), pp. 247–256. Academic Press, New York.

Bryan, W. H., and Hill, D. (1941). Spherulitic crystallization as a mechanism of skeletal growth in the hexacorals. *Proc. R. Soc. Queensl.* **52**, 78–91.

Canu, F., and Bassler, R. S. (1920). North American Early Tertiary Bryozoa. *U.S. Natl. Mus., Bull.* **106**, 1–879.

Cheetham, A. H. (1954). A new early Cretaceous cheilostome bryozoan from Texas. *J. Paleontol.* **28**, 177–184.

Cheetham, A. H. (1968). Morphology and systematics of the bryozoan genus *Metrarabdotos. Smithsonian Misc. Coll.* **153**, 1–121.

Cheetham, A. H., and Cook, P. L. (in press). General features of the class Gymnolaemata. *In* "Treatise on Invertebrate Paleontology" (R. Robison, ed.), Vol. G1 (rev.) (Bryozoa). Geological Society of America, Lawrence, Kansas.

Cheetham, A. H., Rucker, J. B., and Carver, R. E. (1969). Wall structure and mineralogy of the cheilostome bryozoan *Metrarabdotos. J. Paleontol.* **43**, 129–135.

Cook, P. L. (1973). Preliminary notes on the ontogeny of the frontal body wall in the Adeonidae and Adeonellidae (Bryozoa, Cheilostomata). *Bull. Br. Mus. (Nat. Hist.) Zool.* **25**, 243–263.

Cutler, J. F. (1973). Nature of "Acanthopores" and related structures in the Ordovician bryozoan *Constellaria. In* "Living and Fossil Bryozoa" (G. P. Larwood, ed.), pp. 257–270. Academic Press, New York.

Dodd, J. R. (1967). Magnesium and strontium in calcareous skeletons: A review. *J. Paleontol.* **41**, 1313–1328.

Folk, R. L. (1973). Carbonate petrology in the Post-Sorbian age. *In* "Evolving Concepts in Sedimentology" (R. N. Ginsburg, ed.), pp. 118–158. Johns Hopkins Univ. Press, Baltimore, Maryland.

Folk, R. L. (1974). The natural history of crystalline calcium carbonate: Effect of magnesium content and salinity. *J. Sediment. Petrol.* **44,** 40–53.

Gautier, T. G. (1973). Growth in bryozoans of the order Fenestrata. *In* "Living and Fossil Bryozoa" (G. P. Larwood, ed.), pp. 271–274. Academic Press, New York.

Goreau, T. F. (1959). The physiology of skeleton formation in corals. 1. A method for measuring the rate of calcium deposition by corals under different conditions. *Biol. Bull. (Woods Hole, Mass.)* **116,** 59–75.

Greely, R. (1969). Basally "uncalcified" zoaria of lunulitiform Bryozoa. *J. Paleontol.* **43,** 252–256.

Håkansson, E. (1973). Mode of growth of the Cupuladriidae (Bryozoa, Cheilostomata). *In* "Living and Fossil Bryozoa" (G. P. Larwood, ed.), pp. 287–298. Academic Press, New York.

Hansen, H. J. (1972). Two species of Foraminifera of the genus *Turrilina* with different wall structure. *Lethaia* **5,** 39–45.

Harmer, S. F. (1902). On the morphology of the Cheilostomata. *Q. J. Microsc. Soc. [N.S.]* **46,** 263–350.

Harmer, S. F. (1957). The Polyzoa of the Siboga Expedition, pt. 4, Cheilostomata Ascophora II. *Siboga Exped. Rep.* **28d,** 641–1145.

Hay, W. W., Wise, S. W., and Stieglitz, R. D. (1970). Scanning electron microscope study of fine grain size biogenic carbonate particles. *Trans. Gulf Coast Assoc. Geol Soc.* **20,** 287–302.

Hincks, T. (1880). "A History of the British Marine Polyzoa." J. VanVoorst, London.

Hinds, R. W. (1975). Growth mode and homeomorphism in cyclostome Bryozoa. *J. Paleontol.* **49,** 875–910.

Kobayashi, I. (1969). Internal microstructure of the shell of bivalve molluscs. *Am. Zool.* **9,** 663–672.

Kobayashi, I. (1971). Internal shell microstructure of Recent bivalvian molluscs. *Sci. Rep. Niigata Univ., Ser. E.* **2,** 27–50.

Levinsen, G. M. R. (1909). "Morphological and Systematic Studies on the Cheilostomatous Bryozoa." Nat. Forfatt. Forlag, Copenhagen.

Loeblich, A. R., Jr., and Tappan, H. (1964). Protista 2. *In* "Treatise on Invertebrate Paleontology" (R. C. Moore, ed.), Vol. C1 (Sarcodina), Geological Society of America, Lawrence, Kansas.

Lowenstam, H. A. (1954). Factors affecting the aragonite-calcite ratios in carbonate-secreting marine organisms. *J. Geol.* **62,** 284–322.

McKinney, F. K. (1969). Organic structures in a late Mississippian trepostomatous ectoproct (Bryozoan) *J. Paleontol.* **43,** 285–288.

Mutvei, H. (1972). Ultrastructural relationships between the prismatic and nacreous layers in *Nautilus* (Cephalopoda). *Biomineralization Res. Rep.* **4,** 80–86.

Nye, O. B. (1968). Aspects of microstructure in post-Paleozoic Cyclostomata. *Atti Soc. Ital. Sci. Nat. Mus. Civ. Stor. Nat. Milano* **108,** 111–114.

Pohowsky, R. A. (1973). A Jurassic cheilostome from England. *In* "Living and Fossil Bryozoa" (G. P. Larwood, ed.), pp. 447–461. Academic Press, New York.

Pohowsky, R. A. (1974). Notes on the study and nomenclature of boring Bryozoa. *J. Paleontol.* **48,** 556–564.

Pohowsky, R. A. (1976). Boring Bryozoa (abstr). *In* "Bryozoa 1974" (L. David, ed.). Doc. Lab. Geol. Fac. Sci., Lyon.

Poluzzi, A., and Sartori, R. (1973). Carbonate mineralogy of some Bryozoa from Talbot Shoal (Strait of Sicily, Mediterranean). *Gior. Geol.* **39,** 11–15.

Poluzzi, A., and Sartori, R. (1976). Report on the carbonate mineralogy of Bryozoa. *In* "Bryozoa 1974" (L. David, ed.). Doc. Lab. Geol. Fac. Sci., Lyon.

Raup, D. M. (1960). Ontogenetic variation in the crystallography of echinoid calcite. *J. Geol.* **34,** 1041–1050.

Raup, D. M. (1962). The phylogeny of calcite crystallography in echinoids. *J. Paleontol.* **36,** 793–810.

Rucker, J. B. (1967). Carbonate mineralogy of Recent cheilostome Bryozoa. *Geol. Soc. Am., Program Annu. Meet.* Abstract, pp. 191–192.

Rucker, J. B. (1968). Skeletal mineralogy of cheilostome Bryozoa. *Atti Soc. Ital. Sci. Nat. Mus. Civ. Stor. Nat. Milano* **108,** 101–110.

Rucker, J. B., and Carver, R. E. (1969). A survey of the carbonate mineralogy of cheilostome Bryozoa. *J. Paleontol.* **43,** 791–799.

Ryland, J. S. (1970). "Bryozoans." Hutchinson University Library, London.

Sandberg P. A. (1971). Scanning electron microscopy of cheilostome bryozoan skeletons; techniques and preliminary observations. *Micropaleontology* **17,** 129–151.

Sandberg, P. A. (1973). Degree of individuality in cheilostome Bryozoa: Skeletal criteria. *In* "Animal Colonies" (R. S. Boardman, A. H. Cheetham, and W. A. Oliver, Jr., eds.), pp. 305–315, Dowden, Hutchinson & Ross, Inc., Stroudsburg, Pennsylvania.

Sandberg, P. A. (1975). Bryozoan diagenesis: Bearing on the nature of the original skeleton of rugose corals. *J. Paleontol.* **49,** 587–606.

Sandberg, P. A. (1976). Ultrastructural clues to skeletal development in cheilostome Bryozoa. *Stockholm Contr. Geol.* **30,** 1–14.

Sandberg, P. A., Moore, M. C., Stieglitz, R. D., and Worsley, T. R. (1969). Relationship between ultrastructure and mineralogy in some cheilostome bryozoans. *Geol. Soc. Am., Program Annu. Meet.* Abstract, Part 4, pp. 72–73.

Schneider, D. (1957). Orientiertes Wachstum von Calcit-Kristallen in der Cuticula mariner Bryozoen. *Zool. Anz.* **21** (Suppl.), 250–255.

Schneider, D. (1963). Normal and phototropic growth reactions in the marine bryozoan *Bugula avicularia. In* "The Lower Metazoa" (E. C. Dougherty, ed.), pp. 357–371. Univ. of California Press, Berkeley.

Schopf, T. J. M. and Manheim, F. T. (1967). Chemical composition of Ectoprocta (Bryozoa). *J. Paleontol.* **41,** 1197–1225.

Silén, L. (1941). Cheilostomata Anasca (Bryozoa) collected by Professor Dr. Sixten Bock's expedition to Japan and Bonin Islands 1914. *Ark. Zool.* **33a**(12), 1–130.

Silén, L. (1942). Origin and development of the cheilo-ctenostomatous stem of Bryozoa. *Zool. Bidr. Uppsala* **22,** 1–59.

Silén, L. (1944). On the formation of the interzoidal communications of the Bryozoa. *Zool. Bidr. Uppsala* **22,** 433–488.

Söderqvist, S. T. (1968). Observations on extracellular body wall structures in *Crisia eburnea* L. *Atti Soc. Ital. Sci. Nat. Mus. Civ. Stor. Nat. Milano* **108,** 115–118.

Sorauf, J. E. (1971). Microstructure in the exoskeleton of some Rugosa (Coelenterata). *J. Paleontol.* **45,** 23–32.

Sorauf, J. E. (1974). Growth lines on tabulae of *Favosites* (Silurian, Iowa). *J. Paleontol.* **48,** 553–555.

Tavener-Smith, R. (1968). Wall structure and acanthopores in the bryozoan *Leioclema asperum. Lethaia* **2,** 89–97.

Tavener-Smith, R. (1969). Skeletal structure and growth in the Fenestellidae (Bryozoa). *Palaeontology* **12,** 281–309.

Tavener-Smith, R., and Williams, A. (1970). Structure of the compensation sac in two ascophoran bryozoans. *Proc. R. Soc. London, Ser. B* **175,** 235–254.

Tavener-Smith, R., and Williams, A. (1972). The secretion and structure of the skeleton of living and fossil Bryozoa. *Philos. Trans. R. Soc. London, Ser. B* **264,** 97–159.

Taylor, J. D., Kennedy, W. J., and Hall, A. (1969). The shell structure and mineralogy of the Bivalvia. Introduction, Nuculacea-Trigonacea. *Bull. Br. Mus. Nat. Hist., Zool., Suppl.* **3,** 1–125.

Thomas, H. D., and Larwood, G. P. (1956). Some "uniserial" membraniporine polyzoan genera and a new American Albian species. *Geol. Mag.* **93,** 369–376.

Thomas, H. D., and Larwood, G. P. (1960). The Cretaceous species of *Pyripora* d'Orbigny and *Rhammatopora* Lang. *Palaeontology* **3,** 370–386.

Towe, K. M., and Cifelli, R. (1967). Wall ultrastructure in the calcareous Foraminifera: Crystallographic aspects and a model for calcification, *J. Paleontol.* **41,** 742–762.

Utgaard, J. (1973). Mode of colony growth, autozooids, and polymorphism in the bryozoan order Cystoporata. *In* "Animal Colonies" (R. S. Boardman, A. H. Cheetham, and W. A. Oliver, Jr., eds.) pp. 317–360. Dowden, Hutchinson, & Ross, Inc., Stroudsburg, Pennsylvania.

Voigt, E. (1966). Die Erhaltung vergänglicher Organismen durch Abformung infolge Inkrustation durch sessile Tiere. *Neues Jahrb. Geol. Palaeontol., Abh.* **125,** 401–422.

Wada, K. (1972). Nucleation and growth of aragonite crystals in the nacre of some bivalve molluscs. *Biomineralization Res. Rep.* **6,** 141–159.

Wells, J. W. (1969). The formation of dissepiments in zoanthinarian corals. *In* "Stratigraphy and Paleontology, Essays in Honor of Dorothy Hill" (K. S. W. Campbell, ed.), pp. 17–26, Aust. Natl. Univ. Press, Canberra.

Wilbur, K. M., and Simkiss, K. (1967). Calcified shells. *Comp. Biochem.* **26A,** 229–293.

Wilbur, K. M., and Watabe, N. (1967). Mechanisms of calcium carbonate deposition in coccolithophorids and molluscs. *Stud. Trop. Oceanogr.* **5,** 133–154.

Williams, A. (1968). A history of skeletal secretion among articulate brachiopods. *Lethaia* **1,** 268–287.

Williams, A. (1971). Scanning electron microscopy of the calcareous skeleton of fossil and living Brachiopoda. *In* "Scanning Electron Microscopy: Systematics and Evolutionary Applications" (V. H. Heywood, ed.), pp. 37–66. Academic Press, New York.

Woollacott, R. M., and Zimmer, R. L. (1972). Origin and structure of the brood chamber in *Bugula neritina* (Bryozoa). *Mar. Biol.* **16,** 165–170.

6

Polymorphism

LARS SILÉN

The phenomenon known as polymorphism takes many peculiar and characteristic expressions in the bryozoans. For a long time it has held the interest of scientists, attracting them for functional, taxonomic, and phylogenetic considerations. An almost unmanageable amount of specific information on polymorphs has been amassed, mainly in the vast taxonomic literature, whereas serious attempts at a more comprehensive treatise are rather few and incomplete, directed as they are generally along one, or a few special lines. The terminology has tended to be partly illogical and our knowledge of structure too often is incomplete. Concerning the solid facts for judging the biological importance of polymorphism for the species, a statement by Busk (1854), in regard to the avicularia, still is valid to a deplorable degree. "With respect to their . . . functions, nothing but conjectures have been offered." In fact, our knowledge of the functions of the polymorphs consists to too large an extent, to this day, of inferences drawn from their structure. One thing safely can be presumed, however; in the bryozoans, polymorphism is represented to such an extent and takes such a multitude of often very specific expressions, that great biological importance must be ascribed to it. And it should be stressed that in the last few years—in a number of interesting papers—new observations and new points of view have been introduced rendering it possible to at least see the first signs of "light."

I. SOME ELEMENTARY DEFINITIONS

The bryozoans are represented by two types of units on different levels: the colony (zoarium) composed of fundamentally equivalent units, the zooids. The common type of zooid constituting the bulk of the colony contains all, or at least most, of the nutritive, reproductive, and other organs rendering it capable of functioning as one complete organism. Their property of being self-supporting makes these common zooids the unit model of the bryozoan colony and the basis of comparison when judging the deviating types of zooids often occurring with it in the same colony.

"Polymorphism is discontinuous variation in the morphology of zooids," an abbreviated version of the definition given by Boardman *et al.* (1969) seems to me the most neutral and satisfactory expression for the phenomenon. I say this because there appear in numerous bryozoans morphologically distinct zooid types within one and the same colony. The fact must be stressed that a number of organizational patterns may exist within a species, and that a zooid can be organized according to one or the other of these patterns, but nothing in between.

It can be noted that the term polymorphism is used sometimes in the sense of colonial polymorphism in the bryozoans to express the colonial variability produced by different environments (see, for instance, Harmelin, 1973c). This phenomenon will not be dealt with here.

II. DISTRIBUTION OF POLYMORPHISM AMONG BRYOZOANS

It is a striking phenomenon that in the major groups of bryozoans polymorphism is richly developed in the stenolaemates and particularly in the gymnolaemates, but is absent in the phylactolaemates. Certain structural features of the phylactolaemates have made it a generally held opinion that they represent the most primitive stage within the bryozoans. Could it be that the absence of polymorphism is one more expression of the general primitiveness of the group? Or is this absence of polymorphism dependent on the exclusive restriction of the phylactolaemates to freshwater (in contrast with the practically total restriction of remaining groups to marine environments)? If so, what is there in the marine and the limnal environments, respectively, which favors the appearance of polymorphism in one case, but prevents it in the other? Schopf (1973) has advanced the hypothesis that among marine cheilostomes stable environmental conditions favor polymorphism, while unstable ones do not. Freshwater, it is true, represents an extremely unstable environment in many respects. So the absence of polymorphism in the phylactolaemates might be explained by using the principles proposed by Schopf.

However, even in the cheilostomes, for example, where polymorphism is highly developed as a rule, species exist lacking one or more types of the polymorphs present in the group. This fact is valid not only for single species, but also sometimes for genera, or even entire families such as Membraniporidae and Electridae. A plausible conclusion seems to be that among the cheilostomes, although characterized as a group by the ability to develop polymorphs, forms exist lacking this hereditary capacity (primarily or secondarily?) where the absence of one or more polymorphs is not due to an absence of the conditions inducing the formation of polymorphs (see Section XVII and Fig. 44).

Another problem is presented by the facultative presence or absence of a certain polymorph (avicularia) in certain widespread cheilostome species from various regions (cf. survey by Schopf, 1973). Could this difference between different populations of one and the same species be attributed to a genetic difference that has not been expressed (at least as far as has been observed) in other morphological features? Schopf and Gooch (1971) have demonstrated that identical morphologies do not preclude genetic differences, even in closely proximate populations of *Schizoporella errata.* Another explanation might be that the environmental differences affect such intraspecific differences as polymorphism. A third, and possibly the most reasonable assumption, might be that these differences are due to interactions between gene pool and environment. It might be tempting to speculate about the nature of such possible interactions. However, I think such speculations are premature at this point, as the present speculations have already far surpassed our existing knowledge.

III. TYPES OF POLYMORPHS: GENERALITIES

In view of the numerous types of polymorphs described, an attempt to systematize them is a task as ambiguous as necessary. It is not made easier by the confusing terminology in use, based as it originally was on opinions often more than a hundred years old.

As indicated above, the basis for polymorphs is represented naturally by the anatomically complete zooid capable of feeding itself. This zooid traditionally is called the *autozooid.* Its counterpart is the *heterozooid* in which the feeding apparatus is incomplete or absent; consequently heterozooids must be nourished by autozooids.

The term heterozooid is often used synonymously with the term polymorph. Although it is true that polymorphism takes its most drastic expressions in the heterozooids, polymorphism exists among the autozooids as well. Observations of such cases have been advanced particularly in the last few years. This, together with earlier known facts the significance of which in the present respect has not always been fully understood, makes me suspect that polymorphism will prove a far more common phenomenon within the autozooids in the future than yet realized. Because of these circumstances, it is necessary to clarify the pertinent definitions.

The term autozooid, when used in the strictest sense, should designate, in all respects, a self-supporting zooid. The two most significant prerequisites are the presence of a complete apparatus for feeding and metabolism, as well as a hermaphroditic propagation apparatus. Among the autozooids, there can exist autozooidal polymorphs, or autopolymorphs–zooids which possess a functional feeding apparatus but deviate in one respect or the other from the ideal autozooid mentioned above.

The heterozooid is, of course, always a polymorph, by definition. However, there may be circumstances where it might be preferable, perhaps, to use the term heterozooidal polymorphs, or heteropolymorphs (making too much of a good thing though it may seem) to avoid confusion when autozooidal polymorphs are considered simultaneously.

Schematically, the polymorphs can be grouped into four main morphologically defined categories:

1. Autozooidal polymorphs, which possess a functioning feeding apparatus but are specialized in one or several respects. An extremely heterogeneous group of polymorphs functionally and presumably also phyletically.

2. Avicularia (vibracula included), in which the feeding apparatus is transformed into a sensory body, and the operculum, with its pertaining muscles and adjacent parts of the body, is enlarged and transformed into a grasping or sweeping apparatus. Certainly a phyletically homogeneous group of heterozooidal polymorphs with fundamentally similar functions.

3. Kenozooids, in which there is no feeding apparatus or other interior organs. Thus, the fundamental definition is totally negative. A very heterogeneous group of heterozooidal polymorphs functionally and, above all, phyletically.

4. Gonozooids, a quite heterogeneous group of autozooidal and heterozooidal polymorphs specialized for different reproductive functions.

Some further reflections are called for before attempting to systematize the numerous types of polymorphs hiding behind this preliminary sketch. When choosing the main criterion for the final classification of polymorph types, several courses are theoretically possible. To use a purely morphological criterion is unmanageable. It would lead to an arrangement where, to choose only one example, the autozooidal polymorphs would be placed in one group without any regard to the high number of unrelated types represented among them, nor to the relations of these types, at least functionally, to different types of heterozooidal polymorphs.

An arrangement along phyletic lines would lead to similar unsatisfactory results. We can presume that the polymorphs of the gymnolaemates and the stenolaemates have originated quite independently. The gonozooids of the stenolaemates, for instances, are too different in every respect from those seen in the gymnolaemates to draw an assumption of a common origin. Further, polymorphism of the gymnolaemates and stenolaemates, respectively, is the result of a fundamentally different astogeny. Even within the gymnolaemates, the gonozooid types found are, in all probability, heterogeneous. Many other examples could be added. Therefore, I have chosen function as the main criterion, in spite of the fact that it certainly has its shortcomings as well.

The functional aspects considered will be active defense, passive defense, colony strengthening, colony support, survival of unfavorable conditions, interzooidal connections, sexual reproduction, and ancestrulae. Further aspects certainly could be added. However, I hope the most characteristic expressions of polymorphism will be included in this arrangement.

IV. ACTIVE DEFENSE: AVICULARIA AND VIBRACULA

The prerequisites for realizing this particular level of heterozooidal organization are met only within the autozooids of the cheilostomatous groups of gymnolaemates; the operculum found in them is the organ hypertrophying into the mandible of the avicularium or the seta of the vibraculum. As all other bryozoans lack an operculum, this polymorph is consequently absent there. [Similar structures observed in the extinct stenolaematous Meliceritidae offer a problem still unresolved, according to Levinsen (1912) and Ryland (1970).] These zooids occur in a great many shapes often of a curious appearance. This makes them the heterozooids studied most intensely over a long period of time, especially since their structure and position in the colony offer characters very useful in taxonomy.

However, whatever the variations, the morphological kernel of the avicularia—

the sensory body, the mandible and its muscles and functions, the rostrum—is so identical in all of them that they certainly must be presumed to represent one heterozooidal group of common origin (Fig. 1). The vibracula as a group differ from the avicularia in the prolongation of the operculum into a narrow seta, pivoting against the front of the zooid in a different way, and performing horizontal, sweeping movements instead of the vertical opening and closing movement of the avicularian mandible. Nevertheless, the fundamental similarities are great, and obvious transitional types exist between avicularia and vibracula so that there should be hardly any hesitation in considering them one homogeneous group of heterozooids (Silén, 1938; Hastings, 1963).

Besides the well-known morphological characteristics of the avicularia/vibracula briefly referred to above, one structural detail calls for a closer scrutiny, viz., the body volume. As a general rule this heterozooid is smaller than the autozooid of the same species (Silén, 1938). Even in the few cases of avicularia seemingly larger than the corresponding autozooids, a close scrutiny will disclose that this seeming superiority in size of the avicularium is explained always by the hypertrophied development of one or more of the structures always enlarged in the avicularium as parts of the grasping apparatus. The principally smaller body size of this heterozooid will prove of importance in the discussion (Section XVII) of the factors inducing the development of a zooid into exactly this type of heterozooid.

The avicularia/vibracula take defined positions in the colony (Fig. 2). The laws regulating their formation in exactly these positions (and no other ones!) likewise will be discussed in another context later (Section XVII).

Three main types of avicularia are commonly distinguished, viz., *vicarious, interzooidal,* and *adventitious,* terms referring to their position in the zoarium. The vicarious avicularium is formed by the same distolateral budding process as the autozooid; consequently, it holds the same position in the zoarium and is as able as the autozooid to form autozooids by further budding. The adventitious avicularia are situated usually as small bodies on the free (frontal and lateral) walls of the autozooids and are, as a rule, incapable of further budding, giving them an appearance of mere appendages. The term interzooidal is used for comparatively small avicularia wedged in between the autozooids. These are formed by lateral budding, as are the vicarious ones. However, they do not give rise to new autozooids. Sometimes they replace pore chambers. The vicarious and interzooidal avicularia intergrade, and the latter term may be superfluous (Hastings, 1963; Powell, 1966).

Two or even more distinct types of avicularia can be present in one and the same species [cf. for instance, the Reteporidae; see Buchner (1924)]. On the other hand, there are species and entire groups of species within the cheilostomes totally lacking avicularia, in spite of the presence in them of the same structural prerequisites for the formation of avicularia as in the aviculiferous

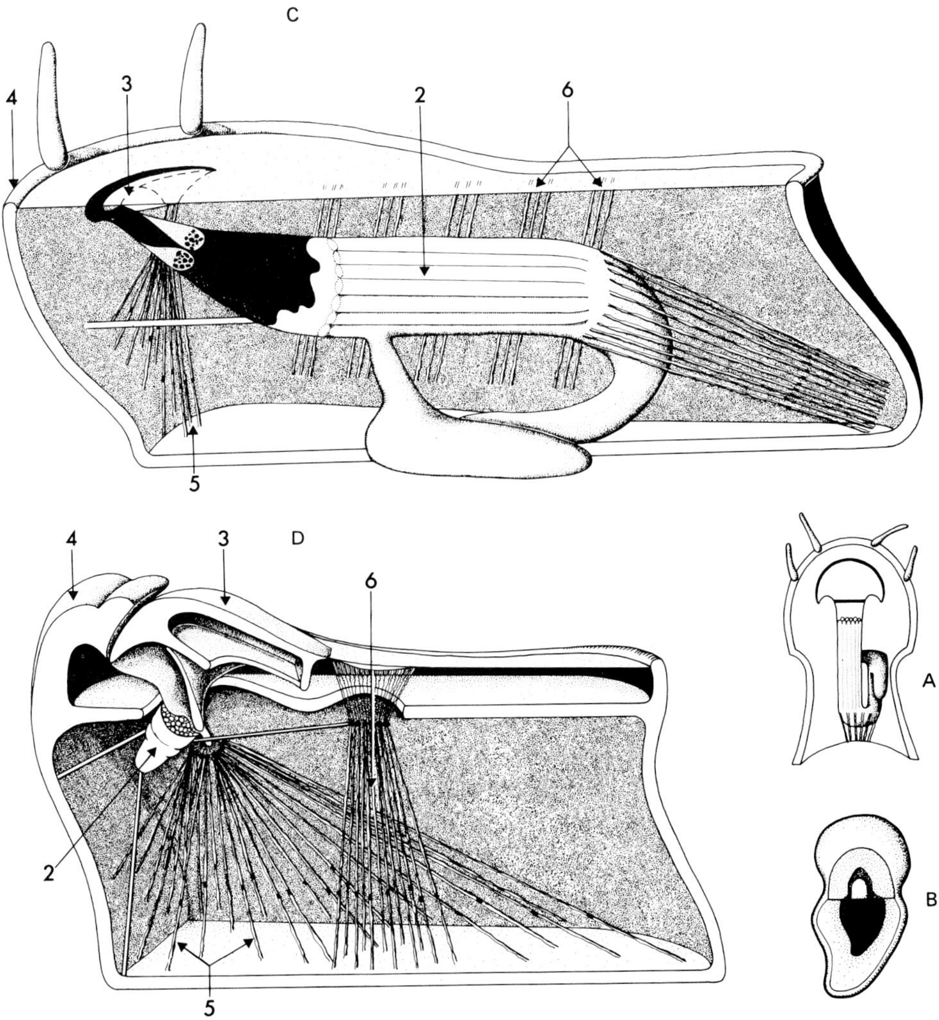

Fig. 1. *Flustra foliacea.* Autozooid (A) and vicarious avicularium (B) in frontal view. Diagrammatic sagittal sections of autozooid (C) and avicularium (D). Redrawn from Silén (1938). Main structural deviations of avicularium from autozooid. (1) Body volume reduced; (2) polypide reduced into sensory body; (3) operculum enlarged into mandible; (4) walls of distal part strengthening into rostrum; (5) mm occlusores operculi enlarged into mm adductores mandibuli; (6) mm parietales concentrated into mm abductores mandibuli; (7) sexual organs disappeared. (1) and (7) Not shown.

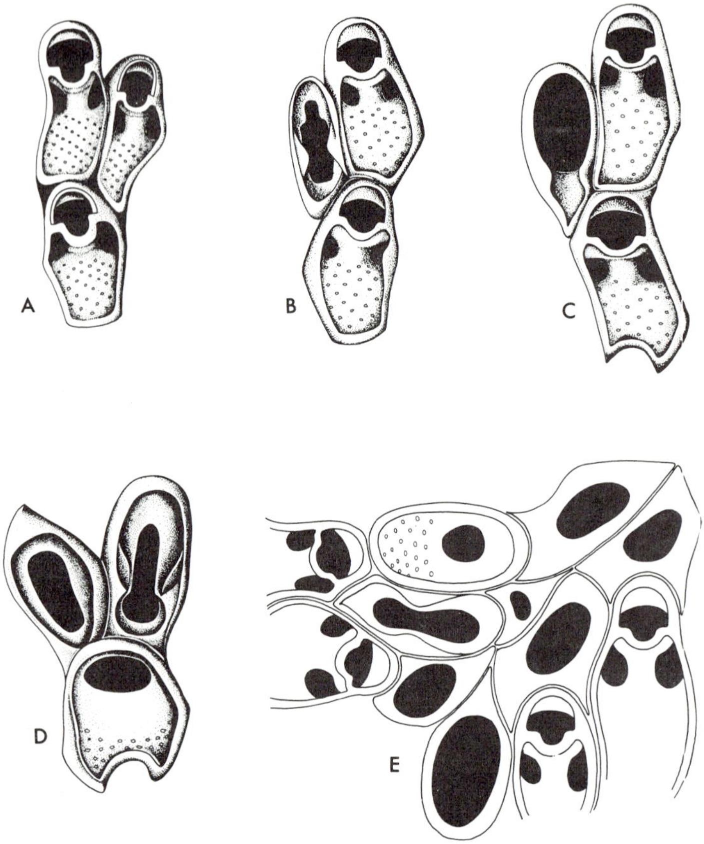

Fig. 2. *Thalamoporella rozieri.* Induction of heterozooid formation by factor in colony structure, here consisting in size reduction of distolateral bud at bifurcation of zooid row, resulting in formation of autozooid (A), avicularium (B), or kenozooid (C); D, both buds reduced, larger one avicularium, smaller one kenozooid; E, disturbance of colony structure resulting in irregular formation of avicularium and kenozooids. A hierarchy autozooid–avicularium–kenozooid is seen directly related to the gradation of the reduction of body size or the deformation of body. Redrawn from Powell and Cook (1966).

forms. D. F. Soule (1973) has made a preliminary study of the morphogenesis of the giant adventitious avicularia occurring in certain ascophoran bryozoans, especially Smittinidae, which either replace frontal avicularia of a "normal" smaller size or are present with them.

The special structure of the avicularium has always been explained as expressing a special function of this zooid and several hypotheses have been advanced concerning the precise nature of this function (cf. the survey of Kaufmann,

1971). However, the most commonly accepted hypothesis always has been that the avicularium has a grasping function, in addition to the protection of the colony against foreign organisms. Recently, this interpretation was confirmed by Kaufmann's study of the faculties and functions of the avicularium of *Bugula*. This study is the more valuable as Kaufmann succeeded in defining the function more exactly as constituting a preferential defense against tube-building crustaceans about 1–2 mm in size. He also managed to define the effect of the *Bugula* avicularium qualitatively, in that a first step was taken toward a quantitative approximation of the work performed by the avicularium. This is, to my knowledge, the first attempt to reach a distinct expression of the biological importance of a polymorph in the bryozoans and represents a certainly not easy, albeit very promising, way of attacking the problem (cf. Section XVIII). Obviously, the functions found in the *Bugula* avicularium are not expected to be necessarily identical in all avicularia. The fact that in certain species one and the same colony possesses more than one type of avicularia may be interpreted as indicating differences in their mode of functioning. Such differences probably indicate no fundamental deviation from the main task of defense against foreign organisms, but may indicate differences in type and/or size of the "prey."

Wyer and King (1973) found that the avicularia of *Bugula* and *Flustra* do not seem to be a defense against pycnogonids, neither such as *Achelia echinata* directly feeding on *F. foliacea,* nor those solely walking on the colonies or feeding on the epiphytes, epizoites, or accumulated detritus on the colonies.

Some avicularian types are provided with a rounded mandible, others with a pointed one. Buchner (1924) and Schopf (1973) considered the rounded mandible more primitive, as its precursor, the operculum, is always rounded. Further, Schopf considered the pointed mandible a specialization increasing the effectiveness of the avicularium; it possibly may be so. However, the hypothesis ought to be given closer scrutiny before being accepted. Certain species of Reteporidae, for instance, possess two or more avicularian types, some provided with rounded, others with pointed mandibles. The immediate assumption is that the "rounded types" and the "pointed types" do not represent different levels of effectiveness per se, but rather different degrees of effectiveness against different preys, thus differentiations for meeting different demands.

Most avicularia, as entire bodies, are immobile. They sit waiting for the prey to enter the open jaws and touch the trigger, i.e., the sensory body. The so-called bird's-head avicularia of the Bicellariellidae are an exception to the rule, performing as they do movements which have to be interpreted certainly as increasing the effectiveness of the grasping action. (See Figs. 3–6.)

The present author assumed (1950), for certain morphological reasons, that the avicularia of the bicellariellidan genera *Beania* and *Dendrobeania* are immobile, though in the bird's-head type the "head" is joined to the mother zooid by a slender stalk. However, this is not true. In 1971, I had the opportunity to observe living specimens of *Beania magellanica* at Laboratoire Arago, France (Silén, unpublished). If undisturbed, the avicularium remains immobile, its front facing frontally and its beak pointing distally, thus fundamentally taking the same position as the autozooid; its stalk is straight, its mandible open (Fig. 4). When the sensory body is touched, the mandible snaps shut and the avicularium simultaneously jerks forward and then backwards again into the starting position. The entire movement is so sudden and swift that it can be missed easily. After awhile, the mandible snaps open again and remains open without any movement of the avicularium as a whole. No muscles were observed in the stalk, and no explanation of the mechanism can be ventured upon for the present. No doubt, the forward jerk might be assumed to improve the possibilities of grasping animals passing by.

The mode of function of the bird's-head avicularia of the likewise bicellariellidan *Bugula* and *Bicellariella* (Fig. 3), as analyzed by the author (1950) and Kaufmann (1971), may be interpreted as a further improvement of the action seen in *Beania.* Here the avicularia are swaying rhythmically and incessantly to and fro, with the mandible open. When the sensory body is touched, the "head" instantly jerks forward and the mandible simultaneously shuts, remaining so for a while. Then the mandible slowly opens again and the rhythmic movement resumes. As to the muscular actions effecting the swaying movement, the final answer must be left to the future, the quoted authors being of different opinions. The difficulty is that no muscles have been hitherto observed in the stalk, and thus it can hardly be imagined that the bending action is performed by direct muscular action.

Rey (1928) made an attempt to examine the influence of mechanical and chemical stimuli on the actions of the *Bugula* avicularium. The results were too preliminary to allow distinct conclusions. To my knowledge further attempts have not been made in this direction though they might possibly be fruitful.

There exist certain morphological indications that the function is not exclusively a grasping one in all avicularia. Thus, in some species the mandible is prolonged to such a degree that it protrudes far beyond the rostrum (for instance, in *Microporella* and *Mastigophora pesanseris;* Figs. 5 and 6). This elongation of the mandible and the absence of a fully rigid maxilla preclude an effective grasping function when the mandible is closed and indicate that these avicularia have a different function. However, the nature of this function is an open question and further study is highly desirable.

In this connection, one further question is of interest: What is the significance of the large glands characteristic of the avicularia of the Reteporidae? There seems to be no answer to this query at present.

The vibracula are especially characteristic of the Scrupocellariidae (Fig. 7), and certain lunulitiform genera such as *Discoporella* and *Cupuladria,* among others (Fig. 8). In the Scrupocellariidae, a stimulation of the sensory body starts

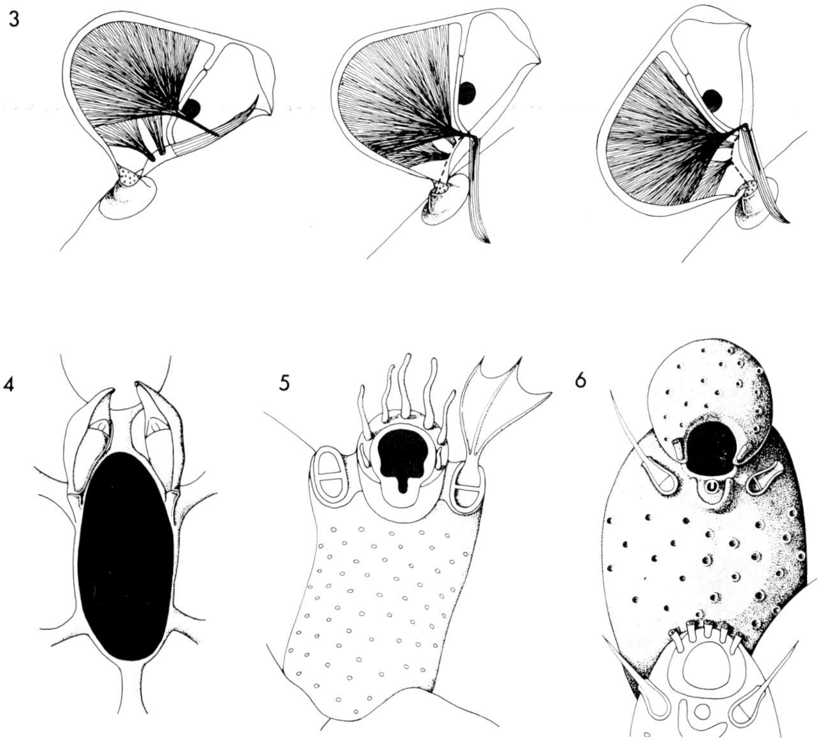

Figs. 3–6. Avicularia.

Fig. 3. *Bugula.* Movements of avicularian head and mandible. Redrawn from Silén (1950).

Fig. 4. *Beania magellanica.* Pair of stalked avicularia facing frontally. Redrawn from Osburn (1950).

Figs. 5 and 6. *Mastigophora pesanseris* and *Microporella marsupiata.* Avicularia with mandible protruding beyond rostrum. Redrawn from Osburn (1952).

a short series of movements of the seta: it sweeps to and fro closely over the surfaces of the zoarium. The movements of the setae are not coordinated among the vibracula of a colony (Silén, 1950). The action is interpreted, traditionally and as it seems reasonably, as a cleaning function by which detritus and/or small organisms are removed. The cup-shaped colony of *Cupuladria* rests free on a substratum of sand or the like. It is raised a little above the bottom by the setae of the peripheral vibracula; if covered by sand, the colony is raised to the surface of the substratum again by the work of the setae which then clean the surface of remaining particles; if occasionally turned upside down, it sometimes can regain its normal position by the action of the setae (Cook, 1963). Though the effects of the vibracular activity may seem rather different in the case of *Scrupocellaria*

7

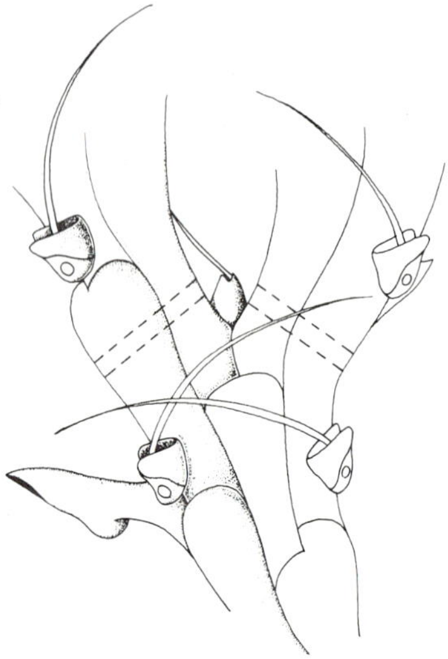

8

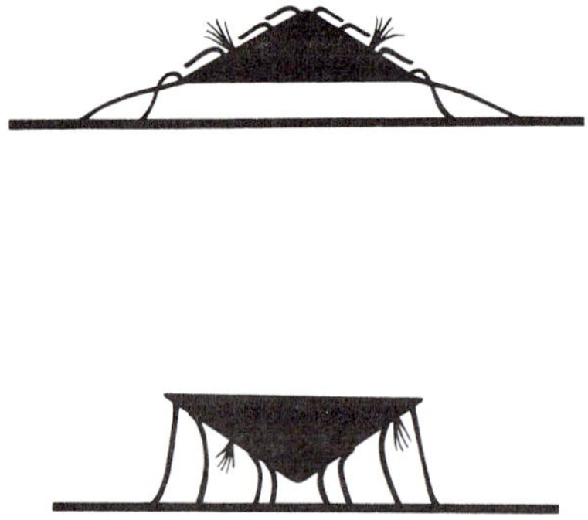

9

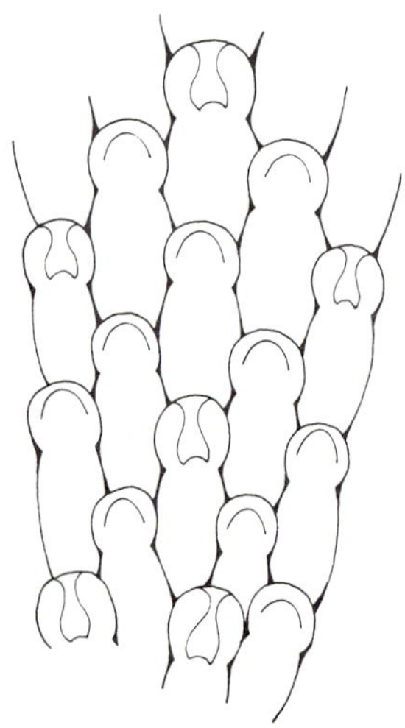

Fig. 7. *Scrupocellaria varians.* Lateral and axial vibracula, lateral giant avicularium. Redrawn from Osburn (1950).

Fig. 8. *Discoporella umbellata.* Colony raised above substrate by setae of marginal vibracula; basal side downward (top) and upward (bottom). Redrawn from Cook (1963).

Fig. 9. *Steginoporella magnilabris.* Size and position of A-zooids (operculum simple) and B-zooids. Redrawn from Silén (1938).

and of *Cupuladria,* it is easily seen that they can be referred to the same basal function, viz., a sweeping away of foreign objects.

Banta (1973) has examined the possible ways along which nourishment can be delivered to frontal (adventitious) avicularia in ascophoran bryozoans by their mother zooids. In the majority of cases the avicularia were found to be connected with the mother zooids by the marginal pores of the fronts of the latter. However, there seem to exist exceptions for which, at present, only more or less theoretical explanations can be advanced.

A treatise on the avicularia is not complete without a few words on the zooids described as B-zooids in the Steginoporellidae (and reported from a few other genera as well, cf. Hyman 1959, p. 327) (Fig. 9). These zooids possess a fully developed feeding apparatus, and thus are autozooidal polymorphs and do not differ, as far as is known, from the "normal" A-zooids in function, but only in their enlarged opercula and some modification of the relevant muscles. According to the nomenclature proposed in the present paper, they are, consequently, autozooidal polymorphs. According to the traditional opinion (Harmer, 1926; Marcus, 1949; Hyman, 1959; Ryland, 1970), the B-zooids represent a developmental stage on the way to the formation of true "vicarious" avicularia. This opinion is based mainly on the modification of the opercular apparatus indicated above, as well as on the fact that they are lacking in the single known species, *S. mandibulata,* which possesses characteristic avicularia. However, this interpretation of the B-zooids does not seem convincing (Silén, 1938). In all other features they coincide with autozooids and differ from the avicularia. Possibly the most distinct difference from the avicularia, including those of *S. mandibulata,* is the fact that they take a quite normal position in the zoarium (none of the positions is characteristic of the avicularia) and their body volume has not diminished at all. If the attempt to explain the trigger mechanisms effecting the avicularium formation holds true (cf. Section XIII), it cannot include the B-zooids. At least until the special function, if any, of the B-zooids is revealed, they ought to be placed as a special category within the autozooidal polymorphs (cf. also Section XIII).

V. CLEANING: NANOZOOIDS

The vibracula serve as cleaning structures, as related above. Nevertheless, their close relations to the avicularia make it necessary to treat both vibracula and avicularia in the same category. However, there remains one distinct type of polymorph that can hardly be related to another category. It is represented by the nanozooids of the cyclostome genus *Diplosolen* which have been well-known for a long time (survey in Borg, 1926) but the function of which has long remained an enigma. Nanozooids are identical in shape with the normal zooids among which they are scattered according to a certain pattern, only being much smaller. The polypide is greatly reduced and does not possess a feeding function; thus the nanozooids are heterozooids. They have no sexual functions. However, Silén and Harmelin (1974) have shown that the nanozooids have a cleaning function, their single tentacle sweeping away detritus from the colony surface.

"True nanozooids" are regularly formed in the common bud, whereas "occasional nanozooids" have been seen in *Plagioecia sarniensis* and *P. dorsalis,* and are formed irregularly in connection with disturbances of colony growth and are possible indications of the mode of origin of the "true nanozooids." In *P. sarniensis* and related species the old zooids are closed by a calcareous diaphragm provided with a small tubule; they contain a reduced polypide with one tentacle. These "secondary nanozooids" represent a unique phenomenon within the bryozoans as they start life as autozooids and then turn into heterozooids, the original feeding polypide degenerating and being replaced by a much reduced one. The functions, if any, of the "occasional" and "secondary nanozooids" are obscure.

The facts now related concerning the nanozooids may prove of importance for the understanding of at least some of the numerous structures described as heterozooids, sometimes even called nanozooids, in the stenolaemates.

VI. PASSIVE DEFENSE: SPINOZOOIDS

Kenozooids occur in the majority of gymnolaemates and in numerous stenolaemates. They appear in a quite confusing number of morphologically differing forms, certainly with quite different functions and probably often of separate origin. It is highly desirable to grasp at least some basic arrangement of this mass of types. Such an attempt will be made here. This undertaking demands an introduction of some new terms for the sake of uniformity.

The spines commonly appearing in the gymnolaemates and certain stenolaemates (Crisiidae) are called spinozooids (Figs. 10–18). In the gymnolaemates the spinozooids appear as slender processes emanating from the free walls of the autozooids. They are usually shorter and always considerably narrower than the autozooids and have a much smaller volume. They lack the ability to bud further and thus each corresponds to a single zooid. Their counterpart among the avicularia is the adventitious type. The spinozooids of *Flustrella* (Fig. 16) probably are related to the interzooids (cf. Section X).

That the nature of the spines was zooidal was introduced by the present author (1944a) and was confirmed by several authors later. However, there is the possibility that there also exist spine-shaped processes which are mere protuberances from the autozooids and to which one cannot ascribe a zooidal nature. Further research is necessary to determine if such nonzooidal spines actually exist and, if so, to draw the exact limits between the two fundamentally quite separate categories of spines.

When dealing with dwarfed zooids of this and other types, we touch upon a problem seemingly easy in theory to solve but intricate in practice: How does one distinguish between zooids and mere evaginations from a zooid? There seems to be one objective way to do this, viz., to decide whether or not the body in question is separated from its mother zooid by a pore plate. However, this is not always an easy task. If the body has a narrow connection with its mother zooid, an existing pore plate can be reduced to an often quite trifling constriction. The possibility also exists that the pore plate is totally lacking even in cases when the body should be interpreted as a dwarfed zooid. An indirect, though more subjective and ambiguous method, is to try to decide whether the body studied, for instance, a spine, has a position where, in the same or other species, is found a body of undoubted zooidal nature, for instance, an autozooid (Fig. 38) or an avicularium (Figs. 39 and 40). Nevertheless, cases exist in which a definite decision on this point cannot be reached as yet.

Concerning the function of the spinozooids, their structure and position, in the majority of cases at least, are such as to make the quite reasonable assumption that they form a passive defense—as compared with the active defense by the avicularia/vibracularia—against foreign objects and organisms for the zoarium as a whole and/or especially for the single zooids. The most illustrative example is in the spines which in numerous anascan gymnolaemates are situated around the thin-walled, and thus particularly damageable, aperture field, or at least at the aperture (Figs. 10–13). Sometimes they form a veritable palisade. For example in the Cribrimorpha and others, they grow horizontally over the aperture field and fuse into a protective cover that in most species forms a perforated armor (Figs. 14 and 15).

Sometimes the spinozooids take very special forms, making the protective function obvious. Such is the case in many Scrupocellariidae, for instance, where one spine widens into a flat shield, the scutum, more or less covering the frontal membrane (Fig. 17), and in some Chapperiidae, where one spine branches to form a similar cover (Fig. 15). Stebbing (1973) has demonstrated that "spines" (spinozooids or spiny protuberances) assist in preventing overgrowth by other bryozoans and organisms. However, perhaps one reservation should be added. In some species a special function is not obvious for the spinozooids. Theoretically, at least, such spinozooids might be remnants of a more ancient type of budding, lacking any current function—remnants so neutral as to advantage or disadvantage that they have not been obliterated wholly during evolution.

A protective function, though of a special kind, probably should be ascribed to certain spinozooids, viz., for the protection of embryos. There are indications that the ectooecium, the exterior fold of the hyperstomial ooecium common in

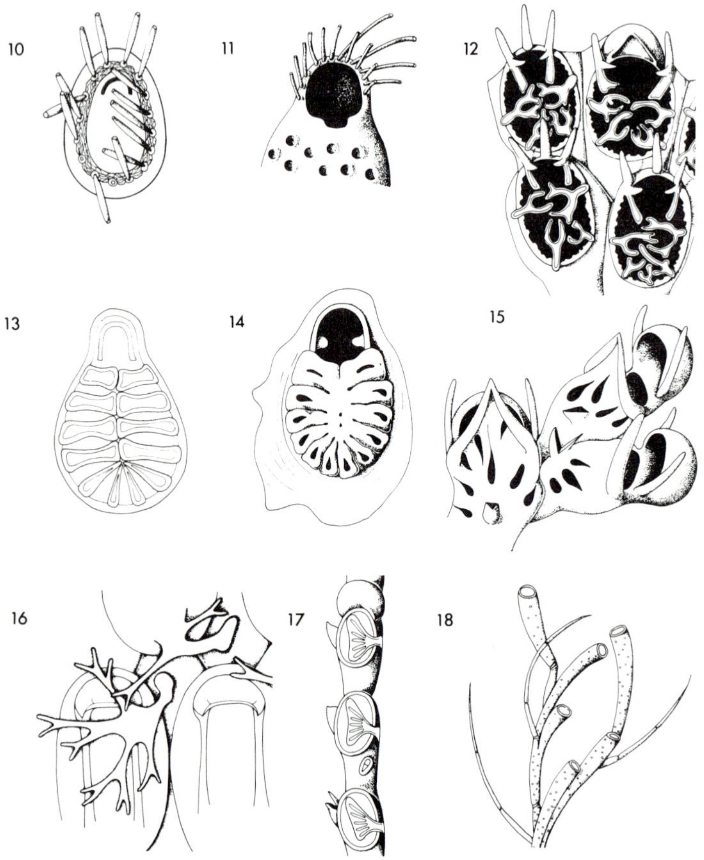

Figs. 10–18. Spinozooids.

Fig. 10. *Callopora whiteavesi.* Circumopesial row of spinozooids. Modified from Levinsen (1916).

Fig. 11. *Dakaria biserialis.* Circumapertural row of spinozooids. Redrawn from Osburn (1952).

Fig. 12. *Membraniporella aragoi.* Branching circumopesial spinozooids. Redrawn from Marcus (1938).

Figs. 13–15. Formation of frontal shield by fusion of spinozooids, in *Membraniporella nitida* [redrawn from Smitt (1868)], *Figularia figularis* [redrawn from Silén (1941)], *Chapperia frontalis* [redrawn from Osburn (1950)].

Fig. 16. *Flustrella gigantea.* Spinigerous kenozooids (spinozooids) between autozooids. Redrawn from Silén (1947).

Fig. 17. *Scrupocellaria scabra.* Flattened spinozooids, scuta, shielding opesiae. Redrawn from Silén (1938).

Fig. 18. *Crisidia cornuta.* Redrawn from Osburn and Soule (1953).

the gymnolaemates, is formed by one or two processes from the autozooid, processes closely reminiscent of spinozooids (Fig. 19). This interpretation possibly gets some support from the fact that in certain species where the frontal membrane or its distal end is surrounded by typical spinozooids, the two distomedial ones are missing in the ooeciferous zooid, the inference being that they have been absorbed in the formation of the ooecium (Silén, 1944b; Harmelin, 1973b).

I would be inclined to consider this picture of the formation of the ectooecium as generally valid—disregarding the statements by Calvet (1900) and Levinsen (1909) that it is formed from the zooid distal to the maternal one—were it not for the paper by Woollacott and Zimmer (1972) on the formation of the ooecium in *Bugula neritina* (Fig. 20). These authors demonstrated with indisputable accuracy that the exterior fold is represented by a kenozooid originating from the distal zooid. One direct consequence of this finding is that the picture first given here of the formation of the exterior fold cannot be held generally valid. Thus the problem must be: Are the conditions found in *Bugula* generally valid and the former picture wholly invalid? So similar do the hyperstomial ooecia look all through the gymnolaematous system that it would not seem a likely assumption that they could exhibit different constructions. On the other hand, there exist species with typical hyperstomial ooecia, such as *Scruparia* and *Hippothoa flagellum* (Figs. 21 and 22), where the ooeciferous zooids are separated widely from other zooids and are not followed by a distal zooid; it is difficult to imagine how in this case heterozooids emanating from zooids other than the maternal ones can take part in the formation of the ectooecial fold.

According to a recent preliminary investigation by D. F. Soule (1973) the ooecia of certain Smittinidae are formed partly from the maternal zooid, partly from the distal one. The latter "is connected to the ovicell by communication pores." It is not wholly clear whether this fact indicates a heterozooidal nature of the parts derived from the distal zooid. For the time being, we apparently have to be content with the possibility that in the hyperstomial ooecium, the ectooecium is formed in different ways in different species. This is obviously, however, a point on which further research is needed badly.

One fact remains invariant in these otherwise controversial descriptions: The exterior fold of the ooecium is made up of kenozooids, possibly modified spinozooids. The ooecial vesicle emanating from the maternal zooid "is a simple evagination of the body wall, not a modified individual" (Woollacott and Zimmer, 1972).

In the peculiar "*Electra*" or *Tendra zostericola*, the ovum is transferred to an embryo chamber formed between the frontal membrane and a cover of what is undoubtedly a row of spinozooids above the membrane. This case seems to be

19

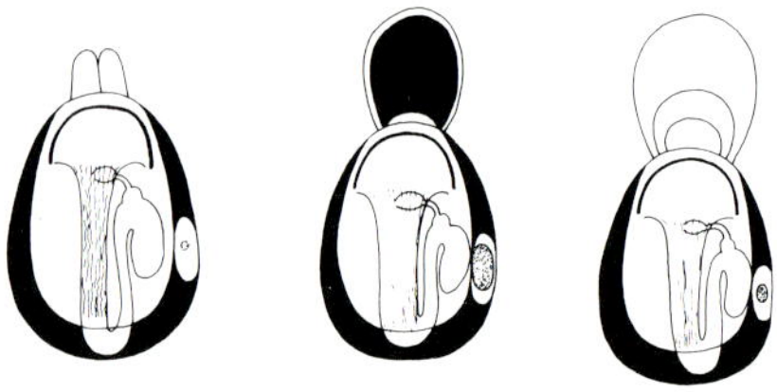

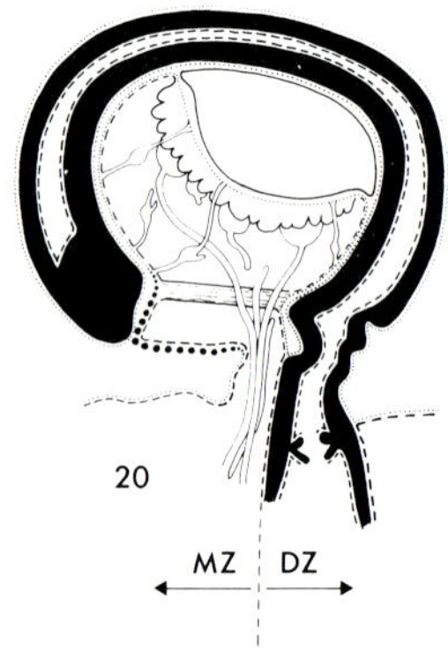

20

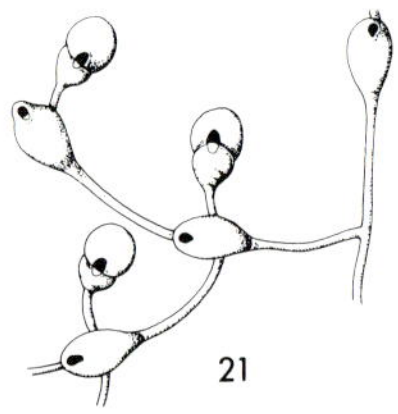

21

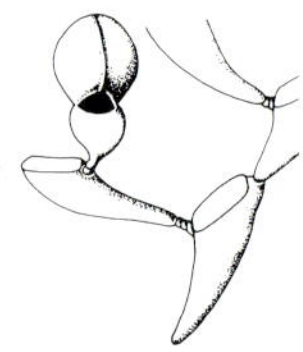

22

Figs. 19–22. Formation of ectooecium.

Fig. 19. *Callopora dumerili.* Formation of ectooecium by paired protrusions (spinozooids?) from maternal zooid. Redrawn from Silén (1944b).

Fig. 20. *Bugula.* Section through hyperstomial ooecium. Ectooecium (thick black line) a heterozooid emanating from zooid (DZ) distal to maternal zooid (MZ). Redrawn from Woollacott and Zimmer (1972).

Figs. 21 and 22. *Scruparia chelata* and *Hippothoa divaricata:* hyperstomial ooecia formed by gynozooids though distal zooids are absent. Redrawn from Hincks (1880).

unique and obviously is not connected with the true ooecia. The function of spinozooids for the protection of the embryo is, nevertheless, the same (cf. further Braiko, 1967).

As to their positions, there are two kinds of spinozooids in the gymnolaemates: some of them situated frontally and remnants of an ancient, nearly obsolete budding pattern, and others situated along the distolateral sides of the mother zooids and, in colonies of a strict astogeny, constituting remnants of potential, though suppressed, zooid rows (Silén, 1944a). Thus, these two categories should be diphyletic in origin, if these two budding patterns are perceived as strictly separate.

The spinozooids of many Crisiidae (the processus spiniformes of Borg) "may be regarded as transformed internodes consisting of one single zooid" (Borg, 1926). Considering the wholly different astogeny of the stenolaemates, they hardly can be considered equivalent to the spinozooids of the gymnolaemates. In forms such as *Crisidia* (Fig. 18) they are numerous and long and are situated in such a way that it may be assumed that they serve some kind of function, for instance keeping away foreign bodies from the colony as a whole.

Emschermann (1972, 1973) has suggested an ion regulatory function for the spines in the entoprocts. Despite the entirely different origins of these structures and of the spinozooids of the bryozoans *sensu strictu* it might be worthwhile to examine the spinozooids, or at least some types of them, from this point of view as well.

VII. COLONY STRENGTHENING: INTERCALATED AND MARGINAL KENOZOOIDS

These kenozooids exhibit such a variety of types of morphology, position, and probable origin that it seems premature at present to try to introduce a specific system of nomenclature suitable for them. The introduction of a new term such as vicarious kenozooids, or even vicariozooids, might be adequate in a great many gymnolaemates, in analogy with the term vicarious avicularia and with the same reservations as advanced above concerning that term. The term would refer to kenozooids taking the place of autozooids in the colony and being capable of further budding. These often appear in encrusting species with a comparatively free mode of growth (for example, species of *Electra*) in places where the growth is disturbed by irregularities of the substratum or by other causes. However, in numerous other forms with a more regular, often strongly accentuated zoarial pattern, they appear regularly in more or less exactly defined positions. This is especially the case in erect species, as is well demonstrated by members of the Flustridae. Their thin, leaflike colonies possess, in many species, a marginal zone of vicariozooids, which grows, gradually broadening, downward into the ramification points where there may be a zone of three or more vicariozooids with particularly thick walls.

Generally, a vicariozooid corresponds to the autozooid not only in position but also in size in the sense that a single vicariozooid clearly corresponds to a single autozooid. Usually, the vicariozooid is smaller than the autozooid, at least in volume. Vicariozooids are often variable in shape, thus differing more or less from the autozooids and, as already indicated, their walls, particularly free ones, may be thickened. The aperture field exhibits all sorts of variations, being vast to wholly obliterated.

Sometimes marginal vicariozooids (*Securiflustra*) may surpass the autozooids two or more times in length. Each vicariozooid no longer corresponds to one autozooid but to a number of autozooids in a zooid row between which the transverse septa have failed to appear. Such zooids are, so far, transitional to rhizozooids as defined in Section VIII. The function, if any, of the vicariozooids is that in erect colonies they contribute to the rigidity of the colony as a whole and render its margins more resistant to mechanical disturbances from the environment, as shown, for instance, by the flustrid case mentioned above.

In the stenolaemates, a considerable number of kenozooid types have been described under different names judged superfluous by Brood (1972) in his valuable survey of them. "The supposed main function of the kenozooids is in fixing the zoarium to its substratum or in forming a dorsal thickening in order to strengthen the stems." Some of them seem intercalated between the autozooids and, if a special function should be ascribed to them, it would be for rendering the colony more rigid (cf. also Astrova, 1973; Viskova, 1973; Morozova, 1973).

VIII. COLONY SUPPORT: RHIZOZOOIDS, STYLOZOOIDS, AND STOLONS (?)

Rhizozooids is a modified term for the kenozooids appearing in the literature under such names as rhizoids (Borg, 1926), rootlets, or radical fibers. They are circular in transverse section and tubular. Except when traces of an aperture field are seen, the body wall is uniform. Rhizozooids form tubes connected with the colony body only by their areas of origin and otherwise extend free from the colony to the substratum, or in some forms, to another part of the same colony. They are longer than the corresponding autozooid and each represents a short or long zooid row in which traverse walls—septa—have failed to appear. Rhizozooids usually show a positive geotropic or negative phototropic mode of growth, contrasting with the remaining parts of the colony.

Rhizozooids occur commonly in certain types of colonies in the gymnolaemates as well as in the Crisiidae in the stenolaemates, viz., in erect colonies that are not rigid but are either entirely flexible or, perhaps, provided with flexible joints. Thus, there seems to exist an interrelation between the type of structure of the colony and the occurrence of rhizozooids, the two factors cooperating to form a structure resisting disturbances from the environment, not by rigid reinforcement but by plasticity. Generally speaking, the rhizozooids obviously serve as holdfasts to secure a suitable position for the main bulk of the colony. The role of rhizozooids in the life of different species of *Scrupocellaria* is commented on by Stebbing (1971). In the species where a rhizozooid joins another branch of the same colony, it exhibits an orientated growth to a pore plate of a zooid of the latter branch (Silén, 1941; *Dendrobeania* and *Camptoplites;* Hastings, 1943; Ryland, 1970; Harmelin, 1973a) (Fig. 23). The factors causing this orientation are still unexplained. These interconnective rhizozooids obviously add to the firmness of otherwise finely branched and brittle colony types.

Stebbing (1971) shows that the distal ends of the rhizozooids in some *Bugula* species penetrate into the colonies and zooids of other bryozoans. He hints at the possibility that the rhizozooids play another role than mere holdfasts: "It is also possible that they are capable of absorbing nutrients from the host colonies." If this idea is correct, it is of great importance, indeed. Among other things, it means that certain bryozoans are partly parasitic. Such rhizozooids still should be termed heterozooids, since they have a nonautotrophic mode of obtaining nutrients. However, they would differ from all other heterozooids in representing not a drain but an asset for the nourishment of the colony. Further research is badly needed on this point.

The term stylozooids ("stem zooids") is created in an attempt to establish some preliminary categories, at least for the numerous and morphologically variable types of zooids, probably of heterogeneous origins, found in the basal parts of many erect colonies (Figs. 24–26). The variety of forms is so great that only a few examples are chosen here from the Bicellariellidae. The most extreme conditions possibly are seen in *Kinetoskias* (Fig. 25). Here the colony consists of a narrow, cylindrical stem up to 15 cm long (Koren and Danielssen, 1877) and on top of it, a bush-shaped portion a few centimeters high containing the autozooids. The detailed structure of the stem does not seem to have been examined closely; it is evident, however, that it is undivided and represents one basal zooid of the colony (possibly the ancestrula?). In *Semikinetoskias* (Silén, 1941) the colony is also raised on a peduncle. This consists of a single zooid about 7.5 mm in length fixed to the substratum by its basal end. An aperture field is indicated but internal organs are lacking, so the zooid (the ancestrula?) is thus a kenozooid. In *Caulibugula* (Harmer, 1962), the stem is composed of a single row of cylindrical kenozooids (occasionally broken by a region of autozooids). In *Bugula, Bicellariella,* and other genera the colony base (as well as the bases of certain of its branches) is built up of a single row of autozooids. Morphologically, they differ distinctly from the autozooids constituting the bulk of the colony (Figs. 26 and 43); thus they are autozooidal polymorphs (cf., for example, Boardman *et al.,* 1969, and also Section XVII concerning astogeny and polymorphism.)

On the basis of present knowledge, it is uncertain whether the examples cited are directly connected. In any case, they demonstrate a general tendency in common, viz., to form a stalk constituted of autozooidal or heterozooidal polymorphs. And the effect of that stalk is that it raises the productive portions of the colony above the substratum into a position more favorable for feeding and other functions.

Numerous other cases of polymorphism occur, primarily and/or secondarily,

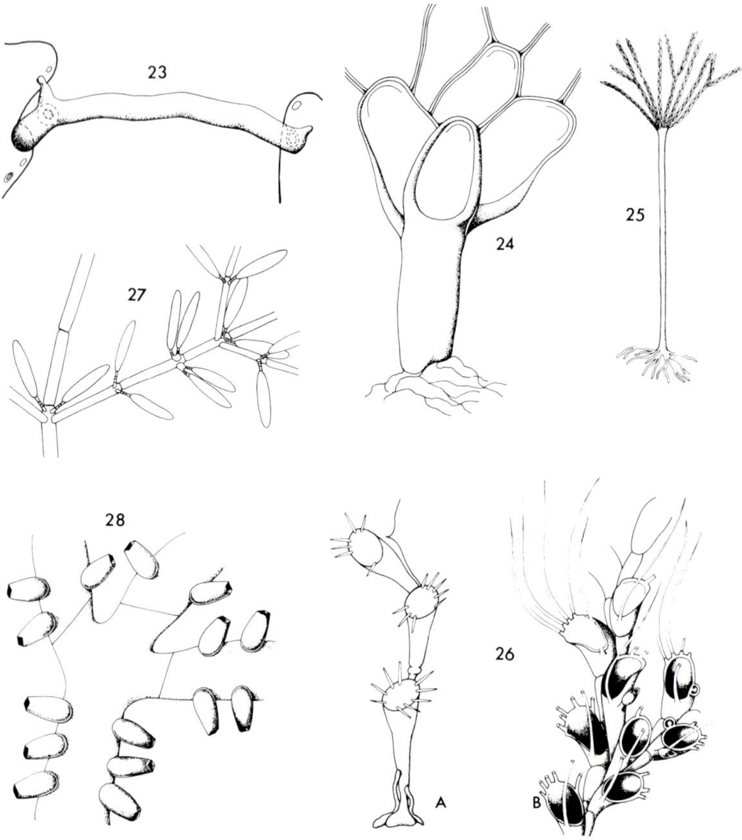

Figs. 23–28. Rhizozooids, stylozooids, and stolons.

Fig. 23. *Dendrobeania japonica.* Rhizozooid connecting left branch of colony with right one. Redrawn from Silén (1941).

Fig. 24. *Sarsiflustra japonica.* Autozooidal stylozooid (probably ancestrula). Redrawn from Silén (1938).

Fig. 25. *Kinestoskias smitti.* Sketch indicating elevating effect of stylozooid on productive portion of colony. Redrawn from Silén (1950).

Fig. 26, A and B. *Bicellariella ciliata.* Ancestrula and first zooids (stylozooids) [redrawn from Ryland (1969)] compared with normal autozooids [redrawn from Hincks (1880)].

Fig. 27. *Mimosella bocki.* Stolonial tree supporting autozooids. Redrawn from Silén (1942).

Fig. 28. *Zoobotryon verticillatum.* Stolonial tree supporting autozooids. Redrawn from Osburn and Soule (1953).

in erect colonies, especially at their bases, but also elsewhere; these heterozooids certainly play a role in securing fixation and strengthening of the colony.

The stolons which are present in large groups of ctenostomatous gymnolaemates, offer a special, intricate problem. In the colonies of these groups, the stolons constitute the foundation supporting the autozooids (Figs. 27–28). They form tubes that extend over the substratum or unite into veritable erect trees. The tubes are subdivided by transverse septa. They might be interpreted as kenozooid rows. However, Jebram (1973b) has advanced arguments for the hypothesis that each stolon portion does not represent an entire zooid but instead part of an originally unitary autozooid bud from which the other part develops into the definite autozooid placed on the stolon portion. If this is correct, it follows that the stolons are formed in a unique way without any known counterparts elsewhere in the bryozoans. In any event, the function of the stolons seems obvious, viz., to serve as supports for the definite autozooids and to raise them into favorable positions in relation to the substratum and the surrounding water. They thus have a function similar to that indicated above for the rhizo- and stylozooids.

IX. SURVIVAL OF UNFAVORABLE CONDITIONS: RHIZOZOOIDS AND STOLONS, HIBERNACULA, STATOBLASTS, AND SACCULI

In discussing rhizozooids and stolons, their special function in survivability ought to be stressed. In many species with erect colonies founded on rhizozooids and stolons, in the gymnolaemates and the Crisiidae, it is a regularly recurring phenomenon that the erect portions die off in the winter but the rhizozooids or stolons survive and demonstrate the ability to bud new zooid rows again in the spring, thus regenerating the portions lost. This phenomenon was studied by Numakanai (1960, 1967) and Stebbing (1971) in *Bugula,* Zirpolo (1933) in *Zoobotryon,* and Braem (1951) in *Bowerbankia* (cf. also Nielsen, 1971). It is not known whether this de- and regeneration can be repeated several seasons by the same colony. If such repetitions take place infinitely, this would mean, theoretically at least, eternal life achieved by the method described—a rather optimistic assumption.

Nothing seems to be known about the theoretical possibility that the rhizozooid system might lead to asexual reproduction outside the limits of the primary colony. Rhizozooid tubes crawling along the substratum can be seen giving rise to new erect tufts in *Bugula neritina,* for instance. When the new tufts still are connnected to the original one, they could be said to form a colony complex. However, if the basal connections are broken in a regular way, it would be in order to talk of the asexual reproduction of colonies, not just zooids. However it may be, indications are clear that the rhizozooids and stolons add considerably to the distribution over wider areas in many species.

Among the heterozoids, the hibernacula of the fresh and brackish water gymnolaemates *Victorella* and *Paludicella* should also be included. They are special resting zooids formed in adverse environmental conditions including temperature and salinity (Ryland, 1970). The statoblasts characteristic of the phylactolaemates serve the same purpose but are formed in a quite different way, viz., as internal buds. Certain statoblast types (floatoblasts and spinoblasts) become free and have the function of distributing the species. Thus they are asexual propagative bodies. The case of *Aetea sica* reported by Simma-Krieg (1969) is of particular interest. In certain, as yet unknown conditions, the lateral porechambers produce, instead of the adnate autozooids normally emanating from them, erect tubes ending in small sausage-shaped bodies called sacculi, which are detached and give rise to new colonies. The sacculi are evidently homologous to zooids, and thus we find here a clear case of free, asexual heterozooids serving the distribution of the species.

X. INTERZOOIDAL CONNECTIONS: INTERZOOIDS

The term interzooids, if really needed, is intended to designate the structures well-known as pore plates and pore chambers, or dietellae, characteristic of the gymnolaemates. These quite small bodies situated along the distolateral sides of the autozooids and elsewhere are of a zooidal nature as generally is accepted nowadays (Silén, 1944c; Powell, 1967; Banta, 1969b; Ryland, 1970; Gordon, 1971; Harmelin, 1973a,b; Pohowsky, 1973). They originate as buds along the walls of zooids directly attached to the corresponding walls of neighboring zoöids, with space generally lacking for their further growth. Their distal walls, as well as the corresponding spots of the adjacent zooids, are dissolved, whereby communications are established between the zooids (Fig. 29).

The process described does not take place at random, nor does it occur, in any species, in structures varying at random morphologically. The interzooids of a particular species can be precisely defined, chiefly as to number and structure. So it is a significant fact that these bodies are often very useful in taxonomy. This phenomenon of development of special interconnecting zooids must be assumed to have originated as an occasional occurrence, but now has become standardized along rigid laws. Gordon (1971) has raised the question of whether all the bodies, here called interzooids, are really homologous.

The function of the interzooids is obvious, viz., to serve as intercommunication between the "more ordinary" zooids within the colony. However, certain

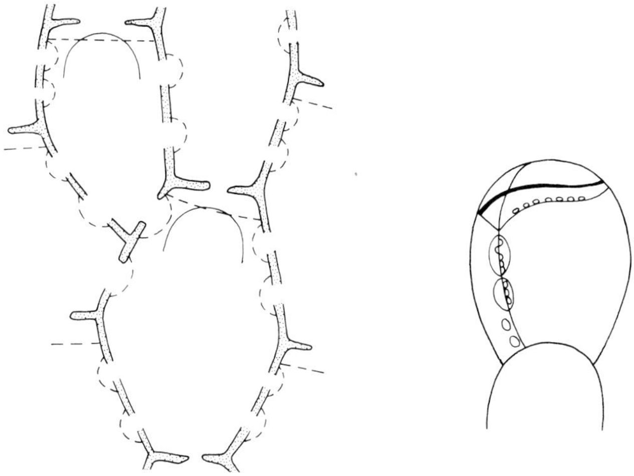

Fig. 29. *Callopora dumerili.* Pore chambers in an autozooid (right) and diagram of their arrangement (left). Redrawn from Silén (1944c).

questions remain to be answered. Why could not the same effect be reached in a more simple way, one autozooid directly merging with an adjacent one at defined spots? Why then these intermediary zooids? Why are the interzooids sometimes developed as pore plates, sometimes as pore chambers? Does some special physiological process take place within the more isolated cavity of the pore chamber? Or are all these complicated arrangements with interzooids expressions purely of laws of colonial architecture? To speculate: Might such expressions represent a primary necessity of the ordinary zooid to form buds along the sides of its distal half? Whereby the possibility remains that each bud is capable of forming a complete zooid if space is available.

In special circumstances it has been observed that the interzooid has the potential to develop into a more complete zooid. Particularly in species with an irregular colony growth, such as *Electra pilosa,* the lateral walls of a zooid row are primarily free, and there autozooids often are developed in the positions taken, as a rule, by interzooids. The interzooidal avicularia have been demonstrated to replace interzooids (Hastings, 1963). In some erect, cellularine gymnolaemates, rhizozooids emanate from pore plates in such a way that it may be possible that they do not represent separate zooids, but in fact are direct prolongations of the interzooid itself; further research is necessary on this point. *Flustrella* possesses spinozooids that probably should be interpreted as interzooids (Silén, 1947a).

XI. SEXUAL REPRODUCTION: GONOZOOIDS

The term gonozooid, introduced a long time ago, commonly is used in a restricted sense, viz., to designate heterozooids with exclusively female functions. However, we shall see that there also exist autozooidal polymorphs with exclusively male or female functions. It is a fundamental fact that the basic bryozoan zooid is hermaphroditic and serves the sexual functions, together with all other basic functions. The conventional use of the term gonozooid leads to certain logical difficulties when attempting to survey all kinds of sexual polymorphs in the bryozoans, difficulties so severe that it seems necessary to propose a new nomenclature in spite of the inappropriateness per se in changing a term so well-established in use (cf. Ryland, 1968, p. 229).

With respect to sexual reproduction, the following zooid types can be discerned:

1. *Hermaphroditic autozooids*
2. *Gonozooids,* autozooidal or heterozooidal polymorphs specialized for sexual reproduction
   a. *Gynozooids,* exclusively female polymorphs, autozooidal or heterozooidal; the term gonozooids in the sense hitherto used is in the main synonymous with heterozooidal gynozooids
   b. *Androzooids,* exclusively male polymorphs; to my knowledge only autozooidal androzooids are known
3. *Sterile zooids,* all heterozooids, except gonozooids, as well as, possibly, some yet undefined autozooidal polymorphs which are permanently sterile; "seasonal sterility" also appears, of course, in otherwise sexually active zooids

Naturally, the actual explanation of the gonozooids should begin with the conditions met within the stenolaemates in which sexual differentiation assumes drastic expression. So far as is known, all of the stenolaemates possess heterozooidal gynozooids (formerly called gonozooids) of very characteristic appearance and structure. These have been well-known for a long time and were particularly well-analyzed by Borg (1926) (Figs. 30 and 31). This demonstrates how unsatisfactory is the application of the term gonozooid only to female polymorphs; if there exist exclusively female zooids, then it follows that there must be male zooids as well! The reason why the term gonozooid has been used solely for female zooids is to be found in the striking appearance and obvious reproductive functions of these zooids; in contrast, the male functions demand so small a space that neither the interior organs nor the exterior body shape are changed. Thus, it is not possible, at least at present, to discern androzooids from sterile autozooids, except by the presence or absence of male reproductive organs and cells. Nevertheless, there is reason to assume that all autozooids do not possess male functions, not even potentially; rather, there actually exists a

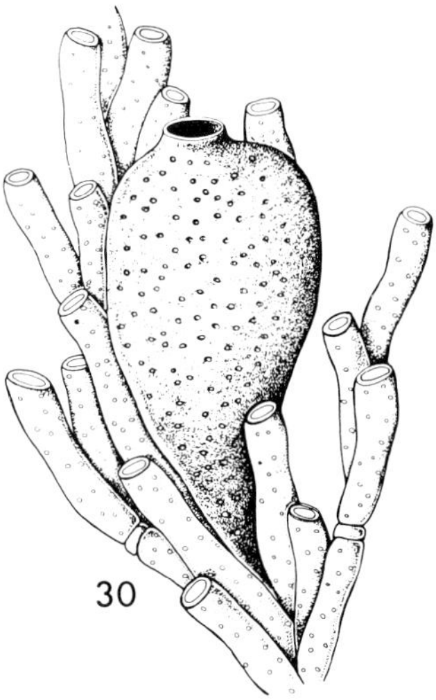

30

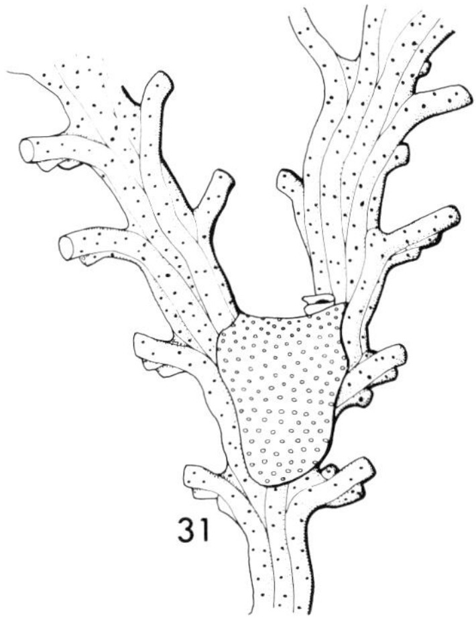

31

Figs. 30 and 31. Gynozooids in Cyclostomata.

Fig. 30. *Crisia.* Gynozooid among autozooids (possibly androzooids). Redrawn from Osburn and Soule (1953).

Fig. 31. *Tervia brevovicella* gives an example of exact location of gynozooid in stenolaemates, viz., to branching points of colony. Redrawn from Silén (1951).

dimorphism between androzooids and sterile autozooids. As one indication can be mentioned the fact that colonies of certain stenolaemates are reportedly dioecious (cf. survey by Borg, 1926). If true, this obviously means that, in such cases, the autozooids are sterile in the colonies possessing gynozooids.

Consequently, we probably have to consider, with respect to sexual conditions, the following polymorphs:

1. Autozooids
   a. Sterile autozooids
   b. Androzooids
   c. Gynozooids?
2. Heterozooids
   a. Sterile heterozooids
   b. Gynozooids

The explanation for the strongly modified structure of the gynozooids of the stenolaemates, particularly if compared with the lack of transformation of the androzooids, certainly is to be associated with the occurrence of polyembryony peculiar to this bryozoan group. The enormous number of embryos often formed could not be accommodated in a space produced solely by either a reduction of other organs or an increase of the body volume—both of these processes are necessary. On the other hand, the formation of male cells influences neither of these structures.

An interesting phenomenon, representing traces of the presumably original autozooidal and hermaphroditic conditions of the predecessors of the stenolaematous gynozooids of today, is that polypides really are formed in the gynozooids, though their development usually stops and is followed by a reduction. However, in several cyclostomes such as the Lichenoporidae, the polypides regress only after having reached a functional state and having formed primordial testes (Borg, 1926). It may be that, before their reduction, the polypides serve to collect the sperm in a similar way as in the cheilostomes (Silén, 1966) and to transfer them to the ovum.

The problem of whether all the gynozooids of the stenolaemates are homologous or not has been discussed in the literature. Harmer (1898) took the former position, but Borg (1926, 1933) and Brood (1972) considered the so-called colonial brood chamber of the suborder Cerioporina (Lichenoporidae) a structure separate from other gynozooids because it incorporates in itself portions of zooids adjacent to the primary gynozooid. If a nonexpert on stenolaemates such as I should dare put forth an opinion on this matter, it would be that the definite gynozooid of the Lichenoporidae as an entity is not homologous to the remaining gynozooids, but that in the complex there is a primary zooid that is homologous to other zooids. Thus, largely seen, there should be one original gynozooid type homologous to all stenolaemates.

The picture of sexual conditions in the gymnolaemates traditionally has been regarded as distinctly different from that in the stenolaemates. Gymnolaemates' colonies have been considered to be constituted essentially of autozooids, all of which are capable of functioning as hermaphrodites. However, recent observations indicate that the real conditions perhaps are not as simple as that, especially if correlated with some other features that have long been well-known. Possibly, we are now just at the beginning of a reevaluation of the sexual conditions of this bryozoan group.

Dioecious autozooids have been reported for several cheilostomes (Vigelius, 1884; Stach, 1938; cf. the survey by Banta, 1969a). However, morphologically defined sexual polymorphs seem to be rare. *Hippothoa hyalina* (as well as *H. bougainvillei:* Gordon, 1968) represents an unusually clear case of sexual polymorphism (cf. the survey of Marcus, 1938). Here male, female, and sterile zooids are morphologically distinct (Fig. 33). The andro- and gynozooids are both dwarfed, but they differ from each other in body shape. All three types possess polypides, and consequently should be termed autozooidal polymorphs. The reservation here is that it does not seem established whether the gonozooids still possess a feeding capacity.

Gynozooids characterized by a special body shape have been described in several gymnolaemate species with different systematic positions, for instance,

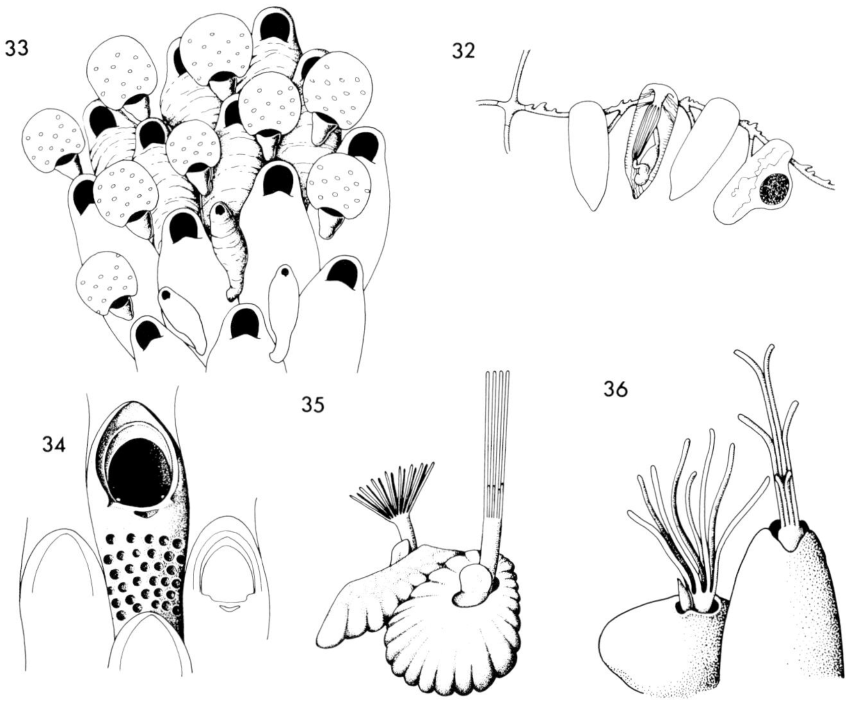

Figs. 32–36. Gonozooids in Gymnolaemata.

Fig. 32. *Penetrantia densa.* Gynozooid and autozooids. Redrawn from Osburn and Soule (1953).

Fig. 33. *Hippothoa hyalina.* Dwarfed gynozooids and androzooids, and presumably sterile autozooids. Redrawn from Marcus (1938).

Fig. 34. *Hippodiplosia insculpta.* Difference of apertural shape in fertile (left) and infertile zooids. Redrawn from Osburn (1952).

Fig. 35. *Hippopodinella adpressa.* Structure of tentacle crown in normal autozooid (left) and androzooid. Redrawn from Gordon (1968).

Fig. 36. *Hippoporidra senegambiensis.* Structure of tentacle crown in normal autozooid (left) and cortical zooid (possibly androzooid). Redrawn from Cook (1968).

other species of *Hippothoa, Thalamoporella, Didymozoum, Synnotum, Penetrantia,* among others (Figs. 32 and 34) (Levinsen, 1909; Harmer, 1926; Marcus, 1938; Silén, 1947b; Ryland, 1970). Among the forms mentioned, indications are that distinctive androzooids occur in *Thalamoporella* and *Synnotum* (Marcus, 1941). Otherwise, androzooids, morphologically distinguishable from the exterior, are apparently not known. It can be added in this respect that in *Steginoporella* (cf. Section II), embryos have been observed only in the A-zooids. Could it be that the A- and B-zooids have something to do with sexual dimorphism?

Clear sexual polymorphs, thus, are known from many different parts of the class. Usually, this phenomenon has been dealt with only in passing in the literature. The question naturally arises: How widespread really is this phenomenon in the gymnolaemates?

The problem is strengthened by the interesting report by Gordon (1968) on the existence in *Hippopodinella adpressa* (Fig. 35) of androzooids which cannot be discerned from other zooids in their external morphology, but can be in the structure of the polypide—the tentacles deviate from the normal condition in number and size. Moreover, the tentacles are devoid of cilia. This fact is of primary interest; a mobile extrusable polypide exists, which cannot have a feeding function as it lacks the appropriate cilia. Consequently, in spite of the presence of a well-developed, though modified, polypide, this male polymorph in fact ought to be classified among the heterozooids, if we adhere strictly to the definitions of autozooids as zooids capable of feeding themselves, and heterozooids as unable to do so. A somewhat similar case is reported by Cook (1968) for *Hippoporidra senegambiensis* (Fig. 36). There the so-called cortical zooids deviate from the normal autozooids by being larger, possessing thicker walls, and having only about half the number of tentacles which are devoid of cilia and capable of protruding but not expanding. In addition to their apparent strengthening function, these zooids may have a male function. If really true, this polymorph represents a unique example of a simultaneous specialization for two wholly different functions (cf. also Cook, 1973).

Thus, we have in the gymnolaemates sexual autozooidal polymorphs of two kinds, those recognizable by their exterior morphology and those distinguished by their tentacle structure exclusively. But this statement is incomplete in one way. One morphological and functional characteristic is always present, though not directly stressed, viz., that represented by the germ cells themselves. Then, logically, we find that there must exist a third kind of autozooidal polymorph, the dioecious autozooids indicated above which can be exclusively recognized by the type of germ cells produced.

In at least Electridae and Membraniporidae there is what could be called a seasonal sexual dimorphism. In sexually active zooids the "normal" polypides are replaced by polypides possessing the well-known and striking intertentacular organ serving as transport organ for ova (cf., for instance, Silén, 1966). After reproduction the polypides are replaced again by "normal" polypides without intertentacular organs. This "momentary dimorphism" probably should be ascribed to the sexual state of the zooids and cannot be associated at present with real polymorphism. Could the *Hippopodinella* case referred to above be analogous with the above phenomenon, the special polypide being a transitory one appearing only during sexual seasons? But even in such a case, the question remains unanswered whether the "normal polypides" are replaced temporarily by "sexual polypides" only in specific zooids, which as yet are indistinguishable from possibly permanently sterile zooids.

Similar considerations perhaps can be applied to the presence/absence of the ooecia (ovicells) so characteristic of most cheilostomatous gymnolaemates. Frequently the ooecia are situated in restricted zones or otherwise characteristic positions in the colonies. Is this related only to the lag between the times their mother zooids are formed and become sexually mature, or are there other factors involved?

Powell (1967) has demonstrated the presence in *Cribrilina annulata* of dimorphic female zooids. Both types are recognized by the presence of ooecia, but whereas one of them does not differ from other autozooids in position or general structure, the other type is dwarfed and frontally situated on other zooids. The latter thus represent sexual polymorphs, gonozooids (or possibly gynozooids, if the male function is absent).

So-called embryo sacs are formed in the phylactolaemates and in certain ctenostomes and cheilostomes in the gymnolaemates. Possibly they are all homologous to one another and homologous as well to the central embryo chamber of the ooecia. They have been interpreted tentatively as internal zooid buds (Braem, 1897; Silén, 1944a; cf. also Jebram, 1973a). However, their true nature is still so obscure that an attempt to evaluate them is premature. Obviously, wide fields are open to research before the problem of the occurrence and role of sexual polymorphs in the gymnolaemates is solved.

It has been demonstrated that the ooecia of the gymnolaemates are partly of zooidal nature, kenozooids (spinozooids?) partaking in their formation. However, in spite of the fact that they play an important role in reproduction, they should not be classed among the gonozooids in my opinion. This term should be used exclusively for polymorphs containing germ cells.

XII. FROM FREE TO SEDENTARY LIFE: ANCESTRULAE

In the gymno- and stenolaemates, the primary zooid of the colony is formed by metamorphosis of the larva, producing the ancestrula which differs clearly in structure from the remaining zooids of the colony. It usually possesses a functional polypide and could be called an autozooidal polymorph. It apparently has no special function in the life of the developed colony except constituting the primary or sometimes the only holdfast. Its structural pecularities certainly are to be seen as the result of the transformation of the free larva into a sedentary individual, as well as of its role as a starting point for astogeny. Nevertheless, considering its structure is characteristically modified, the ancestrula should be classed among the polymorphs.

XIII. INDUCTION AND ORIGIN OF POLYMORPHISM: GENERALITIES

The prime goal of this discussion is to analyze the concept of the individual in the bryozoans—a concept that is twofold, as will be seen in Section XIV—in order to define the units that are acted on by the factors inducing the formation of polymorphs as well as by the factors that have fixed in the gene pool the "programs" for all existing polymorph possibilities, but for only those possibilities.

The inducing factors are contained in the astogeny of the colony. The factors fixing polymorphism in the gene pool and causing the selection of the existing polymorph categories are to be looked for, presumably, in the biological role of the polymorphs in the life and evolution of the polymorph types.

XIV. THE INDIVIDUAL CONCEPT

What is the individual in the bryozoans? The answer, at first, might seem a simple one: the zooids of the obviously basic type—the autozooids—generally appear as well-defined units with a complete complement of organs for their provision and propagation, in other words with typical animal bodies. The polymorphs represent modifications of that fundamental pattern, including modifications also in the genetic sense of the word. This aspect of the problem was emphasized in an earlier paper by the present author (1938). However, it was overemphasized there. Though containing a kernel of truth, that aspect does not represent the entire picture.

We must concern ourselves here with an individual on another, higher level, viz., the colony. A colony is the structure resulting from the development of one zygote, and is thus, as a whole, a genetically singular unit, one individual or clone (Silén, 1944c, pp. 483–485; Schopf, 1973), disregarding, for the moment, the possibility of somatic mutations within a colony (Boardman *et al.,* 1969; cf. also Jebram, 1973a).

This fact is demonstrated clearly by certain behavior of the colony that can be defined in immunological terms. In a previous paper of mine (1944c, p. 454), it was stressed that different parts, branches, etc., of one and the same colony of a particular species are able to fuse, not only attaching superficially to each other, but also establishing quite intimate, organic connections with one another. Many examples can be given of this phenomenon in the bryozoans.

The most significant expression of this rule can be seen in the gymnolaemates. In practically all members of this large group, the zooids form direct connections between each other through their lateral walls, fusing secondarily in small areas. The resulting openings, traversed by living tissues, form parts of the pore-plate/pore-chamber systems. Thus, in the gymnolaemates, this ability for different parts of the colony to fuse is fundamental to the whole internal structure of the colony. This experiment—nature's own—certainly can be repeated in the laboratory with the use of suitable material (cf. Theodor's 1970 experiments on Gorgonacea). However, such laboratory experimentation on bryozoans remains to be done.

If we now turn to behavior of one colony of a certain species in respect to other colonies of the same species, it is observed that they do not fuse on contact (Silén, 1944c; Knight-Jones and Moyse, 1961; Stebbing, 1973). One exception to this rule was reported by Moyano (1967) in *Membranipora hyadesi.* Despite extensive search, Stebbing only found two exceptions in *M. membranacea* while he found such a fusion between different colonies common in *Parasmittina trispinosa.* However few they may be, the exceptions are surprising and remain still unexplained. Such exceptions could be expected in the stenolaemates where polyembryony effects the formation of possibly hundreds of genetically identical larvae from one zygote considering the realistic possibility that such larvae with their short, free life might settle close to one another. Dr. Harmelin, Marseille, kindly informed me that, in fact, fusions between colonies of the same size, thus of approximately the same age and possibly issued from the same mother colony, are not rare in *Diplosolen obelium* and have been observed in *Diastopora compacta* as well. However, the histological examination necessary to confirm the actual fusion has not been performed yet.

The phenomena described above are obviously reflections of the ability of members of a certain species to recognize what is "self" and what is "nonself." In other words, these phenomena (disregarding the apparent exceptions noted) are expressions of allogenic incompatibility, the foundation of immunology.

And it is obvious that within the bryozoans, the colonies, exhibiting this ability as they do, are the individuals. From another point of view, the colony as a whole is subdivided into zooids; when we state that a colony is able to fuse within itself, we could as well express that phenomenon by stating that its zooids can fuse with one another. However, as the zooids are able to fuse within the colony, this fact demonstrates that genetically and immunologically they do not represent separate individuals.

To sum up, in the bryozoans there are "individuals" of two different levels, the colony and the zooid, which can be affected by modificative powers. The mutations must be located in the genetic individual, the colony, and "natural selection operates at the colonial level of organization" (Schopf, 1973).

Boardman and Cheetham (1973) have recently published a treatise of great importance in this connection: "Degrees of colony dominance in stenolaemate and gymnolaemate Bryozoa." Here the reader will find an excellent analysis of the balance, different in different types of bryozoans, between the two forces represented by the zooid-individual and the colony-individual, as elucidated by colony structure. The papers of Banta (1973b), Sandberg (1973), Utgaard (1973), and Blake (1973) should also be considered in this connection.

XV. ONTOGENETIC DEVELOPMENT OF POLYMORPHS

A comparison of the modes of formation of different structures in different polymorphs with corresponding ontogenetic patterns in the autozooids is of interest for further discussion. Unfortunately, there is little such research, and what there is has been confined mainly to the avicularia of gymnolaemates and the gynozooids of stenolaemates.

An investigation into the development of the vicarious avicularia of *Flustra foliacea* was done by the present author (1938) (Fig. 37). If the conditions of the autozooid are taken as normal, as is natural to do, the development of the avicularium takes place in the following sequence and with the following results: (a) Body volume is reduced; (b) polypide rudiment develops centrally instead of proximally; (c) differentiation of polypide remains rudimentary; and (d) specialized, further differentiation of operculum and appendages is exaggerated. In the

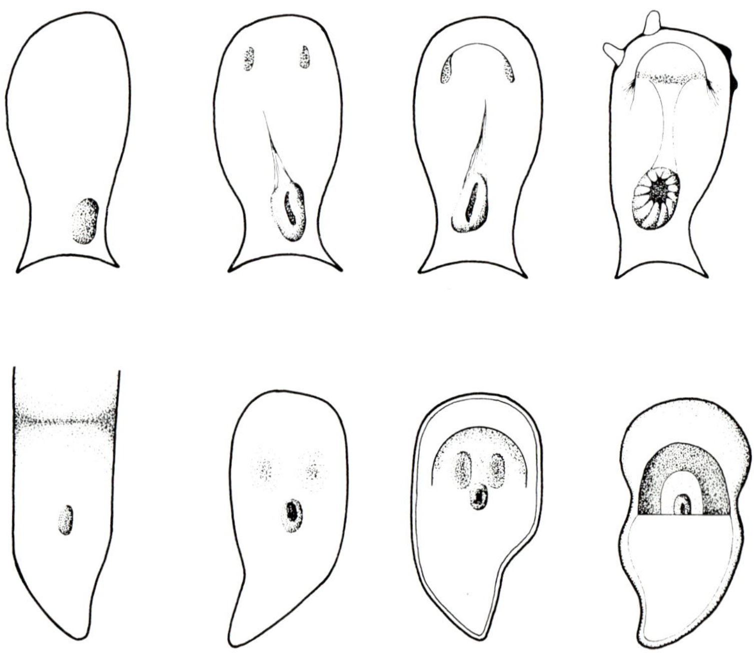

Fig. 37. *Flustra foliacea.* Differentiation of polypide and operculum in autozooid (above) and avicularium (below). Occlusor muscles indicated. From Silén (1938).

kenozooid, the body is even smaller and more modified; the polypide, operculum, and other structures are not differentiated even rudimentarily. Apparently, these characteristics of the ontogeny can be applied to all avicularia and kenozooids.

The ontogeny of the stenolaematous gynozooids offers a different picture. The body volume is increased, often enormously, above the normal zooid size. A polypide is differentiated, but sooner or later, sometimes even after having reached a functional state, it undergoes regression until it disappears. The arrest of development and onset of regression are correlated with the initiation of the reproductive activity.

The processes briefly related here indicate the need to discern two series of polymorphs, one comprising sexual polymorphs, the other sterile polymorphs—series with different selective pressures influencing them.

XVI. SUBSTITUTION OF DIFFERENT POLYMORPHS

It has been indicated for *Flustra* (Fig. 1) that the avicularium, without exception, assumes a characteristic position in the colony. However, all the zooids assuming that position are not differentiated into avicularia—some turn into autozooids. The difference between the buds that are to be avicularia and those to be autozooids can be recognized at an early stage when the body has reached its definitive size but its differentiation has not started; all of the zooids in the position in question are smaller than the autozooids in general, but nevertheless those above a certain size invariably turn into autozooids and those smaller always into avicularia (Silén, 1938).

In *Thalamoporella* (Silén, 1938; Powell and Cook, 1966) (Fig. 2) zooids in certain defined positions can turn into autozooids, avicularia, or kenozooids, according to their size, thus forming an even longer series from one possible extreme, the complete autozooid, to the other extreme, the heterozooid without any interior differentiation, the kenozooid.

At this point, one and the same inducing condition can be suspected to lie behind the formation of the various zooid types mentioned. The strength of this inducing factor is graduated. However, its effect is discontinuous. It results in two (*Flustra*) or three (*Thalamoporella*) distinct zooid types and no intermediate ones. The different possibilities are realized on specific levels along a graduated scale.

There are several examples of such series of morphs alternatively appearing in defined positions, though not always in one and the same species (see also Buchner, 1924) (Figs. 38–41):

Autozooid–vicarious avicularium–vicarious kenozooid, at divisions of zooid rows (cf. above)

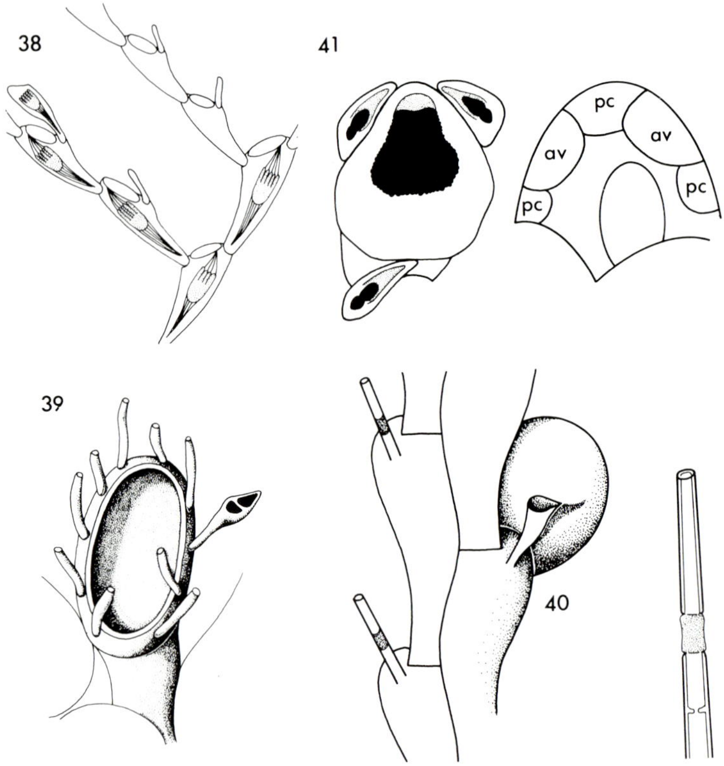

Fig. 38–41. Facultative development of identically situated buds into different zooid types.

Fig. 38. *Scruparia.* Facultative formation of autozooids or spinozooids in identical frontal position. Redrawn from Silén (1938).

Fig. 39. *Cauloramphus cymbaeformis.* Adventitious avicularium in same position as spinozooid. Redrawn from Osburn (1950).

Fig. 40. *Tricellaria ziczac.* Colony in basal view showing avicularium and spinozooids in corresponding position. Spinozooid more enlarged to show position of pore plate. Redrawn from Silén (1942).

Fig. 41. *Ellisina canui.* Autozooid and (interzooidal) avicularia from frontal (left) and basal side. Avicularia replace porechambers. Redrawn from Silén (1938).

Autozooid–adventitious avicularium–spinozooid, formed by frontal budding at the proximal end of the aperture field [can the dwarfed gynozooids described by Powell (cf. Section XI) possibly be added to this series?]

Adventitious avicularium–spinozooid–possibly ectooecium; formed by frontal budding at the distal of the aperture field

Interzooid (pore plate or pore chamber)–interzooidal avicularium–interzooidal spinozooid–rhizozooid? (cf. Section X)

These examples, certainly there are additional ones, are excellent illustrations, I think, of the thesis that the colony, and, consequently, each zooid of that colony have inherited certain specific and potentially different possibilities for the differentiation of the zooid and no other ones.

XVII. CONDITIONS INDUCING POLYMORPHISM

Which factors induce the development of a zooid in one or the other direction, that is, its modification into one or the other morph? These factors, of course, are contained in the environment of the developing zooid bud. In this connection two categories of environment are to be discerned, viz., the environment associated with its particular position in the specific colony of which it is a part, and environment in its usual sense represented by the substratum and water conditions. The colonial environment, as we might designate it, has a regular, constitutional influence on the growing bud; the environment represented by the surroundings, if strong enough, results in modifications, or even direct disturbances, of the colonial astogeny, provoking deviations from the normal formation of the zooid.

Many bryozoans, particularly those with erect colonies, exhibit quite strict laws within their colonial structure, often so precisely determinate that formulas can be evolved for them. This has been done for the Bicellariellidae and Scrupocellariidae (Harmer, 1923; Silén, 1938) and the Crisiidae (cf. survey in Borg, 1926), for example. In encrusting forms, these laws are generally less rigid. In forms such as several Electridae, they even must be said to be rather loose and restricted to the most elementary principles; this is only natural, considering the more intimate dependence of the substratum within the encrusting species. Recently, the principles of astogeny and, consequently, of colonial structure, have been discussed fully by Boardman *et al.* (1969) in a most instructive treatise, referred to for those who might want more detailed and precise information on the subject (Figs. 42 and 43).

The cheilostomes are selected for offering the most illustrative examples for further discussion. In this group, two types of astogeny apparently are to be discerned (Silén, 1944a). One is represented by the distolateral budding which produces the great bulk of the colony and establishes its primary characteristics. The other budding emanates from the frontal sides of the mother zooid (cf. further Banta, 1972). The appearance of the frontal budding is such that an interpretation seems in order: It represents, in actual form, the remnants of a mode of budding that is mainly suppressed today and usually results in dwarfed zooids that form as spinozooids and (adventitious) avicularia.

42

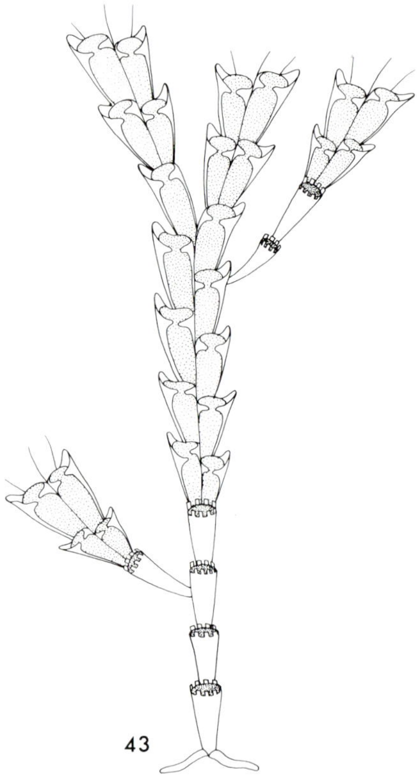

43

Figs. 42 and 43. Astogenetic changes in encrusting (*Escharoides*) and erect (*Bugula*) cheilostomes. Redrawn from Boardman *et al.* (1969).

In the predominant distolateral budding system (the primogenial budding of Banta, 1972), two phenomena are to be observed. The first is the quincunx arrangement of the zooids, i.e., each zooid of each zooid row is situated so that its distal half is on a level with the proximal halves of the two adjacent zooids of the zooid rows situated on each side of it. This consistent pattern is broken, potentially, when the growing colony widens in the distal direction; consequently, there is sufficient space for a zooid row to divide into two (sometimes three). The terminal zooid of the undivided row gives rise, at this point, to a directly terminal daughter zooid and a more laterally situated one. The former is of the length normal to the species, whereas the latter is dwarfed to a varying degree. By this process, the terminal septum of the dwarfed zooid becomes situated between the levels of the terminal septum of its sister zooid on one side and of the terminal septum of the corresponding zooid of the adjacent row on the other side, and the quincunx arrangement is restored (Fig. 44). The crucial point, in the present discussion, is that the process making possible the restoration of the quincunx pattern always results in the formation of one dwarfed zooid. In numerous cheilostomes it is seen that precisely this zooid is transformed into a polymorph. Such are the cases referred to above, of the vicarious avicularia or kenozooids in *Flustra* and *Thalamoporella* (cf. for full information Silén, 1938).

The second phenomenon of interest is the fact (particularly discussed by Boardman *et al.,* 1969) that in the series of zooids extending from the ancestrula to the distal margin of the colony, not all the zooids are identical in structure. Rather the colony consists of subsequent generations of zooids, all those on the same level being identical in structure, but those on one level differing more or less in structure in a characteristic way from those on other levels. This is the case within a restricted region around the ancestrula, the zone of astogenetic change; this region is followed by zones of astogenetic repetition.

These laws, covered very briefly here, become most strikingly expressed in cheilostomes with erect zoaria. For instance, many Bicellariellidae and Scrupocellariidae possess erect, strictly biserial colonies divided into separate branches. Fundamentally, the zooid rows of the cheilostome colony have the potential to form new lateral rows in a number regulated by space only. As this potential is not realized in the mentioned forms, this obviously means that the number of zooid rows formed in reality is the effect of still another force: a genetic "blueprint" for an exact colony shape. The specified colony form is achieved by the suppression of a number of potential zooid rows. There are two ways in which suppression can occur: The potential zooid rows are so totally suppressed that no trace of them can be perceived at their would-be emanation points. Or they start forming but the initial zooid bud comes to a standstill long before it reaches the length normal for the complete zooid; dwarfed, it is differentiated into a polymorph, such as an avicularium or a spinozooid.

I think we can safely conclude from this exposition, incomplete though it may be, that asexual polymorphs resulting in dwarfed zooids are induced regularly by genetically fixed laws of astogeny. In some ways, hitherto unrevealed, this restriction of the body volume influences the formation of the primordium of the interior organs, the polypide bud, so that it is either not formed at all or is hampered in further development and transformed. This absence, or transformation of the polypide bud, has an inductive effect on the further differentiation of the zooid as a whole, the end result being a kenozooid, an avicularium, etc.

It is a fact of considerable interest that in many species avicularia, kenozooids, and other polymorphs may also appear outside the exact positions indicated. This phenomenon has not been observed in species with an erect and precise colony structure, but in species with a more variable colony shape, particularly

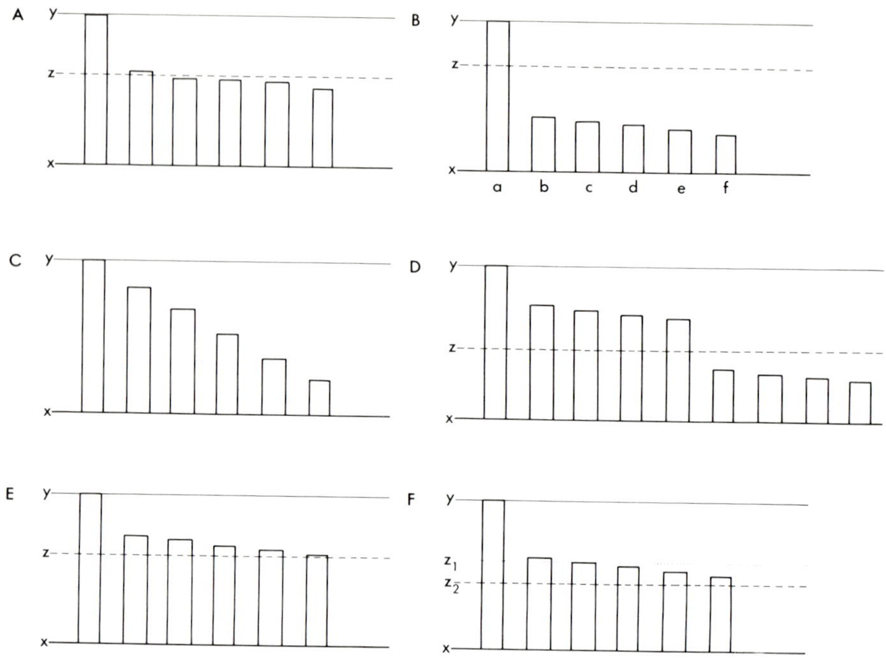

Fig. 44. Theoretical diagrams to demonstrate the relation between on one side the presence or absence of avicularia (or other heterozooids), on the other the genetic basis for formation of avicularia. The line x indicates the proximal zooid wall. The line y is drawn at a distance from x corresponding to the normal zooid length. The line z is the genetically fixed size limit above which the zooid develops into an autozooid and below which it develops into an avicularium. The zooid (a) represents in all the diagrams the normal autozooid size. In A the size-reducing factor (for instance, the quincunx pattern, see also Fig. 2) causes a reduction of certain zooids b–f (for example, distolateral buds at the bifurcations of zooid rows) making them vary in size around the line z; some of them (b) fall above that line and develop into (unusually small) autozooids, others (c–f) below it and develop into avicularia. In otherwise identical conditions the size-reducing factor is changed in E so that the size reduction is less strong: All zooids fall above the line z and, consequently, develop into autozooids. Thus, the species in question has "lost" its avicularia owing to a change in the reducing factor in spite of the genetic basis for the formation of avicularia still remaining. In F, in conditions otherwise identical with A, the line z is moved further downward (from the original position z 1 to z 2) illustrating a change in the genetic base demanding a stronger size reduction for the formation of avicularia; all the zooids fall above the line z and the species has "lost" its avicularia owing to a genetic change though the potential possibility of avicularia formation in general is not lost. Diagram B demonstrates the case of a stronger size reduction making all reduced zooids fall below the line z and develop into avicularia; in this species the avicularia are small though there is a genetical basis for the formation of bigger avicularia. In case C the variation in size is strong but no avicularia are formed as the genetic basis for avicularia formation (the line z) is lacking. In D there are two categories of size reduction of which only one falls below the line z and develop into avicularia, whereas all other zooids develop into autozooids of two size categories. The diagrams can be imagined altered so that another "genetic line" is added to z (situated below it) whereby the third category kenozooids could be taken in account as well. Redrawn from Silén (1938).

encrusting ones (*Thalamoporella* and many others), it is rather common. Forces outside the encrusting colony, above all irregularities of the substratum or collisions with colonies of different species (Gordon, 1972), can induce local modifications as is seen easily because of disturbances in the extension of the zooid rows (Fig. 2E). In such areas of disturbance, polymorphs appear if the species normally forms polymorphs. These polymorphs are characterized by a deformation of the zooid shape in one way or the other, often resulting in dwarfism. In *Thalamoporella,* avicularia and kenozooids often occur under such circumstances. As to the process inducing their formation, I think the same theory can be applied as advanced above, with the only difference being that in the present case the effective stimulus is from the exterior environment, while in the former case the determinant is from the colonial environment.

It may be added here, perhaps a little prematurely, that in my opinion it is reasonable to assume that this formation of polymorphs by extrinsic forces is a secondary phenomenon, occurring as it does only in forms also possessing the same polymorph types induced by the regular forces inherent in the colony structure. If so, the origin of these polymorph types must be looked for in the colony pattern, and the irregularly induced polymorphs have no evolutionary significance.

In turning to the sexual polymorphs—the gonozooids—it is obvious that the same laws cannot be applied to explain their induction. It is true that a certain regularity, but only an approximate one, can be discerned in the Crisiidae. An exact positioning occurs in *Tervia (Tubuliporidae)* (Fig. 31) and occasionally in the cheilostomes (cf. Borg, 1926, etc.), but as yet no general rule has been discerned, even less a rule comparable to those valid for the asexual polymorphs. The hypothesis appears reasonable that in sexually polymorphic species the development of a particular zooid into a gonozooid polymorph is caused by the presence of one or more germ cells in that specific zooid. Consequently the germ cells would represent the force inducing the formation of sexual polymorphs. However, then the problem immediately arises: Where do the germ cells originate and along which lines are they directed to the incipient gonozooids? Borg (1926, pp. 336 ff. and 409 ff.) has given an interesting survey of the process taking place in the Crisiidae. The germ cells are formed below the terminal membrane of the common bud typical for the stenolaemates, male and female cells not being distinguishable at first. When the young polypide buds are formed, some of the germ cells are distributed among them, the other ones degenerating. However, though female primordial germ cells are much superior in number to the polypide buds, each bud does not receive such a cell; those few buds which do presumably later develop into gynozooids. Thus, the distribution of the germ cells to and hence determination of the incipient gonozooids is localized in the growing zone of the colony in the stenolaemates at least. (Though the growth takes place in a different way in the gymnolaemates, no common bud existing there, it does not seem improbable that the germ cells are formed in an analogous way there, viz., in the growing edge of the colony.) If this distribution is regulated by distinct rules, further investigations should be concentrated upon this portion of the colony in search for the solution of the problem of induction of gonozooid formation.

An interesting fact already referred to above is the formation of primordial testes, disappearing later on, in the gynozooids of the Lichenoporidae. One might make the following speculation. In the stenolaemates the male cells are distributed to all the zooids; the female cells only to a few zooids. Male cells arriving into a zooid with a female cell are suppressed, as are those that arrive in kenozooids. The colony of a monoecious species would thus be constituted by a few gynozooids, numerous autozooidal androzooids, and sterile kenozooids. The female colony of a dioecious species would consist of a few gynozooids and sterile autozooids (and kenozooids), the male colony of male autozooids exclusively (and kenozooids).

It can be annotated in this connection that the induction of the formation (or modification) of the kenozooids which constitute the ectooecial fold of the hyperstomical ooecia (whether they are formed from the maternal zooid or the zooid distal to it), (cf. Section VI), as well as of the ooecial vesicle, is to be ascribed presumably to the female germ cells of the maternal zooid. Concerning a few other types of polymorphs, the B-zooids of the Steginoporellidae and the nanozooids of *Diplosolen* among others, no indication of the nature of their induction has been found as yet.

XVIII. ORIGIN AND EVOLUTION OF POLYMORPHISM

As seen from the aforementioned, it seems appropriate to relate the induction of the polymorphs, in many cases at least, to precise forces inherent in the colonial or external environment of the polymorph or in the distribution of the germ cells. However, the fact has been stressed repeatedly that even in those cases where the influence of that force is graded, the result is discontinuous; the zooid can develop only into one of a restricted number of existing distinctly defined polymorphs, and not into some morph belonging to an endless, theoretical series of transitional variations.

How have the existing types of polymorphs originated historically, and how have they been retained and evolved by means of natural selection? A colony, functioning perfectly in all respects and wholly composed of identical autozooidal monomorphs can be imagined quite easily. In fact, such colonies exist, as demonstrated by all the phylactolaemates and certain gymnolaemates. A colony wholly composed of heterozooidal polymorphs is obviously impossible. Polymorphs can exist only in the simultaneous presence of autozooids. From the aspect of nutrition, the possession of heterozooid polymorphs is a disadvantage to the colony, and thus to the species. The production of each heterozooid means not only the loss of a feeding individual to the colony, but also represents a drain on the productive members of the colony, the autozooids, which must nourish them. [The possibility that certain rhizozooids may absorb nutrients from other bryozoans, as suggested by Stebbing (1971), is disregarded until confirmed.]

However, if the presence of heterozooids in one way means a burden to the colony, the simple fact that they exist must mean, nevertheless, that they represent an asset to the colony and to the species—advantages at least as big as the loss represented by their incapability of feeding, and preferably bigger than that loss! If the assets did not balance the liabilities, the heterozooid mutations obviously could not have been retained during natural selection.

As yet, no attempts have been made to appreciate the energy drain on the autozooids effected by the heterozooids. However, in many cases it must be supposed to be considerable, for instance, in regard to such active polymorphs as the avicularia, especially when incessantly active as in *Bugula,* or when very numerous and large, as in certain species of the Reteporidae. The energy consumption is presumably lower per body volume in the inactive kenozooids. Nevertheless, the total energy cost is probably great in such forms as *Zoobotryon* and similar ctenostomes where the total kenozooid volume equals or surpasses the total autozooid volume (Fig. 25).

It should not be entirely impossible to reach, in a suitable material, at least some approximate quantification of the total energy consumed by the heterozooids as compared with the energy produced by the autozooids. In this way we could obtain one form of estimation of the biological importance of the heterozooids, postulating perhaps that their advantage to the colony is at least as large as the disadvantage represented by their drain on the production of the colony. Obviously, however, such an approach would mean that an entirely new and wide field of investigation be opened.

Schopf (1973) recently has published an interesting treatise on the "Ergonomics of Polymorphism. Its Relation to the Colony as the Unit of Natural Selection in Species of the Phylum Ectoprocta." Schopf proposes "a model for the development of polymorphism in ectoprocts derived from the model E. O. Wilson (1968) used to predict the characteristics of caste formation in the social insects." This theory is based on the advantage to the colony of distributing special tasks to separate types of polymorphs (exemplified by avicularia and vibracula), the promotion of polymorphism by the repeated stimulus of stable environments, and the counteraction of polymorph formation by unstable environments. The theory is illustrated by comparing the occurrence of avicularia/vibracularia in tropical, arctic, deep-sea, and estuarine cheilostome faunas.

In turning from theory-searching to hard facts for making a balance sheet for the polymorphs, we are confronted unfortunately with an almost complete absence of knowledge in the present situation. A qualitative determination of the functions of the polymorphs is possible to a certain degree, as is seen from the above survey. The reservation must be repeated, however, that conclusions regarding the modes of functions are still largely interpretations of the structural characteristics of the polymorphs and their positions in the colony. There is a desperate need for direct observations into the mode of working of the polymorphs. This, I think, will prove a wide and fruitful field for future investigators, though such research by no means will be an easy task. However, even more necessary, though even more difficult, is the matter of obtaining a quantitative measure of the advantage supplied to the colony and to the species by each type of polymorph. So long as there are no such data, a firm basis cannot be established for judging the biological importance of the polymorphs, and this is a prerequisite for understanding the origin and evolution of the polymorphs.

If qualitative investigations on the functions of polymorphs are sparse, quantitative investigations of their effects are almost nonexistent. A very interesting exception is the investigation performed recently by Kaufmann (1971) on the avicularia of *Bugula.* This study, it is true, pertains only to the faculties involved in the avicularia. However, Kaufmann was able to demonstrate the potential of the avicularia to reduce the harm done by tube-building and predatory animals, to such an extent that it follows that the investigation if expanded in the same direction would lead almost automatically to a quantitative measuring of the effect of the work done by the avicularia. Though investigations of this kind meet with many intricate obstacles in practice, it should not be impossible to quantify in analogous ways, and even experimentally, the effects of at least some other types of polymorphs as well, such as different types of avicularia, vibracula, and stylozooids. It would be of particular interest to know the consequences of the presence in one and the same colony of several morphologically distinct types of such specialized polymorphs as the avicularium.

In this chapter the idea has been pursued that the heterozooids mean an extra drain on the direct feeding resources of the colony—a drain that is compensated or rather overcompensated for by the favors rendered to the colony by their special activities. However, here as well as in other areas touched on above, the gonozooids, particularly the gynozooids, seem to represent an exception. They consume a large amount of energy, of course. However, we do not have to have special reasons for stating that it hardly can mean noteworthy differences in the energy balance of the colony whether its sexual propagation is performed by normal hermaphroditic autozooids or by polymorphs, just as no differences can be expected, in this respect, if a species propagates by numerous small embryos poor in yolk, by a few big embryos rich in yolk, or by polyembryony. These different modes of propagation represent exclusively different techniques for reaching the same goal; the gain, if any, is dependent on the environmental situation of each species, today and/or in ancient times.

In a most interesting study, Cheetham (1973) has followed the evolution of polymorphism, progressive or retrogressive, that has really taken place during geological periods of considerable length in evolutionary lineages in the two cheilostomatous genera *Poricellaria* (samples from late Cretaceous to Holocene) and *Vincularia* (samples from early Eocene to middle Oligocene). "The hypothesis that continuous variation in zooid morphology has been parceled out as polymorphic differences can be tested biometrically in cheilostomes in which groups of zooids are regularly segregated in identifiable regions of colonies. The numerous attributes by which polymorphs differ can be condensed by principal component analysis to a few size and shape factors. Application of this method to evolutionary lineages in *Poricellaria* and *Vincularia* has revealed similar series of gradual changes in both a monomorphic to polymorphic transformation and one in which polymorphism is reduced."

To briefly conclude, the big task for bryozoologists in the future, in my opinion, is the intense and detailed study of all manifestations of bryozoan life *in vivo* as well as *in vitro* material. This task could be extremely tedious, to be sure, but it is certainly unavoidable if a full comprehension of the functions, importance, and history of the bryozoans is to be reached. At present, our knowledge of these topics consists only of fragments, loosely related. Though this chapter offers more of a sketch than a detailed picture, it is my hope that this very lack of knowledge will serve as a stimulus and a guide to future research.

REFERENCES

Astrova, G. G. (1973). Polymorphism and its development in the trepostomatous Bryozoa. *In* "Living and Fossil Bryozoa" (G. P. Larwood, ed.), pp. 1–10. Academic Press, New York.

Banta, W. C. (1969a). *Uscia mexicana,* new genus, new species, a watersiporid bryozoan with dimorphic autozooids. *Bull. S. Calif. Acad. Sci.* **68,** 30–35.

Banta, W. C. (1969b). The body wall of cheilostome Bryozoa. II. Interzooidal communication organs. *J. Morphol.* **129,** 149–170.

Banta, W. C. (1972). The body wall of cheilostome Bryozoa. V. Frontal budding in *Schizoporella unicornis floridana. Mar. Biol.* **15,** 63–71.

Banta, W. C. (1973a). The significance of areolae in cheilostome Bryozoa. *In* "Living and Fossil Bryozoa" (G. P. Larwood, ed.), pp. 209–219. Academic Press, New York.

Banta, W. C. (1973b). Evolution of avicularia in cheilostome Bryozoa. *In* "Animal Colonies" (R. S. Boardman, A. H. Cheetham, and W. A. Oliver, eds.), pp. 295–303. Dowden, Hutchinson & Ross, Stroudsburg, Pennsylvania.

Blake, D. B. (1973). Coloniality and polymorphism in Bryozoa of the families Rhabdomesidae and Hyphasmoporidae (order Rhabdomesonata). *In* "Animal Colonies" (R. S. Boardman, A. H. Cheetham, and W. A. Oliver, eds.), pp. 361–376. Dowden, Hutchinson & Ross, Stroudsburg, Pennsylvania.

Boardman, R. S., Cheetham, A. H., and Cook, P. L. (1969). Intracolony variation and the genus concept in Bryozoa. *Proc. North Am. Paleontol. Conv., 1969* pp. 294–320.

Boardman, R. S., and Cheetham, A. H. (1973). Degrees of colony dominance in stenolaemate and gymnolaemate Bryozoa. *In* "Animal Colonies" (R. S. Boardman, A. H. Cheetham, and W. A. Oliver, eds.), pp. 121–220. Dowden, Hutchinson & Ross, Stroudsburg, Pennsylvania.

Borg, F. (1926). Studies on Recent cyclostomatous Bryozoa. *Zool. Bidr. Uppsala* **10,** 181–507.

Borg, F. (1933). A revision of the Recent Heteroporidae (Bryozoa). *Zool. Bidr. Uppsala* **14,** 253–394.

Braem, F. (1897). Die geschlechtliche Entwicklung von *Plumatella fungosa. Zoologica (Stuttgart)* **10,** 1–96.

Braem, F. (1951). Über *Victorella* und einige ihrer nächsten Verwandten sowie die Bryozoenfauna des Ryck bei Greifswald. *Zoologica (Stuttgart)* **102,** 1–59.

Braiko, V. D. (1967). Biology of reproduction of *Membranipora zostericola* Nordm. (Bryozoa.) *Zool. Zh.* **46,** 1119–1121.

Brood, K. (1972). Cyclostomatous Bryozoa from the Upper Cretaceous and Danian in Scandinavia. *Stockholm Contrib. Geol.* **26,** 1–464.

Buchner, P. (1924). Studien über den Polymorphismus der Bryozoen. 1. Untersuchungen an japanische Reteporiden. *Zool. Jahrb., Abt. Syst. Oekol. Geogr. Tiere* **48,** 155–216.

Busk, G. (1854). Remarks on the structure and function of the avicularian and vibracular organs of the Polyzoa. *Trans. Am. Microsc. Soc.* **2,** 26–33.

Calvet, L. (1900). Histoire naturelle des Bryozoaires Ectoproctes marins. *Trav. Inst. Zool. Univ. Montpellier* [2] **8,** 1–488.

Cheetham, A. H. (1973). Study of cheilostome polymorphism using principle components analysis. *In* "Living and Fossil Bryozoa" (G. P. Larwood, ed.), pp. 385–409. Academic Press, New York.

Cook, P. L. (1963). Observations on live lunulitiform zoaria of Polyzoa. *Cah. Biol. Mar.* **4,** 407–413.

Cook, P. L. (1968). Observations on living Bryozoa. *Atti Soc. Ital. Sci. Nat. Mus. Civ. Stor. Nat. Milano* **108,** 155–160.

Cook, P. L. (1973). Preliminary notes on the ontogeny of the frontal body wall in the Adeonidae and Adeonellidae (Bryozoa, Cheilostomata). *Bull. Br. Mus. (Nat. Hist.), Zool.* **25,** 243–263.

Emschermann, P. (1972). Cuticular pores and spines in the Pedicellinidae and Barentsiidae (Entoprocta), their relationship, ultrastructure and suggested function, and their phylogenetic evidence. *Sarsia* **51,** 7–16.

Emschermann, P. (1973). Cuticular pores and spines, ion regulatory organs in Entoprocta. *In* "Living and Fossil Bryozoa" (G. P. Larwood, ed.), pp. 261–270. Academic Press, New York.

Gooch, J. L., and Schopf, T. J. M. (1970). Population genetics of marine species of the phylum Ectoprocta. *Biol. Bull. (Woods Hole, Mass.)* **138,** 138–156.

Gordon, D. P. (1968). Zooidal dimorphism in the polyzoan *Hippopodinella adpressa* (Busk). *Nature (London)* **219,** 633–634.

Gordon, D. P. (1971). Zooidal budding in the cheilostomatous bryozoan *Fenestrulina malusii* var. *thyreophora. N. Z. J. Mar. Freshwater Res.* **5,** 453–460.

Gordon, D. P. Biological relationships of an intertidal bryozoan population. *J. Nat. Hist.* **6**, 503–514.

Harmelin, J. G. (1973a). *Callopora minuta* n. sp., nouvelle espèce de Bryozoaires Chilostome (Alderinidae) des côtes françaises de Méditerranée. *Cah. Biol. Mar.* **14**, 29–37.

Harmelin, J. G. (1973b). Les Bryozoaires des peuplements sciaphiles de Méditerranée: Le genre *Crassimarginatella* Canu (Chilostomes Anasca). *Cah. Biol. Mar.* **14**, 471–492.

Harmelin, J. G. (1973c). Morphological variations and ecology of the Recent cyclostome bryozoan "*Idmonea*" *atlantica* from the Mediterranean. *In* "Living and Fossil Bryozoa" (G. P. Larwood, ed.), pp. 95–106. Academic Press, New York.

Harmer, S. F. (1898). On the development of *Tubulipora,* and on some British and Northern species of this genus. *Q. J. Microsc. Sci.* **41**, [N.S.] 73–157.

Harmer, S. F. (1923). On cellularine and other Polyzoa. *J. Linn. Soc. London* **35**, 293–361.

Harmer, S. F. (1926). The Polyzoa of the Siboga Expedition. II. Cheilostomata Anasca. *Siboga Exped.* **28b**, 183–501.

Hastings, A. B. (1943). Polyzoa. I. *'Discovery' Rep.* **22**, 301–510.

Hastings, A. B. (1963). Notes on Polyzoa (Bryozoa). VI. Some setiform heterozooecia. *Ann. Mag. Nat. Hist.* [13] **6**, 177–184.

Hincks, T. (1880). "A History of the British Marine Polyzoa," I-CXLI, pp. 1–601. Van Voorst, London.

Hyman, L. H. (1959). Ectoprocta. *In* "The Invertebrates," Vol. V, pp. 275–515. McGraw-Hill, New York.

Jebram, D. (1973a). Zooid individuality and brooding organs (Bryozoa). *Z. Morphol. Tiere* **75**, 255–258.

Jebram, D. (1973b). Stolonen-Entwicklung und Systematik bei den Bryozoa Ctenostomata. *Z. Zool. Syst. Evolutionsforsch.* **11**, 1–48.

Kaufmann, K. W. (1971). The form and functions of the avicularia of *Bugula* (phylum Ectoprocta). *Postilla* **151**, 1–26.

Knight-Jones, E. W., and Moyse, J. (1961). Intraspecific competition in sedentary marine animals. *Symp. Soc. Exp. Biol.* **15**, 72–95.

Koren, J., and Danielssen, D. C. (1877). "Fauna littoralis Norvegiae," Vol. 3. Bergen, Norway.

Levinsen, G. M. R. (1909). "Morphological and Systematic Studies on the Cheilostomatous Bryozoa." Nationale Forfatterers Forlag, Copenhagen.

Levinsen, G. M. R. (1912). Studies on the Cyclostomata Operculata. *Kgle. Dansk, Vidensk. Selsk. Skr.* **7**, 1–52.

Levinsen, G. M. R. (1916). Bryozoa. Dan.-Eksp. Grönlands Nordöstkyst 1906–1908 **3**, 16. *Medd. Grönland* **43**, 433–472.

Marcus, E. (1938). Bryozoarios Marinhos Brasileiros. II. *Univ. Sao Paulo, Fac. Filos. Cienc. Let. Bol., Zool.* **2**, 1–137.

Marcus, E. (1941). Sôbre Bryozoa do Brasil. *Univ. Sao Paulo, Fac. Filos. Cienc. Let. Bol., Zool.* **5**, 3–208.

Marcus, E. (1949). Some Bryozoa from the Brazilian coast. *Commun. Zool. Mus. Hist. Nat. Montevideo* **3**, 1–34.

Morozova, I. P. (1973). New and little-known structures of Fenestelloidea. *In* "Living and Fossil Bryozoa" (G. P. Larwood, ed.), pp. 327–333. Academic Press, New York.

Moyano, H. I. G. (1967). Sobre la fusión de los colonias de *Membranipora hyadesi* Jullien, 1888. *Not. Mens. Mus. Nac. Hist. Nat. (Santiago)* **126**, 1–4.

Nielsen, C. (1971). Entoproct life-cycles and the entoproct/ectoproct relationship. *Ophelia* **9**, 203–341.

Numakunai, T. (1960). An observation on the budding of the stolon of a bryozoan, *Bugula neritina* Linné. *Bull. Biol. Stn. Asamushi* **10**, 99–101.

Numakunai, T. (1967). Histological observations on the budding of the overwintering stolon of a bryozoan, *Bugula neritina* L. *Sci. Rep. Tohoku Univ., Ser. 4* **33**, 499–508.

Osburn, R. C. (1950). Bryozoa of the Pacific Coast of America. 1. Cheilostomata-Anasca. Allan Hancock Pacific Expedition 14, Vol. 1, pp. 1–269. Allan Hancock Foundation, University of Southern California, Los Angeles.

Osburn, R. C. (1952). Bryozoa of the Pacific Coast of America. 2. Cheilostomata-Ascophora. Allan Hancock Pacific Expedition 14, Vol. 2, pp. 271–611. Allan Hancock Foundation, University of Southern California, Los Angeles.

Osburn, R. C., and Soule, J. D. (1953). Bryozoa of the Pacific Coast of America. 3. Cyclostomata, Ctenostomata, Entoprocta, and Addenda. Allan Hancock Pacific Expedition 14, Vol. 3, pp. 613–841. Allan Hancock Foundation, University of Southern California, Los Angeles.

Pohowsky, R. A. (1973). A Jurassic cheilostome from England. *In* "Living and Fossil Bryozoa" (G. P. Larwood, ed.), pp. 447–461. Academic Press, New York.

Powell, N. A. (1966). Note on the relationship between kenozooecia and avicularia in some Cribrilinidae (Polyzoa). *Trans. R. Soc. N. Z.* **7**, 155–158.

Powell, N. A. (1967). Sexual dwarfism in *Cribrilina annulata* (Cribrilinidae-Bryozoa). *J. Fish. Res. Board Can.* **24**, 1905–1910.

Powell, N. A., and Cook, P. L. (1966). Conditions inducing polymorphism in *Thalamoporella rozieri* (Audouin) (Polyzoa, Anasca). *Cah. Biol. Mar.* **7**, 53–59.

Rey, P. (1928). Action de divers agents physiques ou chimiques sur les mouvements des aviculaires de *Bugula*. *Mem. Fac. Sci. Univ. Paris* **335**, 13–20.

Ryland, J. S. (1968). Terminological problems in Bryozoa. *Atti. Soc. Ital. Sci. Nat. Mus. Ci. Stor. Nat. Milano* **108**, 225–236.

Ryland, J. S. (1969). A nomenclatural index to "A History of the British Marine Polyzoa" by T. Hincks (1880). *Bull. Br. Mus. (Nat. Hist.), Zool.* **17**(6), 207–260.

Ryland, J. S. (1970). "Bryozoans." Hutchinson University Library, London.

Sandberg, P. A. (1973). Degree of individuality in cheilostome Bryozoa: Skeletal criteria. *In* "Animal Colonies" (R. S. Boardman, A. H. Cheetham, and W. H. Oliver, eds.), pp. 305–315. Dowden, Hutchinson & Ross, Stroudsburg, Pennsylvania.

Schopf, T. J. M. (1973). Ergonomics in polymorphism: Its relation to the colony as the unit of natural selection in species of the phylum Ectoprocta. *In* "Animal Colonies" (R. S. Boardman, A. H. Cheetham, and W. A. Oliver, eds.), pp. 247–294. Dowden, Hutchinson & Ross, Inc., Stroudsburg, Pennsylvania.

Schopf, T. J. M., and Gooch, J. L. (1971). Gene frequencies in a marine ectoproct: A cline in natural populations related to sea temperature. *Evolution* **25**, 286–289.

Silén, L. (1938). Zur Kenntnis des Polymorphismus der Bryozoen. Die Avicularien der Cheilostomata Anasca. *Zool. Bidr. Uppsala* **17**, 149–366.

Silén, L. (1941). Cheilostomata Anasca (Bryozoa) collected by Dr. Sixten Bock's expedition to Japan and the Bonin Islands 1914. *Ark. Zool.* **33A**, 1–130.

Silén, L. (1942). Carnosa and Stolonifera (Bryozoa) collected by Prof. Dr. Sixten Bock's expedition to Japan and the Bonin Islands 1914. *Ark. Zool.* **34A**, 1–33.

Silén, L. (1944a). The anatomy of *Labiostomella gisleni* (Bryozoa Protocheilostomata). *K. Sven. Vetenskapsakad. Handl.* [3] **21**, 1–111.

Silén, L. (1944b). The main features of the development of the ovum, embryo and ooecium in the ooeciferous Bryozoa Gymnolaemata. *Ark. Zool.* **35**, 1–34.

Silén, L. (1944c). On the formation of the interzooidal communications of the Bryozoa. *Zool. Bidr. Uppsala* **22**, 433–488.

Silén, L. (1947a). On the spines of *Flustrella* (Bryozoa). *Zool. Bidr. Uppsala* **25**, 134–140.

Silén, L. (1947b). On the anatomy and biology of Penetrantiidae and Immergentiidae (Bryozoa). *Ark. Zool.* **40A,** 1–48.

Silén, L. (1950). On the mobility of entire zooids in Bryozoa. *Acta Zool. (Stockholm)* **31,** 349–386.

Silén, L. (1951). Bryozoa collected by the "Skagerak" expedition, 1946. *Göteborgs K. Vet.-och Vitterh.-Samhälles Handl.* **6,** Ser. B 6, 1–9.

Silén, L. (1966). On the fertilization problem in the gymnolaematous Bryozoa. *Ophelia* **3,** 113–140.

Silén, L., and Harmelin, J.-G. (1974). Observations on living Diastoporidae (Bryozoa Cyclostomata), with special regard to polymorphism. *Acta Zool. (Stockholm)* **55,** 81–96.

Simma-Krieg, B. (1969). On the variation and special reproduction habits of *Aetea sica* (Couch). *Cah. Biol. Mar.* **10,** 129–137.

Smitt, F. A. (1868). Kritisk förteckning öfver Skandinaviens Hafs-Bryozoer. 3. Öfvers. K. Vet.-Akad. Förh. 1867, 279–429. Stockholm 1868.

Soule, D. F. (1973). Morphogenesis of giant avicularia and ovicells in some Pacific Smittinidae. *In* "Living and Fossil Bryozoa" (G. P. Larwood, ed.), pp. 485–495. Academic Press, New York.

Stach, L. W. (1938). Observations on *Carbasea indivisa* Busk (Bryozoa). *Proc. Zool. Soc. London, Ser. B* **108,** 389–399.

Stebbing, A. R. D. (1971). The epizoic fauna of *Flustra foliacea* (Bryozoa). *J. Mar. Biol. Assoc. U. K.* **51,** 283–300.

Stebbing, A. R. D. (1973). Observations on colony overgrowth and spatial competition. *In* "Living and Fossil Bryozoa" (G. P. Larwood, ed.), pp. 173–183. Academic Press, New York.

Stebbing, A. R. D. "Self-recognition" in colonial marine animals–the Bryozoa. (unpublished.)

Theodor, J. L. (1970). Distinction between "self" and "not-self" in lower invertebrates. *Nature (London)* **227,** 690–692.

Utgaard, J. (1973). Mode of colony growth, autozooids, and polymorphism in the bryozoan order Cystoporata. *In* "Animal Colonies" (R. S. Boardman, A. H. Cheetham, and W. A. Oliver, eds.), pp. 317–360. Dowden, Hutchinson & Ross, Stroudsburg, Pennsylvania.

Vigelius, W. J. (1884). Morphologische Untersuchungen über *Flustra membranaceo-truncata. Biol. Zentralbl.* **3,** 705–721.

Viskova, L. A. (1973). On the morphology and systematics of some Late Cretaceous cyclostomatous Bryozoa. *In* "Living and Fossil Bryozoa" (G. P. Larwood, ed.), pp. 497–502. Academic Press, New York.

Woollacott, R. M., and Zimmer, R. L. (1972). Origin and structure of the brood chamber in *Bugula neritina* (Bryozoa). *Mar. Biol.* **16,** 165–170.

Wyer, D. W., and King, P. E. (1973). Relationships between some British littoral and sublittoral bryozoans and pycnogonids. *In* "Living and Fossil Bryozoa" (G. P. Larwood, ed.), pp. 199–207. Academic Press, New York.

Zirpolo, G. (1933). *Zoobothryon verticillatum* (Delle Chiaje). *Mem. Accad. Nuovi Lincei* [2], **17,** 109–442.

7

Feeding in Marine Bryozoans

JUDITH E. WINSTON

I. INTRODUCTION

Bryozoans capture particulate food by means of a funnel of ciliated tentacles—the lophophore. The method they use has been called by Bullivant (1968b) impingement feeding because the cilia of the tentacles create a current which throws the food particles toward the mouth.

While a few workers have cultured bryozoans successfully on several species of algae (see Chapter 8), as yet little is known about the kinds and numbers of organisms captured and utilized by bryozoans in nature. Also important is an aspect of feeding which, as Ryland (1970) remarks, has received no attention, namely, the influence of the dimensions of the tentacular funnel on the types and sizes of particles captured. The relationship of feeding behavior and type of food utilized to the environment in which the various species of bryozoans live is a third parameter about which there is little information.

It is probably true that bryozoans, like other suspension feeders such as barnacles (Crisp, 1964), must be able to utilize a fairly broad diet as most probably depend on phytoplankton, which can show great variations in diversity and species composition at different localities, and at different seasons in the same locality. They must also be capable of adjusting to fluctuations in the abundance of suitable food. Nevertheless, the presence of variations in (a) polypide size, morphology, and tentacle number; (b) mechanisms for selection, e.g., sensory bristles, rejection tracts; of differences in (c) behavior in ingesting particles even in closely related species; and (d) the differential availability of foods in different habitats (some rich in detritus, some in diatoms, etc.) suggests that some degree of specialization has occurred and may be considered important in understanding the diversity of the group.

II. REVIEW OF FUNCTIONAL MORPHOLOGY OF FEEDING

A. The Food-Handling System

As illustrated (Fig. 1) by a zooid of *Conopeum tenuissimum,* the gymnolaemate digestive system is a U-shaped tube enlarged into three distinct sections: it begins with a mouth surrounded by tentacles and ends with an anus located outside the circle of tentacles.

The lophophore in gymnolaemates is a food-gathering organ constructed of a circle of tentacles attached to a basal ridge surrounding the mouth. When the lophophore is extended from the zooecium (as in Fig. 1) it expands into a funnel shape. When it is retracted the tentacles are pulled together into a tight bundle inside the zooecium. Each tentacle is a hollow protuberance and contains a coelomic lumen that is continuous with a ring coelom in the lophophore base (Hyman, 1959).

On each tentacle two lateral bands of cilia move in a laeoplectic fashion; that is, when observed from within the funnel, the metachronal waves pass up the left

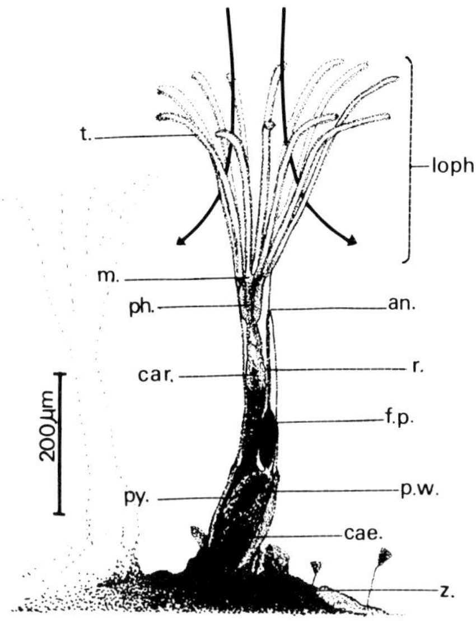

Fig. 1. Polypide portion of the estuarine bryozoan *Conopeum tenuissimum* showing the various regions of the food-handling system. Arrows indicate direction of feeding currents. Abbreviations: loph, lophophore; t, tentacle; m, mouth, ph, pharynx; car, cardia; cae, cecum; py, pylorus; pw, pyloric whirl; fp, fecal pellet; r, rectum; an, anus.

side and down the right side of each tentacle. The wave probably travels around each crotch between tentacles, but not over the tips of the tentacles. Cilia at one point on a tentacle may stop beating without affecting the beat of those further along, and reversals of ciliary action may occur (Bullivant, 1968b; Strathmann, 1973).

The median cells of the inner surface of the tentacle also possess short cilia, but these frontal cilia are sparse and are considered to lack motility (Hyman, 1959; Bullivant, 1968b). Latero-frontal cilia borne by the two cell rows adjacent to the median cell row may function as sensory organelles to detect particles in the water. In addition, short tufts of cilia and longer solitary cilia have been described on the abfrontal surface (Gordon, 1974). Stimulation of these receptors by contact with objects in the water may initiate retraction, and flicking of the tentacles, or the reversal of ciliary motion (Lutaud, 1955, 1973; Bullivant, 1968b; Gordon, 1974).

Tentacles can move independently by means of longitudinal muscle fibers located beneath the peritoneum on their inner and outer surfaces (Hyman, 1959; Smith, L., 1973). Each is provided with sensory and motor nerves connected to the main nerve ring encircling the pharynx (Smith, S., and Watabe, 1971; Smith, L., 1973; Lutaud, 1973; Gordon, 1974).

The mouth, at the bottom of the cup formed by the lophophoral base, opens into a triangular pharynx through a ciliated buccal region. The pharynx walls are made of a vacuolated columnar epithelium whose cells are continuous with the inner epidermis of the tentacles. These cells are true striated myoepithelial cells, lined on their lateral walls with striated myofibrils (Henneguy, 1909; Bullivant and Bils, 1968; Matricon, 1973). Among them in a line down the center of the neural wall of the pharynx occur tall narrow cells, each with a dense central nucleus and one to three bristles projecting into the lumen. These are believed to be sensory cells and could be stimulated by particles in the pharynx and so serve to initiate swallowing or rejection of particles.

The pharynx sucks in water and food particles by means of the contraction of the myofibrils of the myoepithelial cells. Their contraction causes the cell walls to spread, increasing the lumen of the pharynx, and drawing water and food particles just above the mouth into its cavity. This sucking action is repeated until a small ball of food particles forms. By peristalsis of the pharynx this ball is passed into the adjacent unciliated region or esophagus. The foregut region ends with a sphincter which may sometimes project into the lumen of the esophagus (Hyman, 1959).

The mid-gut region begins with a long tubular section, the cardia, whose proximal part may be altered into a gizzard in some ctenostomes. Gizzard development may range from a thickened hardened lining of the proximal part of the cardia as in *Hislopia* (Annandale, 1916) to the complex grinding plates of *Cryptopolyzoon* (Dendy, 1888) or the rows of large and small denticles typical of the Vesciculariidae. Bobin and Prenant (1952), who studied the development of the gizzard in a member of this family (*Bowerbankia*), found that the chitinous teeth which are large dorsally and ventrally and small laterally are each secreted by single epithelial cells.

The cecum is a saclike expansion of the stomach following the cardia. It originates at the turn of the U of the digestive system and extends proximally in species with long narrow zooids (cyclostomes, stoloniferous ctenostomes), but is displaced basally or laterally in those with short broad zooids (carnose ctenostomes, encrusting cheilostomes) (Silén, 1944; Borg, 1926).

Both cardia and cecum have a nonciliated columnar and glandular epithelium. Externally the cardia has both circular and longitudinal muscle fibers, while those of the cecum are mainly circular. As polypides age the cells of these regions appear to accumulate waste materials in inclusions, giving them a darker coloration than the rest of the polypide (see Chapter 10). Most often this region is a reddish or brownish color, though blue, black, and banded tissues have been reported (Waters, 1909, 1918).

The third region of the stomach is the pylorus, a cupuliform area lined with a densely ciliated epithelium and surrounded by circular muscle fibers. It lies between the cecum and the rectum and is separated from the latter by a sphincter (Hyman, 1959).

The rectum (or intestine) is a tubular structure lined with a glandular epithelium and provided with both longitudinal and circular muscle fibers. It leads to the anal opening on the side of the tentacle sheath, outside the lophophore.

B. The Mechanism of Feeding

The mechanism of feeding of the bryozoans and other lophophorates was termed "impingement feeding" by Bullivant (1968b) in analogy with a means of particle separation used in industry. In gymnolaemate bryozoans the lateral cilia of the tentacles in the cone-shaped lophophore beat outward, causing water to flow in through the open end of the inverted cone (see Fig. 1). The current of water is directed into the shallow cup at the base of the cone in which the mouth is located and is deflected upward and outward to exit between the tentacles (Atkins, 1932; Bullivant, 1968b). Bullivant suggested that the sharp deflection of the current over the mouth causes particles to be thrown toward the mouth or into an eddy of comparatively still water above the mouth from which they can be sucked into the mouth by the rapid muscular dilation of the pharynx.

Recently this theory has been challenged by Strathmann (1971, 1973), who believes that impingement could not account for particle capture because the density of the food particles is too close to the density of the seawater. Strathmann has suggested instead that the lophophorates capture particles by means of brief reversals of the lateral cilia brought about by contact of the cilia with passing particles. According to this mechanism the ciliary beat produces a current across the tentacle away from the frontal or inner side (as described for impingement feeding). Particles may either be swept past or retained by local reversals of the beat of the lateral cilia which bounce them toward the mouth in 100 μm jumps. Since the component of the current toward the base of the lophophore is greater near the center of the lophophore, the inward flicking of the tentacles often noted in bryozoans may serve to throw particles into the central current where they can be carried toward the more proximal part of the lophophore. It is possible, of course, that both impingement and ciliary reversal mechanisms may be functioning in bryozoan feeding, but laboratory demonstration of the importance of these processes would depend on sophisticated microcinematographic techniques and has not yet been attempted.

Once inside the pharynx cilia in the buccal region prevent particles from escaping. There is no evidence that mucus plays any role in feeding as ciliates and flagellates were observed by Bullivant (1967) to move freely in the pharynx of *Zoobotryon verticillatum* after ingestion. No secretory cells have been identified on the tentacles; however, Smith (1973) did report that the tentacles were coated with a layer that is "mucoid, probably even mucus." Particles may be rejected from the tract by contraction of the pharynx or in some species (e.g., *Bugula neritina, Farella repens*) by a rejection tract of cilia in the mouth that beats outward toward the abneural pair of tentacles (Bullivant, 1968b).

It is doubtful that particles could travel any distance toward the mouth on the sparse frontal cilia, though Bullivant (1968b) on one occasion observed cells of the diatom *Phaeodactylum* to travel a short distance toward the mouth of *Zoobotryon* in this manner. However, the speed of the feeding current and the rapidity with which particles are usually ingested means that marine bryozoans are able to capture active protistans as large as will enter their mouths (Bullivant, 1968b).

C. Passage through the Digestive System

Once a group of particles enters the pharynx, they are carried into the cardia by successive contractions of the pharynx and esophagus. According to Silén (1944), this swallowing motion occurs at irregular intervals depending on the supply of food. Two or more swallowings may be seen in one minute, or several minutes may pass between swallowings in a given animal. I observed that the rate of swallowing also differs between species. For example, in *Bugula turrita* (fed carmine particles at a high concentration in seawater) particles pass rapidly into the stomach, and clumps of particles do not remain in the pharynx for any observable length of time, while in *Schizoporella biaperta* a clump may stay in the pharynx for a minute or more before descending into the stomach.

Food may remain for some time in the cardia until it is carried further along the digestive tract by the contraction of the circular muscles of the cardia and the relaxation and rounding of the cecum (Silén, 1944). If the species has a gizzard, its grinding action may operate at this time to break down the rigid walls of diatoms and other armored protistans.

Contractions of the circular muscles of the cecum, although they serve to carry particles from one part of the stomach to another, are too simple to provide mixing. Such mixing occurs instead through action of the strong cilia of the pylorus which create a whirling current of stomach fluid (see Fig. 1). In the axis of this current food particles form a solid rotating mass that acts as a stirring rod to mix the other particles in the stomach. The rod of particles is so coherent that it seems there must be some matrix holding the particles together. The composition of this matrix has not been investigated, although Silén (1944) suggested that the substances produced by the secretory cells lining the stomach could have adhesive as well as dissolving qualities. The length of time of rotation of each portion of food in the rod seems to indicate that the process must function in the enzymatic breakdown of the food particles as well as in forming feces (Silén, 1944).

The ctenostome species which have a gizzard have not been observed to possess the digestive whirl, and it may be that development of a gizzard with grinding action makes this type of breakdown and digestion unnecessary (Silén, 1944).

Digestion is both intra- and extracellular. Extracellular digestion occurs at a pH of 6–7 in the cardia and cecum. Intracellular digestion takes place through the absorption of particles by the cells of the stomach epithelium (Ryland, 1970). Ries (1936) followed the digestion of various particles by the ctenostome *Zoobotryon* and found that starch is hydrolyzed in the stomach lumen and the products absorbed by the digestive epithelium. Protein particles are probably digested in the same manner. Fat spheres are phagocytized by the epithelial cells in all regions of the stomach. His studies indicate also that carbohydrate, fat, and protein are stored in the stolon to provide a colony food reserve.

When the food has been digested the remains are carried from the pylorus into the rectum by a sudden depression of the pylorus and opening of the canal into the rectum. The ovoid fecal pellet thus formed passes through the rectum by means of a slow peristalsis of the rectal walls. Normally a fecal pellet is defecated when the polypide is fully protruded, but it can be evacuated by the pressure of a sudden retraction of the polypide.

The rate of passage of material through the alimentary system appears to be dependent on (1) condition of the animal (i.e., recently fed, starved, etc.), (2) quality of the food, (3) concentration of particles (up to a satiation concentration), and (4) behavioral, physiological, and other species-specific differences.

In recent studies of the comparative rates of food passage in several species of bryozoans I found consistent differences between species and orders as to the length of time necessary for particle passage through the alimentary system. In these observations carmine dye suspended in seawater was used as the "food." These nonnutritive particles seemed to be passed through the alimentary system somewhat more rapidly than digestible particles of algae but were useful in studying the comparative rates of food passage.

The stoloniferous ctenostomes, *Bowerbankia gracilis* and *Bowerbankia imbricata,* had passage times of 58 and 45 minutes, respectively. These figures agree fairly well with Bullivant's (1968a) observation that food passage in *Zoobotryon verticillatum,* another stoloniferan, took about 60 minutes. In the carnose ctenostome studied, *Flustrellidra hispida,* passage time was similar, averaging about 45 minutes. It might be expected that *Flustrellidra,* with the pyloric whirl-type of food processing, would be able to pass food more rapidly than the gizzard-bearing ctenostomes in which food seems to be moved along only by weak muscular contractions. However, the polypide of *Flustrellidra* is about twice as big as those of *B. gracilis* and *B. imbricata*, and this may be the reason food takes a relatively long time for passage.

None of the cheilostomes observed took longer than 40 minutes for food passage. The primitive cheilostome *Aetea* passed food through the gut in about 30–40 minutes. Although this species has a polypide similar in appearance to that of a stoloniferan ctenostome, it lacks a gizzard and food is processed by being rotated in the pylorus.

Bugula turrita had the fastest passage time of any species observed, averaging about 21 minutes. Though *Bugula stolonifera* has a polypide very similar in size and morphology, and the two species may occur side by side on a piling, the passage time for *B. stolonifera* was consistently longer, averaging 27 minutes. Polypides of these two species quickly recover from water changes and remain expanded most of the time. Both of the species swallowed the carmine particles with only rare rejection of a particle, though *B. turrita* appeared to consume particles even less discriminantly than *B. stolonifera.*

Zooids of *Electra pilosa* have a longer food passage time—about 37 minutes. This slowness may be due to behavioral factors, as, in the laboratory at least, this species was quite sensitive to motion of the water or too high a concentration of carmine particles, and the polypides remained expanded less of the time than those of some other cheilostome species.

Passage time for *Schizoporella biaperta,* an ascophoran cheilostome with medium-sized polypides, was about 28 minutes, though passage to the cecum was quite slow, and the cardia of some polypides were still lined with red when the first red fecal pellets had appeared in the rectum.

Schizoporella biaperta also ingested carmine particles unselectively, but a larger ascophoran species, *Cryptosula pallasiana,* behaved quite differently in that its polypides seemed to select against the carmine particles. Though dye particles entered the pharynx, most were ejected back into the lophophoral funnel and removed by the streams of water flowing out between the tentacles. Since the polypides did retain a few particles it was possible to get an estimate of food passage time (about 34 minutes).

I was able to examine only one cyclostome, *Crisia eburnea.* The polypides of cyclostomes are more difficult to observe than those of cheilostomes and ctenostomes as they can project from the zooecium only enough to bring the mouth level with the orifice (Borg, 1926). Thus, it is difficult to follow the passage of food particles. In addition, at least under laboratory conditions, the polypides are retracted more than half the time. Estimates of food passage time for *Crisia* ranged from 37 to 57 minutes, but in one case the polypide I observed required almost 2 hours to produce a red fecal pellet. Probably 37 minutes would be fairly close to the minimum time necessary in *Crisia,* considering the behavioral characteristics and the fact that the polypides process food by whirling it in the pylorus.

D. Clearance Rates

The rate at which a suspension feeder can clear a suspension can be an important parameter for the determination of its success in the marine ecosystem. Bullivant (1968a) studied the rate of feeding by the ctenostome *Zoobotryon verticillatum* on different concentrations of algal suspensions over a period of 48 hours. By placing a small portion of the colony in a closed container in which there was a suspension of algae of known concentration and by measuring the change in concentration with time, he found that *Zoobotryon* was able to clear suspensions of four species of algae at rates ranging from 0.152 to 1.05 ml/zooid/hr. These clearance rates are high compared with those of *Daphnia,* oyster veligers, and rotifers, but higher rates are to be expected in smaller animals because of their higher rates of respiration. Furthermore, the high clearance rate suggests that although the colony of *Zoobotryon* may be relatively large the zooids behave metabolically as individuals (Bullivant, 1968a). One could also point out that (1) especially in *Zoobotryon* there is a large nonzooidal component of the colony to be nourished, and (2) there is active asexual growth in all bryozoans for which nutrients are needed. The clearance rates obtained show that this bryozoan has a maximal ingestion rate at algal cell concentrations which approach the highest it experiences in nature.

Strathmann (1973) attempted to measure clearance rates in bryozoans by an indirect method. He used dissected tentacles (kept on a cooled microscope stage) and measured the velocity of particles past the lateral cilia by means of rapid (50/sec) multiple exposure photographs. According to Strathmann, ". . . If particles are retained from a 20 μm wide zone on each side of a tentacle and enter this zone at a velocity of 1 cm/sec, then the maximum clearance rate is about 1.4 ml/h/cm of tentacle. . . . As an upper estimate of the maximum clearance rate at this current velocity, one can take the product of the tentacle length times the tentacle number times 1.4 ml/h/cm." He offers theoretical maximum clearance rates for two species *Membranipora villosa* (0.9 to 1.6 ml/hr/zooid) and *Schizoporella unicornis* (2.0 to 2.5 ml/hr/zooid) (see Table II). Interestingly enough, although Strathmann does not give a theoretical rate for *Zoobotryon,* if this parameter is calculated, using a tentacle number of 8 and a tentacle length of 320 μm [estimated from diagram given by Bullivant (1967)] the maximum rate turns out to be 0.358 ml/hr/zooid. This figure is similar to the experimental average of 0.368 ml/hr/zooid determined by Bullivant for *Zoobotryon* colonies fed *Monochrysis* (the algal food on which they grew best). Bullivant's results also showed that *Zoobotryon,* like many other suspension feeders, feeds fairly constantly.

III. VARIATION IN THE BRYOZOAN POLYPIDE

A. Tentacle Number

Hyman (1959) stated that the number of tentacles varies in different species of marine bryozoans from 8 to about 34 and noted also that the number is not always constant within individuals of the same species. I compiled from the literature information on tentacle number in order to learn (1) range of variation in each of the orders and in selected families, genera, and species, and (2) whether tentacle number can be related to morphological or ecological parameters.

The distribution of tentacle number counts within each major subdivision of marine bryozoans is shown in Fig. 2. In calculating the distributions, a single tentacle number was used for each species; in species with a range of tentacle numbers, the average or median number was calculated if only the range was given, but the modal count was chosen if known.

As Fig. 2 shows, range of tentacle number is least in the cyclostomes (Fig. 2a), varying from 8 to 16. Tentacle number is consistent in some species and

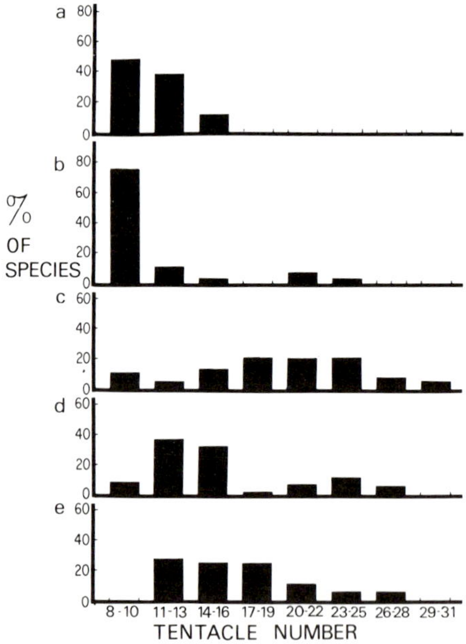

Fig. 2. Distribution of tentacle number in marine bryozoans. (a) Cyclostomes (31 species). (b) Stoloniferous ctenostomes (29 species). (c) Carnose ctenostomes (39 species). (d) Anascan cheilostomes (60 species). (e) Ascophoran cheilostomes (53 species).

genera (especially in those with few tentacles), while in the species where variation occurs it is generally on the order of only 2 or 3 tentacles.

The stoloniferous ctenostomes (Fig. 2b) also have the majority of polypides with a small number of tentacles (8–10). The range of variation in most species is not great (e.g., polypides of the family Vesciculariidae almost always have 8 tentacles); however, two species of the Mimosellidae, *Farella repens* and *Farella gigantea,* show large ranges of 11 and 7 tentacles, respectively.

The carnose ctenostomes (Fig. 2c) present quite a different picture. This group is most variable with respect to the number of tentacles present within a species; many have a range of 10 or 11 tentacles. The group has few species with 8 to 13 tentacles, and most have between 17 and 35 tentacles. Carnose ctenostomes have the highest tentacle numbers of any of the groups; *Flustrellidra hispida* may have up to 34.

Tentacle number in the anascan cheilostomes (Fig. 2d) has a bimodal distribution with a large peak between 11 and 16 and a smaller one between 20 and 28. Nine is the smallest number of tentacles found in the anascans.

None of the ascophoran cheilostomes has fewer than 10 tentacles (Fig. 2e). In this group there seems to be one peak in tentacle number between 11 and 19. In both the anascans and the ascophorans a range of 3 to 5 in tentacle number within a species is common. Some families, such as the Reteporidae, are quite invariable with respect to tentacle number (tentacle numbers of 11 to 15 occur), while others, e.g., Bugulidae (12–30) and Membraniporidae (9–25) are more highly variable.

The evidence cited above, though scanty (based on information from 31 species of cyclostomes, 68 species of ctenostomes, and 113 species of cheilostomes) supports the statement of Waters (1904) that while tentacle number ". . . is not likely to be one of the most useful characters of classification, . . . it is certainly a character which should be recorded; and we are now seeing that in many genera there is always approximately the same number, so that a great variation from the normal should cause us to carefully examine other characters to make sure that our classification is correct." Waters cited tentacle number wherever possible in the many papers and monographs he produced, but the trend among later systematists has been to ignore almost entirely polypide morphology and concentrate on the anatomy of hard parts. This is unfortunate, for such information is not only valuable in pointing out a discrepancy in classification, but may be useful also in elucidation of the ecology and feeding behavior of the various species and even of the higher groups.

Theoretically a larger and/or more efficient food-collecting device might be expected in suspension feeders which live at great depths where food particles are less abundant (Bullivant, 1968b). Deepwater species of *Bugula,* for example, have longer and more numerous tentacles than shallow-water species of the same genus (Braem, 1940). Of course, some families living in shallow water have a large number of tentacles, and some which live at considerable depths have a small number; ecological explanations must be found for these species also. For example, the erect fan- or cup-shaped colonies of the Reteporidae may be such effective particle traps that the autozooids lining their surfaces do not need long or numerous tentacles to obtain sufficient food. Tentacle number, while it may provide a clue to feeding efficiency, must be correlated with morphology and ecology in evaluating a species' strategy for maintaining itself.

B. Size of Mouth and Lophophore

To learn more about the role of differences in morphology, especially the morphology of the soft parts, with respect to feeding biology, I measured various morphological parameters of living individuals of 10 shallow water species of bryozoans from the Woods Hole, Massachusetts, region. Table I gives the results of these measurements.

It can be seen from this table that there exists considerable difference between these species regarding such features as tentacle length, lophophore diameter, mouth size, as well as zooecial length and width. Tentacle number varies from 8 to 30, tentacle length from 0.2 to 1.0 mm, and the diameter of the expanded lophophore from 0.2 to 1.4 mm. While the absolute differences may not seem very great, one can see that the largest species observed, *Flustrellidra hispida,* had a tentacle length over 5 times that of the smallest, *Crisia eburnea,* a factor which must surely influence the food-collecting efficiency of the lophophore.

All of these variables are related. Figure 3 shows the relationship of tentacle number to tentacle length and lophophore diameter. This evidence suggests that a large number of tentacles is able to form a more efficient feeding cone as it is able to mourish a larger animal. As mentioned in Section II,D, Strathmann (1973) measured the velocity of particles past the lateral cilia of bryozoan tentacles and used that information to calculate a maximum clearance rate of approximately 1.4 ml/hr/cm of tentacle. An upper estimate of the maximum clearance rate for a bryozoan species is then given by the product of the tentacle length times the number of tentacles times 1.4 ml/hr/cm:

$$\text{Clearance rate (ml/zooid/hr)} = [\text{T\#}]\ [\text{TL (cm)}]\ [1.4]$$

This theoretical maximum clearance rate can then be used to compare the efficiency of the lophophores of bryozoans of different sizes and with various numbers of tentacles. Doing this for the species from the Woods Hole area on which measurements were made, we find that maximum clearance rate does increase with size of lophophore and number of tentacles (Table II).

Figure 4 shows the relationship between the diameter of the mouth and the diameter of the lophophore. This graph suggests that animals with lophophores

TABLE I
Measurements on Ten Species of Woods Hole Area Bryozoans

Species	Tentacle number[a]	Tentacle length (mm)	Lophophore diameter (mm)	Length of zooecium (mm)	Width of zooecium (mm)	Length of orifice (mm)	Width of orifice (mm)	Diameter of mouth (mm)	Size of fecal pellet (mm)
Cyclostome									
Crisia eburnea	8	0.193	0.213	0.582	0.102	0.079	0.079	0.019	0.023 × 0.053
Cheilostome									
Aetea sp.	12	0.291	0.220	0.675	0.095	0.068	0.047	–	–
Electra pilosa	12–14	0.336	0.360	0.535	0.286	0.076	0.108	–	0.036 × 0.095
Bugula stolonifera	13	0.386	0.536	0.552	0.167	0.100	0.109	0.042	0.070 × 0.157
Bugula turrita	14	0.421	0.488	0.686	0.159	0.095	0.095	0.038	0.044 × 0.097
Schizoporella biaperta	13–14	0.504	0.433	0.427	0.250	0.079	0.103	0.047	0.040 × 0.087
Cryptosula pallasiana	16	0.749	0.796	0.742	0.479	0.230	0.188	0.069 × 0.047	0.080 × 0.127
Ctenostome									
Bowerbankia gracilis	8	–	–	0.711	0.175	–	–	–	–
Bowerbankia imbricata	10	0.410	0.364	0.897	0.174	0.047	0.047	0.033	0.009 × 0.085
Flustrellidra hispida	27–30	1.01	1.42	1.06	0.372	0.246	0.246	0.090	0.100 × 0.240

[a]Numbers represent means of measurements on 3–6 individuals of the species.

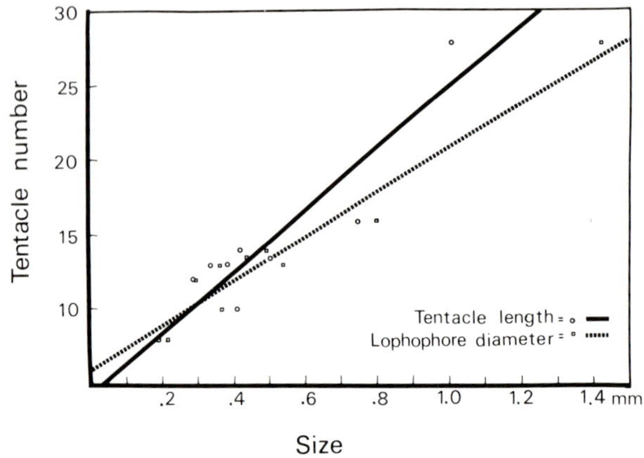

Fig. 3. Relationship of tentacle number to tentacle length and lophophore diameter for nine species of Woods Hole area bryozoans.

of increasing sizes also have mouths of increasing sizes and thus might be utilizing different size fractions of the phytoplankton. *Crisia,* for example, could not ingest any of the larger diatoms and dinoflagellates and would probably feed on nannoplankters, while the two species of *Bugula* and *Schizoporella biaperta* with larger mouth diameters could be expected to handle a more varied diet.

TABLE II
Theoretical Maximum Clearance Rates[a] for Woods Hole, Chesapeake Bay, and West Coast Bryozoans

Species	Tentacle number	Tentacle length (cm)	Maximum clearance rate (ml/hr/zooid)
Crisia eburnea	8	0.019	0.21
Zoobotryon verticillatum	8	0.032	0.358
Bowerbankia imbricata	10	0.040	0.56
Aetea sp.	12	0.030	0.50
Conopeum tenuissimum	11–12	0.029	0.45–0.49
Electra pilosa	12–14	0.033	0.56–0.64
Bugula stolonifera	13	0.039	0.71
Bugula turrita	14	0.042	0.83
Schizoporella biaperta	13–14	0.050	0.91–0.98
Membranipora villosa	16	0.04–0.07	0.9–1.6
Cryptosula pallasiana	18	0.075	1.68
Schizoporella unicornis	18	0.08–0.10	2.0–2.5
Flustrellidra hispida	27–30	0.10	3.78–4.20

[a]At current velocity of 1 cm/sec.

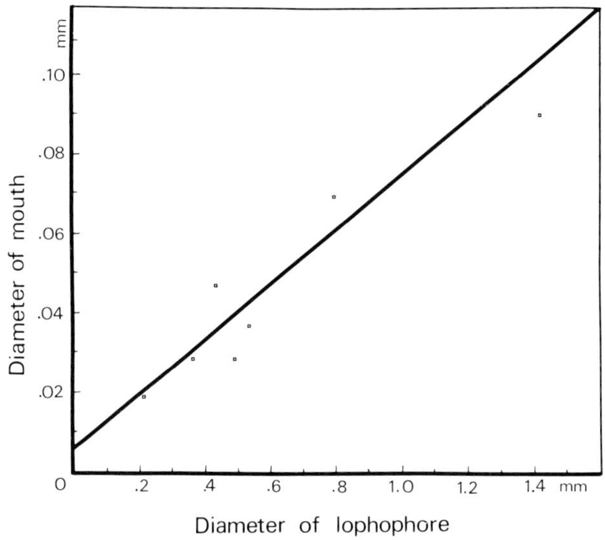

Fig. 4. Relationship between diameter of mouth and diameter of lophophore in seven species of Woods Hole area bryozoans.

C. Lophophore Shape and Other Possibilities for Increasing Current Flow

While the lophophore of marine bryozoans has been described as a circular funnel, not all species have lophophores with tentacles of the same length. Among the species common around Woods Hole, Massachusetts, for example, *Schizoporella biaperta* has a lophophore in which the adneural tentacles are shorter (and more curving) than the neural tentacles so that the funnel appears slanted in side view.

As shown in Plate I, shape of the funnel also varies. In cyclostomes such as *Crisia* and ctenostomes such as *Bowerbankia* and *Victorella* the tentacles are held straight and remain rather close together when the lophophore is expanded. In forms with larger lophophores the tentacles may remain straight at the tips as in *Membranipora tenuis,* or they may curve outward at tips and be less tightly adherent so that the funnel has more of a bell shape, as in *Conopeum tenuissimum, Alcyonidium verrilli,* or *Bugula* spp., among others. In the largest species I have observed, the ctenostome *Flustrellidra hispida,* the bell shape is extremely pronounced and the tentacles are capable of a great range of movement. These polypides can bring the tentacles of the lophophore together at the tips in a purselike arrangement, can wave the tentacles independently of each other, or with an extremely rapid motion can flatten the bell completely into a disc, then bring it back into a bell shape again.

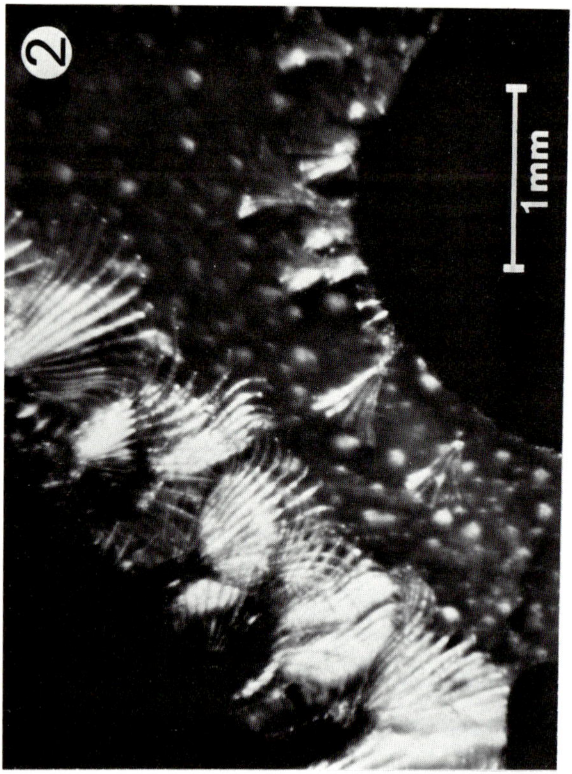
2
1 mm

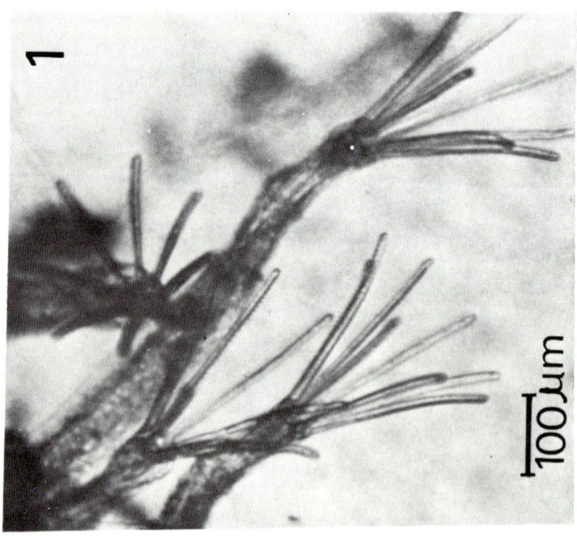
1
100 µm

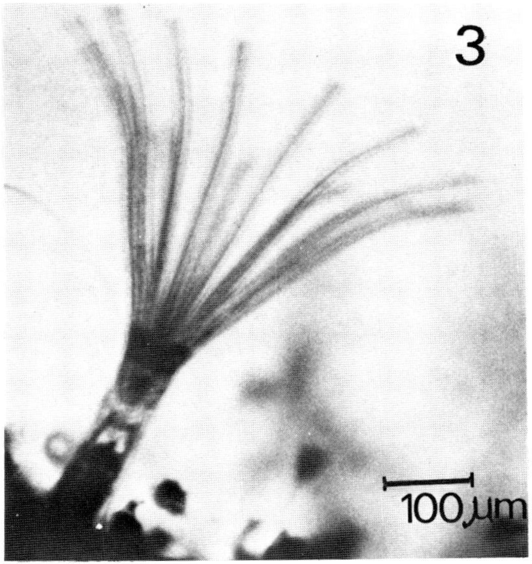

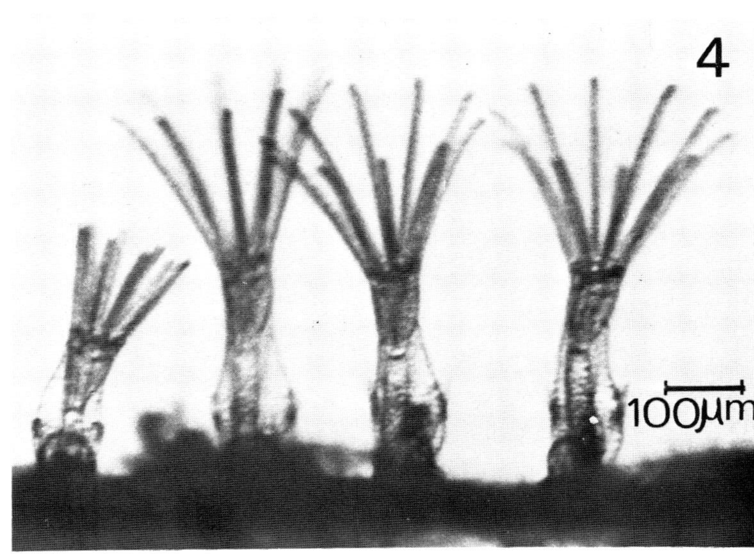

Plate I. Examples of variation in size and shape of the lophophore in marine bryozoans. (1) *Bowerbankia gracillis,* a small simple lophophore, with 8 tentacles. (2) Upper left: *Flustrellidra hispida,* an extremely arge bell-shaped lophophore with 30 or more tentacles. Lower right: *Conopeum tenuissimum.* (3) *Conopeum tenuissimum,* a medium-sized slightly bell-shaped lophophore. (4) *Membranipora tenuis,* a medium-sized stiffly held lophophore.

Other sedentary colonial or gregarious suspension feeding organisms have evolved methods of increasing the flow of water over the feeding surfaces of their aggregations. For example, Knight-Jones and Moyse (1961) described the ridging and hummocking of the miniature reefs of the serpulid *Filograna,* and of crowded populations of barnacles which result in bringing additional food within the reach of all individuals by increasing the turbulence of water currents, and by increasing the surface area available for feeding.

This same aggregation effect may obtain in bryozoan colonies. Where zooids are small and widely spaced—as in *Bowerbankia* or the *Crisia*-like cyclostomes—small simple lophophores are sufficient for gathering food, but in massive colonies some method of increasing current flow may be necessary, either through increasing the size of the lophophore and the number of tentacles, or creating a pattern in the arrangement of the lophophores that would draw a greater current, or by ridging and hummocking of the surface of the skeleton of the colony to create greater turbulence. Banta *et al.* (1974) have observed that lophophores of large encrusting colonies of a *Membranipora* species from California are arranged in a distinct pattern when expanded so that water passing through them is channeled into excurrent chimneys. They also suggest that raised skeletal protuberances in Recent and fossil cheilostomes and monticules of fossil stenolaemate Bryozoa may be skeletal expressions of such excurrent channels which would have aided efficient water flow through the colonies. Branching colonies like those of the Cellarioidea also permit each polypide a slightly different position so that the colony can encounter food-bearing water in all directions, and as mentioned previously, the fan- or funnel-shaped colonies of the Reteporidae may likewise be assumed to be highly efficient in creating feeding currents.

Although the fact that food resources can presumably be shared by all members of a colony might make the need for efficient food gathering seem less critical, making food available to as many zooids as possible may have a considerable effect on the colony's survival and particularly its ability to reproduce.

D. Behavioral Differences

Behavioral parameters relevant to the feeding biology of bryozoans range from the amount of time the zooids keep the lophophore expanded, to the range of movement of which the polypide portion is capable, and to differences in the ingestion of food materials. As mentioned in Section II,C, some bryozoan species ingest carmine dye readily while others actively reject it. Those species which I have observed to be most selective in rejecting the dye include *Flustrellidra hispida, Cryptosula pallasiana,* and *Conopeum tenuissimum*—all species characteristic of estuarine areas where large quantities of suspended particles might be expected to cause difficulty for suspension feeders not adapted to deal with them. This situation is similar to the adaptation described by Menzel (1955) for *Crassostrea virginica.* He found that the euhaline oyster *Ostrea equestris* feeds indiscriminately on carmine, charcoal, plankton, detritus, silt, etc., until its stomach is completely full. However, in the estuarine species *Crassostrea virginica* particles are carried to the palps and formed into a ball; if edible they are ingested, but if inedible they are conveyed to the mantle to become pseudofeces. Thus *Crassostrea* is able to live in silty turbid waters which are fatal to *Ostrea.*

Differences in the rate of ingestion of particles may occur even in closely related species such as *Bugula turrita* and *B. stolonifera.* While using micronic beads (plastic spherules several μm in diameter) to study the size of particles available to these two species, I noted a consistent difference with respect to the number of beads swallowed (Dudley, 1970). When fed a seawater suspension of beads ranging in size from 1 to 30 μm the average number of beads ingested by *B. stolonifera* polypides (counting all those that swallowed at least one bead) was 2.2, the average number swallowed by *B. turrita* polypides was 7.8. Whether this difference is attributible to more active rejection or capture of beads by one species (note that *B. turrita* should have a slightly higher maximum clearance rate according to Table II), or to differences in physiology (e.g., food passage time is faster in *B. turrita* than in *B. stolonifera*), it is most interesting, for these results show that there can be a significant difference in the feeding behavior even of two species with similar ecology and polypide morphology.

IV. BRYOZOAN FEEDING AND CULTURE

A. Review of the Literature on Feeding and Culture

There are three possible methods by which suspension feeders could achieve some degree of trophic separation: by means of selection based on (1) physical characteristics of the particles such as size, shape, or density; (2) chemical characteristics of the particles such as substances present in or released by them; and (3) behavioral characteristics based on the different behavior patterns of the various species or groups of suspension feeders.

Behavioral selection and size selection no doubt play a part, but also important is possible preference for different portions of the phytoplankton based on such physical considerations as presence of a siliceous, cellulose, or calcareous wall structure, and chemical considerations based on the chemical composition of the species of phytoplankton algae themselves.

Until recently the literature on the food items of bryozoans was sparse. Hasper (1912) mentioned feeding young zooids of *Bowerbankia pustulosa* on the diatom *Nitschia closterium* forma *minuta* and the cyanophyte *Pleurococcus mucosus.* While there is no mention of culturing marine bryozoans in the literature again until recently, there have been some observations on gut contents of polypides.

Hentschel (1922) described the gut contents of *Membranipora tuberculata* encrusting *Sargassum* in the Sargasso Sea as containing diatoms including *Coscinodiscus,* coccolithophores–especially *Coccolithophora leptopora,* peridinian dinoflagellates–mostly gymnodinian, and surprisingly the nematocysts of *Physalia.*

Hunt (1925) examined the gut contents of *"Lepralia" foliacea* and *Cellaria* spp. from localities on the Plymouth fishing grounds and noted that they contained small diatoms, silicoflagellates, peridinians, coccolithophores, algal cysts, and detritus. He also observed the capture of small flagellates by bryozoans and considered that they and perhaps bacteria might also serve as food for them.

Marcus (1941) illustrated a horizontal section through a polypide of *Thalamoporella evelinae* showing diatoms in the gut and in cells of the cecal walls. This figure, reproduced in Hyman (1959) has been taken by many as evidence that diatoms are the main food source of marine bryozoans.

Schneider (1959, 1963) cultivated *Bugula avicularia* on the colorless dinoflagellate *Oxyrrhis marina,* which itself was fed on a green flagellate, *Dunaliella* sp., but noted that while the polypides ingested both the *Oxyrrhis* and the *Dunaliella,* the *Dunaliella* cells left the gut intact.

Jebram (1968) also used *Oxyrrhis marina* (fed upon *Dunaliella*) and *Cryptomonas* sp. (Cryptophyceae) to culture several bryozoan species. Species he found to grow well on these food sources included *Alcyonidium* sp., *Bowerbankia gracilis, Farella repens, Electra crustulenta, E. monostachys, Conopeum seurati, C. reticulum,* and *Bugula stolonifera.* In later work (1970, 1973) Jebram concentrated on the effect of growing Bryozoa in an apparatus in which water currents could be produced and varied (see Chapter 8). In his 1970 paper, he noted qualitative effects of food on growth: larger zooids with larger and differently proportioned guts–changes in the portion of the gut were produced under favorable nutritional conditions. His 1970 experiments included the use of *Monochrysis* as food. On a *Monochrysis* diet *Farella repens* grew slowly, but colonies did not become sexually mature. *Bowerbankia gracilis* colonies did become mature, but had a growth form containing comparatively few zooids spaced along very long stolons. In contrast, the use of *Cryptomonas* as food caused rapid and regular maturity of *Farella* colonies, as well as dense growth of mature zooids along more slowly growing stolons in both *Bowerbankia gracilis* and *B. imbricata.*

Jebram (1975) also studied both qualitative and quantitative effects of several foods (*Cryptomonas* sp., *Monochrysis lutheri, Cyclotella nana, Phaeodactylum tricornutum, Oxyrrhis marina,* and *Dunaliella marina*) upon the growth of *Conopeum seurati* and *Bowerbankia gracilis.* Of the single food diets *Oxyrrhis marina* and *Cryptomonas* sp. proved to be most favorable, and the two diatoms *Cyclotella* and *Phaeodactylum,* least favorable. Mixtures of two foods also were found to induce good growth. For *Conopeum seurati* growth on a mixture of *Cryptomonas* and *Dunaliella* was better than on either of the foods alone, while growth on a mixture of *Oxyrrhis* and *Dunaliella* was initially more slow than on *Oxyrrhis* alone, but eventually increased above even that for the *Oxyrrhis* diet.

Bullivant (1967, 1968a) cultured the ctenostome *Zoobotryon verticillatum* on several species of algae and found that the colonies grew well on the chrysophyte flagellate *Monochrysis lutheri* and the aberrant diatom *Phaeodactylum tricornutum,* showed some growth on the diatom *Cyclotella nana* and the coccolithophore *Cricosphaera carterae,* but were unable to grow on the dinoflagellate *Amphidinium carterae,* the green flagellate *Dunaliella tertiolecta,* and the diatom *Thalassiosira fluvaiatilis.* His results indicate that, as for other marine invertebrates which have been cultured (e.g., clam and oyster larvae, barnacle larvae, copepods), different algae apparently have different food values to *Zoobotryon.* In addition, Bullivant pointed out that although the ctenostome *Zoobotryon,* which has a gizzard, is able to grow when fed the diatom *Phaeodactylum,* the cheilostome *Bugula neritina,* which lacks a gizzard, cannot.

B. Growth of *Conopeum tenuissimum*

In order to assess better the effects of different unicellular algae on the growth of a bryozoan without a gizzard, I carried out a series of culture experiments using the estuarine cheilostome *Conopeum tenuissimum* (Canu) (Winston, 1976). These experiments took place in 1972 at the Virginia Institute of Marine Science and the Marine Biological Laboratory, Woods Hole, Massachusetts.

Slides on which young (1–4 days old) colonies of *C. tenuissimum* had settled were cleaned of all but one central colony and placed in rectangular glass dishes (1 dish/treatment). The dishes were aerated continuously, and three times a week the feeding medium was replaced using fresh filtered seawater and a measured amount of phytoplankton suspension; at each feeding the colonies were carefully cleaned of debris.

Experiments ran for about 40 to 42 days for two reasons: In this amount of time reproduction would have occurred in nature and it was desired to see if some foods supported reproductive activity; and by this time the colonies could grow to the edge of the slides.

1. Foods Producing Good Growth

Two single algal foods, *Dunaliella tertiolecta* and *Gymnodinium simplex,* were found capable of supporting good growth of *Conopeum tenuissimum* (Fig. 5).

Dunaliella tertiolecta is a chlorophyte flagellate about 6 × 11 μm in size. It has been used several times before in the culture of invertebrates, but the results have been variable.

Colonies fed *Dunaliella* showed extremely good growth over the 42 day period; mean zooid number increased from 4 to 2484 (N = 7). In addition to having six times as many zooids as colonies raised on fair foods (e.g., *Monochrysis*), colonies fed *Dunaliella* were different in shape. As well as growing along the major axes that developed in the young colony, buds also developed distolaterally and distally to fill in the space between the branches, giving a *Dunaliella*-fed colony a more circular shape in contrast to the double-fan shape of a *Monochrysis*-fed colony. At the end of the culture period all colonies

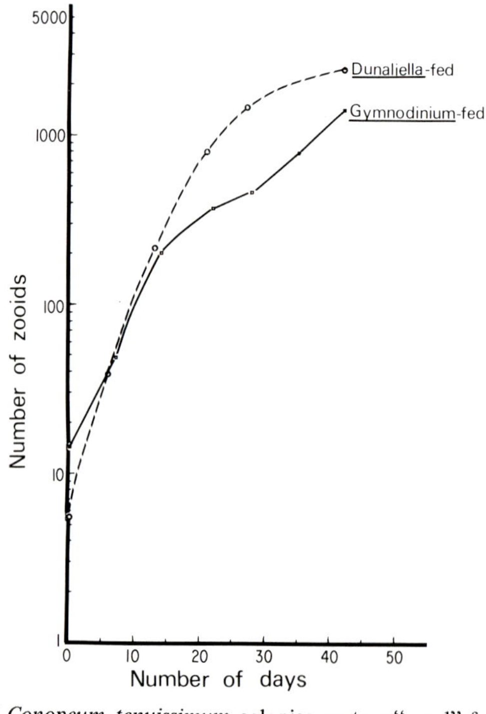

Fig. 5. Growth of *Conopeum tenuissimum* colonies on two "good" foods—the chlorophyte flagellate *Dunaliella tertiolecta* and the dinoflagellate *Gymnodinium simplex.*

appeared healthy, and observations of the zooids at this time showed some with polypides with intertentacular organs and ovaries containing developing eggs.

Gymnodinium simplex is a small (8 × 10 μm) unarmored dinoflagellate which occurs in the York River waters with some frequency (McKiernan, 1968), and thus appeared to be a good choice as a naturally occurring food for *Conopeum.* Due to difficulties with the culture of this alga, the *Gymnodinium* experiments had to be begun late in August at VIMS and continued in September at Woods Hole, and the detrimental effects of travel on the algae and the colonies can be seen in the middle dip in the growth curve. In spite of this problem the colonies grew quite well.

Total mean zooid increase (N = 6) was 1394 from 15 to 1409 individuals. When colonies were examined for evidence of reproduction two of the six contained zooids with sperm as well as some with developing eggs; thus in spite of complications due to transporting the cultures, *Gymnodinium simplex* appeared to be an adequate food for the growth and maturation of *Conopeum tenuissimum.*

It is also of interest that only the *Conopeum* colonies grown on *Gymnodinium* showed the reddish-brown color in the polypides which is characteristic of this species when taken from its natural environment. This color presumably derives from the pigment contained in the chloroplasts of the dinoflagellate. *Conopeum* colonies also showed good growth on a two-food and a three-food combination (Fig. 6).

Colonies fed a mixture of *Monochrysis lutheri* and *Dunaliella tertiolecta* grew quite well. Mean zooid number (N = 8) increased from 9 to 1291. Two colonies appeared to be developing reproductive structures at the end of the period. Growth on this two-food mixture was better than on any single food except *Gymnodinium* and *Dunaliella.*

A mixture of three foods, *Monochrysis–Dunaliella–Cyclotella,* supported slightly better growth than did the mixture of two foods. The growth curve was similar in shape, but a somewhat higher final mean number of zooids was reached–1629, an increase of 1623, from 6 (N = 6). When these colonies were examined at the end of the experimental period two were found to have zooids containing ovaries, indicating that this mixture of three foods was adequate for both growth and reproduction of *Conopeum.*

2. *Foods Producing Fair Growth*

Fair growth of the *Conopeum* colonies on two single foods is shown in Fig. 7. *Monochrysis lutheri* is a small (2.5 × 7 μm) chrysophyte flagellate which has been used in many invertebrate culture experiments. During 42 days of culture the mean zooid number (N = 5) increased from 8 to 370. The colonies were straggling and star-shaped in form with growth occurring mainly from the distal

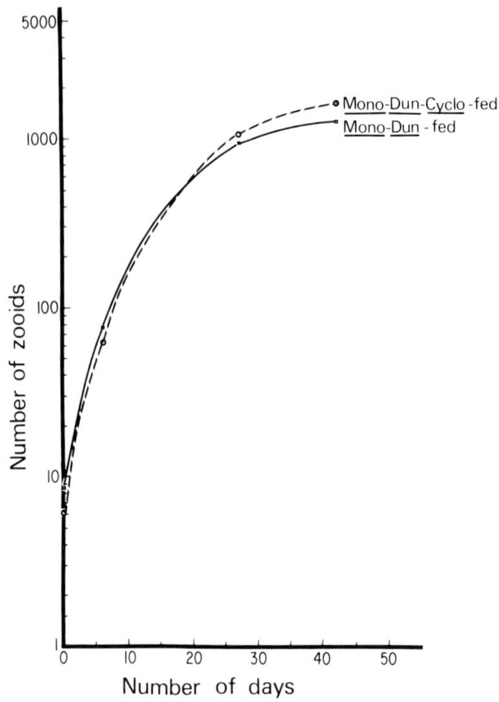

Fig. 6. Growth of *Conopeum tenuissimum* colonies on two combinations of foods–a two-food mixture (*Monochrysis–Dunaliella*) and a three-food mixture (*Monochrysis–Dunaliella–Cyclotella*).

ends of chains of zooids. No signs of readiness for reproduction (zooids with sperm, ovaries, or polypides with intertentacular organs) were found in any of the colonies at the end of the experimental period.

Va-12 is an unidentified chrysophyte flagellate (4.5 × 4.2 μm) which has been used in the culture of oyster larvae at the Virginia Institute of Marine Science. It proved to be only a "fair" food for *Conopeum*. Final mean zooid number ($N = 7$) was 201, an increase of 194 from 7 at the start of the culture period. No evidence of reproductive activity was observed in these colonies.

3. Foods Producing Poor Growth

Results of the culture of *Conopeum tenuissimum* on foods that produced only poor growth are shown in Fig. 8. *Cyclotella nana* is a small (6.4 μm in diameter) centric diatom which has also frequently been utilized in the culture of marine invertebrates. This species, however, failed to support good growth of *Conopeum tenuissimum*. Mean zooid number ($N = 8$) increased only from 6 to

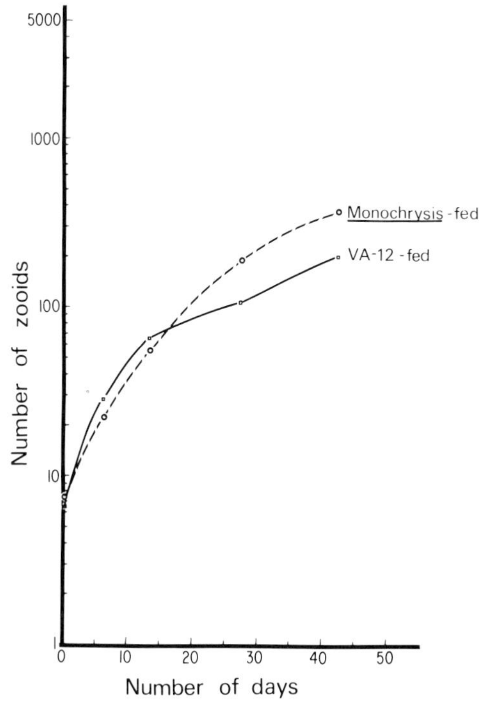

Fig. 7. Growth of *Conopeum tenuissimum* colonies on two "fair" foods–the chrysophyte flagellates *Monochrysis lutheri* and Va-12.

35. These colonies seemed in poor condition at the end of the experimental period, with functional polypides only in the outmost zooids.

Nannochloris occulata (1 – 2 μm), a nonmotile chlorophyte, has been used successfully in larval culture work at VIMS, but turned out to be a poor food for *Conopeum* colonies. Final mean zooid number (N = 5) was 42 individuals, an increase of 35. As the mean growth curve (Fig. 8) shows, growth on this alga was comparable to that on *Cyclotella* and very different from growth on the other chlorophyte species *Dunaliella tertiolecta.*

V. GENERAL DISCUSSION

A. The Ecology of Suspension Feeders

Suspension feeders comprise a large group of animals from many phyla, which remove particulate matter, detritus or small organisms from the water by

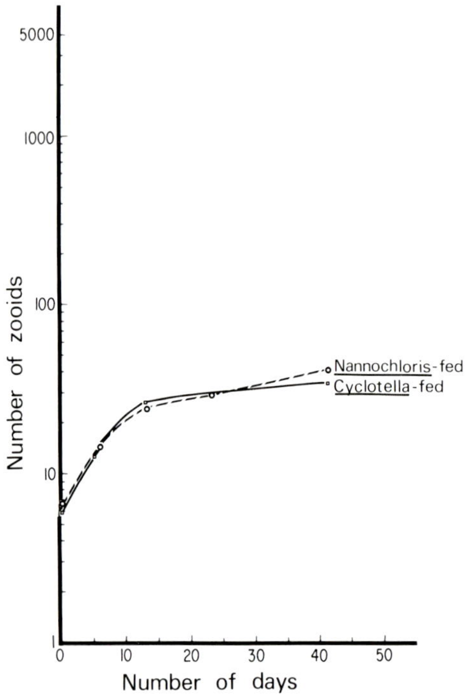

Fig. 8. Growth of *Conopeum tenuissimum* colonies on two "poor" foods–the non-motile chlorophyte *Nannochloris occulata* and the centric diatom *Cyclotella nana.*

means of cilia, mucous nets, setal filters, or other mechanisms. It has been generally believed that suspension feeders do not show selection of food items, but feed in a quantitative fashion, removing particles from the water at a uniform rate regardless of their food content. However, while this may hold true for many large filter feeders, particularly bivalve mollusks, there is increasing evidence that smaller organisms employ selection for various foods, as well as showing marked differences in filtering rate, depending on diet used, or even when one species of phytoplankton food is used at different concentrations.

Trophic parameters for benthic and planktonic suspension feeders may differ with respect to size, feeding organ, method of feeding, and nature of food ingested. Clearance rates may also be important, but they have been measured in different ways and show so much variation that it is difficult to make generalizations about them that are applicable to major groups of organisms. In addition, clearance rates themselves do not tell much about the actual food of the organisms in question, as many suspension feeders will clear particles on which they cannot grow. Although information on the feeding biology of many of these groups is far from complete, it is possible to make a few generalizations about the fraction of the plankton that each one is able to utilize.

The sponges (Reiswig, 1971) appear to be mainly dependent upon particles less than 50 μm in size, particularly the very smallest particles, bacteria, and unresolvable particulate organic matter. Protozoans (foraminiferans and ciliates), and meioinvertebrates are also unable to use anything but relatively small particles–nannoplankton and the smaller members of the microplankton (less than 100 μm). Though little is known of the food preferences of protozoans, Rapport *et al.* (1972) found that *Stentor* preferred protozoan to algal prey.

Edmondson (1965) found that the relative significance of algal food organisms to the rotifer *Keratella* was partly, but not entirely, related to size, and that some algae supported reproduction of the rotifer while others did not.

The crustacean setose suspension-feeding mechanism apparently precludes ingestion of the smallest particles (lower limit of 2–10 μm depending on species), but otherwise shows a range of variation with respect to particles accepted. Crisp (1964) has shown that feeding is not indiscriminate in barnacles. While they feed on a wide size range of particles from flagellates to small crustaceans, unwanted particles can be rejected; they take nutritious particles more readily than nonnutritious ones, and the nutritious particles stimulate greater cirral activity.

The setigerous suspension feeders seem, in general, to be more selective than the bivalves. Much of the evidence on selection by filter feeders has come from this group. For example, *Calanus helgolandicus* has been shown (Corner, 1961) to select preferentially a diet of high organic content from the particulate food available. Richmond and Rogers (1969) have further suggested that *C. helgolandicus* actively hunt paired (dividing) cells of algae and passively filter the single cells.

In general, copepods appear to filter cells less than 50 μm in size while larger cells are separately seized. Wilson (1973) postulated that copepods can "scan" the distribution of particles, capturing particles a little larger than the last ingested, until the peak of abundance is reached. Cladocerans have a different mechanism which involves the algae passing between their shells, limiting them to small particles and ensuring that the maximum-size particle ingested increases in direct proportion to the length of the animal, making it possible to predict what fraction of the phytoplankton community would be available to them (Burns, 1968).

Size appears to be the primary criterion in considering the behavior of filter feeders with respect to food. If, as Slobodkin (1966) has stated, "The true filter feeder takes an organism as food which is effectively infinitely small with respect to itself", probably only sponges, mollusks, and other large organisms can satisfy the definition.

Figure 9 shows a polypide of *Conopeum tenuissimum* in relation to several

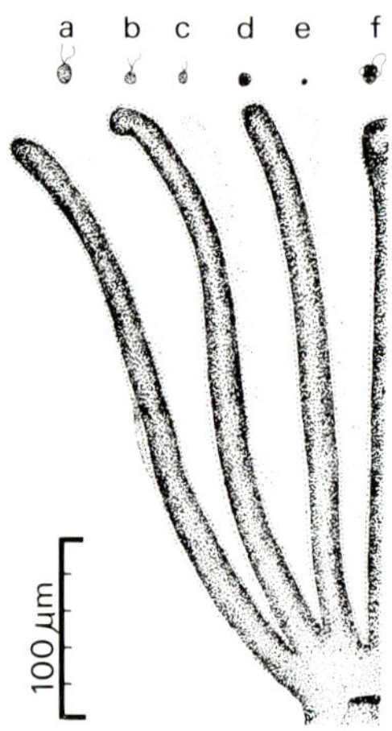

Fig. 9. Schematic representation of the lophophore region of *Conopeum tenuissimum* in relation to several species of phytoplankton algae on which it has been cultured. (a) *Dunaliella tertiolecta*; (b) *Monochrysis lutheri*; (c) Va-12; (d) *Cyclotella nana*; (e) *Nannochloris occulata*; (f) *Gymnodinium simplex*.

food organisms used in growth experiments with this species. It is obvious that the differential between the size of the polypide and the size of its food is far less than that between a suspension-feeding coral polyp, for example, and the plankton it collects, much less than that between an oyster and its food organisms.

B. The Food of Suspension Feeders

The possible food types available to suspension feeders are phytoplankton, detritus, bacteria, and dissolved organic matter.

1. Dissolved Organic Matter

Dissolved organic matter is of questionable importance as a food source for suspension feeders. The conclusion of Krogh (1931) and Bond (1933) that dissolved organic matter has little if any importance as a food for aquatic organisms has, in general, been supported by the results of later workers. While it may serve as an accessory food source, its chief importance to filter feeders probably lies in its function as a source of nutrients for their planktonic food. Stephens and Schinske (1961) found *Bugula turrita* to be one of many marine invertebrates able to remove amino acids from seawater, but the possible nutritional significance of such an uptake is not yet known.

2. *Bacteria*

Bacteria have been found to be capable of sustaining growth in a number of filter feeders, although different species of bacteria may vary in nutritional value, and some are directly toxic. The herbivorous copepods have filtering mechanisms that are relatively permeable to particles less than 5 μm in size, but most other filter feeders are capable of retaining particles down to the size of bacteria. The only documented instance of bacteria serving as food for bryozoans is in the cyphonautes-type larva (Devèze, 1953).

3. *Detritus*

There have been few direct experiments aimed at ascertaining the value of detritus as food for various organisms, as there is no method available for separating detritus from similar-sized particles suspended in water, and experimenters must use artificial detritus which can differ in content and food value from the natural product. Fenchel (1970) studied the feeding of the amphipod *Parahyalella phelpleyi* upon detritus derived from *Thalassia testudinum* (turtle grass). He discovered that *Parahyalella* uses only the microorganisms living on the dead plant material and passes the plant material itself undigested through its system.

Although experiments carried out on *Conopeum tenuissimum* (Dudley, 1974) showed that the species could not be cultured on detritus from laboratory oyster tanks, I have observed that polypides of *Flustrellidra hispida* from Great Bay, New Hampshire, contain many irregular particles that could be organic detritus, and have, as well, sharp-edged particles of sand or silt and some phytoplankton within their guts. It is possible that this species may have some means of digesting the bacteria and other coating particles from the inorganic particulates.

4. *Phytoplankton*

Phytoplankton no doubt provides the most important food source for suspension feeders, including bryozoans. Algal diets have been shown to support the growth of numerous suspension feeders including freshwater cladocerans, and marine barnacles, barnacle nauplii, copepods, bivalves, bivalve larvae, and euphausiids. Different algae may vary greatly in food value, and some species (e.g., *Chlorella*) may be disliked or even toxic to the suspension feeder.

All of the bryozoan culture work to date indicates that algae are the major food source for bryozoans, with the following qualifications: (1) The food is limited, by the size of the bryozoan mouth, mainly to the portion of the phytoplankton less than 50 μm in diameter. (2) Small flagellates appear to be the best food source. These can belong to several groups of algae—chrysophyte, pyrrophyte, chlorophyte—but not all algae even from the same group and of the same size have equal food value for a species. (3) This may be related to the chemical composition of the algae for a mixture of foods is found to be better than most single foods in the case of *Conopeum tenuissimum.* In cultures of other invertebrates a combination of foods has been shown to overcome the nutritional deficiencies (e.g., limiting amounts of certain essential amino acids, vitamins, etc.) of any one algal species, resulting in improved growth, and in some cases, achievement of reproductive maturity. (4) While diatoms seem to be suitable food for bryozoans with gizzards (e.g., *Bowerbankia* grows well on *Nitschia,* and *Zoobotryon* on *Phaeodactylum* and *Cyclotella*), they do not appear to support growth in bryozoans without gizzards (*Bugula neritina* was unable to grow on *Phaeodactylum* and *Conopeum tenuissimum* was unable to grow on *Cyclotella*). (5) The Pyrrophyta are important, as the cultivation of bryozoans on *Oxyrrhis, Cryptomonas,* and *Gymnodinium* indicates. Although bryozoans probably cannot use those species with a thick wall structure or those which are known to contain toxic metabolites, there are many lightly armored or naked forms—particularly in the blooms which occur in summer temperatures in enclosed or inshore waters—that could serve as suitable food. Important evidence which points to the utilization of these organisms by bryozoans is the observation that when cultured on certain pyrrophytes [e.g., *Conopeum tenuissimum* cultured on *Gymnodinium* and *Conopeum reticulum* cultured on *Cryptomonas* (Jebram, 1968)], the gut contains a brownish-orange to reddish color that is characteristic of many species of bryozoans when collected in their natural environment.

C. Possibilities for Differentiation of Trophic Structure in Suspension-Feeding Communities

The prevalent opinion concerning separation or structuring of trophic resources by suspension-feeding communities is that, in the words of Knight-Jones and Moyse (1961), "their competition for food is rather like the competition of plants for CO_2." In other words, since plankton is available to all, there should be no need of competition and hence of trophic specialization in suspension-feeding groups.

Levinton (1972) compared stability and trophic structure between deposit-feeding and suspension-feeding communities and concluded that food supplies for suspension feeders are unpredictable and fluctuating, while those for deposit feeders are relatively constant. Hence members of a suspension-feeding community should not be trophically specialized because of competition for food. He noted that particles carried toward a suspension-feeding animal can pass through its feeding apparatus (as some particles pass between the tentacles of a bryozoan), be rejected, be ingested and digested, or be ingested and passed through without being digested. Missed, rejected, or undigested particles will be carried at least some distance away in the current; thus it cannot be said that another individual (of the same or a different species) in the community will have access to the same food particle. Levinton concluded that the different species in the same area would not be exploiting the same food resource, thus competitive exclusion by trophic resource exploitation does not exist.

Nevertheless, while food may not be a limiting factor in most marine habitats, it is true that at least partial separation of the trophic resource does exist among suspension feeders in the same community. If the resource is thought of as the same kind of particle or the same size of particle (rather than the same particle) it can be seen that due to structural limitations and differences in feeding methods alone different groups of suspension feeders will share overlapping but partially separated portions of the food resource. In addition, evidence is accumulating that the operation of "selection" by these animals is not as unlikely as was once assumed (e.g., Paulson and Scheltema, 1968; Mullin, 1963; Edmondson, 1965).

Sponges (Reiswig, 1971) are apparently specialized for retention of small particles such as bacteria and URPOC (unresolvable particulate organic carbon), though they may take in particles up to 50 μm in size depending on the structure of the species. Mariscal (1965) found the entoproct *Barentsia gracilis* to feed mainly on detritus, and Dales (1957) suggested that this feeding preference was also true for serpulid polychaetes. Ascidians and bivalves can filter a wide size range of plankton, but ectoprocts are limited to members of the nannoplankton. The lower size limit ranges from less than 1 μm, for mucous sheet and choanocyte feeders, to an intermediate range for ciliary and impingement feeders, and a higher limit (3–4 μm or more) for the setal suspension feeders. Thus, it can be seen that possibilities for partial separation of trophic resources do occur and are being taken advantage of by suspension feeders.

D. Differences in Trophic Structure in Different Habitats

The food utilized and the degree of trophic separation will depend upon the environment in which the organism lives. The most important trophic parameters are the amount, sizes, and kinds of particles present. Variation of the particle size spectrum in different environments has only recently become the object of investigation as the use of new techniques (such as electronic particle counting) have allowed the quantitative investigation of the smaller elements. Probably the most important finding of these investigations has been the discovery of the role of those phytoplankton components too small to be collected with ordinary plankton nets (Wood and Davis, 1956; Yentsch and Ryther, 1959).

However, geographic variations in the relative importance of net vs. nannoplankton are still poorly understood. It has been documented for both temperate and tropical areas that nannoplankton may be responsible for 80–90% of the observed phytoplankton productivity (Malone, 1971). The studies that have been made, however, have dealt primarily with oceanic areas. To properly assess the value of net and nannoplankton as food for suspension feeders, we still need studies of these parameters: size distribution, composition, productivity, and fluctuation of the various components of the plankton, for inshore areas, to be linked with studies of actual gut contents of the animals over the same time period.

Many size classifications have been put forth (Dussart, 1965), but as Sheldon *et al.* (1973) state, ". . . particle range from 1–100 μ . . . includes most of the particles likely to be involved in primary production. . . . However . . . this size range, often referred to as 'phytoplankton,' covers two orders of magnitude in terms of diameter and six orders of magnitude in terms of volume. The relative difference in size between a bacterium and a flagellate at 1 μ and a diatom or protozoan as 100 μ is roughly the same as the difference between a mouse and an elephant." It is apparent that any attempt to deal with particle size distribution in assessing the possible food available to a suspension-feeding organism must involve a finer subdivision of the 1–200 μm portion of the particle size spectrum than is usually made.

Using particle size analysis Parsons and LeBrasseur (1970) found that while nannoplankton could serve as food for both small and, to some extent, large zooplankton organisms, the larger organisms also required food that was larger in size or that could readily be concentrated (e.g., chain-forming species). If this type of analysis were to be applied to a bryozoan community (or other benthic suspension-feeding community), and the particle spectrum available to them, a certain degree of trophic separation would be expected on that basis alone. For example, in an estuarine locality like Great Bay, New Hampshire, several bryozoan species may be found inhabiting a single subtidal piece of rockweed (*Fucus* sp.). Figure 10 shows the range of particles each species would be capable of ingesting according to the size of its mouth. While no information is available on the actual size spectrum of particles occurring in that environment, one can see from the figure that if three or four species of algae formed abundance peaks, there would be a difference both qualitatively and quantitatively in the food available to each bryozoan species. In all the types of habitats discussed below, the occurrence of such a continuous spectrum of particles must be taken into consideration.

1. Shallow Marine

In the waters of the nearshore region where phytoplankton algae are most available they probably provide the main food resource for bryozoan populations. From the size restrictions on feeding and from experimental work on bryozoan culture thus far, it appears that most of the food resource would come from phytoplankton under 50 μm in size. Microflagellates are likely to be of special value in this regard. Naked dinoflagellates may also be important, and diatoms

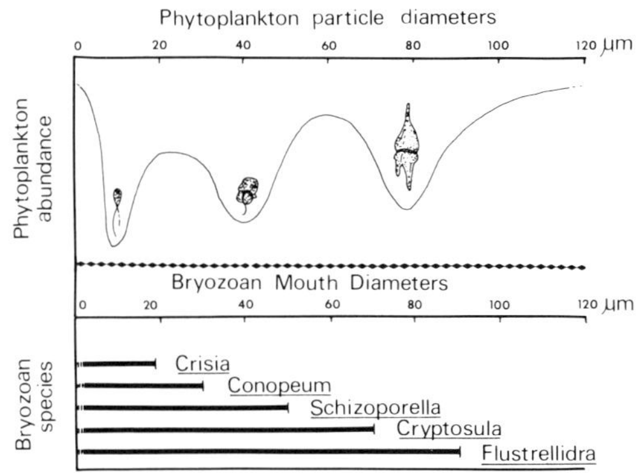

Fig. 10. Mouth diameters of five species of bryozoans in relation to a hypothetical phytoplankton particle size spectrum.

and other armored forms may provide food at least for ctenostomes with gizzards.

It must be kept in mind that the composition and concentration of the plankton varies not only with season, but with habitat, so that the kinds and quantitites of algae available will differ for species living in different environments. Although, in general, the satiation concentrations for most suspension feeders are higher than the algal concentrations that would occur except under bloom conditions, it is obvious that the concentrations of phytoplankton are high enough in inshore habitats to adequately support suspension feeders—at least during the summer months (Jørgensen, 1962; Bullivant, 1968b).

2. *Tropical Marine*

It is in tropical waters where phytoplankton food resources are low but constant that the greatest differentiation with respect to trophic resources could be expected to occur. Reiswig (1971) found that although three tropical Demospongiae all fed primarily on unresolvable particulate organic matter, they differed in their response to armored cells. One species showed apparent size selection of armored cells; another retained them at low efficiencies, but without size selection; and the third showed high retention of armored cells due to a behavioral adaptation. These adaptations appear to make it possible for the sponges to become the dominant filter feeders in the relatively nannoplankton-poor reef waters.

Like sponges, bryozoans probably will be found to display their greatest degree of trophic diversification in tropical waters where the supply of nannoplankton is scarce—though at this time no work exists on the culture or feeding biology of tropical species.

3. Deep Sea

The problem of defining the food resource of deep-sea suspension feeders remains unresolved. Various authors have criticized the possibilities of detritus and bacteria as food for benthic suspension feeders on the grounds that such food is not abundant enough to provide for their nutritive requirements. However, the fact remains that many suspension-feeding organisms including bryozoans manage to survive there. Perhaps a lowering of the rate of metabolism (as had been generally postulated for deep-sea organisms and seems to be supported by the data of Jannasch *et al.*, 1971) allows organisms to make use of a slow but steady supply of detrital or bacterial material.

Schopf (1969) analyzed the gut contents of some deep-sea cheilostomes. He found the digestive system to be as well-developed as in shallow-water genera. The fore- and mid-gut regions were usually empty (probably due to the time elapsed in raising and processing the samples), but the intestine region contained fecal pellets. These did not include recognizable phytoplankton cells, but did contain detritus, including birefringent particles of $CaCO_3$. Schopf also described a structural modification which may have arisen in response to detrital feeding. The curved cylindrical colonies of *Euginoma* are oriented parallel rather than perpendicular to the bottom, a modification which would allow the polypides maximum contact with the bottom particles—perhaps for feeding on organic matter and bacteria in the bottom sediments.

4. Estuarine

It is generally recognized that the strongly physically controlled estuarine environment favors organisms with high reproductive rates, small body size, and relative lack of specialization with respect to food and habitat preferences (Dudley, 1973). Among estuarine bryozoans then, trophic resources could be expected to be largely overlapping. Yet even here differences can exist. Estuarine ctenostomes like *Bowerbankia* or possibly *Victorella* which possess gizzards could digest armored forms such as diatoms which appear to be unavailable to cheilostomes like *Bugula* or *Conopeum.* The size of particles ingested would be limited by the size of the mouth of the polypides; thus *Bowerbankia,* for example, would be limited to those under 30 μm while *Flustrellidra* has the potential for ingesting those up to 90 μm. As mentioned previously, several estuarine species show great tenacity in rejecting particles; this capacity may have adaptive value for a suspension feeder living in turbid estuarine water.

VI. SUGGESTIONS FOR FUTURE RESEARCH

It is obvious from the preceding pages that our knowledge of the feeding biology of marine and estuarine bryozoans is fragmentary. Research in a variety of areas is needed in order to fill in these gaps; of special significance are the following:

1. Studies of the feeding parameters (lophophore size, mouth size, number of tentacles, etc.) for species in various communities in order to learn what degree of trophic separation is maintained by them, and how different degrees of separation might be adaptive in various habitats.

2. A detailed study of feeding behavior. While food placed in the culture water acts as a stimulus—usually causing the organisms to expand their lophophores—no one has examined the ways in which the individual zooid can react to food.

3. Studies of resource regimes in various microhabitats such as Reiswig (1971) conducted for tropical sponges. Much more information is needed on the composition and abundance of the nannoplankton in various areas.

4. Studies of tropical species are especially needed; theoretically, this is where the greatest degree of trophic diversification should occur.

VII. CONCLUSIONS

Marine bryozoans are not alike with respect to structure and size of the feeding apparatus. Variation in tentacle number, lophophore size and shape, mouth size, rate of food passage, and behavior with regard to suspended particles indicate that partial separation of trophic resources must occur.

Recent studies on the culture of marine bryozoans have led to the following conclusions concerning their food:

1. Phytoplankton organisms under 50 μm in size represent the major food source.

2. Different algae may vary in food value to a particular species of bryozoan. Even two algal species of the same size and from the same class may not have the same food value for a species of bryozoan, as work with *Conopeum tenuissimum* has shown.

3. A mixture of algal foods is probably better than most single foods.

4. Diatoms are probably suitable foods for bryozoans with gizzards, but not for those which lack gizzards. The pyrrophyte, chrysophyte, and chlorophyte flagellates seem to be more suitable as food for species lacking gizzards.

5. Food other than phytoplankton (e.g., detritus) may become important in areas where phytoplankton production is limited or phytoplankton is unavailable.

Note added in proof. Ryland (1975) and Winston (in press) demonstrate that some of the morphological and behavioral adaptations speculated upon in this chapter do exist and are important in feeding in nature. Ryland (1975) analyzed lophophore parameters in a single population of bryozoans from one locality, showing that while lophophore dimensions varied within that population, dominance of a few species led to less variation in size range for the community as a whole. Winston (in press), in contrast, studied lophophore parameters and individual and colony behavior for 56 species from varied habitats in Florida and Panama to learn how these related to feeding activities. Polypide morphology, particularly introvert length and lophophore symmetry, varied from species to species; the variation was linked to behavioral strategy. Individual (polypide) behavior ranged from passive filtration, to tentacle feeding, to cage-capture of active protistans, to mechanisms for the concentration of small particles by large individuals. Colony-wide patterns of morphology and behavior reflected various methods of dealing with water currents, ranging from weakly integrated patterns characterized by a high degree of functional independence of individual polypides to highly integrated patterns in which both skeletal morphology and polypide orientation worked to enhance and channel feeding currents.

REFERENCES

Annandale, N. (1916). Zoological results of a tour in the Far-East. Polyzoa Entoprocta and Ctenostomata. *Mem. Asiat. Soc. Bengal* **6**, 13–37.

Atkins, D. (1932). The ciliary feeding mechanism of the entoproct Polyzoa and a comparison with that of the ectoproct Polyzoa. *Q. J. Microsc. Sci. [N.S.]* **75**, 393–423.

Banta, W. C., McKinney, F. K., and Zimmer, R. L. (1974). Bryozoan monticules: Excurrent water outlets? *Science* **185**, 783–784.

Bobin, G., and Prenant, M. (1952). Structure et histogénèse du gésier des vescicularines (Bryozoaires Cténostomes). *Arch. Zool. Exp. Gén.* **89**, 175–202.

Bond, R. M. (1933). A contribution to the study of the natural food-cycle in aquatic environments. *Bull. Bingham Oceanogr. Collect.* **4**, 1–89.

Borg, F. (1926). Studies on Recent cyclostomatous Bryozoa. *Zool. Bidr. Uppsala* **10**, 181–507.

Braem, F. (1940). Über die Querstreifung im Pharynx der gymnolämen Bryozoen und über den Bau des Munddarms. *Z. Morphol. Oekol. Tierre* **36**, 668–676.

Bullivant, J. S. (1967). Aspects of feeding of the bryozoan *Zoobotryon verticillatum.* Ph D. Thesis, University of Southern California, Los Angeles.

Bullivant, J. S. (1968a). The rate of feeding of the bryozoan *Zoobotryon verticillatum. N. Z. J. Mar. Freshwater Res.* **2**, 111–134.

Bullivant, J. S. (1968b). The method of feeding of lophophorates (Bryozoa, Phoronida, Brachiopoda). *N. Z. J. Mar. Freshwater Res.* **2**, 135–146.

Bullivant, J. S., and Bils, R. F. (1968). The pharyngeal cells of *Zoobotryon verticillatum* (Del Chiaje), a gymnolaemate bryozoan. *N. Z. J. Mar. Freshwater Res.* **2**, 438–446.

Burns, C. W. (1968). The relationship between body size of filter-feeding Cladocera and the maximum size of particle ingested. *Limnol. Oceanogr.* **13**, 675–678.

Corner, E. D. S. (1961). On the nutrition and metabolism of zooplankton. I. Preliminary observations on the feeding of the marine copepod *Calanus helgolandicus* (Claus). *J. Mar. Biol. Assoc. U. K.* **41**, 5–16.

Crisp, D. J. (1964). An assessment of plankton grazing by barnacles. *In* "Grazing in Terrestrial and Marine Environments" (D. J. Crisp, ed.), pp. 251–264. Blackwell, Oxford.

Dales, R. P. (1957). Some quantitative aspects of feeding in sabellid and serpulid fan worms. *J. Mar. Biol. Assoc. U. K.* **36,** 309–316.

Dendy, A. (1888). Anatomy of an arenaceous polyzoan. *Proc. R. Soc. Victoria [N.S.]* **1,** 1–11.

Devèze, L. (1953). Utilisation possible du contenu bactérien des eaux par les larves cyphonautes. *Rec. Trav. Stn. Mar. Endoume* **8,** 49–54.

Dudley, J. (1970). Differential utilization of phytoplankton food resources by marine ectoprocts (Abstr.). *Biol. Bull. (Woods Hole, Mass.)* **139,** 420.

Dudley, J. (1973). Observations on the reproduction, early larval development, and colony astogeny of *Conopeum tenuissimum* (Canu). *Chesapeake Sci.* **14,** 270–278.

Dudley, J. (1974). The culture of *Conopeum tenuissuimum* in relation to feeding, growth and distribution of estuarine ectoprocts. Ph.D. Thesis, University of Chicago, Chicago, Illinois.

Dussart, B. M. (1965). Les différentes catégories de plancton. *Hydrobiologia* **26,** 72–74.

Edmondson, W. T. (1965). Reproductive rate of planktonic rotifers as related to food and temperature in nature. *Ecol. Monogr.* **35,** 61–111.

Fenchel, T. (1970). Studies on the composition of organic detritus derived from the turtle grass *Thalassia testudinum. Limnol. Oceanogr.* **15,** 14–20.

Gordon, D. P. (1974). Microarchitecture and function of the lophophore in the bryozoan *Cryptosula pallasiana. Mar. Biol.* **27,** 147–163.

Hasper, M. (1912). On a method of rearing larvae of Polyzoa. *J. Mar. Biol. Assoc. U.K. [N.S.]* **9,** 435–436.

Henneguy, M. F. (1909). Sur un épithélium à fibres musculaires striées. *C. R. Hebd. Seances Acad. Sci.* **148,** 134–138.

Hentschel, E. (1922). Uber den Bewuchs auf den treibenden Tangen der Sargassosee. *Mitt. Zool. Mus. Hamburg* **38,** 1–26.

Hunt, O. D. (1925). The food of the bottom fauna of the Plymouth fishing grounds. *J. Mar. Biol. Assoc. U. K.* **13,** 560–598.

Hyman, L. H. (1959). "The Invertebrates," Vol. V. McGraw-Hill, New York.

Jannasch, H. W., Eimhjellen, K., Wirsen, C. O., and Farmanfarmaian, A. (1971). Microbial degradation of organic matter in the deep sea. *Science* **171,** 672–675.

Jebram, D. (1968). A cultivation method for saltwater Bryozoa and an example for experimental biology. *Atti Soc. Ital. Sci. Nat. Mus. Civ. Stor. Nat. Milano* **108,** 119–128.

Jebram, D. (1970). Preliminary experiments with Bryozoa in a simple apparatus for producing continuous water currents. *Helgoländer Wiss. Meeresunters.* **20,** 278–292.

Jebram, D. (1973). Preliminary observations on the influences of food and other factors on the growth of Bryozoa with a description of a new apparatus for the cultivation of sessile plankton feeders. *Kieler Meeresforsch.* **29,** 50–57.

Jebram, D. (1975). Effects of different foods on *Conopeum seurati* (Canu) (Bryozoa Cheilostomata) and *Bowerbankia gracilis* Leidy (Bryozoa Ctenostomata). *Docum. Lab. Géol. Fac. Sci. Lyon H. S.* **3,** (1), 97–108.

Jørgensen, C. B. (1962). The food of filter feeding organisms. *Rapp. Cons. Explor. Mer* **153,** 99–107.

Knight-Jones, E. W., and Moyse, J. (1961). Intraspecific competition in sedentary marine animals. *Symp. Soc. Exp. Biol.* **15,** 72–95.

Krogh, A. (1931). Dissolved substances as food for aquatic organisms. *Rapp. Cons. Explor. Mer.* **75,** 7–36.

Levinton, J. (1972). Stability and trophic structure in deposit-feeding and suspension-feeding communities. *Am. Nat.* **106,** 472–486.

Lutaud, G. (1955). Sur la ciliature du tentacule chez les Bryozoaires Chilostomes. *Arch. Zool. Exp. Gén.* **92,** 13–19.

Lutaud, G. (1973). L'innervation du lophophore chez le Bryozoaire Chilostome *Electra pilosa* (L.). *Z. Zellforsch. Mikrosk. Anat.* **140,** 217–234.

McKiernan, G. M. (1968). Seasonal distribution of dinoflagellates in the Lower York River, Virginia. M.S. Thesis, Virginia Institute of Marine Science, Gloucester Point, Virginia.

Malone, T. C. (1971). The relative importance of nannoplankton and net plankton as primary producers in tropical oceanic and neritic plankton communities. *Limnol. Oceanogr.* **16,** 633–639.

Marcus, E. (1941). Sôbre Bryozoa do Brasil. *Univ. Sao Paulo, Fac. Filos. Cienc. Let., Bol. Zool.* **22,** No. 5, 3–208.

Mariscal, R. N. (1965). The adult and larval morphology and life history of the entoproct *Barentsia gracilis* (M. Sars, 1835). *J. Morphol.* **116,** 311–338.

Matricon, I. (1973). Quelques données ultrastructurales sur un myoepithelium: Le pharynx d'un Bryozoaire. *Z. Zellforsch. Mikrosk. Anat.* **136,** 569–578.

Menzel, R. W. (1955). Some phases of the biology of *Ostrea equestris* Say, and comparison with *Crassostrea virginica* (Gmelin). *Publ. Inst. Mar. Sci., Univ. Tex.* **4,** 69–153.

Mullin, M. (1963). Some factors affecting the feeding of marine copepods of the genus *Calanus. Limnol. Oceanogr.* **8,** 239–250.

Parsons, T. R., and LeBrasseur, R. J. (1970). The availability of food to different trophic levels in the marine food chain. *In* "Marine Food Chains" (J. H. Steele, ed.), pp. 325–343. Univ. of California Press, Berkeley.

Paulson, T. C., and Scheltema, R. S. (1968). Selective feeding on algal cells by the veliger larvae of *Nassarius obsoletus* (Gastropoda, Prosobranchia). *Biol. Bull. (Woods Hole, Mass.)* **134,** 481–489.

Prenant, M., and Bobin, G. (1956). "Bryozoaires. Première Partie: Entoproctes, Phylactolèmes, Cténostomes," Faune de France, 60. Lechavalier, Paris.

Rapport, D. J., Berger, J., and Reid, D. B. W. (1972). Determination of food preference of *Stentor coeruleus. Biol. Bull. (Woods Hole, Mass.)* **142,** 103–109.

Reiswig, H. M. (1971). Particle feeding in natural populations of three marine demosponges. *Biol. Bull. (Woods Hole, Mass.)* **141,** 568–591.

Richmond, S., and Rogers, J. N. (1969). The feeding of *Calanus hegolandicus* on synchronously growing populations of the marine diatom *Ditylum brightwelli, Limnol. Oceanogr.* **14,** 701–709.

Ries, E. (1936). Fütterungsversuche bei *Zoobothryon* (Bryozoa). *Z. Vergl. Physiol.* **23,** 64–99.

Ryland, J. S. (1970). "Bryozoans." Hutchinson University Library, London.

Ryland, J. S. (1975). Parameters of the lophophore in relation to population structure in a bryozoan community. *In* "Proceedings of the 9th European Marine Biology Symposium, 1975" (H. Barnes, ed.), pp. 363–393. Aberdeen University Press, Aberdeen, Scotland.

Schneider, D. (1959). Der Aufbau der *Bugula*-Tierstocke und seine Beeinflussung durch Aussenfaktoren. *Biol. Zentralbl.* **78,** 250–283.

Schneider, D. (1963). Normal and phototrophic growth reactions in the marine bryozoan *Bugula avicularia. In* "The Lower Metazoa" (E. C. Dougherty, ed.), pp. 357–371. Univ. of California Press, Berkeley.

Schopf, T. J. M. (1969). Geographic and depth distribution of the phylum Ectoprocta from 200 to 6,000 meters. *Proc. Am. Philos. Soc.* **113,** 464–474.

Sheldon, R. W., Sutcliffe, W. H., and Prakash, A. (1973). The production of particles in the surface waters of the ocean with particular reference to the Sargasso Sea. *Limnol. Oceanogr.* **18,** 719–733.

Silén, L. (1944). On the division and movements of the alimentary canal of the Bryozoa. *Ark. Zool.* **35A,** 1–40.

Slobodkin, L. (1966). *In* "Marine Biology III; Proceedings of the Third International Interdisciplinary Conference" (W. T. Edmondson, ed.), p. 85. N. Y. Acad. Sci., New York.

Smith, L. W. (1973). Ultrastructure of the tentacles of *Flustrellidra hispida* (Fabricius). *In* "Living and Fossil Bryozoa" (G. P. Larwood, ed.), pp. 335–342. Academic Press, New York.

Smith, S. G., and Watabe, N. (1971). The ultrastructure of the tentacles of *Electra monostachys,* a marine bryozoan. *Am. Zool.* **11,** 696 (abstr.).

Stephens, G. C., and Schinske, R. A. (1961). Uptake of amino acids by marine invertebrates. *Limnol. Oceanogr.* **6,** 175–181.

Strathmann, R. R. (1971). The feeding behaviour of planktotrophic echinoderm larvae: Mechanisms, regulation and rates of suspension feeding. *J. Mar. Biol. Ecol.* **6,** 109–160.

Strathmann, R. R. (1973) Function of lateral cilia in suspension feeding of lophophorates (Brachiopoda, Phoronida, Ectoprocta). *Mar. Biol.* **23,** 129–136.

Waters, A. W. (1904). Bryozoa. Expéd. Antarct. *Belge. Resulte Voy. Belg. Rapp. Sci. Zool.* **2,** 1–114.

Waters, A. W. (1909). Marine biology of the Sudanese Red Sea. Bryozoa. Pt. I. Cheilostomata. *J. Linn. Soc. London, Zool.* **31,** 123–181.

Waters, A. W. (1918). Some collections of the littoral marine fauna of the Cape Verde Islands, made by Cyril Crossland in the summer of 1904–Bryozoa. *J. Linn. Soc. London, Zool.* **34,** 1–44.

Wilson, D. S. (1973). Food size selection among copepods. *Ecology* **54,** 909–914.

Winston, J. E. (1976). Experimental culture of the estuarine ectoproct *Conopeum tenuissimum* from Chesapeake Bay. *Biol. Bull.* (*Woods Hole, Mass.*) **150,** 318–335.

Winston, J. E. (in press). Polypide morphology and feeding behavior in marine ectoprocts. *Bull. Mar. Sci.*

Wood, E. J. F., and Davis, P. S. (1956). Importance of smaller phytoplankton elements. *Nature (London)* **177,** 438.

Yentsch, C. S., and Ryther, J. H. (1959). Relative significance of the net phytoplankton and nannoplankton in the waters of Vineyard Sound. *J. Cons. Explor. Mer.* **24,** 231–238.

8

Experimental Techniques and Culture Methods

DIETHARDT JEBRAM

I. INTRODUCTION

Many aspects in scientific work on the Bryozoa require observations of live animals – not only in physiology, ontogeny, and ecology, but for anatomy and morphology as well; laboratory observations may also help find new facts and arguments for systematics and phylogenetics (Jebram, 1970b, 1973*).

*Read as paper "Advances in Cultivation of Bryozoa" on Sept. 7, 1971, at the 2nd International Conference of the International Bryozoology Association in Durham, England.

Scientists working with freshwater animals were the first to bring bryozoans into the laboratory for studying food uptake, growth, and development. Braem (1888, 1890), Otto (1921), Marcus (1926b), and Rogick (1937, 1938) performed the first significant experiments with freshwater bryozoans. Marcus (1926a) provided the first important experimental research on marine bryozoans, but utilized simple methods. Grell (1950 in Tübingen, unpublished) and Hauenschild (1962, 1968, 1970) developed culture techniques for many invertebrates and were also the first who found suitable methods for rearing marine bryozoans. Schneider (1959) and Kaissling (1963) studied photic responses using Grell's cultivation methods. Lynch (1947, 1949) and Ryland (1959, 1960) evaluated the behavior and taxes of bryozoan larvae. Wayss (1968) conducted original experiments on the physiology and regenerative growth of freshwater bryozoans. Borg (1926), Atkins (1932), Ries (1936), Becker (1937), Rüsche (1938), Silén (1944), Bullivant (1968a,b), Jebram (1968, 1970b, 1973, 1975b), Strathmann (1973), and Winston (1976) studied food uptake and effects of different food species on various bryozoans. Jebram developed an apparatus for simple experiments with water currents (1970a) and another one for long-term rearing experiments (1973).

In working with live animals, various difficulties have to be puzzled out. In comparison with normal aquarium techniques, the cultivation of animals that feed continuously on plankton, like bryozoans, raises special questions. For this reason this chapter will not simply give suggestions on methods, procedures, and recipes, but will point out various methodological problems as well.

II. EXPERIMENTAL WORK IN THE FIELD

For some ecological studies bryozoans may be collected in their habitat by a variety of methods (e.g., manual sampling at low-tide, or with scraping nets, bottom nets or dredges, skin diving) (Friedl, 1925; Sloane *et al.*, 1957; and many other authors).

Individual colonies can be marked *in situ* for observations about colonial development, regenerative growth, production of larvae, and competitive interactions (Gordon, 1968).

One may undertake similar studies by using pieces of artificial substratum submerged in the water: plates of different types of materials may be used, fastened within a frame and fixed at a pier or tethered between an anchor and a float (Grave, in Needham et al., 1937). This method has long been utilized in synecological work, especially pollution research, and is suitable for all types of sessile organisms. Menon and Nair (1967) and Menon (1973) investigated the vertical and horizontal distribution of estuarine bryozoans in waters near Cochin (India) using submerged panels, while Geraci and Relini (1970a,b,c) studied the fouling of bryozoans in coastal waters of Italy with a related method. Test panels of submerged artificial substratum are useful for studies on the settlement of larvae, preferential substratum selection, seasonal occurrence of larvae, and the sequences of colonization of new substrata.

Influences of newly settled larvae, predators, or competitors may be evaluated by elimination of selected species from an area in the natural environment and/or by covering areas with fine nets or other devices (perhaps by using diving equipment). Initiating isolation in different seasons may show possible influences of seasonal newcomer species which are excluded by this isolation.

III. SIMPLE REARING IN THE LABORATORY

Bryozoan colonies on their natural substratum may simply be stored, for some time, in tanks with running water from the habitat and containing the natural plankton as food (Marcus, 1926a,b; Bronstein, 1938a,b, 1940; Ryland, 1959, 1960, 1962; Cook, 1961, 1962, 1963, 1964, 1968; Lutaud, 1961; Cook and Hayward, 1966; and many other authors). This method is suitable for several kinds of research (e.g., preliminary studies of the methods of food uptake, budding of new zooids, brooding, embryo development, and function of the nervous system).

Unfortunately, experimental conditions are not well known when using this method of rearing. Food will be scarce in most cases; bacteria and indigestible protistans may quickly increase. If the natural substratum is seaweed, it will degenerate with time. Colonies on stones or shells may remain growing there for longer times, but often such colonies can be observed only with difficulty because of their rugosity.

If there is no running habitat-water in the laboratory, one must use smaller vessels and add suitable food. The food may be plankton from the environment of the bryozoans. Culture media in limited volumes must be renewed frequently (see Sections IV and V).

The use of natural plankton is possible only at stations near the native habitat of the bryozoans. In most cases an artificial propagation of the food for the experimental animals would also be necessary. Moreover, a defined food source will provide for more rigorously controlled experiments. (Cultivation of food for bryozoans will be described in Section V.)

Elucidation of many problems require intensive experiments under well-defined conditions and over longer periods of time.

IV. GENERAL CONDITIONS FOR LABORATORY WORK

A. Materials Used for Cultivation Facilities

All metals, even stainless steel, introduce toxic ions into the medium and for this reason should not be used as culture equipment. Red or black rubber may also have toxic effects on the cultivated organisms after prolonged exposure.

Synthetic plastic materials may be more or less stable. Boxes of polyethylene or polystyrol may be used. Perspex (polymethyl metacrylate) has proved to be very suitable material for building various experimental apparatuses (Jebram, 1973). Perspex may easily be cut, drilled, and cemented (with chloroform or Acrifix 92).

Polyvinylchloride or polyethylene may be used for tubes, but glass or silicon rubber are more suitable. Where an absolutely stable concentration of diluted gases in the medium should be maintained, however, silicon rubber is undesirable because it is permeable to gases.

Glass is the best material for rearing vessels, but even it is not absolutely inert chemically. All types of glass leach metal ions into the medium, but in very small amounts. Glass mainly supplies sodium ions. Sodium is one of the main components of the salt in the seawater and is also a common "pollutant" in freshwater and in various sorts of materials; therefore, this additional sodium in such small amounts bears no real complications. Even the other ions of trace elements, e.g. Si or B, leached into the medium from the glass walls, may be ignored in most cases.

Generally, vessels or utensils that were previously used with toxic agents, like formalin or mercuric chloride, should never be used for rearing of organisms, or for storage of rearing water or agents used for preparation of culture media.

B. Use of Experimental Animals

When rearing bryozoans for the first time, one should work with physiologically robust and eurytopic species. Generally in bryozoans, as well as in other groups of animals, species from brackish water or polyhaline ones are easier to rear than those from fresh or stenohaline waters. Uncalcified animals are less delicate for rearing than forms with calcified walls. For these reasons, in many cases, ctenostomes and membraniporid cheilostomes are cultured with the least difficulty. Phylactolaemates seem likewise tolerant to laboratory conditions to some extent. Several ascophoran cheilostomes have been found to be more stenotopic and fastidious in rearing. Stenolaematous bryozoans present the most difficulties and demand the greatest attention to the quality of culture conditions.

In many cases for experimental work which requires a statistical basis, the genetic diversity of colonies from different origins should be eliminated by use of strain material. Bryozoans easily multiply asexually by division of individuals within colonies. Although bushy and stoloniferous forms are simply propagated by cutting off branches of colonies, the crustous species strongly adhere to the substratum and can only be subcultured by separating a part of the substratum with a portion of the adhering colony. Jebram (1968) developed a simple but practical method for subculturing. Colonies are allowed to grow not on the slide itself but on coverslips, which are held fast on the slides by slit pieces of polyvinyl tubing. The bryozoans grow from one coverslip to the other (Fig. 1). Afterward, the colonies can be subdivided along the borders of the coverslips and the separate pieces used for further subculturing.

To avoid confusion, each slide with a bryozoan colony should be marked by a diamond pencil with the essential data including the strain of the bryozoan species, food species, salinity, temperature, perhaps colony number, and speed and direction of water currents (indicated by an arrow). Changes in the conditions during the growth of the colony are also to be marked on the slide and indicated by arrows. For example: A_{Tm}/F:O (6) 15°→10°, means strain A of *Tanganella muelleri;* food: *Oxyrrhis marina;* 6‰ salinity; temperature 15°C changed to 10°C.

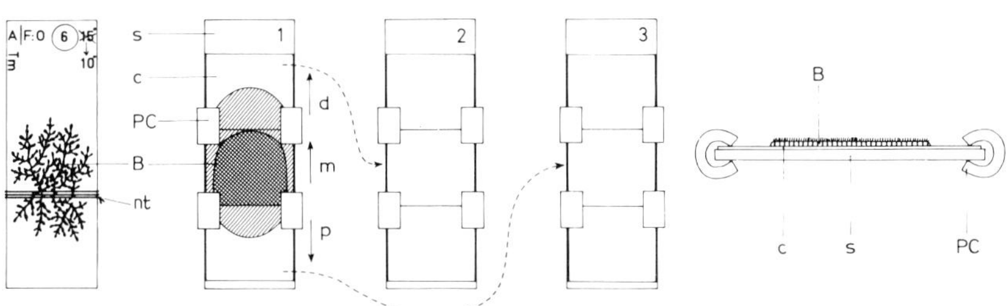

Fig. 1. Culturing bryozoans on glass slides and asexual multiplying of crustous colonies. Left slide with a serial (or stolonial) colony originating from a primary branch fastened by a nylon thread to slide with the essential data (from left to right: bryozoan strain, food species, salinity, and temperature). Far right slide in a cross-section view shows the growth of incrusting bryozoans on a coverslip. The middle slides 1, 2, and 3 demonstrate the kind of asexual multiplying of incrusting colonies by replacement of coverslips. Abbreviations: B, bryozoan colony; c, coverslip; d, distal; m, middle; p, proximal part of the growing colony; nt, nylon thread; PC, clamps of slit pieces of polyvinyl tubing; s, glass slide (76 × 26 mm). Modified from Jebram (1968).

C. Minimization of Eutrophication and Bacterial Growth

Destruction of incompletely digested parts of the feces, as well as products of protein metabolism and excretion, in time enrich the rearing medium with ammonia and other agents. Ammonia is toxic. It will be oxidized in steps to nitrate, which is not toxic, but for rearing of most organisms the medium should not contain more than 0.3–0.5 gm/liter of nitrate (Hauenschild, in Schlieper, 1968). During the process of oxidation of ammonia other toxic reagents such as nitrite and phenolic agents are produced. In higher concentrations nitrate may easily be reduced to nitrite (Hückstedt, 1968). Eutrophication of some degree may be desirable for propagation of bacteria or algae, but extensive eutrophication is more or less toxic to experimental animals like bryozoans.

Controlling bacterial growth is the most difficult factor in rearing bryozoans, because many species of bacteria produce toxic agents, and because bacteria have high rates of propagation. Bryozoa may eat few bacterial cells, but a high concentration of bacteria in the food has been observed to reduce the vitality of bryozoans, and perhaps even cause their death (Jebram, 1973). Some of the bacteria tend to adhere to the walls of the vessels and to the bryozoans. When the bacterial coating has attained greater extents it interferes with respiration of the bryozoans and may hinder evagination of the polypides. Toxic excretions of the adhering bacteria may diffuse into the bryozoan bodies. For these reasons colonies occasionally must be cleaned of bacteria and feces by gentle brushing.

For preparation of media used for rearing in the laboratory one should pay attention to some basic facts. During the retention of organisms in cultivation vessels of limited volume, the exchange of agents between the organisms and the water alters the chemical composition of the medium. The organisms take up agents through mouth or body wall from the water, e.g., vitamins, amino acids, and oxygen (if the organism is heterotrophic) or cabonic acid (if the organism is autotrophic), and perhaps nitrate, phosphate, and trace elements. On the other hand, the medium will receive agents from the organisms, including excretory products, products of the breakdown of feces, carbonic acid (if the organism is heterotrophic) or oxygen (if the organism is autotrophic), and various other agents diffusing through the body wall into the water. The uptake of carbonic acid by assimilating alga (or the consumption of oxygen by heterotrophs) may drastically alter the pH of the medium even in a short time!

Therefore, maintenance of the chemical stability is a main task during cultivation and experimental work. A high redox potential for the oxidation of toxic agents has to be maintained by adjusting the chemical composition of the medium and by aeration. Agents must be added for chemical stabilization and for nutrition of autotrophic organisms.

All cultivation media must be sterilized before use to eliminate unknown protistans and to prevent a rapid growth of bacteria. Antibiotics and bactericides may not be generally useful, because they may also cause negative effects in the physiology of small-sized experimental animals like bryozoans.

In the common aquarium technique, there are some simple methods for delaying the eutrophication of the water, but most of them are not sufficient for long-term rearing of bryozoans.

First, the medium may be filtered; however, the greater part of the planktonic food organisms for the bryozoans will also be mechanically eliminated from the water. Filtration of the medium cannot be used, if one works with planktonic food species.

Second, bacteria swimming in the medium may be killed by UV-light, but the suspended food protistans will be killed as well; on the other hand, the bacteria adhering to all walls of the rearing vessels and to the enclosed bryozoans cannot be killed by UV-light without also killing the bryozoans. UV-light cannot solve this problem in culturing bryozoans.

Third, adding large amounts of medium dilutes the dissolved and suspended toxic products for some time, when comparatively few experimental animals will live in it; on the other hand, voluminous containers are difficult for handling. Large water masses must be circulated, because in different parts of a big volume of still water regional variations in the chemical conditions will develop, and the buffering effect of a large amount of water cannot be maintained.

Planktonic food organisms for bryozoans are, in most cases, autotrophic algae which can incorporate eutrophying substances like nitrate, phosphate, and sulfate. Thus algal cells, added as food to the bryozoan cultures, also may partly clean the medium of contaminants which enter the water as waste from the bryozoans. The culture chambers must be illuminated to permit photosynthesis and assimilation of the algae. This arrangement of circulation of chemical components may work for some time, if the number of autotrophic plankton cells is large enough and associated with only a few bryozoan colonies. Maintaining these conditions is a delicate process. Bryozoans are eager feeders and feed as long as suitable food is present. Bryozoans will ingest large amounts but will incompletely digest the food, if it is plentiful. Intensive feeding by bryozoans quickly reduces the concentration of planktonic organisms and produces a large amount of feces and excretory products. Food cells must be added at frequent intervals for the maintainance of a sufficiently large cell number, because the feeding rate of the bryozoans may be greater than the propagation rate of the algal cells. In spite of these precautions, contamination cannot be suppressed absolutely in a closed system. In all culture systems without running water the medium must be renewed from time to time. One must search for ways to overcome these difficulties and how to expand the span of time until the concentration of contaminants attains a physiologically unsuitable level. For practical work, one must find a compromise: the volume of the medium should be large, but the vessels should not be too voluminous for handling; the number of bryozoan colonies per vessel should be small, but in an experimental series the number of colonies should be large enough to provide statistically reliable data; the colonies should remain as long as possible in a vessel, but not too long because of the accumulation of toxic substances.

Generally, in culture technique, the importance of the chemical composition of the medium and of the nutritive conditions increases with a decrease in the body size of the experimental organisms. Therefore, composition of the medium and choice of food supply are the most important obstacles in culturing small-sized animals such as bryozoans.

V. CULTIVATION OR LONG-TERM REARING

A. Cultivation Media

Otto (1921), Marcus (1926a,b), Graupner (1929, 1933), and Rogick (1937, 1938) reared bryozoans with "habitat-water" and "old aquarium water," but they could maintain the life of the colonies for only short periods of time, a maximum of about $5\frac{1}{2}$ months. From the point of view of modern aquarium techniques and water chemistry the use of "old aquarium water" is nonsense.

Water from the natural habitat (seawater, brackish, or freshwater) may be used for laboratory work, as long as it is not polluted by oil, phenols, or detergents. Tests by several workers on several species of algae reared in water from various stations of a lake or bay have shown that even neighboring waters may produce important differences in growth and vitality of the inoculated organisms. At any given locality the chemical composition of the water varies with time because of seasonal or other external conditions. Even with the same total salinity a given "natural seawater" will not be the same as another. Besides pollution by man, the most important differences in natural waters occur in their varying contents of organic agents, like vitamins, amino acids, humic acids, and tannins, but differences may also appear in the occurence and concentration of the inorganic components, e.g., trace elements.

For many simple experiments small differences in the components of the water may be ignored. However, even if the differences may be small numerically, in some cases the unknown components can cause difficulties and questionable results in physiological experiments or may be toxic. Additionally, many of the organic agents facilitate the propagation of bacteria.

A large amount of the dissolved organic agents could be extracted from the natural water by filtering through or shaking with charcoal. The unwanted inorganic components would not be extracted, otherwise one would produce distilled water (which is unsuitable even for freshwater organisms).

TABLE I
Simple Artificial Seawater[a,b]

Compound	Conc. (gm)	Compound	Conc. (gm)
NaCl	28.4	$CaCl_2 \cdot 2\ H_2O$	1.6
$MgSO_4 \cdot 7\ H_2O$	7.0	KCl	0.7
$MgCl_2 \cdot 6\ H_2O$	5.1	$NaHCO_3$	0.2

[a]Data from Hauenschild (1962).
[b]Dissolved in 1000 ml distilled water.

Unknown factors should be eliminated from experiments as completely as possible. The experimental conditions could be defined more precisely by using an artificial culture medium produced with distilled water* and defined agents. Hauenschild (1962) provided a simple recipe for artificial seawater (Table I). Wiedemann and Kramer (after Hückstedt, 1971) introduced a more complete mixture for artificial seawater (Table II). Some artificial-salt compositions for rearing marine fishes in aquaria are offered by several manufacturers, but scientists rearing invertebrates should be cautious in the use of commercial sea salts. During the process of preparing bulk mixtures, different components may be incompletely mixed, so that the chemical composition of commercial sea salts may vary to significant extents and may be unsuitable for working experimentally with sensitive animals like bryozoans.

For several experiments one may make a compromise by preparing a medium mixed up with one part of an artificial medium and one part of natural habitat water. This method has proved to be suitable for several experiments with saltwater bryozoans (Jebram, 1968, 1970a, 1973).

All cultivation media must be sterilized. Sterilization may be performed by heating up to 80°–90°C, by autoclavating, or by sterile filtering through fine filtration membranes (e.g., Millipore). Heating or autoclavating should be repeated about 24 hours later to kill young germs of resistant spores.

Besides CO_2 and H_2O, the main nutrient agents for autotrophic algae are nitrate and phosphate, used per liter of medium as 50 mg $Ca(NO_3)_2$ and 12.5 mg KH_2PO_4 for freshwater or 100 mg $NaNO_3$ and 20 mg $Na_2HPO_4 \cdot 12\ H_2O$ for seawater.

Enrichment of an artificial medium with trace elements and stabilizing and chelating agents may be achieved by some simple but practical means, for example, addition of a solution of a commercial powdered algal or yeast extract (about 5–10 mg per liter of medium) or a soil extract solution (about 10–20 ml per liter of

*One should not use freshwater from the tap pipes, because it is served by metal tubes in most cases. This is especially important in the morning as the tap water contains large amounts of metal ions if the water has remained in the metal supply lines through the night.

TABLE II
Artificial Seawater[a,b]

Compound	Conc. (gm)	Compound	Conc. (gm)
NaCl	27.65	$NaHCO_3$	0.2500
$MgSO_4 \cdot 7\,H_2O$	6.92	NaBr	0.1000
$MgCl_2 \cdot 6\,H_2O$	5.51	KI	0.0001
$CaCl_2 \cdot 6\,H_2O$	1.45	$SrCl_2$	0.0150
KCl	0.65		

[a]After Wiedemann and Kramer, data from Hückstedt (1971).
[b]Dissolved in 1000 ml distilled water.

medium). Soil extract will be produced by extraction of soil by a Soxhlet extraction apparatus or by autoclaving of about 1 liter dry soil with 2 liters distilled water in a 3-liter Erlenmeyer flask for about 2 hours (50-ml milled volcanic stones may be added). Naturally, the soil for the solution should be taken from an area which is not polluted by artificial manure, insecticides, fungicides, etc. After cooling the soil is removed by filtration. Each time, after a part of these extracts is removed from the storage bottles, the extracts must be sterilized again, otherwise bacteria, fungi, or protozoan populations will develop.

The composition of the soil, yeast, or algal extracts may vary, and their effects on the physiology of the algae are not known in detail. Therefore, when using these extracts the experimental conditions are not defined as well as may be necessary in some cases. One of the main effects of these extracts is that they seem to include agents that reduce surface tension and enhance the permeability of the cell membrane and thus support assimilation. Another effect of soil extracts is the buffering and the chelation of metal ions by the humic acid complexes.

Despite our incomplete understanding of their detailed physiological functions, these extracts are suitable for the propagation of autotrophic or mixotrophic protistans and may be used in many experiments in which the growth conditions of the food source are not considered to be critical variables. Soil extract used in small amounts has been found to be tolerated by all brackish and many polyhaline water bryozoan species.

In experiments dealing with the problems of nutrition and metabolic requirements one should substitute chemically defined components for the extracts mentioned above. Na_2EDTA (disodium ethylenediamine-tetraacetate) (20 mg per liter of medium) or Tris [tris(hydroxyl-methyl)-aminomethane = amino-methylidinetrimethanol] are used as stabilizer and chelater agents in rearing media for various invertebrates by many authors. My own long-term tests have proved that Tris is toxic for saltwater bryozoans, even when present in small amounts.

The pH of freshwater media should be about 6–7; freshwater may be buffered with Sørensen's phosphate buffer. Seawater should have a pH of about 7.6–8.6, in general 8.3; seawater media may be buffered with $NaHCO_2$ and Na_2CO_3 (6:1) (Geisler, 1964; Hückstedt, 1971). Addition of stabilizing agents may be necessary for artificial freshwater media as well as for artificial seawater in many cases. Unfortunately, the carbonate buffer as well as the phosphate buffer will be destroyed after time through its assimilation by autotrophic algae.

A mixture of trace elements, P II metals (McLaughlin and Zahl, 1959), has proved useful for propagation of several algae in artificial media (dosage: 10–30 ml per liter of medium). P II metals mixture contains in 1 ml: 1 mg Na_2EDTA, 10 μg Fe (as Cl), 40 μg Mn (as Cl), 5 μg Zn (as SO_4), 1 μg Co (as SO_4), 0.2 mg B (as H_3BO_3).

Another mixture of trace elements, SL 4, has been developed by Pfennig (1966, personal communication) for cultivation of some bacteria and has also been found to be useful for the propagation of algae (dosage: 1 ml per liter of medium). SL 4 contains in 1 liter distilled water: 500 mg Na_2EDTA, 200 mg $FeSO_4 \cdot 7\ H_2O$, 10 mg $ZnSO_4 \cdot 7\ H_2O$, 3 mg $MnCl_2 \cdot 4\ H_2O$, 30 mg H_3BO_3, 20 mg $CoCl_2 \cdot 6\ H_2O$, 1 mg $CuCl_2 \cdot 2\ H_2O$, 2 mg $NiCl_2 \cdot 6\ H_2O$, 3 mg $Na_2MoO_4 \cdot 2\ H_2O$.

Several algae (e.g., *Cryptomonas* species) need or at least grow better in artificial media supplemented with vitamins; the most important vitamin is B_{12}. Some algae require mixtures of several vitamins and/or amino acids in artificial media (Provasoli et al., 1957, 1970; Lee et al., 1970; and many other authors).

For long-term rearing of freshwater bryozoans (*Plumatella repens*) together with autotrophic food protistans Wayss (1968) used a medium (pH 7.0 ± 0.1) mixed up 1:1 from a Knop solution and a soil solution: Knop solution contained in 1 liter distilled water 100 mg $Ca(NO_3)_2$, 25 mg $MgSO_4 \cdot 7\ H_2O$, 25 mg KH_2PO_4; soil solution was prepared by dilution of a soil extract 1:10 with distilled water.

Hauenschild (1962) introduced a synthetic medium for cultivation of autotrophic seawater organisms which is prepared with a reduced amount of magnesium. This medium is intended to prevent precipitation of $MgHPO_4$ by the normally increased content of phosphate in artificial media. This medium (Table III) may be used for rearing of some eurytopic bryozoan species together with autotrophic food protistans. For some delicate algae Hauenschild's medium should be augmented with organic agents like vitamins and/or amino acids or replaced by other media.

The nutrient agents, especially the extracts from yeast, algae, or soil, necessary for the propagation of the food protistans, eutrophy the medium. Eutrophication is tolerated by bryozoans from brackish water or polyhaline ones. Therefore, for cultivation of such bryozoans one can feed 1 part of algal culture diluted with 2 parts of seawater lacking any enrichments. However, typical marine bryozoans do not tolerate eutrophication.

For some experiments the food algae should be separated from their eutrophic propagation medium before they are added to the experimental animals.

TABLE III
Synthetic Seawater Medium[a,b]

Compound	Conc. (mg)
Main salts	
$NaCl$	26,700
$MgSO_4 \cdot 7\ H_2O$	3,200
$MgCl_2 \cdot 6\ H_2O$	2,200
$CaSO_4 \cdot 2\ H_2O$	1,600
KCl	700
$NaHCO_3$	200
Trace elements	
Cation sources	
$LiCl$	0.006
$RbCl$	0.060
$SrCl_2$	3.800
$AlCl_3$	0.030
$FeCl_3$	0.200
$ZnSO_4$	2.300
$MnSO_4$	0.600
$CoSO_4$	0.006
$CuSO_4$	0.001
Anion sources	
KBr	22.000
KI	0.020
Na-silicate	20.000
H_3BO_3	2.000
Na_2MO_4	0.200
Chelating agent	
Na_2EDTA	20.000
Nutrient salts for autotrophic food algae	
$NaNO_3$	100
$Na_2HPO_4 \cdot 12\ H_2O$	20

[a]Modified from Hauenschild (1968).
[b]Dissolved in 1000 ml distilled water.

Various methods of filtration have proved to be impracticable for this. Since 1966 centrifugation has been found to be the best method for separating algae from their culture medium, especially for feeding stenohaline bryozoans (Jebram, 1968). One should use a centrifuge with large centrifugation vessels (500–100 ml) and with a speed of 2000–2700 rpm or a continuous flow centrifuge, because bryozoans require large amounts of food. The algae are washed from the centrifugation vessels into calibrated tubes with sterilized, unenriched culture medium, and from there defined quantities may be added to the experimental vessels. Thus, bryozoans may be reared in a sterilized medium without nutrient agents.

B. Nutrition of Bryozoans

Bryozoans may consume various kinds of "food" including both live protistans and nonliving particles. Marcus (1926b) reported that Piscidin 000 (commercial, dry, finely powdered fish food) was accepted by different species of freshwater bryozoans. Ries (1936) observed that *Zoobothryon verticillatum* may eat various digestible materials, which will not normally be the food in the natural habitat (e.g., fat globules from milk or white of egg, blood cells from man or fishes, sperm from sea urchins, particles of starch from potatoes), but may also take up fine particles of indigestible materials [like India ink, carmine, soot, Norit (charcoal), melanine from *Sepia,* pyrrole blue, Fe-saccharate, etc.].

Marcus (1926a,b) described the "disgusted attitude" (*Ekelstellung*) of the tentacles, if the animals are (strongly) mechanically stimulated or if the chemical conditions, salinity, or temperature vary to unaccepted degrees. Ries (1936) also reported the "disgusted attitude" of the tentacles, when *Zoobothryon verticillatum* was fed with some unnatural kinds of "food"; but because initially the mouth took up some of these particles Ries concluded that generally bryozoans do not make selections between different kinds of food and will consume all swallowable particles swimming in the water. Perhaps this may be true with the species *Z. verticillatum.* Dudley (1970) and Ryland (1975) concluded from some experiments on filtration of artificial nonliving particles and from some accounts of specimens from the natural habitats, that bryozoans passively consume all plankton particles which can pass through their mouth. Thus bryozoans should "select" food only passively according to appropriate particle sizes.

On the contrary, my own observations on various species have clearly shown that in most cases gymnolaematous bryozoans do "select" food to some extent—not only for satisfactory size, but also according to the "taste" (Jebram, 1975b). "Selection" of food may be observed by feeding with natural mixed plankton, or when colonies are adapted to one species of food algae (monofood) and then brought into vessels with another food species.

The food "selection" of the bryozoans is in most cases not performed particle-by-particle but by variations of the polypide behavior in general according to the "taste" of the food (mixture) or of the agents diluted into the medium. Various forms of the "disgusted attitude" and various degrees of "nonevagination" of the polypides on one hand, while on the other hand, the evagination of all polypides and "eager feeding," have been found to be the main responses of all tested gymnolaematous bryozoans to the different "tastes" of the surrounding plankton mixtures.

For example, when a colony of *Conopeum seurati,* formerly fed with *Cryptomonas* spp., is newly fed with (old) cultures of the diatom *Phaeodactylum tricornutum,* the polypides starting to evaginate will quickly retreat again. After some time, tentacles may evaginate, adopt the "disgusted attitude" and not collect food; alternatively, evaginated polypides repeatedly reject cells of *P.*

tricornutum accumulated in the tentacle crowns. Colonies of *C. seurati* literally died of starvation, although a sufficient supply of algal cells was present. Specimens of *C. seurati, C. reticulum,* and *Electra pilosa* may eat some of the young cultures of *P. tricornutum* but do not grow well with this diatom as monofood; after some time the colonies stopped the budding of new zooids and more and more polypides underwent reduction to brown bodies.

Some ctenostome bryozoans (e.g., *Bowerbankia gracilis*) accepted *Phaeodactylum tricornutum* as food more readily than did the cheilostomes, but after some days the ctenostomes tested with *P. tricornutum* as monofood stopped their growth as well. Cessation of growth also occured in the ctenostome *Zoobothryon verticillatum*, but required some weeks.

The dislike of old cultures of *Phaeodactylum tricornutum* as food by *Conopeum seurati* and other bryozoans could only depend on its "taste" (or perhaps structure) but not on its size, because the cells of *P. tricornutum* are significantly smaller than the cells of the dinoflagellate *Oxyrrhis marina.* Additionally, *O. marina* is a powerfully moving protistan; nevertheless, it is well accepted as food by *C. seurati* and many other bryozoans. Several bryozoan species are able to attain sexual maturity with *O. marina* as a monofood.

Even comparatively voluminous particles may be accepted as food by bryozoans, if these particles "taste good." For example, the brackish water membraniporids *Electra crustulenta* and *Conopeum seurati* may eat planula larvae of the hydroid *Laomedea loveni* (Jebram, 1975b). The smaller diameter of the planula larvae of *L. loveni* is about twice the diameter of the "normally" open mouth of these bryozoans. Another example is the dinoflagellate *Gymnodinium splendens,* which is larger than the "normally" open mouth of *Conopeum seurati*; nevertheless, this membraniporid appeared "eager" to catch and ingest these voluminous protistans, while centric diatoms of comparable sizes were more or less disliked or ignored.

My observations described above agree with those more superficially reported by Marcus (1926b), who observed that freshwater bryozoans preferred some sorts of food more than others. Bullivant (1968a,b) reported that some marine bryozoans grew better with some food species than with others. Jebram (1968, 1970b, 1973, 1975b) found that variations of the food species (see Table IV) may cause changes in budding pattern, modification of growth forms, and different growth rates of several saltwater bryozoan species and may also influence the general vitality of the colonies and the attainment of maturity. Cummings (1975) and Winston (1976) made related observations on other species of bryozoans.

The colors of the bryozoan embryos are suggested to be diagnostic characters by Ryland (1958). I have proved by experiments that this trait is influenced by the kind of nutrition (Jebram, 1975b). Careful observations on native specimens of *Bowerbankia gracilis* (and of cf. *B. gracillima*) revealed that the color of the reserves may vary to some degree in the natural habitats as well (Jebram, in press). On the other hand, the appearance of "approximately specific" colors of the ova, embryos, and lasting buds in many bryozoan species coupled with my own observations on food uptake led to the assumption that the "specific" reserve colors may be partly caused by a selective food-uptake behavior of each gymnolaematous bryozoan species (Jebram, 1975a,b).

Several tests (D. Jebram, unpublished) revealed, on the other hand, that "favor" of a food species is no evidence for its sufficiency. In several cases protistans, which were more or less accepted by the bryozoans, did cause definite toxic effects (histolysis of polypides, death of colonies); some of these toxic protistans are included in Table V.

Additionally, research about the sufficiency and physiological effects of single protistan species as food for bryozoans is complicated by an interaction of a "persevering effect" and a "summation effect." Several agents from the food (perhaps needed in smaller quantities) may be stored in the zooids for long periods of time and can be used up during periods of absence of such foods. If in experiments one monofood (or mixture) is replaced by another one, reserves from the former nutrition can compensate for deficiencies in the contents of the new food. This complication hinders early detection of the effects of the new food. On the contrary, the new food is an addition to the reserves remaining from the former source and causes by the "summation effect" an increase in vitality and growth. In summary, in general mixtures of protistan species (so far as none is toxic) have proved to enhance vitality, health, and growth rates of bryozoans. For example, the effects of the good monofood *Oxyrrhis marina* or *Cryptomonas* spp. will be essentially improved by addition of the less good monofood *Dunaliella marina* or by trace quantities of (young) *Phaeodactylum tricornutum* (which was the least effective as monofood in long-term tests) (Jebram, 1975b, and unpublished results).

In addition to species or size of food, age and physiological conditions of the food organisms are also important factors. Autotrophic algae should be used as food during their logarithmic growth phase, because they will be healthy and in best physiological condition at this time. The nutritive worth will also be influenced by the components of the culture medium of the food algae.

The heterotrophic dinoflagellate *Oxyrrhis marina* must be fed with autotrophic basic foods, e.g., *Dunaliella* spp. or *Phaeodactylum tricornutum*. Cultures of *O. marina* proved to be a very good food for many species of bryozoans at the time when *O. marina* has digested its food and when the color of the dinoflagellate culture changes from green (if *Dunaliella* is being fed) or brown (if *Phaedactylum* is being fed) to clear light brown. When the suspension of *O. marina* becomes too old, about 2 or 3 days later, it becomes toxic to various species of bryozoans (Jebram, 1975b). Colonies fed with overaged *O. marina* resorb more and more polypides. When most zooids of a colony have eaten overaged *Oxyrrhis* cells,

TABLE IV
Foods Reported to Be More or Less Sufficient for Bryozoans[a]

Species		Medium[b]	Reference
Cryptophyta			
Cryptomonas sp.[c]		S	Emschermann (1961); Hauenschild (1968); Jebram (1968, 1970b, 1973, 1975b)
Cryptomonas sp.			Menon (1972)
Cryptomonas maculata		S	Jebram (1975, unpublished)
Cryptomonas calceiformis		S	Jebram (1975, unpublished)
Hemiselmis brunnescens		S	Jebram (1975, unpublished)
Hemiselmis aff. *virescens*		S	Jebram (1975, unpublished)
Euglenophyta			
Euglena sp.		F	Marcus (1926b); Rogick (1937)
Eutreptia marina	(X)	S	Bullivant (1968a, 1968b); Jebram (1975, unpublished)
Chrysophyta (incl. Haptophyceae)			
Monochrysis lutheri		S	Bullivant (1968a, 1968b); Jebram (1970b, 1973, 1975b)
Isochrysis galbana		S	Jebram (1973, unpublished)
Chrysochromulina			
pringsheimii	(X)	S	Jebram (1975, unpublished)
minor	(X)	S	Jebram (1975, unpublished)
Coccolithus pelagicus		S	Jebram (1975, unpublished)
Cricosphaera carterae		S	Bullivant (1968a, 1968b); Jebram (1975, unpublished)
Dinophyta			
Gymnodinium splendens	(X)	S	Jebram (1975, unpublished)
Gymnodinium simplex		S	Jebram (1975, unpublished); Winston (1976)
Amphidinium carteri	(X)	S	Bullivant (1968b), Jebram (1975; toxic, unpublished)
Oxyrrhis marina	(X)	S	Grell (1950, unpublished); Schneider (1959); Kaissling (1963); Hauenschild (1968); Jebram (1968, 1970a, 1973, 1975b); Eiben (1976)
Bacillariophyceae			
Cyclotella nana		S	Bullivant (1968b); Jebram (1975b)
Thallassiosira fluviatilis		S	Bullivant (1968b)
Phaeodactylum tricornutum	(X)	S	Bullivant (1968a, 1968b); Jebram (1975b, and unpublished)
Chlorophyceae			
Chlamydomonas sp.		F	Marcus (1926b); Rogick (1937)
Chlamydomonas pomiformis		F	Rüsche (1938)
Chlorogonium sp.		F	Wayss (1968)
Haematococcus pluvialis		F	Wayss (1968)
Dunaliella tertiolecta		S	Bullivant (1968a, 1968b)
Dunaliella marina		S	Jebram (1975b)
Prasinophyceae			
Pyramimonas obovata		S	Jebram (1975, unpublished)
Pyramimonas grossii		S	Jebram (1975, unpublished)
Micromonas pusilla		S	Jebram (1975, unpublished)
Micromonas squamata		S	Jebram (1975, unpublished)

continued

Table IV (*Continued*)

Species	Medium[b]	Reference
Rhizopoda		
Arcella sp.	F	Rogick (1937)
Ciliata		
Colpoda sp.	F	Marcus (1926b)
Paramecium sp.	F	Rogick (1937)
Paramecium trichium	F	Wayss (1968)
Paramecium bursaria	F	Wayss (1968)
Paramecium aurelia	F	Wayss (1968)
Paramecium caudata	F	Wayss (1968)
Rotifera	F	Rogick (1937); Rüsche (1938)
Philodina sp.	F	Wayss (1968)
Adineta sp.	F	Wayss (1968)
Macrotrachela sp.	F	Wayss (1968)

[a]Problematical species (from my own tests) are indicated by (X); some of these are discussed in the text. Compare also Table V.

[b]S, seawater; F, freshwater

[c]This strain of *Cryptomonas* sp. is subcultivated from the stock of Prof. D. K. Grell, Zoological Institute, Tübingen.

TABLE V
Protistan Species Unable to Sustain Growth or with Toxic Effects if Eaten by Bryozoans as Monofood for Prolonged Periods[a]

Chrysophyceae
Olisthodiscus luteus
"Monas" sp. (heterotrophic, with monadal and amoeboid phase)
Haptophyceae
Prymnesium parvum
Phaeocystis pouchetii (motile phase)
Chrysochromulina kappa
Chrysochromulina strobilus
Chrysochromulina brevifilum
Ochrosphaera napolitana
Dinoflagellata
Amphidinium carteri
Amphidinium klebsii
Bacillariophyceae
Cyclotella nana
Phaeodactylum tricornutum
Skeletonema costatum

[a]After my unpublished long-term tests mainly with the cheilostomatous species *Electra pilosa.* Most of the protistan species listed here "taste bad" to the bryozoans, however, a few are well accepted.

even polypide buds and completed polypides which have not evaginated and therefore not eaten will become histolyzed. In such colonies all tissues will be injured so much that no regenerative growth is possible, and the colony will die after some time. For these reasons, *O. marina* may be a good food for many bryozoan species, but one has to pay attention to several factors: First, it must be used in the right stage of culture, about 2 days after the inoculation of a portion of *O. marina* into a suspension of its basic food (ca. 1:4); second, the food *O. marina* presented to the bryozoans should be renewed every 2 or 3 days with newly fed *Oxyrrhis* cultures to prevent deleterious effects. A similar kind of negative "overaging effect" was observed when *O. marina* was fed on overold cultures of a basic food source (older than about 3 weeks).

The important influences of the food species is corroborated by another observation. Although several of the laboratory-treated bryozoan species fed well on both *Oxyrrhis marina* and *Cryptomonas* sp., and grew and matured well with either as a monofood, every time the type of food was changed, all active polypides were reduced and replaced by new ones (Jebram, 1968). Similar observations have been made after changes between other food species. The physiology of the polypides seems to be regulated by the food as well as by other factors. A change of the kind of food requires a change in the physiology of digestion, which often results in a replacement of the polypides. Polypides seem to be incapable of adapting their physiology to drastic changes of external factors, e.g., quick change of the kind of food or salinity. Probably some of the bryozoan species that grew poorly under the laboratory conditions provided, did so because of this phenomenon. Surely there are bryozoan species which require other kinds of food than previously used in laboratories.

Bacteria may be eaten by bryozoans. Small amounts of digested bacteria seem to have no negative effects, but larger quantities have been found to be toxic for some bryozoans.

My own observations over several years have generally shown that (1) even if nonliving particles may be consumed by bryozoans, most species of this animal group will grow well and may become sexually mature only with sufficient quantities of live food of a suitable quality, and (2) different bryozoan species require mixtures of different species of protistans as their main food and may survive with only a few food species as monofoods.

For some quantitative experiments the number of food cells in the culture medium are to be determined. Food cells may be counted microscopically in a particle-counting chamber for blood cells or electronically by the different types of counters (e.g., Coulter Counters). Bullivant (1968a) found that the bryozoans in a medium exposed to the electrodes of a Coulter Counter did not behave normally; the polypides took longer to evaginate than usual. This observation is alarming considering the wide use of Coulter Counters in experimental work with planktonic organisms and therefore needs special consideration and further investigation.

C. Substrata for Laboratory Work

For detailed precise observations, statistical analysis, and making photographs, incrusting bryozoans should be grown on plane, transparent substrata. It has been found that most bryozoan species, which may survive laboratory conditions, regardless of their natural substratum in the habitat (algae, hydroids, woods, rocks, shells of mollusks or arthropods), are also able to grow on artificial substrata like glass or Perspex.

Glass is a hard material and will remain clear and transparent for longer times. The surface of glass has very limited exchange of substances or ions with the incrusting animals or the medium. Additionally, glass may easily be cleaned.

The transparency of glass allows observations of the experimental animals from below; this is very useful because in most bryozoans, even the Ascophora with strongly calcified frontals, the basal walls are thin and likewise transparent. One may look through the slide and the basal wall into the body cavities of the bryozoans and thus ascertain their physiological conditions by observing the color of the guts and the development of polypides and ovaries (Jebram, 1973). With the polypides hanging downwards most of the feces will fall away and not dirty the colonies.

Perspex and several related artificial materials also may be used as substrata, but their surfaces will easily be scratched, reducing their transparency. Therefore, it is impossible to make detailed microscopic observations through a Perspex plate that has been used as the substratum for bryozoans for some time. On the other hand, Perspex may be used for such species which dislike the absolute smoothness and/or the (very weak) acidic conditions of the surface of glass.

Two sizes of slides have been found useful for laboratory work with bryozoans: Slides 76 × 26 mm, as normally used for histological specimens, when used for culturing incrusting colonies may be directly prepared as material for microscopic examination. Square slides 76 × 76 mm are very well suited for statistical experiments which require information such as "mean zooid numbers" or "minimum zooid numbers of colonies" (Jebram, 1973, 1975b).

Slides used for propagation of colonies may be covered with one to three coverslips (26 × 20 mm) or be fitted with a belt of nylon thread. Coverslips are fastened with slit pieces of polyvinyl tubing on the slides. Before the slides are used as substrata they should be stored in a holding device (as used for staining histological specimens), cleaned with a commercial glass cleaner such as Extran or Deconex (1:10 in water) for 4–8 hours, washed with running freshwater for 1 hour, and then repeatedly washed with distilled water; they may be sterilized, if desired.

For wood-penetrating bryozoans, pieces of old coniferous wood are cleaned mechanically, fastened with nylon thread on a slide, and repeatedly washed with boiling distilled water. Some wood-penetrating species may also grow on glass or Perspex slides.

D. Vessels for Rearing

Small glass dishes (diameter about 9 cm) are useful for rearing bryozoan colonies on single slides (76 X 26 mm) (Fig. 2). Such dishes may be used for observations of live colonies through a stereomicroscope. If in one experimental series several colonies, each on a slide, are to be grown under the same conditions, the slides may be suspended with plastic clamps in larger vessels of glass or Perspex (Fig. 3). The larger, square slides may be placed as well in these vessels, with the bryozoans hanging downward (Fig. 4). Larger vessels must be aerated because sufficient oxygen will not diffuse through a water column more than about 5 cm in height. Such aeration also provides the necessary water circulation.

In vessels like those mentioned above, planktonic food species with low swimming activity (like *Phaeodactylum*) settle at the bottom of the container, even if the water is agitated by aeration bubbles. The sedimented cells may die and pollute the medium thus favoring the propagation of bacteria.

This problem has been solved by the development of a new apparatus for long-term rearing of bryozoans and other plankton feeders (Jebram, 1973). This

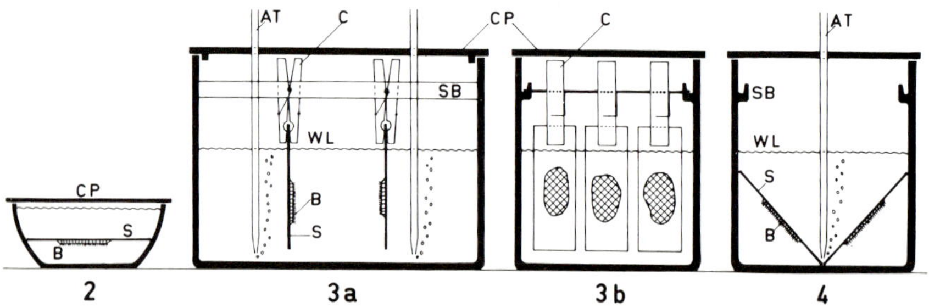

Fig. 2. Glass dish (9 cm in diameter) with a bryozoan colony on the underside of the slide.

Fig. 3. Perspex vessels (about 18 X 12 X 13 cm) with slides hanging on plastic clamps and with aeration tubes. (a) Side view; (b) frontal view. Abbreviations: AT, aeration tube; B, bryozoan colony; C, plastic clamp; CP, coverplate; S, glass slide; SB, side board; WL, water level.

Fig. 4. Perspex vessel (like Fig. 3b) with square slides (76 X 76 mm) and with aeration tubes. For abbreviations see Fig. 3.

new apparatus consists of an experimental vessel and pumping devices. The essential attribute of the new apparatus is the parabolic bottom (made of Perspex) of the experimental vessel. This form sustains a vertical water rotation and hinders sedimentation of the planktonic food organisms (Fig. 5).

The pumping device of the new apparatus consists of a descending tube, bubble pump, ascending tube, container vessel, water-propelling tube, air inlet tube, air outlet tube, and connecting tubes. The different parts of the apparatus are stored in an external box (of Perspex) (Fig. 6). The frontal coverplate over

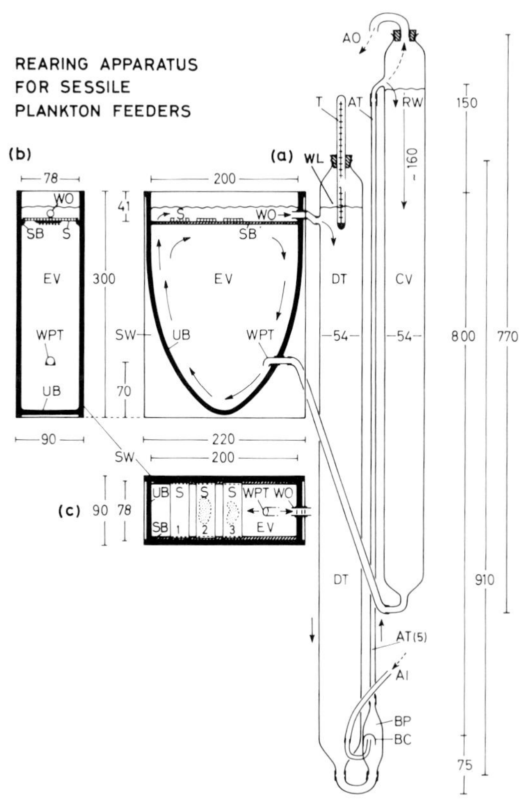

Fig. 5. Apparatus for rearing sessile plankton feeders. (a) The apparatus in side view; (b) cross section; and (c) view from above on the experimental vessel. Abbreviations: AI, air inlet tube; AO, air outlet tube; AT, ascending tube; BC, bubble cup; BP, bubble pump; CV, container vessel; DT, descending tube; EV, experimental vessel; RW, raised water level in CV; S, slides with the experimental animals; SB, side board for S; SW, side walls of EV; UB, U-formed bottom of EV; WL, water level in DT and EV; WO, water-outlet tube; WPT, water-propelling tube. Numbers (except on the slides) represent millimeters; arrows indicate the movements of water and air in the apparatus. (After Jebram, 1973.)

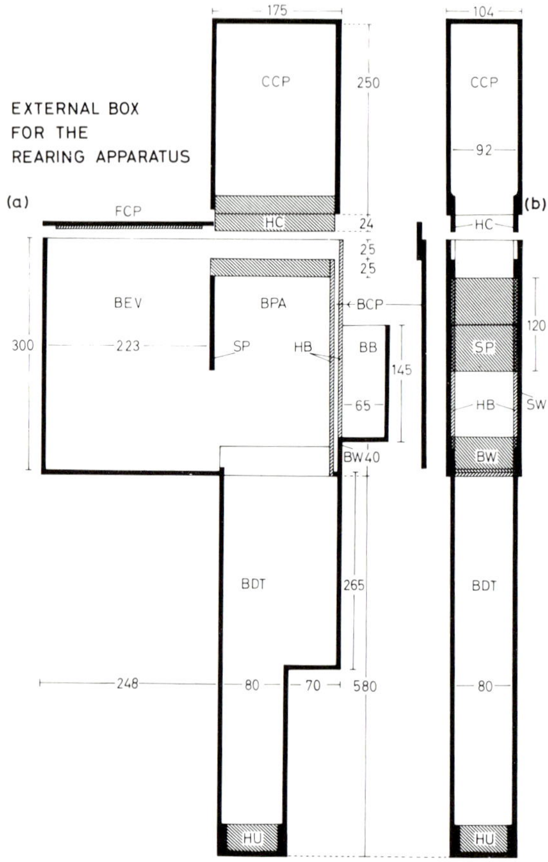

Fig. 6. External box for rearing apparatus. (a) Side view of the box; (b) cross section through the back part. Abbreviations: BCP, back coverplate; BDT, box for the descending tube and the bubble pump; BEV, box for the experimental vessel; BPA, box for (a part of) the pumping apparatus; BW, back wall; CCP, cover cup for the pumping apparatus; FCP, front coverplate (glass); HB, holding device boards for BCP; HC, holding device boards for CCP; HU, holding device for the U-tube connecting the descending tube and bubble pump; SP, separating wall; SW, side wall of the external box. Numbers represent millimeters. (After Jebram, 1973.)

the experimental vessel is made of glass. The other walls and the covercup are made of Perspex. The pumping apparatus and the tubes connecting it with the experimental vessel are made of glass and silicon rubber, respectively. Experimental vessels can also be built with side walls of glass and a parabolic bottom of Perspex. Such vessels may be dry sterilized at 110°C. Sterile work is therefore possible with these experimental vessels.

The culture apparatus has a total volume of about 7 liters. This size is large enough to maintain the stability of the chemical conditions for some weeks, when only few bryozoan colonies live in it. In this apparatus the water circulates in a closed system constantly for some weeks. Any type of mechanical pump would damage the planktonic food organisms during such long periods of time. For this reason the water rotation is produced by the hydrostatic pressure of a raised water level in a container vessel above the level of the experimental vessel. The water is raised by an improved bubble pump (Jebram, 1970a) with a diameter of 22 mm and a bubble cup 10 mm in diameter.

The water level in the experimental vessel is near its water outlet. This arrangement supports a water current parallel to the water surface. The current has a high speed on the parabolic bottom of the experimental vessel, while it moves at only a few centimeters per second at the surface.

Bryozoans will grow attached to glass slides (76 X 26 mm) lying horizontally on small boards suspended a small distance beneath the water surface in the experimental vessel. This working distance allows observation of the animals with a stereomicroscope (up to 4 X objective magnification with a Leitz microscope) without removing the colonies from the water. The colonies are placed in the apparatus on the undersides of the slides so the evaginating polypides extend downward to prevent feces from contaminating the colonies. The larger square slides (76 X 76mm) should stand vertically near the vertical frontal wall (made of glass) of the experimental vessel as otherwise air bubbles may accumulate under these slides.

The high current speed along the parabolic bottom hinders sedimentation of food organisms and of fecal material. Feces and some food cells may settle on the top sides of the horizontally placed slides, especially on the first slide.

The use of food centrifuged from its propagation medium and the prevention of the sedimentation of the planktonic cells delay the eutrophication of the medium and the increase of bacteria. Therefore, cleaning of the colonies by brushing appears to be necessary at less frequent intervals.

The total volume of the medium in the culture apparatus should not be replaced all at once after a long period of time, but in aliquots at several shorter intervals. It has proved useful to replace about 1–2 liters of medium per week; the added medium should be at the culture temperature. So, in these new cultivation apparatus the stability of the chemical conditions may be maintained for several (2–5) weeks.

E. Management of Temperature

Temperature is an important ecological and physiological factor. Both the growth rates and the colony formation of bryozoans are influenced by this parameter.

For simple temperature control the experimental vessels of smaller sizes (as shown in Figs. 2–4) may be placed in normal refrigerators or constant-temperature boxes which are produced in many varieties by several manufacturers. When using such a chamber one should provide for aeration of the experimental vessels and for their illumination to enable assimilation by the autotrophic food algae.

In such commercial growth chambers the temperature in the experimental vessels may vary by small degrees, because the vessels are surrounded by air which will easily be replaced during opening of the thermostat box. A more constant temperature may be achieved if the experimental vessels are placed in a constant-temperature bath. The temperature of the bath can be regulated by a cooling (or heating) coil (e.g., thermostats from Colora or Haake).

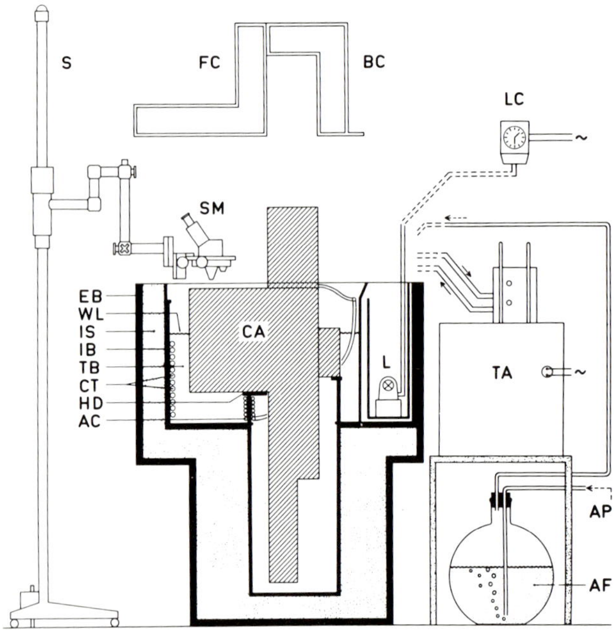

Fig. 7. Constant-temperature box for the long-term rearing apparatus. Abbreviations: AC, air precooling tubes; AF, air filtration through a charcoal bath; AP, air inlet from aeration pump; BC, back cover of Perspex; CA, cultivation apparatus for long-term rearing; CT, primary cooling tubes; EB, external box; FC, front cover of Perspex; HD, holding device for CA; IB, internal box of hard polyethylene; IS, isolation by Styropore; L, lamp in a double-walled box of Perspex; LC, light clock; S, support for SM, stereomicroscope; TA, temperature-regulating aggregate (thermostat); TB, constant-temperature bath for CA; WL, water level of TB.

For the new apparatus for long-term rearing a special constant-temperature box has been designed (Fig. 7). A thermostat controls the temperature in the constant-temperature bath. The bath is isolated by Styropore. The cultivation apparatuses stand on holding devices in the bath and are covered by double-walled cover devices of Perspex. The air is filtered through a bath of charcoal in distilled water and precooled in glass tubes in the temperature bath before its inlet into the bubble pumps of the cultivation apparatuses. The apparatuses are illuminated from above outside and from behind inside the constant-temperature box. The lamp inside the constant-temperature box is isolated by a double-walled box of Perspex.

A light clock is used to control the photoperiod. In normal cases, bryozoans may be reared under long-day conditions with a light:darkness regimen of 16:8 hours.

F. Still or Agitated Waters

Water movements have several different properties, each of which may affect living organisms (Riedl, 1964, 1966); three of these seem to be of primary importance for the growth of bryozoans.

1. Speed of water particles (cm/sec); important for problems of current exposure, rheotaxis, and rheotropism.
2. Current volume, rate of change of water ($m/cm^2/sec$); important for the provision of adequate food for filter feeders, for problems of gas exchange in respiration, and ion exchange through the body walls.
3. Pressure of current (kp/cm^2); important for problems of current exposure and stability of the body walls, especially of the erect colony forms, and for rheotropism.

The first two parameters have been tested by several experimenters, but the third one has been more or less ignored, perhaps because of the technical difficulties involved in assessing its importance.

During rearing of animals the movement of water is of general importance, because it prevents development of inhomogeneities in the medium, supports the diffusion of oxygen and, to some extent, distributes suspended food.

Simple agitation of the water in the rearing vessel is produced by aeration bubbles or by shaking devices such as shaking tables or shaking cars (Hauenschild, 1968) (Fig. 8).

Most bryozoans live in coastal waters, where oscillating water movements predominate. For experiments with continuous oscillating water movements Svoboda (1970) developed an apparatus which consists mainly of a large piston pump, the experimental chamber, and a compensatory cylinder (Fig. 9). The movement velocity may be varied from 70 to 160 cm/sec. The exposure of

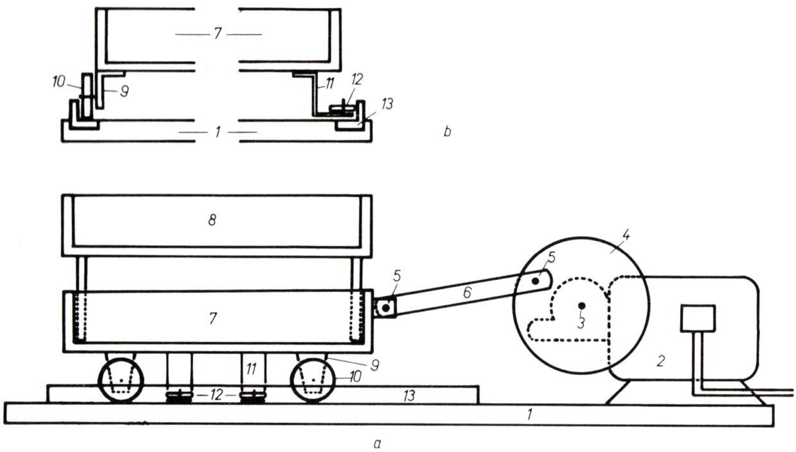

Fig. 8. Shaking apparatus. Abbreviations: 1, base plate; 2, electric motor; 3, axle of transmission; 4, large transmission wheel; 5, trunnion; 6, connecting rod; 7, shaking car; 8, replaceable top device; 9, receptacle of ball-bearing axle; 10, ball-bearing; 11, holder for axle of conducting wheel; 12, conducting wheel; 13, rail of steel. (Original after Hauenschild, 1968.)

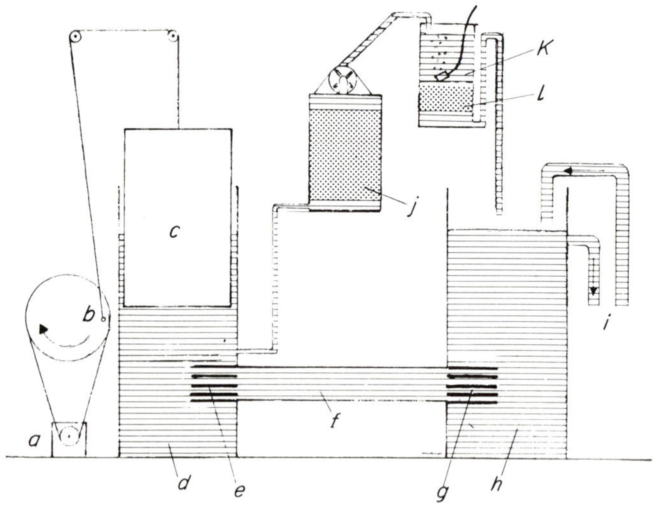

Fig. 9. Apparatus for producing oscillating water movements. Abbreviations: a, electric motor; b, excentric wheel; c, piston; d, pumping cylinder; e and g, tubes divided vertically and horizontally by Perspex plates; f, experimental chamber of Perspex; h, compensating container; i, tubes connected to refrigerator; j, Eheim filter loaded with Hydraffine charcoal; k, ozonization contact tube; l, Hydraffine charcoal for removing surplus ozone. (Original after Svoboda, 1970.)

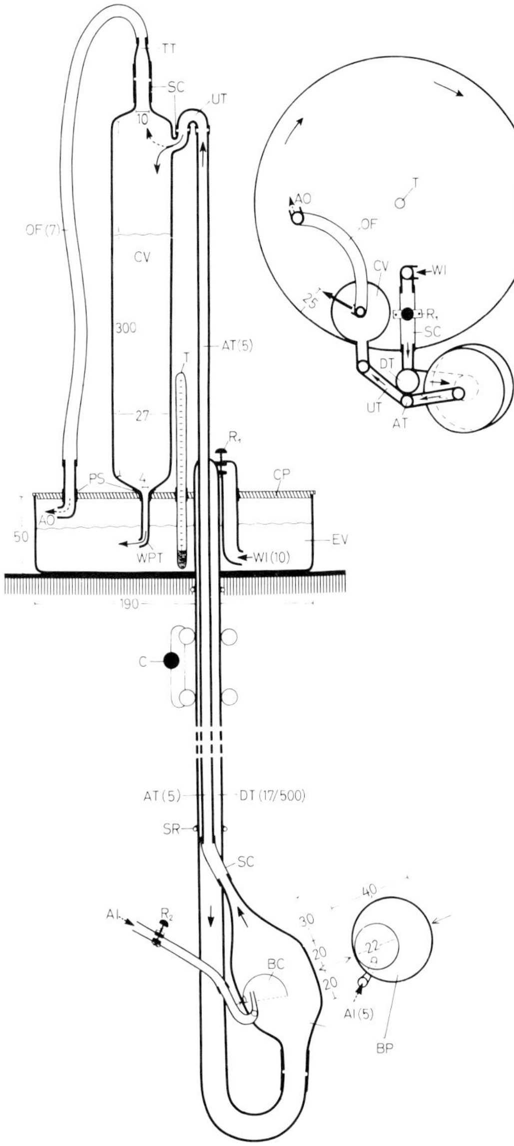

Fig. 10. Apparatus for producing continuous water currents. Left hand, front view; right hand, view from above. Abbreviations: AI, air inlet tube; AO, air outlet tube; AT, ascending tube; BC, bubble cup; BP, bubble pump; C, clamp; CP, coverplate; CV, container vessel; DT, descending tube; EV, experimental vessel; OF, overflow tube; PS, packing sleeve; R_1, regulator of WI; R_2, regulator of AI; SC, silicon rubber connecting tube; SR, rubber securing ring; T, thermometer; TT, tapered tube; UT, U-tube; WI, water inlet tube; WPT, water-propelling tube. Arrows indicate the movements of water and air; numbers represent millimeters. (After Jebram, 1970a.)

TABLE VI
List of Bryozoan Species Reared in Laboratories[a]

Species	Reference
Phylactolaemata	
Cristatella mucedo	Braem (1912); Otto (1921); Marcus (1926b); Oda (1954)
Pectinatella magnifica	Kammerer (1907); Braem (1912)
Lophopus crystallinus	Kammerer (1907); Otto (1921); Marcus (1926b, 1934)
Lophopodella carteri	Rogick (1937, 1938); Oda (1954)
Plumatella fungosa	Braem (1897); Marcus (1926b); Rogick (1937); Rüsche (1938)
Plumatella repens	Marcus (1926b); Bullivant (1968b); Wayss (1968)
Plumatella fruticosa	Otto (1921); Wiebach (1952, 1953)
Fredericella sultana	Braem (1908); Otto (1921); Marcus (1926b)
Stenolaemata	
Crisiella producta	Jebram (1968, unpublished) (x)
Crisia eburnea	Jebram (1968, unpublished) (x)
Crisia aculeata	Jebram (1968, unpublished) (x)
Crisia sp.	Schneider (1959)
Tubulipora sp.	Schneider (1959)
Tubulipora sp.	Jebram (1968, 1970b) (x)
Lichenopora sp.	Schneider (1959)
Disporella separata	Bullivant (1968b)
Ctenostomata	
Flustrellidra hispida	Ryland (1959, 1960); Jebram (1973, and unpublished)
Alcyonidium hirsutum	Ryland (1959, 1960); Jebram (1968, 1970b)
Cf. *Alcyonidium gelatinosum*	Jebram (1968, 1970b, 1973)
Cf. *Alcyonidium polyoum*	Ryland (1959, 1960); Jebram (1968, 1970b, 1973)
Paludicella articulata	Marcus (1926b); Jebram (1968, unpublished)
Farrella repens	Marcus (1926a); Jebram (1968, 1970a, 1970b, 1973)
Triticella koreni	Jebram (1970b, 1973) (x)
Valkeria uva	Jebram (1970b, 1973) (x)
Bulbella abscondita	Braem (1951); Jebram (1969a, 1970b)
Tanganella muelleri	Braem (1951); Jebram (1969a, 1970b)
Victorella pavida	Braem (1951); Jebram (1969a) (x)
Bowerbankia imbricata	Jebram (1968, 1970b, 1973)
Bowerbankia gracilis	Braem (1951; Schneider (1959); Jebram (1968, 1969a, 1970b, 1973, 1975b); Eiben (1976)
Cf. *Bowerbankia gracillima*	Jebram (in press)
Zoobothryon verticillatum	Zirpolo (1933); Ries (1936); Bullivant (1968a, 1968b); Jebram (1975, unpublished)
Buskia nitens	Jebram (1969b, 1970b, 1973)
Cheilostomata	
Aetea sp.	Schneider (1959); Emschermann (after Hauenschild, 1968)
Electra verticillata	Bobin and Prenant (1960)
Electra pilosa	Bobin and Prenant (1960); Marcus (1926a); Jebram (1968, 1970a, 1970b, and unpublished); Menon (1972)

continued

TABLE VI (*Continued*)

Species	Reference
Electra posidoniae	Jebram (1968) (x)
Electra monostachys	Cook (1964); Jebram (1968, 1969b, 1970a, 1970b)
Electra crustulenta	Cook (1961, 1962, 1964); Jebram (1968, 1969a, 1969b, 1970b)
Conopeum seurati	Cook (1962); Cook and Hayward (1966); Jebram (1969a, 1969b, 1970a, 1970b, 1973, 1975b)
Conopeum reticulum	Cook (1964); Jebram (1968, 1970a, 1970b); Menon (1972)
Conopeum tenuissimum	Winston (1976)
Membranipora membranacea	Bronstein (1938a, 1938b, 1940); Lutaud (1961); Jebram (1968, unpublished) (x), Menon (1972)
Scrupocellaria scruposa	Marcus (1926a); Ryland (1960)
Scrupocellaria reptans	Lutaud (1953); Schneider (1959)
Bugula simplex	Grave (1930, 1937)
Bugula turrita	Grave (1930, 1937)
Bugula neritina	Lynch (1947, 1949); Mawatari (1951); Schneider (1959); Ryland (1960); Bullivant (1968b)
Bugula fulva	Ryland (1960)
Bugula plumosa	Ryland (1960)
Bugula stolonifera[b]	Grell (1950, unpublished); Schneider (1959); Kaissling (1963); Jebram (1968, 1970b)
Aspidelectra melolontha	Jebram (1968, unpublished) (x)
Cribrilina punctata	Jebram (1968, 1970b) (x)
Crassimarginatella papulifera	Gordon (1971a)
Eurystomella foraminigera	Gordon (1971a)
Steganoporella buskii	Cook (1968)
Discoporella umbellata	Cook (1963)
Cupuladria doma	Cook (1963)
Hippoporidra senegambiensis	Cook (1968)
Hippothoa hyalina	Ryland (1959, 1960); Jebram (1968) (x)
Fenestrulina malusii	Gordon (1971a, 1971b)
Microporella ciliata	Jebram (1968) (x)
Escharella immersa	Jebram (1968) (x)
Cryptosula pallasiana	Jebram (1968, 1970b)
Schizoporella sanguinea	Schneider (1959)
Schizoporella unicornis	Bullivant (1968b); Jebram (1968, unpublished)
Schizoporella unicornis floridana	Cummings (1975)
Porella sp.	Schneider (1959)
Holoporella brunnea	Bullivant (1968b)

[a]A small selection only could be presented here from the large number of papers dealing with phylactolaemates, because in most cases scientists working with freshwater animals rear them for some time, and complete references concerning reared phylactolaemates would infalte and distort this list. The list of gymnolaemate species reared is largely complete. If in my own research rearing could be maintained for more or less short times only, it is indicated by (x).

[b]At first erronously described as *Bugula avicularia.* Cultivated colonies from the stock of Prof. K. Grell have been determined to be really *Bugula stolonifera* by Dr. J. S. Ryland (personal communication).

sessile experimental animals to the direction of the water oscillations can be regulated by special turning devices in the wall of the experimental chamber.

Oscillations are complex movements of the water particles. Before investigations of oscillatory factors are initiated the effects of continuous water currents should be analyzed. The kinds of influences which water movements have on the growth of bryozoans can only be discerned by comparison with the growth of colonies in still water. For these control experiments the colonies must be reared in dishes with a small water height to enable a sufficient quick diffusion of oxygen through the medium as other modes of aeration would produce some movements of the water.

Continuous nonoscillating water currents can easily be produced in a circular experimental vessel (Jebram, 1970a). The circular form reduces friction, and comparatively high current speeds may be created by low-energy expenditure produced by the hydrostatic pressure of a raised water level in a container vessel above the experimental vessel. The circular shape causes a gradient of the current speed in the experimental vessel. The experimental animals are placed on slides or glass rods near the external wall of the experimental vessel, where the current attains its maximum speed. In this simple and small apparatus the current speed may be increased up to about 0.5 m/sec (Fig. 10).

Preliminary experiments by Marcus (1926a) and Jebram (1970a) have demonstrated that several bryozoan species have positive rheotropic growth reactions. Smaller current speeds, such as near the water surface in the new apparatus for long-term rearing, are especially favorable for the growth of Bryozoans. (See Table VI for a list of bryozoan species reared in the laboratory.)

ACKNOWLEDGMENTS

I am indebted to Prof. Dr. K. Grell for his kindness in introducing me some years ago to the basic methods of cultivation in the Protozoological Laboratory in the Zoological Institute, University of Tübingen, and for giving me some stock cultures of food organisms. I very much thank Dr. J. C. Green, The Laboratory of the Marine Biological Association of the U.K., Plymouth, and Dr. W. Koch, Institute for Plantphysiology of the University of Göttingen, for providing me with stock cultures of food protistans, and Prof. Dr. C. Hauenschild and Prof. Dr. W. A. Müller, Zoological Institute of the Technical University of Braunschweig, for suggestions and critical discussions concerning special problems of rearing and experimental work. My research on saltwater bryozoans is supported by the Deutsche Forschungsgemeinschaft (Grant Nos. Je 61/1-7).

REFERENCES

Atkins, D. (1932). The ciliary feeding mechanism of the entoproct Polyzoa and a comparison with that of the ectoproct Polyzoa. *Q. J. Microsc. Sci.* 75, 393–423.

Becker, G. (1937). Untersuchungen über den Darm und die Verdauung von Kamptozoen, Bryozoen und Phoroniden. *Z. Morph. Oekol. Tiere* **33**, 72–127.

Bobin, G., and Prenant, M. (1960). *Electra verticillata* (Ellis et Solander 1786) Lamouroux 1816 (Bryozoaire Chilostome). *Cah. Biol. Mar.* **1**, 121–156.

Borg, F. (1926). Studies on recent cyclostomatous Bryozoa. *Zool. Bidr. Uppsala* **10**, 181–507.

Braem, F. (1888). Untersuchungen über die Bryozoen des süssen Wassers (Vorl. Mittlg.). *Zool. Anz.* **11**, 503–509; 533–539.

Braem, F. (1890). Untersuchungen über die Bryozoen des süssen Wassers. *Bibl. Zool.* (Kassel) **2**, H.6, 1–134.

Braem, F. (1897). Die geschlechtliche Entwicklung von *Plumatella fungosa*. *Zoologica* (Stuttgart) **23**, 1–96.

Braem, F. (1908). Die geschlechtliche Entwicklung von *Fredericella sultana*. *Zoologica* (Stuttgart) **52**, 1–38.

Braem, F. (1912). Die Keimung der Statoblasten von *Pectinatella* und *Cristatella*. *Zoologica* (Stuttgart) **67**, 35–64.

Braem, F. (1951). Über *Victorella* und einige ihrer nächsten Verwandten sowie über die Bryozoenfauna des Ryck bei Greifswald. *Zoologica* (Stuttgart) **102**, 1–59.

Bronstein, G. (1938a). Mécanisme de la formation du polypide chez *Membranipora membranacea* (L.). *C.R. Acad. Sci.* **207**, 506–508.

Bronstein, G. (1938b). Note sur la croissance résiduelle des fragments de colonies chez *Membranipora membranacea* (L.). *C.R. Soc. Biol.* **128**, 65.

Bronstein, G. (1940). Sur les gradients physiologiques dans une colonie de Bryozoaires. *C. R. Acad. Sci.* **209**, 602–603.

Bullivant, J. S. (1968a). The rate of feeding of the bryozoan *Zoobotryon verticillatum*. *N. Z. J. Mar. Freshwater Res.* **2**, 111–134.

Bullivant, J. S. (1968b). The method of feeding of Lophophorates (Bryozoa, Phoronidea, Brachiopoda). *N. Z. J. Mar. Freshwater Res.* **2**, 135–146.

Cook, P. L. (1961). The development of *Electra crustulenta* (Pallas) (Polyzoa, Ectoprocta). *Essex Nat.* (1960) **30**, 258–266.

Cook, P. L. (1962). The early larval development of *Membranipora seurati* (Canu) and *Electra crustulenta* (Pallas), Polyzoa. *Cah. Biol. Mar.* **3**, 57–60.

Cook, P. L. (1963). Observations on live lunulitiform zoaria of Polyzoa. *Cah. Biol. Mar.* **4**, 407–413.

Cook, P. L. (1964). The development of *Electra monostachys* (Busk) and *Conopeum reticulum* (Linaeus), Polyzoa, Anasca. *Cah. Biol. Mar.* **5**, 391–397.

Cook, P. L. (1968). Observations on living Bryozoa. *Atti Soc. Ital. Sci. Nat. Mus. Civ. Stor. Nat. Milano* **108**, 155–160.

Cook, P. L., and Hayward, P. J. (1966). The development of *Conopeum seurati* (Canu), and some other species of membraniporine Polyzoa. *Cah. Biol. Mar.* **7**, 437–443.

Cummings, S. G. (1975). Zooid regression in *Schizoporella unicornis floridana* (Bryozoa Cheilostomata). *Chesapeake Sci.* **16**, 93–103.

Dudley, J. W. (1970). Differential utilization of phytoplankton food resources by marine ectoprocts. *Biol. Bull.* (*Woods Hole, Mass.*) **139** (2), 420.

Eiben, R. (1976). Einfluß von Benetzungsspannung und Ionen auf die Substratbesiedlung und das Einsetzen der Metamorphose bei Bryozoenlarven (*Bowerbankia gracilis*). *Mar. Biol.* **37**, 249–254.

Emschermann, P. (1961). Über die Brutkörper bei dem Kamptozoon *Barentsia gracilis*. *Zool. Jahrb., Abt. Allg. Zool. Physiol. Tiere* **69**, 333–338.

Friedl, H. (1925). Koloniebildung, Besiedlung und Wachstum bei marinen Bryozoen. *Arb. Zool. Inst. Univ. Insbruck* **2**, 137–168.

Geisler, R. (1964). "Wasserkunde für die Aquaristische Praxis," 143 pp. Alfred Kernen Verlag, Stuttgart, Germany.

Geraci, S., and Relini, G. (1970a). Osservazioni sistematico-ecologiche sui Briozoi del fouling portuale di Genova. *Bul. Mus. Inst. Biol. Univ. Genova* **38**, 103–139.

Geraci, S., and Relini, G. (1970b). Fouling di zone inquinante. Osservazioni nel Porto di Genova, I Briozoi. *Pubbl. Stn. Zool. Napoli* **38**, suppl., 21–32.

Geraci, S., and Relini, G. (1970c). Insediamento su panelli atossici immersi nella Rada vi Vado Ligure: I Briozoi. *Pubbl. Stn. Zool. Napoli* **38**, suppl., 19–33.

Gordon, D. P. (1968). Growth, regeneration and population biology of cheilostomatous polyzoans. Thesis Ph.D., University of Auckland.

Gordon. D. P. (1971a). Colony formation in the cheilostomatous bryozoan *Fenestrulina malusii* var. *thyreophora. N. Z. J. Mar. Freshwater Res.* **5**, 342–351.

Gordon, D. P. (1971b). Zooidal budding in the cheilostomatous bryozoan *Fenestrulina malusii* var. *thyreophora. N. Z. J. Mar. Freshwater Res.* **5**, 453–460.

Graupner, H. (1929). Haltung und Aufzucht von Süßwasserbryozoen. *In* "Handbuch der biologischen Arbeitsmethoden" (E. Abderhalden, ed.), Vol. 9, Part 2, No. II, pp. 1419–1423. Urban & Schwarzenberg, Berlin.

Graupner, H. (1929). Haltung und Aufzucht von Süßwasserbryozoen. *In* "Handbuch der biologischen Arbeitsmethoden" (E. Abderhalden, ed.), Vol. 9, Part 5, pp. 420–430. Urban & Schwarzenberg, Berlin.

Grave, B. H. (1930). The natural history of *Bugula flabellata* at Woods Hole, Mass., including the behavior and attachment of the larva. *J. Morphol.* **49**, 355–383.

Grave, B. H. (1937). *Bugula flabellata* and *B. turrita. In* "Culture Methods for Invertebrate Animals" (J. G. Needham et al., eds.), pp. 178–179. Dover, New York (reprint 1959).

Hauenschild, C. (1962). Die Zucht mariner Wirbelloser im Laboratorium (Methoden und Anwendung). *Kiel. Meeresforsch.* **18**, 28–37.

Hauenschild, C. (1968). Hälterung und Laboratoriumszuchten von Invertebraten. *In* "Methoden der meeresbiologischen Forschung" (C. Schlieper, ed.), pp. 192–209. VEB Gustav Fischer Verlag, Jena.

Hauenschild, C. (1970). Die Zucht von niederen marinen Wirbellosen und ihre Anwendung in der experimentellen Zoologie. *Helgol. Wiss. Meersunters.* **20**, 249–263.

Hückstedt, G. (1968). "Aquarientechnik," 3rd ed. Franckhsche Verlagshandlung, Stuttgart.

Hückstedt, G. (1971). "Aquarienchemie," 4th ed. Franckhsche Verlagshandlung, Stuttgart.

Jebram, D. (1968). A cultivation method for saltwater Bryozoa and an example for experimental biology. *Atti Soc. Ital. Sci. Nat. Mus. Civ. Stor. Nat. Milano* **108**, 119–128.

Jebram, D. (1969a). Bryozoen als Holzschädlinge im Brackwasser. *Kiel. Meeresforsch.* **25**, 224–231.

Jebram, D. (1969b). Zur Bryozoen-Fauna der deutschen Meeresgebiete und Brackwässer. I. Neue Funde. *Kiel. Meeresforsch.* **25**, 336–345.

Jebram, D. (1970a). Preliminary experiments with Bryozoa in a simple apparatus for producing continuous water currents. *Helgol. Wiss. Meeresunters.* **20**, 278–292.

Jebram, D. (1970b). Stolonen-Entwicklung und natürliches System bei den Bryozoa Ctenostomata. Inaugural-Dissertation, University of Kiel, 1970, 106 pp.

Jebram, D. (1973). Preliminary observations on the influences of food and other factors on the growth of Bryozoa with the description of a new apparatus for cultivation of sessile plankton feeders. *Kiel. Meeresforsch.* **29**, 50–57.

Jebram, D. (1975a). Dauerknospen ("Hibernacula") bei den Bryozoa Ctenostomata in mesohalinen und vollmarinen Gewässern. *Mar. Biol.* **31**, 129–137.

Jebram, D. (1975b). Effects of different foods on *Conopeum seurati* (Canu) (Bryozoa Cheilostomata) and *Bowerbankia gracilis* Leidy (Bryozoa Ctenostomata). *Docum. Lab. Geol. Fac. Sci. Lyon,* H. S.3 (1), 97–108.

Jebram, D. (in press). Bryozoa Ctenostomata in the lakes near Naples. *Pubbl. Stn. Zool. Napoli* **40.**

Kaissling, K.-E. (1963). Die phototropische Reaktion der Zoide von *Bugula avicularia* L. *Z. Vgl. Physiol.* **46,** 541–594.

Kammerer, P. (1907). Über Schlammkulturen. *Arch. Hydrobiol. Planktonkd.* **2/4,** 500–526.

Lee, J. J., Tietjen, J. H., Stone, R. J., Muller, W. A., Rullman, J., and McEnery, M. (1970). The cultivation and physiological ecology of members of salt marsh epiphytic communities. *Helgol. Wiss. Meeresunters.* **20,** 136–156.

Lutaud, G. (1953). Premiere stades de la croissance zoariale chez le Bryozoaire Chilostome *Scrupocellaria reptans* Thompson. *Arch. Zool. Exp. Gén.* **90,** 42–58.

Lutaud, G. (1961). Contribution à l'étude du bourgeonnement et de la croissance des colonies chez *Membranipora membranacea* (Linné), Bryozoaire Chilostome. Thesis Ph.D., Fac. Sci., University of Paris.

Lynch, W. F. (1947). The behaviour and metamorphosis of the larvae of *Bugula neritina* (Linnaeus): Experimental modification of the length of the freeswimming period and the responses of larvae to light and gravity. *Biol. Bull. (Woods Hole, Mass.)* **92,** 115–150.

Lynch, W. F. (1949). Modification of responses of two species of *Bugula* larvae from Woods Hole to light and gravity: Ecological aspects of the behavior of *Bugula* larvae. *Biol. Bull. (Woods Hole, Mass.)* **97,** 302–310.

McLaughlin, J. J. A., and Zahl, P. (1959). Axenic zooxanthellae from various invertebrate hosts. *Ann. Acad. Sci. (N. Y.)* **77,** 55–72.

Marcus, E. (1926a). Beobachtungen und Versuche an lebenden Meeresbryozoen. *Zool. Jahrb., Abt. Syst. Geogr. Biol. Tiere* **52,** 1–102.

Marcus, E. (1926b). Beobachtungen und Versuche an lebenden Süßwasserbryozoen. *Zool. Jahrb., Abt. Syst. Geogr. Biol. Tiere* **52,** 279–350.

Marcus, E. (1934). Über *Lophopus crystallinus* (Pall.). *Zool. Jahrb., Abt. Anat. Ont. Tiere* **58,** 501–606.

Mawatari, S. (1951). The natural history of a common fouling bryozoan *Bugula neritina* (Linnaeus). *Misc. Rep. Res. Inst. Nat. Resour.* **19–21,** 47–54.

Menon, N. R. (1972). Heat tolerance, growth, and regeneration in three North Sea bryozoans exposed to different constant temperatures. *Mar. Biol.* **15,** 1–11.

Menon, N. R. (1973). Vertical and horizontal distribution of fouling bryozoans in Cochin backwaters, southwest coast of India. *In* "Living and Fossil Bryozoa" (G. P. Larwood, ed.), pp. 153–164. Academic Press, New York.

Menon, N. R., and Nair, N. B. (1967). Observations on the structure and ecology of *Victorella pavida* Kent (Bryozoa) from the South West coast of India. *Int. Rev. Ges. Hydrobiol.* **52,** 237–256.

Needham, J. G., Galtsoff, P. S., Lutz, F. E., and Welch, P. S. (1937). "Culture Methods for Invertebrate Animals." Dover, New York (reprint 1959).

Oda, S. (1954). On the monsters of polypides in freshwater bryozoa. *Collect. Breed.* **16,** 15–18.

Otto, F. (1921). Studien über das Regulationsvermögen einiger Süßwasserbryozoen. *Arch. Entwicklungsmech. Org.* **47,** 399–442.

Provasoli, L., McLaughlin, I. I. A., and Droop, M. R. (1957). The development of artificial media for marine algae. *Arch. Mikrobiol.* **25,** 392–428.

Provasoli, L., Conklin, D. E., and D'Agostino, A. S. (1970). Factors inducing fertility in aseptic Crustacea. *Helgol. Wiss. Meeresunters.* **20,** 443–454.

Riedl, R. (1964). Die Erscheinungen der Wasserbewegung und ihre Wirkung auf Sedentarier im mediterranen Felslitoral. 4. Meeresbiolog. Symposium. *Helgol. Wiss. Meeresunters.* **10,** 155–186.

Riedl, R. (1966). "Biologie der Meereshöhlen." Verlag Paul Parey, Hamburg und Berlin.

Ries, E. (1936). Fütterungsversuche bei *Zoobotryon* (Bryozoa). *Z. Vergl. Physiol.* **23,** 64–99.

Rogick, M. (1937). Culturing freshwater bryozoa. *In* "Culture Methods for Invertebrate Animals" (J. G. Needham *et al.,* eds.), pp. 179–181. Dover, New York (reprint 1959).

Rogick, M. (1938). Studies on Freshwater Bryozoa. VII. On the viability of dried statoblasts of *Lophopodella carteri* var. *typica. Trans. Am. Microsc. Soc.* **57,** 178–199.

Rüsche, E. (1938). Hydrobiologische Untersuchungen an niederrheinischen Gewässern. X. Nahrungsaufnahme und Nahrungsauswertung bei *Plumatella fungosa* (Pallas). *Arch. Hydrobiol. Planktonkd.* **33,** 271–293.

Ryland, J. S. (1958). Embryo colour as a diagnostic character in Polyzoa. *Ann. Mag. Nat. Hist.* (ser. 13) **1,** 552–556.

Ryland, J. S. (1959). Experiments on the selection of algal substrates by polyzoan larvae. *J. Exp. Biol.* **36,** 613–631.

Ryland, J. S. (1960). Experiments on the influence of light on the behaviour of polyzoan larvae. *J. Exp. Biol.* **37,** 783–800.

Ryland, J. S. (1962). The effect of temperature on the photic responses of polyzoan larvae. *Sarsia* **6,** 41–48.

Ryland, J. S. (1975). Parameters of the lophophore in relation to population structure in a bryozoan community. *In* "Proceedings of the 9th European Marine Biology Symposium, 1975" (H. Barnes, ed.), pp. 363–393. Aberdeen University Press, Aberdeen, Scotland.

Schlieper, C. (1968). "Methoden der meeresbiologischen Forschung." VEB Gustav Fischer Verlag, Jena.

Schneider, D. (1959). Der Aufbau der *Bugula*-Tierstöcke und seine Beeinflussung durch Aussenfaktoren. *Biol. Zentralblatt* **78,** 250–283.

Silén, L. (1944). On the divisions and movements of the alimentary canal of the bryozoa. *Ark. Zool. Stockholm* **35A**/12, 1–40.

Sloane, J. F., Ebeling, F. J., Kitching, J. A., and Lilly, S. J. (1957). The ecology of the Lough Ine Rapids with special reference to water currents. V. The sedentary fauna of the Laminarian algae in the Lough Ine area. *J. Anim. Ecol.* **26,** 197–211.

Strathmann, R. (1973). Function of lateral cilia in suspension feeding of lophophorates (Brachiopoda, Phoronida, Ectoprocta). *Mar. Biol.* **23,** 129–136.

Svoboda, A. (1970). Simulation of oscillating water movement in the laboratory for cultivation of shallow water sedentary organisms. *Helgol. Wiss. Meeresunters.* **20,** 676–684.

Wayss, K. (1968). Quantitative Untersuchungen über Wachstum und Regeneration bei *Plumatella repens* (L.). *Zool. Jahrb., Abt. Anat. Ont. Tiere* **85,** 1–50.

Wiebach, F. (1952). Über den Ausstoss von Flottoblasten bei *Plumatella fruticosa* (Allman). *Zool. Anz.* **149,** 181–185.

Wiebach, F. (1953). Über den Ausstoss von Flottoblasten bei den Plumatellen. *Zool. Anz.* **151,** 266–272.

Winston, J. E. (1976). Experimental culture of the estuarine ectoproct *Conopeum tennuissimum* from Chesapeake Bay. *Biol. Bull.* (*Woods Hole, Mass.*) **150,** (2), 318–335.

Zirpolo, G. (1933). *Zoobothryon verticillatum* (Delle Chiaje). *Mem. Pont. Acad. Sci. Roma* (2), **17,** 109–442.

9

Interzooecial Communications and the Funicular System

GENEVIÈVE BOBIN

I. INTRODUCTION

While the vertebrates profit, in common, from a certain plan of organization which furnishes both a framework for investigation and possibilities for comparison, each of the various groups of invertebrates offers a diversity of unique features both in structure and function. This observation is also pertinent when one considers animal colonies, the vital functions of which are always complex. In the Bryozoa (or Ectoprocta) these functions are influenced above all by the cystid–polypide duality. Among their essential roles, the cystids are responsible for budding and growth while the polypides provide nutrients to the whole colony.

One would imagine that in the Phylactolaemata, whose colonies are deprived of rigorous compartmentalization and in which the long-lived polypides are unable to regenerate, the colonial life is effected by a direct, reciprocal, and relatively equal cooperation of the zooids; in fact, the *milieu intérieur* is continuous throughout the zoarium.

In contrast, the behavior of the Gymnolaemata is unique. The extreme compartmentalization of the zooids reduces the area of contact between adjacent zooids to minute pores and confers upon each unit a certain autonomy. Moreover, the polypides undergo periodic degeneration and are capable of regeneration and a considerable cystidal polymorphosim often complicates the situation and involves specific supernumerary processes. Otherwise we shall observe a colony of gymnolaemates, whatever is its morphology, exhibiting a sort of "functional polymorphism" of its zooids, the physiological states of which are never synchronous. One would think that the related individuals in such a colony would participate in the collective life either in a nearly equivalent but intermittent manner or by a specialization of roles.

All the authors have described for the bryozoans a funicular system, the importance of which has escaped no one since so many abilities have been attributed to it.

Phylactolaemates are bryozoans that differ notably from the gymnolaemates in their biotope (freshwater), their zoarial structure, their histology, and the expression of their biological functions, especially that of reproduction. Thus, their funicular system consists of a number of individual organs in the form of bands or cables (Brien, 1960), each joined to a single polypide. In the gymnolaemates (from marine or occasionally brackish water), and in the eurystomes in particular, the funicular system consists of a cystidal organ which has zoarial value. This organ consists of more or less branched ribbons, running inside the colony, interrelating all the individuals.

The representatives of the eurystomes (cheiloctenostomes) will be thoroughly studied because of their extreme compartmentalization and the existence of communication plates provided with pores. The rosettes which plug these pores constitute periodic relays along the funicular branches. "Primary plates" are established at the base of each new zooid at the moment of its budding and are incorporated into its transverse partitions or septa. This fundamental category exists in all the eurystomes, but is the only form present in those bryozoans with noncontiguous zooids (for example, the stoloniferous and vesicularine ctenostomes). "Secondary" communication plates are found in addition to the primary ones in forms with adjoining zooids (for example, the alcyonidean ctenostomes and the majority of cheilostomes) and are situated within the coupled lateral walls of neighboring zooids. Cyclostomate gymnolaemates will be briefly examined.

A correlation of results is impossible considering the present state of our knowledge. The problem of the funiculi and the communications is too heterogeneous, varied in its complexity, and insufficiently understood. Furthermore, one should not attempt to make premature generalizations. This is especially true for the bryozoans in which the great number of vital processes are linked to a multiplicity of shapes.

II. CASE OF THE PHYLACTOLAEMATES

Communication among zooids is extensive throughout the colony. The cecum of each polypide is linked to the ventral wall of the cystid by an individual funiculus or "cable" that is hollow and curved, the structure and role of which differ from those of the gymnolaemates. A simple peritoneal epithelium is arranged around the axial cavity. This peritoneum is provided with a strong basal membrane and longitudinal fibers which are situated beneath its convex face. The funiculus serves (with rare exceptions) a double function. Spermatogenesis is an aspect of sexual reproduction (the ovary rising from the ventral peritoneum of the cystid). Spermatogenesis occurs near the cecum in the peduncular region of the funiculus which elongates during the maturation of the gametes. Statoblastogenesis is a form of asexual reproduction, independent from the budding of the growing colony. The colonies of phylactolaemates, flourishing from the beginning of spring, wither at the onset of winter; the preservation of the species, however, is assured by statoblasts which are latent dormant bodies, protected by an encasement and filled with nutrient reserves. Their arrested development starts anew with the onset of warm weather when each statoblast can give rise to a new colony. In fact, these two reproductive functions are interrelated. Indeed (see Brien, 1953, 1960; Brien and Mordant, 1955–1956) the primordium of the young funiculus, which grows from its distal end that is fixed to the wall, contains a type of precocious statoblastic reserve constituted of undifferentiated cystogenic epidermal cells which emigrate into the axial cavity where they are joined by deutoplasmic cells from the peritoneal layer that covers the concave face of the new funiculus. This composite store of basophilic embryonic cells is inhibited in its evolution during the spermatogenetic phase. Afterward, the primordial statoblastic material is reactivated and extends along the whole length of the funiculus, progressively strobilating off globular islets that are graduated in size and age. The largest of these primordia are located near the polypide, while their blastogenetic focus remains localized near the attachment of the funiculus to the cystid. Each primordium contains a "cystogenic vesicle," a form of embryonic disc, which progressively tends to surround as a two-layered epithelium the deutoplasmic mass that is now active in vitellogenesis. The outer layer secretes the shell and the floats, if there are any; the inner one becomes the wall of the future cystid.

These funicular functions require a supply of substances obtained either from the ambient fluid or synthesized *in situ.* We have little information on this matter. The formation of the statoblasts has been studied histochemically only in *Plumatella repens* (Linné) (Gasser, 1962). For completed statoblasts, the various components, including enzymes, were detected by electron microscopy by Lacourt and Willighagen (1966) and Lacourt (1968).

III. CASE OF EURYSTOME GYMNOLAEMATES

The colonies of eurystomes, of varied appearance, have walls which are calcified in varying degrees (cheilostomes) or are uncalcified (ctenostomes). The composition of the walls as well as the differences in zoarial architecture and the proportion of polymorphism lead to the problem of the choice of species. One cannot consider a colony in its totality. It is a morphological whole but a functional mosaic. This is an inextricable sum of life and death, or of rejuvenation and senescence, without permanent size, weight, shape, or structure. Finally, the zoaria are not all visible to the naked eye and cells are of minimal size.

The multiplicity of colonial forms depends on the shape of the zooids, as well as on their patterns of growth and association (that is to say, the distribution of privileged sites of cystidal budding). There exist homogenous colonies all with complete individuals (autozooecia with polypides) and other colonies complicated with polymorphic variations (autozooecia with polypides + various heterozooecia that are entirely cystidal in nature). These circumstances influence the number and topography of the perforated walls and the funicular branches. An arrangement so original obliges us to revise our concepts of physiology and to present in a new way the problem of the exchanges. The known facts will be examined by help of specific examples.

A. Primary Uniporous Communication Plates of *Bowerbankia imbricata* (Ctenostomata: Vesicularina)

The main example chosen, *Bowerbankia imbricata* (Adams), is a ctenostome (that is to say, noncalcified) with plain, bushlike zoaria. Autozooecia are separate and bud directly from the stolon which is a branching file of tubular heterozooecia called coenozooecia. The path of the funiculus is rather simple. Its principal cord is segmented (as is the stolon) by septa and sends a branch toward the basal partition for each autozooecium; inside the autozooecium I have called "funicular tract" that portion of the funiculus which runs from the cecum to the local rosette.

1. Colonial Morphophysiology (Fig. 1)

The vital processes of a colony are so entangled that it is necessary to summarize them briefly. The soft, tuftlike zoarium of *B. imbricata* consists of a basal trunk which is attached to the substrate; though old, it has a latent life. From the trunk originate divided branches extending in all directions; toward the free ends of the branches are located the young regions. The coenozoecia in files form stolonal segments, which are progressively older when one looks back along their axis. They grow, at their free, closed extremity which bears autozooecial buds, in a rather regular order, the more terminal buds being the youngest ones. Each segment, furnished with a group of autozooecia, never has a composition identical to that of its neighbors, and the autozooecia themselves offer a great local diversity of functional states. From such conditions, added to the presence of a single primary communication plate for unity, one should be better able to decipher the topography, age, formation, and the role of the funicular system and of the rosettes which are attached to the sectors that can be rather strictly defined. If the permanent cystid ensures the well being of the colony and the growth by budding, the intermittent polypide irregularly provides for the capture of food from the environment and at least for the prospering of its own cystid.

a. Physiology of the Autozooecium. Nourishment of the polypide, through its digestive tube, is usual. The polypide, active or regressed, also furnishes materials that are used or stored within its cystid. There is no excretory organ, but the cyclical degeneration of the polypide constitutes its own excretory function. The engorged digestive tissue becomes isolated as a brown body. This type of renal purification is completed by the phagocytosis of other debris (from tentacles, muscles) to the advantage of the cystidal unit, which thus benefits from the remnants of the transitory polypide.

Sexual reproduction concerning the autozooecia will not be covered here. An autozooecium, taken individually, can live by itself, but it is certain that the interrelationships between individuals, ensured by the funiculus, pores, and rosettes have a significance which is not only morphological.

b. Physiology of the Cystid. Since the polypides feed the whole group, we must recognize that at any given time and in continuously varying localities, a large number of zooecia are closed and isolated from the external environment. This phenomenon is transitory for autozooecia in the course of their reshaping but definitive for aging individuals and for all coenozooecia. Such entities, degraded or normally closed, are entirely dependent on cystids. Meanwhile, they are living and are the seat of physiological processes. One must ask oneself how these processes are achieved. The interrelations suggest the possibility of mutual reversible help, thanks to individuals momentarily more favored than others both in time and in space.

2. The Funiculus–Rosettes Complex (Figs. 1 and 2)

For more than a century, the funicular cords and the communication plates have been recognized, but often only in a superficial way. Two schools of thought opposed each other. Müller (1860) considered the complex as a colonial nervous system; his view was endorsed temporarily by a few scholars. Nitsche (1871a,b), in turn, considered the funiculus as an element for sustaining the colony while others affirmed that it gave rise to all the cells of the cystidal cavity. Joliet (1877) proved Müller in error; but, although he described the

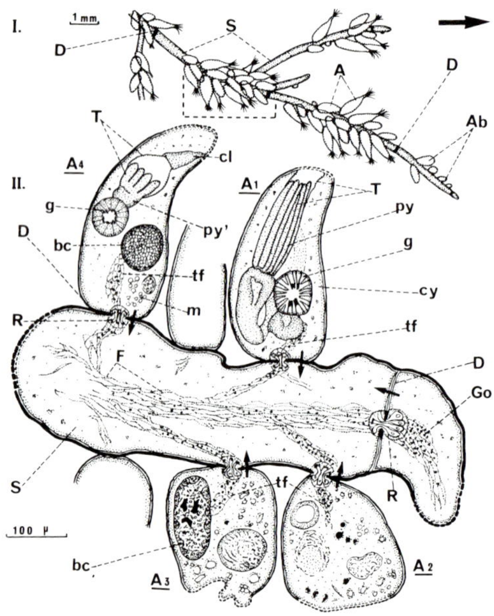

Fig. 1. Boughs of *Bowerbankia imbricata* (after Bobin, 1964). (I) Free end of a living bough. The segment underlined with a dashed line corresponds, in age and position, to (II). (II) Schematized longitudinal section of a stolonal part bearing physiologically different autozooecia (Golgi technique). (Higher arrow, direction of the free extremity of the stolons. Arrows adjacent to the diaphragms; direction of the morphological polarity of the rosettes.) Abbreviations: A, autozooecia (1, normal and functional; 2 and 3, degenerating; 4, at the end of degeneration with neoformation of a polypide); Ab, autozooecial buds; bc, brown body; cl, developing collar; cy, cystid; D, diaphragm; F, stolonal funiculus; g, gizzard; Go, Golgi bodies; m, mesenchyme; py: functional adult polypide; py': polypidean bud; R: rosette; S: stolon; T, tentacles and tentacular sheath; tf, autozooecial funicular tract.

elongated cells of a funicular tissue irregularly charged with reserves, he did not suspect the true structure of the rosette anymore than did Calvet (1900).

Reichert (1869) was the first to point out the communication plates of the ctenostome *Zoobothryon* from which the rosette was later drawn by Zirpolo (1933). However, the only valid histological studies on these elements are due to Davenport (1891) on *Paludicella* and to Brien and Huysmans (1937) on *Bowerbankia pustulosa.* They recognized then, with some precision, that rosette cells cross the pores. Starting from these findings, my research has been carried out on *Bowerbankia imbricata* and, to supplement my studies, with analogous forms (Bobin, 1958a,b, 1962, 1964, 1965, 1971). I should add that these studies concern healthy regions of the colony, without any sexual dominance, to avoid the concomitance of too many processes.

a. Morphology and Staining of the Adult Complex (Figs. 1, II and 2, IV). The stolonal funicular band, branched and flattened, is fractionated along its path and indeterminate in its limits. Connected only to the communication plates, it floats in the cystidal cavity, and so is vulnerable when making histological sections. In the living animal, it appears to be composed of rather large elongated and juxtaposed cells with faint borders and large ovoid nuclei. At the points of bifurcation the appearance of the funicular tissue is reticulated. In the autozooecium, the funicular tract is simple, well trimmed, and composed of smaller cells. *In toto,* the rosettes are visible as so many spherical swellings divided along their equator by a perforated diaphragm.

It is essential to analyze their histology. The two facing hemispheres of a rosette have an only apparent symmetry and together measure between 30 and 40 μm in diameter. The ectocystal diaphragm is sheathed with a continuous fine epidermis that is reflected at the level of the pore (4–5 μm in size).

I have named "special cells" those that are inserted as a "cluster" in the pore. The nucleated lobes of these cells, indented at their summit and provided with a short fringe of microvilli, are lodged on one side of the partition. The cells become thinner as they approach the pore, and, after crossing it, each expands on the other side into a bulging base adorned with a longer fringe of microvilli. Each hemisphere is capped by a cup-shaped layer of limiting cells where the union with the funicular trunks takes place.

The stolonal rosettes are a little larger than those of the autozooecia and their special cells are more numerous and more elongated (about 30 μm rather than 20). The rosettes have a morphological polarity inscribed into the structure of their special cells. Moreover, they have, as does nearly the entire funiculus, a permanent character, which remains constant at the time of the cyclical modifications of the autozooecia. The stainability of the constituents of the complex is rather unusual. The fundamental cytoplasm (reticulated with inclusions of the limiting layers) the principal funiculus, and the funicular tracts, preferentially takes up a very pale coloration of light green or aniline blue or may even tend vaguely toward rose, depending on the dose of the dyes of the trichrome stain employed (Prenant or Masson), without marked preference for any one. This cytoplasm has a sprinkling of granules, which are refringent in the living animal. One finds, among the ordinary cells, some narrower elements of a fibrillar aspect. The funicular nuclei are recognizable by their oblong form, their large size, their paucity of chromatin, and their small nucleoli. Those of the limiting cells are better provided with organized components. The staining affinities of the special cells are very specialized. Their nuclei are small, rich in chromatin, with minute or indiscernible nucleoli. In the adult state, their dense cytoplasm shows, after Prenant's trichrome, a gray-green coloration at the nuclear pole and, due to siderophilia, a blackness at the level of the swollen bases. The bipolar fringes strongly retain the light green and respond positively to tests for mucopolysaccharides. That the fringes are in reality microvilli is confirmed by electron micrographs.

b. Genesis of the Complex: Morphological Polarity (Fig. 2) i. Embryonic Ectomesenchyme. The "funicular mesoderm" (Joliet, 1877) or, more broadly, the mesenchymatous components of the cystid (Calvet, 1900) are derived from elements issued, in the course of budding, from a proliferation of the terminal epidermis. According to Brien and Huysmans (1937) the cystids are "hollow ectomesodermal tubes," and, in the very basophilic blastogenic zones, detached cells of the "embryonic ectoderm" migrate and participate in the formation of the "discontinuous endothelium" of the funiculus and, without doubt, of the "blood" and "renal" cells of the cavity.

Since 1958a, I have been able to confirm the epidermal origin of the primary mesenchyme cells which, once freed into the hollow of the bud, actively divide into an undifferentiated material with multiple potentials, for which I have proposed the term "ectomesenchyme." Whether it migrates beneath the epidermis or remains more central, it evolves, in part, into a reticulum of ameboid vacuolated cells (the "plexus" of Calvet) and various free cells, some of which are characteristic of the species (Bobin and Prenant, 1972). I have confirmed, in the ctenostomes at least, that the cells of the cavity do not take their origin from the funicular tissue as has been sometimes proposed. On the other hand, it is true (Bobin, 1962) that the funiculus is built up at the tip of the stolon, as long as the segment is elongating, by the local and successive addition of small elements that become fusiform while detaching from the epidermis. Although hardly visible within the confusion of the blastema, I have located them with particular clarity in *Amathia lendigera.* I have previously remarked (Bobin, 1958a) that the peritoneal envelope of the primordium of the polypide first appears as a progressive deposit of ectomesenchyme cells, which subsequently undergo mitotic divisions.

Finally, the primitive ectomesenchyme is a malleable reserve material whose diversified utilization is immediate or delayed according to the needs. Even in the fully evolved sectors of the colony, there remain here and there embryonic elements of this type. There are more present, we will see, near the septa undergoing histogenesis.

ii. Partitioning; Establishing of the Complex (Fig. 2). Histogenic partitioning which I have elucidated in *Bowerbankia* (Bobin, 1958a, 1962), takes place in a way comparable to that for the stolons and for the zooecia except for minor morphological differences and variations in time. In effect, in a multisegmented stolon, the terminal coenozooecium during its growth is occupied by the young funiculus, to its apex. At a given moment, evidence of a partition appears behind the important blastogenic zone, initiating the isolation of a new apical, coenozooecial cystid of good size which encloses a mass of embryonic mesenchyme. On the other hand, the deflected funiculus is always positioned parallel to the wall of the future points of budding in the subterminal stolonal segments, without penetrating into the minute cavity of the buds, regardless of whether it is a case of autozooecial buds or of a primordium of the first coenozooecium of a bifurcation of the stolon. The formation of the cystids and the first signs of ectomesenchyme and partitioning are nearly synchronous events. Let us review the sequence of events.

1. A *prepartition* or *mucous mesenchymatous veil* (V) separates quite early the stolonal cavity from the preexisting lumen of the future terminal coenozooecium or from the virtual spaces of the incipient parietal buds which now consist only of an elevation of their epidermis into a dome. This prepartition originates from the migration and the association of the embryonic ectomesenchyme which is already formed or is in the process of separation; these amoeboid elements unite and secrete particles which meet in a mucous network. In the coenozooecium, the veil encloses the eccentric funiculus while in the parietal buds it insinuates itself between the funicular surface and the epidermis. This provisory viscous barrier seems to prevent the intrusion of older elements of the stolon into the young cavity so that the material of a bud, except the modified epidermis, is undifferentiated or entirely new.

2. The *real diaphragm* (D) is elaborated at the base of a bud over the mucous veil in the form of a circular ledge formed by a double epidermal layer which proliferates by mitosis and closes like an iris diaphragm. The two layers, at first thickened, secrete between themselves a layer of ectocyst bound to the general one of the colony. Under the pressure of the constricting diaphragm, the fragile funiculus of the terminal coenozooecium breaks off and there, as in the parietal buds, the veil which is stretched between the walls then obliterates the opening which to this time is quite large.

3. A compact *mucous plug* (C) completes this obliteration when the diaphragm achieves about half of its extension; it is produced from one to three mesenchyme cells that were deposited a little earlier and which disappear when they have produced the plug, the substance of which fuses with that of the veil.

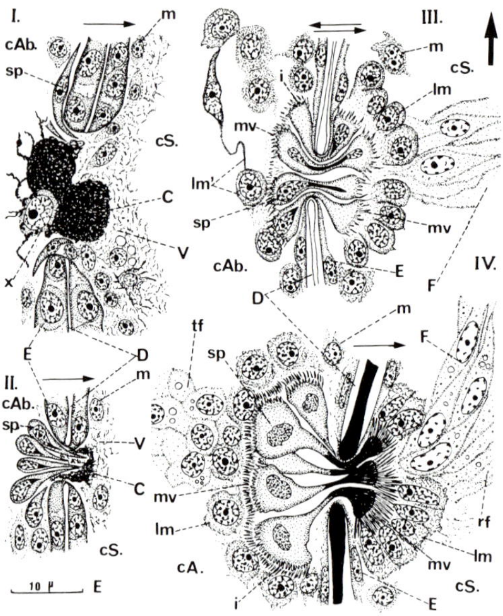

Fig. 2. *Bowerbankia imbricata.* Axial sections of autozooecial rosettes (after Bobin, 1958; trichrome of Prenant). Genesis of the complex and diaphragms. Setting up of the morphological polarity. (I and II) Primitive polarity (plug and mucous veil; setting up of the special cells). (III) Fluctuating polarity (bipolar differentiation of the special cells; moving of the nuclei and setting up of the other elements of the complex). (IV) Fixed polarity (old adult rosette; note the rumpled nuclei of the special cells). (Higher arrow, direction of the free extremity of the stolons. Arrows adjacent to the diaphragms, direction of the polarity.) Abbreviations: C, mucous provisory plug; cA, cavity of the adult autozooecium; cAb, cavity of the autozooecial bud; cS, stolonal cavity; D, ectocyst diaphragm and pore; E, epidermis; F, stolonal funiculus; i, apical indentations; lm, cells of the limiting layers; lm': limiting cells while depositing; m, mesenchymatous cells; mv, microvilli; rf, thin refringent granules, sp, special cells; tf, autozooecial funicular tract; V, transitory mucous veil; x, original cell of the mucous plug.

4. The *special cells* (sp) of the rosette begin their differentiation during the production of the plug; a few small embryonic mesenchyme elements bordering the partition become elongated and, sending out a point, insinuate themselves one after the other against the plug, into the orifice, which narrowing more and more, constricts the special cells at their midpoints and becomes the definitive pore. Thus the primitive rosette, pointed through the pore, is reduced to a single cluster of special cells, still deprived of ornamentation and of links with their surrounding. The mucous plug is resorbed and the septal epidermis becomes more flattened. The special cells then take on their distinctive morphological

characteristics by the indentation of their apices, the elongation and the expansion of their bases, and the growth of the microvillar fringes at both poles.

5. *Limiting layers* (lm) arise concurrently with these last differentiations. On both sides of the partition, but without absolute synchronization, the undifferentiated mesenchyme cells which have accumulated become elongated and deposit themselves one by one against the corresponding pole of the special cells where they become organized into a layer.

6. The *junctions of the funiculi* (F and tf) with the limiting layers during the formation of the latter is also prepared for by migratory phenomena. The funiculi, whether undergoing histogenesis (in parietal buds) or already organized (in stolons) preexist at this junction; their own elements, young or modified amoeboid cells, appear to fulfill this linking role. In the stolonal cavity, additional help is provided by the already deposited local mesenchyme; thus the previously broken pieces of the main funiculus get tied to the rosette or secondary branches are formed in the direction of the new autozooecial septa. When these structures are completed, the epidermis of the septum becomes thin and the ectocyst of the diaphragm reaches its normal height and forms a thickening ring around the pore.

7. In conclusion it has been demonstrated that the funiculi–rosettes complex, formed of components that are originally autonomous but secondarily united, is a cystidal organ the origin of which is exclusively mesenchymatous.

iii. Morphological Polarity (Fig. 2). The morphological polarity of the rosettes inscribed into the structure of the special cells and oriented from their nucleated portions toward the swollen bases is evident, but is not acquired right away. I have previously defined three stages (1958a,b, 1962).

1. The *primitive polarity* (Fig. 2, I and II) expresses the sense of implantation of the new special cells which are inserted into the pore without yet traversing it; their nucleated bodies are grouped on that side of the partition corresponding to their former place: always the cavity of the autozooecial buds and for the stolons it appears to be the cavity of the older of the two terminal coenozooecia. During this setting, the nucleus of each special cell produces a pointed extension into the swollen base and the latter, growing on, reaches the other face of the septum.

2. The *fluctuating polarity* (Fig. 2, III) is significant during the course of construction of the rosettes according to the details already sketched and when funicular relationships are absent or uncertain. This indeterminate phase of polarity, with reversible phenomena, concerns above all the nuclei of the special cells which, by plodding on, going and returning, stretching out or bipartitioning, situate themselves on one side or the other of the young partition (as confirmed by the reaction of Feulgen–Rossenbeck).

3. The *fixed polarity* (Fig. 2,IV) is correlated with the completion of the rosette and its attachments. Because of the final localization of the nuclei, back

to the apex of the special cell, the specifically bipolar ornamentations are confirmed. The structure then is durable and all the adult rosettes are fixed in their morphology and present signs of apparent aging (Bobin, 1964, 1965). The fixed polarity of the rosettes of functional autozooecia is, except for rare cases of inverse polarity, oriented from the zooecia to the stolon and coincides with the natural direction of the primitive polarity. In the stolons, in contrast, the fixed polarity of the coenozooecial rosettes takes a direction much more capricious from one region to another.

c. Cytoplasmic Components and Inclusions of the Complex; Physiological Polarity. (Fig. 1, II and Fig. 3). Bobin proposed (1962) that the complex, a form of intracolonial network, serves for the accumulation and the distribution of substances, particularly nutritive ones. Therefore a search was made for the sources of these products and for evidence of the probable orientation of their transport. The combined results of vital staining, of histology and cytology, and of histochemical tests have provided proof of the incessant activity of the funiculi and of the rosettes and of a constant functional movement at their level between autozooecia and stolon or between successive elements of the stolon. The original tissues of the differentiated complex offer no trait that is strictly glandular and their ribonucleic acids are rare and independent from the inclusions. In the special cells, the specific detection of nucleic acids has permitted me (Bobin, 1958a, 1965) to confirm the migrations of the nuclei (fluctuating and fixed polarities) and to distinguish the three causes of their changing basophilia: constituents of the nuclei, cytoplasmic ribonucleic acids associated with autosynthesis during young stages, and, finally, secondary stainability of the swollen bases due to a superficial sclerous fibrillation that does not create any involution. The distribution, size, form, and density of mitochondria, Golgi bodies, and inclusions vary enormously in time and space. Thus, by turns or at the same time, the seven levels of a rosette and its annexes (two funicular pieces, two limiting layers, bodies of the special cells, and the systems of bipolar microvilli) are cyclically rich or poor. I was able to link the observed facts either to serial stages of budding or to individual physiological phases of the zooecia of a given sector. The direction of useful physiological polarity of the special cells is superimposed upon their morphological polarity (nucleated apical portion → base) for the following principal reasons. The common orientation of nearly all of the autozooecial rosettes (autozooecia → stolon) is a decisive argument reinforced by the fact that the polypides are necessarily responsible for colonial nutrition; they metabolize the digested products in the course of their activity, or become partly a lysable nutrient during their own degradation.

The accumulation, in waves, of variable quantities of funicular inclusions on both sides of a septum evokes a succession of charges and discharges, with paroxysm when the autozooecia are undergoing modification. The parallel cyclical evolution of the mitochondria and of the Golgi bodies corroborates this dynamic activity.

The very small special cells do not permit storage but only moderate transit, perhaps by filtration, but most probably by the absorption of dissociated particles at one pole (the nucleated lobe) and their release at the other pole (swollen bases) by way of the microvillar system. I have furnished conclusive pictures from such a passage (Bobin, 1964) that were largely verified later (Bobin, 1971) in connection with the circulation of the lipids (Fig. 3, I). Lipids are presented in large quantity throughout the funicular complex; their concentration is periodically increased by the lipidic degeneration of the digestive epithelium of the polypide, preceding the final brown body.

d. Scheme of the Colonial Nutrition (Fig. 3, II). The malleability of cellular material, both of organized tissue and of free cells immersed in the fluid of the cavity, allows one to surmise possibilities of direct exchange, at least within the limits of each zooecium. The funicular tissue becomes loaded with products that it transports. Many of them, bound with very active Golgi bodies, shape into diverse inclusions which, subsequently, are dissociated or reconstituted many times according to the level of their progression. The transported inclusions in healthy regions of a colony are usually of nutritive value, such as lipids which are so abundant. In summary, the funicular tract of the autozooecium collects and stores material, particularly from the normal or degenerating stomach with which it is joined. Stored then in the adjacent limiting layer, the inclusions are broken down and make contact with the short, apical microvilli of the special cells. The fine particles are absorbed and traverse the cytoplasm of the special cells without delay and are then released on the other side of the partition at the level of the long microvilli bordering the swollen bases. The minute granules reassemble themselves in the other limiting layer and fuse, forming larger inclusions which diffuse in turn into the adjacent funicular branch to reach finally the principal funicular cord of the stolonal coenozooecium involved.

The useful current of transport toward the stolon is originally determined by the common orientation, in that direction, of the fixed polarity of the autozooecial rosettes. The principal funiculus is provided for because the coenozooecia support a mixture of polypidal units, active or declining. In the rare cases of inverse polarity, the autozooecium, isolated from the colonial economy, lives only for its own benefit and is able to profit from fortuitous gifts from the stolon. The segmental septal rosettes of the stolons have different orientations for no apparent reason. The fluctuating polarity has been fixed at random, but its final orientation has been influenced to some extent by the direction of the richest nutrient current and the temporary regional needs at the time of partitioning. Indeed, the usefulness of the polarity, fixed under transitory influences, risks being altered or even abolished by the intervention of new

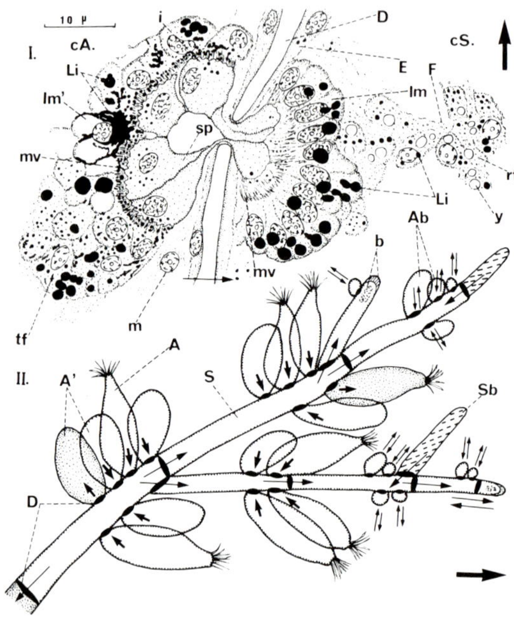

Fig. 3. *Bowerbankia imbricata.* Crossing of the substances through the complex. Physiological polarity. (I) Axial section of an adult rosette and its annexes; the polypide of the corresponding autozooecium is degenerating. In black, lipids (after Bobin, 1971). (II) Scheme of a bough (original) showing the ways of currents determined by the polarity of the local rosettes (the black thickenings show the diaphragms with rosettes and pores). An autonomous physiological sector is shown in white; excluded from currents which are useful to its life are two autozooecia and one stolonal coenozooecium (dotted), the rosettes of which present an inverse polarity. The ends of the stolons (covered with dashes), unfavored by the orientation of the rosettes, have lost their growth blastema. (Arrows to the right, direction of the free extremity of the stolons. Simple arrows adjacent to the diaphragms, direction of the polarity and transit. Double arrows, fluctuating polarity). Abbreviations: A, functional autozooecia (spread-out tentacles); A′, closed autozooecia in course of modification; Ab: autozooecial buds; b: growth blastema of a stolonal bud (epidermis + mesenchyme); cA: cavity of the adult autozooecium; cS: stolon cavity; D, diaphragm; E, epidermis; F, stolonal funiculus; i, apical indentations; Li, lipids; lm, limiting cells, vacuolized in lm′; m, mesenchymatous cells; mv, microvilli; rf, thin refringent granulations; S, stolon; Sb, more or less abortive stolonal bud; sp, special cells; tf, autozooecial funicular tract; y, nonlipidic substances.

factors. Because the polarities vary from place to place and, consequently, the nutrient current may run only a short distance, one should anticipate that the physiological needs of the colony will also be cut up. The area of autonomy of one territory and its functional homogeneity would be defined by the contradictory orientation of a certain number of rosettes. One well-provided coenozooecium can, if the polarity of the rosettes is favorable, nourish its less fortunate neighbors, especially elements totally cut off from the exterior at the free end of the stolon–unfinished elements or elements provided with closed buds. At the same level, unfavorable polarity bodes ill for these units; they survive, but lose the terminal stolon-growing blastema. In such coenozooecia the capacity to bud ceases. In this respect, the example of the rounded or pointed branches of *Vesicularia spinosa* is significant (Bobin, 1962). Aberrant funicular currents and the wasting away which follows them are, in my view, responsible for the premature senescence of certain branches. On the whole, a zoarium incessantly seeks equilibrium and the colonial harmony results from constant competition between the individuals (auto- and cenozooecia) of a given sector and those of the neighboring sectors. The boundaries of a sector depend on the orientation of the surrounding rosettes.

3. Partial Conclusion: **Bowerbankia imbricata** *and Related Forms*

a. Other Roles of the Complex. If the major lines of colonial nutrition are partly clarified by the studies on *B. imbricata,* it is probable that other collective functions are accomplished at the same levels of the funicular and rosette systems, with the possible assistance of cells from the body cavity; some of which are rich in glycogen (spherulous cells, Bobin and Prenant, 1972). I have at the very most presumed (1962) that the funiculus would be able to propagate stimuli which initiate cystidal budding or even intercalary growth of terminal stolonal coenozooecia, subsequent to the partitioning. Less hypothetical is the proposal of Harmer (1891), who attributed to the funiculus the role of a colonial accumulating excretory organ. I believe, in fact, that in the young parts of *B. imbricata,* the stolon, living on substances furnished by the autozooecia, eliminates residues into the closed space of the coenozooecia. For the more one moves back along the older branches and the peduncle of one tuft, the more the funicular tissue is engorged with inert, refringent granules. Eventually, these residues fill the entire cavity of stolonal segments or of old, closed zooecia incapable of producing another polypide. I have often made similar observations, for example, in *Terebripora* (Bobin and Prenant, 1954) as well as in some stoloniferans and "alcyonellines."

b. Vesicularines and Other Ctenostomes. Despite the very deformed ectocyst and the irregular arrangement of the zooids, *Bowerbankia imbricata* has some advantages over the vesicularines which are also tufted and often have more regularly ordered individuals; the dimensions of the zooecia are greater and the diameter of the rosettes are a little larger in *B. imbricata.* An analogous structure and development of the funiculi–rosettes complex can be found in *B. imbricata, B. pustulosa, B. citrina, Vesicularia spinosa,* and *Amathia lendigera* (Bobin, 1958a,b; 1962). The phenomena of cystidal budding, of genesis of polypidal primordium and of histogenesis in general are slower, more coherent, and better sequenced in *B. imbricata;* that has always been necessary for the understanding of the other cases.

The rosettes I have been able to observe in other ctenostomes (1962) have special cells of reduced length (12–15 μm), the microvilli of which are more or less accentuated; the limiting layers which unite the special cells with the funiculi are discrete, discontinuous, or absent in *Valkeria* (Stolonifera), *Nolella* (Paludicellidae), and *Alcyonidium* (Alcyonellidae). The stoloniferans have a zoarium complicated by dwarfed coenozooecia so that the funiculus is further divided by partitions. The Paludicellidae have direct and simple interrelationships among autozooecia; some disjointed zoaria are transitional to the more compact arrangement of *Alcyonidium* where the presence of secondary communication plates anticipates the colonial organization of numerous cheilostomes.

Thus, in *Alcyonidium gelatinosum,* the autozooecia are joined, and each one possesses not only one primary communication plate, but also lateral secondary plates. As these uniporous plates are entirely comparable to one another, it becomes difficult to recognize the primary communication plate and, therefore, the growth rate for the zooecia with variable borders. The direction of the exchanges through the pores and rosettes is uncontrollable, all the more so because the topography of the funicular ramifications is rather hazy.

B. Primary and Secondary Plates of Cheilostomes (Figs. 4 and 5).

The communication plates of the cheilostomes are generally multiporous and, consequently, have many rosettes. The special cells which are few in number (one, two, or four), are very short (12 μm at the most). These plates, incorporated into the more or less calcified walls, have characteristics that are of diagnostic value, as has been recognized since 1896 (Waters, 1896). Thus, one is well informed about their distribution and morphological structure, but rather little is known about their histology. In a given wall, the interzooecial pores may be scattered, grouped, or else all located on thin depressed areas called septulae. In principle, the septulae are delineated by a definite or fuzzy, swollen rim; also, ambiguous intermediates exist between multiporous septulae and groupings of individual pores.

In any case, one must distinguish between primary plates (the fundamental type) established in the budding process during the formation of the "simple" transverse septa and the secondary plates disposed in the common "double" lateral walls of contiguous zooids. In his time, Nitsche (1871a,b) suspected that the two categories existed, the study of which was later developed by Levinsen (1909) and Silén (1944). Histological information is credited to Lutaud (1961) who studied one *Membranipora* and to Banta (1969) who examined a dozen species. Nearly all the examples studied by these authors possess the same type of encrusting zoarium in which the zooids, alternating or arranged in quincunx, are like rigid boxes with six walls (two terminal, two lateral, one frontal, and one basal).

1. Primary Transverse Communication Plates (Fig. 4, I and II).

The transverse ("interior" according to Silén) walls are, above all, terminal proximodistal septa of autozooecia which are arranged in linear files, one arising from the other; homologous, but less constant basal or frontal perforated septa can give access to heterozooecia, including rhizoids. Situated near the base of the partition, the pores are scattered or, more commonly, united into marginal

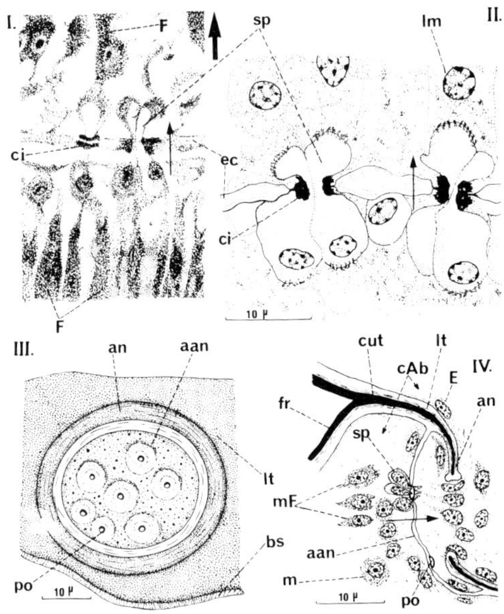

Fig. 4. Communication plates in cheilostomes. (I and II) Transverse plates of *Membranipora membranacea* (I, after Lutaud, 1961; trichrome of Prenant) and of *Watersipora nigra* (II, after Banta, 1969; Toluidine blue); frontal sections of autozooecia. (III and IV) Lateral septulae of *Electra verticillata* (after Bobin and Prenant, 1968); tangential section of adult septula (III, Kossa); transverse section of decalcified buds, showing a making up septula (IV, Mallory).

[Higher arrow, direction of the distal extremity of the autozooecia on (I) and (II). Arrows adjacent to communication plates, direction of the polarity of the rosettes).] Abbreviations: aan, abannular side of the septula; an, "annulus" or annular side of the septula; cAb, cavities of two neighboring autozooecial buds; bs, basal wall; ci, cincture (or ferrule) or the pore; cut, intercalary cuticle of the ectocyst (=external amalgamated pellicles); E, epidermis; ec, ectocyst of the transverse wall; F, adult funicular cells; fr, frontal wall of the bud; lm, cells of limiting layers; lt, double lateral wall; m, young mesenchyme; mF, original mesenchyme of future funicular cells; po, pores (one can see on III their mamillae and ferrules; sp, special cells of the rosettes (young on IV).

groups such as those illustrated by Banta (1969, Figs. 11, 16 and 37) for *Watersipora nigra.* Here, the calcified septum is simply pierced by pores, each being bordered by an organic "cincture." One pore is traversed by one or at most two special cells which are each in the shape of a dumbbell. Their nuclei, confined to the proximal face of the partition, indicate a polarity oriented toward the distal, younger autozooecium; limiting layers of cuboidal cells provide the bipolar junction with the funicular branches.

The explanation, without illustrations, by Banta, of the genesis of these septa seems to correlate with the commentaries and illustrations of Lutaud (1961, text Figs. 9 and 10, Plate VII, Fig. 4) for *Membranipora membranacea.* In summary, there appears on the lateral walls a pair of epithelial thickenings, the progressive extension of which results in the formation of a diaphragm which in closing itself leaves visible sutures, instigating the rupture and the displacement downward of the funicular trunks. In the course of this development, oval areas, at first uncalcified, are established at the future sites of the transverse porous plates. These areas are precociously covered with cells, some of which emanate from the broken funiculi and the others (the future special cells of the rosettes) from the peritoneum of the proximal zooecium.

Allied to the transverse plates are the dietellae ("pore chambers"), generally situated at the distolateral angle of zooecia or at the periphery of the ancestrula in certain species. In reality, these are autozooecial buds with arrested development, in which the partition, established earlier and provided with its perforated communication plates, represents the proximal wall of a daughter zooid and a potential bifurcation site at the free edge of a colony (Silén, 1944). Gordon (1971) has debated the opinion of Silén.

2. Secondary Lateral Communication Plates (Fig. 4, III and IV, Fig. 5)

The lateral ("exterior" of Silén) walls are doubled between contiguous zooids, the two calcified ectocysts being separated by a nonmineralized "intercalary cuticle" (Banta, 1969) formed by the fusion of the two "external pellicles" (Bobin and Prenant, 1968). These walls are pierced by clearly circumscribed plates which are more complex in structure and origin than the former ones. To differentiate these secondary plates, the term "septula" should be reserved for them. A lateral septula is asymmetric. In one of the coupled walls it presents a common, curved, multiporous plate, encircled at times by a false annulus and, in the other, opposite one, a large perforation, the thickened border of which delineates the true annulus, more or less calcified. Banta justifiably speaks of an "abannular" and an "annular" face. Sections of adult septulae have been illustrated by Bobin and Prenant (1968) and Banta (1969); on the whole one can recognize in each pore the usual cellular components of rosettes (Fig. 5, I).

The development of the lateral septulae occurs inside young, continuous, double walls previously organized and, according to Silén (1942, *Electra pilosa*) and Lutaud (1961, Plate V, Fig. 4), completely impregnated with a primary deposit of calcium. These authors presume a secondary dissolution of the calcium. Banta, on the other hand, thinks that the ectocyst of the septular abannular region remains thin and noncalcified and is thus easily resorbed.

Septulae in the distal half of a new cystid (Fig. 5, II and III) being as a "lenticular primordium" (Lutaud, 1961) applied against the lateral basal ectocyst which is still unperforated. Each primordium consists of a high basophilic epithelium and a peritoneal cap, both of which have actively dividing cells; later, an abannular calcified plate is produced and assumes the curvature of the primordium. Each pore of this plate encloses young special cells (Banta, 1969, Fig. 13), the nucleated lobes of which, situated against the convexity of the plate, project their bases toward the annular face which will open into the neighboring zooid. [This sequence, derived from Banta's studies (1969) is not in agreement with that of Lutaud (1961).] Finally, the funicular junctions are perceptible in the two cavities in the form of differentiating cells (Bobin and Prenant, 1968, Fig. 6, IV, m, *Electra verticillata*) (Fig. 4, IV).

3. Funiculi-Rosettes Complex (Fig. 4, I and IV; Fig. 5, IV)

The funicular cells may look like those of the vesicularines (see Banta, 1969, Fig. 6), but frequently they are less adherent, narrower, more fibrous, and more easily stained. According to Lutaud, they contain lipids, mainly, and nonglycogenic PAS-positive granules; but the structure and contents vary throughout even one cystid. The funicular pathways are not constant; however, longitudinal trunks always abut on the transverse rosettes and numerous branches are stretched between these trunks, the digestive loop of the polypide, and the lateral rosettes. In addition, Lutaud described, in *Membranipora membranacea*, a peripheral canal; this continuous collecting circuit unites the parietal ends of the funiculus and all the communication plates of one cystid. Bordered by an "endothelium" of "funicular fibers," the canal carries PAS-positive material interspersed with fat globules. Banta has not recognized this canal in his examples from North America; I have observed in *Electra pilosa* no peripheral organ, but some very similar canalicular patterns. Certain funicular branches exhibit this structure in some autozooecia, in any position whatsoever. I have deduced that such an appearance, without doubt transitory, is the periodic reflection of the functional state that I would especially attribute to colonial nutrition. Concrete information on such physiological problems is lacking for the cheilostomes. The direction of circulation, nevertheless, can be surmised from the orientation of the rosettes, at least in the regular colonies with alternate autozooecia (Fig. 5, IV). In this case, each zooid typically has four (or more) terminal, multiporous communication plates and eight lateral septulae,

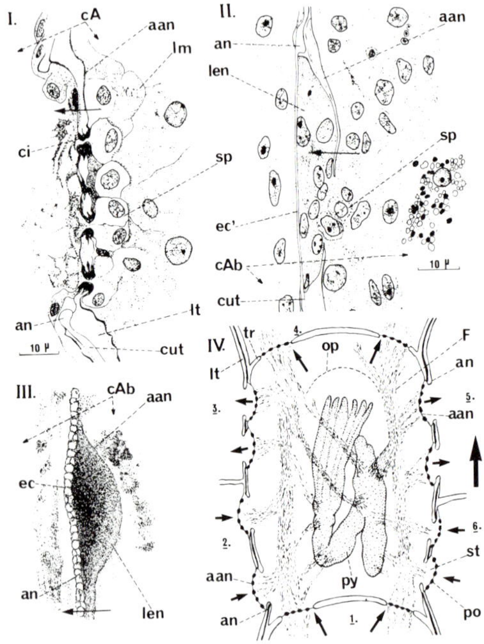

Fig. 5. Communication plates in cheilostomes. (I and II) Sections of lateral septulae of *Watersipora nigra* (after Banta, 1969); adult septula (I, hematoxylin–eosin); lenticular primordium differenciating (II, Richardson staining). (III) *In toto* view of a lenticular primordium of septula in *Membranipora membranacea* (after Lutaud, 1961; Hemalum). (IV) theoretical scheme of an autozooecium of cheilostome, seen from its basal side (orig.); note the branches of the funiculus and the probable physiological polarity of the rosettes. (Arrow on the right on scheme IV, direction of the free edge of the colony. Arrows adjacent to the communication plates, direction of the rosettes polarity). Abbreviations: aan, abannular side of the septula; an, "annulus" or annular side of the septula; cA, cavities of two neighboring autozooecia; cAb, cavities on two neighboring autozooecial buds; ci, cincture (=ferrule) of the pore; cut, intercalary cuticle (=external amalgamated pellicles); E, epidermis; ec, ectocyst temporarily continuous and calcified of the annular side of the lenticular primordium (see the crystals on III, after a photograph of Lutaud); ec', the same ectocyst with a membranaceous nature, according to Banta (on II); F: ramified funicular trunks; len, lenticular primordium; lm, cells of limiting layers; lt: double lateral wall; op: opercular trace; po: place of a pore and a rosette; py, polypide; sp, special cells of the rosettes (during genesis on II); st, lateral septulae; tr, transverse walls with their communication plates; 1 to 6, cavities of the 6 autozooecia with which the figured autozooecium communicates.

but only the rosettes of its distal half properly belong to it, as evidenced from their origin. In this regard and according to the convenient terminology of Lutaud, the zooecia is "donor" in its distal half and "receptor" in its proximal half.

This indicates that one zooecium furnishes, distally (while retaining the nucleated lobes), the special cells of all the transverse plates and of four septulae; proximally the same zooecium collects the swollen bases projecting through the pores of the special cells from neighboring cystids. This polarity is reinforced at the level of the lateral septulae by a morphological feature recognized in part by Silén. In effect, distally the zooid is donor of the lenticular primordium and, afterward, the incurved abannular porous plate of the septula, whose annular face alone is contained in the proximal receptor part of an adjacent individual. Thus one zooecium is joined with six contiguous zooids and it is the very place of distal centrifugal currents and proximal centripetal ones, if one takes into account the morphophysiological polarity of the rosettes as analyzed in the preceding section.

Of course even a quincuncial zoarium undergoes irregularities of growth so that, in some individuals, the number and orientation of the communication plates are subject to alterations. This process will be more perceptible, perhaps, for the zoaria with more capricious growth forms (bushy, bilaminar, whorled, etc.). As in the ctenostomes, one should look for certain functional contradictions.

It should be emphasized that Banta, following Silén, considered the lateral septulae as abortive autozooecial buds, arrested in their morphogenesis because of the lack of space. From her rearing of *Membranipora,* Lutaud (1961) stated that if one zooecial file, at first internal within the colony, becomes marginal following the accidental arrest of the growth of the neighboring files, it continues to grow; however, no rosettes are formed on the denuded margin of the new zooecia, the free wall of which escapes doubling. If, by hypothesis, primordia of septulae were able to exist on the simple lateral limit of such zooecia, they would be, in my view, absent proximally for lack of donor zooecia and, at least, reduced distally to their abannular face.

The question is to know whether the contact between individuals is not indispensable for the genesis of septulae. Lutaud suggested the intervention of determining factors arising from funicular branches localized against the contiguous walls.

A final, curious observation has been made by Bobin and Prenant (1960) in experimental colonies of *Electra verticillata,* obtained from implants of basal stolons on diverse substrates. Sometimes a whorled configuration occurred, sometimes it did not. Zooecial files produced from different stolons regularly may adhere to one another and the cystids may be joined by the normal lateral septular pattern. In certain cases, a slight distance separated the lateral walls of two zooecia; there were then created, in place of septulae, interzooecial bridges, the histological examination of which remains to be studied.

IV. CASE OF CYCLOSTOME GYMNOLAEMATES

The cyclostomes, very calcified, have zoaria either massive or bushy with articular chitinous joints. The tubular and rigid autozooecia have an orifice that is deprived of a closing apparatus. There is an exceptional process of polyembryony, and the incubatory chambers (ooecia) are specific to this group. These characteristics isolate the latter from the eurystomes, as do also the modes of cystidal budding and partitioning, which rather resemble those of the phylactolaemates. As they possess neither flexible walls, elastic areas, nor asci, the zooids are provided with a compensatory "membranous sac" that permits the protrusion of the polypide. Borg (1926a,b) has elucidated the formation and the mechanism of action of this sac. Doubling the cystidal lining and anchored by ligaments, this sac encloses the polypide from the proximal vestibular rim to the base of the cystid where it is attached. It partially partitions the cavity into two hollow cylinders, one inside the other; its wall of thin ectomesenchyme is covered by a homogeneous membrane.

The ectocyst of the majority of species contains scattered "pseudopores" which decorate even the internal zooecial walls which are incorporated in the mass of the zoarium. The pseudopores are, as are those of the gymnocyst of various cheilostomes, excavations developed in the calcified wall and covered superficially by a membranous cuticle that may be impregnated with calcium during aging. Borg indicated that only Robertson (1903) maintained that the pseudopores of Crisiidae are real pores. The general view is that the false pores of cyclostomes are entirely distinct from the true ones which provide interzooecial communications. The only valid investigation of the interzooecial pores, without specifying their number, of cyclostomes are due to Borg, who presented only a single illustration (1926b, Plate 1,1). Sometimes he speaks of a few pores and at other times he indicates they are numerous when he describes the relationship between "nanozooids" and normal zooids.

These pores, formed during the partitioning of the zooids, are small, round orifices crossing the calcareous walls of two neighboring individuals, freely linking their respective body cavities. Thus, they are not filled as in the cheiloctenostomes because of the lack of rosettes and funicular cords. True pores, of a similar nature, pierce the cryptocysts in the two families which are provided with them (Horneridae and Lichenoporidae). Borg thinks that communication pores serve for the circulation of the cavitary fluid and for the exchange of nutrients between zooids.

The methods of partitioning of the zooids are correlated with the specific style of budding in the cyclostomes. Initially, there exists a "common bud" endowed with a "growth zone" from which both the polypides and the primordial sexual cells originate. In Crisiidae, for example, the initiation of the polypide bud gives the signal for the separation by a septum; developing from a lateral basal site, the septum grows obliquely toward the frontal and terminal regions. According to Borg, the few pores set aside in the course of this growth are the only sites of communication between the two new cavities, the larger of which is that of the polypidal zooecium and the smaller that of a new common bud. The process is repeated alternately on one side and then the other. The partitioning of the common bud becomes quite complicated in other genera and species.

Each adult polypide possesses a short funiculus of mesenchymatous origin that is stretched between the gastric cecum and the attachment point of the membranous sac to the cystidal wall at the base of the zooid; no association of this funiculus with any pore whatsoever is mentioned. Each funiculus is, as in the phylactolaemates, an individual differentiation and is without significance for the collective life of the colony; however, it is much more discrete in its structure and its role in the cyclostomes. The germinal cells are located on the funiculus, taken away from the growth zone by the peritoneum of the polypidal primordium. One shall not speak of oocytes, the ultimate destiny of which is associated with a distended ooecium in which the polypide degenerates. The testes, limited by a membrane, develop entirely against or around the funiculus. Although not directly producing the male cells, one may think that the funiculus furnishes them *in situ* with a nutritive supply emanating from the polypide. The funiculus is lacking in the nanozooids, which moreover are sterile.

Borg observed in the cavity of any zooid some spans of mesenchyme, the filiform extensions of which are able to insinuate themselves into the septal pores; this arrangement could permit exchanges. Finally, it should be noted that the polypide with its funiculus is completely enclosed in the internal cavity of the membranous sac. I wonder whether in addition to its mechanical role, the wall of this sac would not permit a filtration of substances or if its membrane would not involve a system of microviili?

V. CONCLUSIONS

This chapter on the funicular system has presented new knowledge that has been obtained in a field that has been slowly enlightened over the course of a century. These exceptional structures, peculiar to the Bryozoa, have aroused justified interest, but their diversity, their dispersion, as well as the delicacy of their components still constitute impediments to their study. Without returning to their morphology, one may insist upon their functional role in transfer of substances necessitated by the communal sharing of resources of any animal colony.

A. Free Transfers

The phylactolaemates appear to obey the usual physiological rules as happens with the coelenterates (including hydroids) and certain ascidians. A common cavitary fluid, through which cells are scattered, is mixed and freely circulates inside the zoarium; this fluid ensures proper physiological conditions for the zooids, which live for one season. The "funicular cable," a strictly individual structure, does not play an apparent collective role; the products that it collects are probably reserved for the development of male gametes and, then, of the statoblasts.

The cyclostomes (gymnolaemates) have with the phylactolaemates some points in common. Notably, communication between zooids is effected through free pores, which are much smaller and more localized than the large intrazoarial orifices of most phylactolaemates. In addition, the rigidity of the calcareous walls of the cyclostomes eliminates the elasticity necessary for the mixing of the cavitary fluid; this inconvenience appears to be compensated for, at the zooecial level, by the presence of the "membranous sac."

In the cyclostomes, the very discrete funiculus is still produced by an individual differentiation, associated with each polypide and without connection, it is believed, with the communication pores. Without an evident role, it might act, not in the production, but in the nutrition of the gametes, especially the male ones. This is the limit of our knowledge.

B. Indirect Transfers

The cheiloctenostomes (eurystomes gymnolaemates) differ from the preceding groups by the possession of a true colonial system specialized for exchanges; this anatomic system appears to be unique in the animal kingdom. Thus, Nature has resolved in a highly original way the problem of the astonishing compartmentalization of a single colony community.

The structure and arrangement of the tissues of the funiculi and rosettes resemble nothing known elsewhere; they are neither nervous nor truly glandular, but serve in aneural conduction by the collection, reconversion, and distribution of diverse substances, thus providing, starting with the polypidal zooecia, for the need of those portions of the colony which are impoverished or deprived of exterior contact. These same tissues also would be, in the older portions of the colony, collectors of wastes. However, the resources as well as the internal exchanges of a colony are subjected in the eurystomes to the vicissitudes of the life of the polypides, which appear, undergo degeneration, and reappear again; consequently, a succession of beneficial phases and interruptions occurs which might dangerously compromise the collective economy if an organized and permanent system (the funiculi–rosettes complex) did not intervene, locally and temporarily, as a regulator with multiple potentials. The marked polarity of the rosettes orients the direction of the currents, and their "special cells," with microvilli, absorb and reject the substances running between neighboring units in a given direction.

However, because of contradictory polarities a colony is deprived of physiological unity, and its general functioning seems subdivided into sectors. From the constant competition of these sectors, precarious equilibrium results, constantly menaced by various predominating factors.

Only one pattern of colonial nutrition had been proposed—that for ctenostomes with a given zoarial style. Other zoarial forms will merit attention in this group as well as in the cheilostomes where the communication between zooids are even more complex. The cyclostomes remain to be explored more closely with respect to their pores.

It is evident that only the further contributions of different researchers will be able to furnish, at length, utilizable arguments for the total comprehension of the colonial physiology of the Bryozoa. Even if a satisfactory scheme is evolved, a diversity of specific expressions must be expected.

REFERENCES

Banta, W. C. (1968a). The body wall of the encrusting cheilostome *Watersipora nigra* (Canu and Bassler) (preliminary report). *Atti Soc. Ital. Sci. Nat. Mus. Civ. Stor. Nat. Milano* **108,** 93–96.

Banta, W. C. (1968b). The body wall of cheilostome Bryozoa. I. the ectocyst of *Watersipora nigra* (Canu and Bassler). *J. Morphol.* **125,** 497–507.

Banta, W. C. (1969). The body wall of cheilostome Bryozoa. II. Interzoidal communication organs. *J. Morphol.* **129,** 149–170.

Bobin, G. (1958a). Structure et genèse des diaphragmes autozoéciaux chez *Bowerbankia imbricata* (Adams) (Bryozoaire Cténostome, Vésicularine). *Arch. Zool. Exp. Gén,* **96,** 59–99.

Bobin, G. (1958b). Histologie des bourgeons autozoéciaux et genèse de leurs diaphragmes chez *Vesicularia spinosa* (Linné). (Bryozoaire Cténostome). *Bull. Soc. Zool. Fr.* **83,** 132–144.

Bobin, G. (1962). Histogénèse des diaphragmes septaux stoloniaux et valeur des rosettes chez les Vésicularines (Bryozoaires Cténostomes). *Arch. Zool. Exp. Gén.* **101,** 14–42.

Bobin, G. (1963). "Bryozoaires et Kamptozoaires. Zoologie 1. Encyclopédie de la Pléiade," pp. 916–942. Gallimard, Paris.

Bobin, G. (1964). Cytologie des rosettes de *Bowerbankia imbricata* (Adams). (Bryozoaire Cténostome, Vésicularine). Hypothèse sur leur fonctionnement. *Arch. Zool. Exp. Gén.* **104,** 1–44.

Bobin, G. (1965). Introduction à l'histophysiologie des rosettes de *Bowerbankia imbricata* (Adams) (Bryozoaire Cténostome, Vésicularine). Les Ribonucléines. *Arch. Zool. Exp. Gén.* **105**, 1–25.

Bobin, G. (1971). Histophysiologie du système rosettes-funicule de *Bowerbankia imbricata* (Adams) (Bryozoaire Cténostome). Les lipides. *Arch. Zool. Exp. Gén.* **112**, 771–792.

Bobin, G., and Prenant, M. (1954). Sur un Bryozoaire perforant (*Terebripora comma* Soule), trouvé en Méditerranée. *Arch. Zool. Exp. Gén.* **91**, 130–144.

Bobin, G., and Prenant, M. (1960). *Electra verticillata* (Ellis et Solander 1786), Lamouroux 1816, (Bryozoaire Chilostome). *Cah. Biol. Mar.* **1**, 121–156.

Bobin, G., and Prenant, M. (1968). Sur le calcaire des parois autozoéciales d'*Electra verticillata* (Ell. et Sol.) Bryozoaire Chilostome *Anasca*. Notions préliminaires. *Arch. Zool. Exp. Gén.* **109**, 157–191.

Bobin, G., and Prenant, M. (1972). Sur les cellules cavitaires de quelques Vésicularines (Bryozoaires Cténostomes). *Cah. Biol. Mar.* **13**, 479–510.

Borg, F. (1926a). On the body-wall in Bryozoa. *Q. J. Microsc. Sci. [N.S.]* **70**, 583–598.

Borg, F. (1926b). Studies on Recent cyclostomatous Bryozoa. *Zool. Bidr. Uppsala* **10**, 181–504.

Brien, P. (1953). Etude sur les Phylactolémates. *Ann. Soc. R. Zool. Belg.* **84**, 301–444.

Brien, P. (1960). Phylactolémates. *In* "Traité de Zoologie" (P.-P. Grassé, ed.), Vol. 5, pp. 1059–1167. Masson, Paris.

Brien, P., and Huysmans, G. (1937). La croissance et le bourgeonnement du stolon chez les Stolonifera (*Bowerbankia* Farre) (Evolution du stolon et de la zoécie chez les Bryozoaires). *Ann. Soc. R. Zool. Belg.* **68**, 13–40.

Brien, P., and Mordant, C. (1955–1956). Relations entre les reproductions sexuée et asexuée à propos des Phylactolémates. *Ann. Soc. R. Zool. Belg.* **86**, 169–189.

Calvet, L. (1900). Contribution à l'histoire naturelle des Bryozoaires Ectoproctes marins. *Trav. Inst. Zool. Univ. Montpellier, Mem [N.S.]* **8**, 1–488.

Davenport, C. B. (1891). Observations on budding in *Paludicella* and some other Bryozoa. *Bull. Mus. Comp. Zool.* **22**, 1–114.

Gasser, F. (1962). Contribution à l'étude du statoblaste chez *Plumatella repens* (Linné) (Bryozoaire Ectoprocte Phylactolème). *Arch. Zool. Exp. Gén.* **101**, 59–72.

Gordon, D. P. (1971). Zoidal budding in the cheilostomatous bryozoan *Fenestrulina malusii* var. Thyreophora *N. Z. J. Freshwater Res.* **5**, 454–460.

Harmer, S. F. (1891). On the nature of the excretory processes in marine Polyzoa. *Q. J. Microsc. [N.S.]* **33**, 123–167.

Joliet, L. (1877). Contribution à l'histoire naturelle des Bryozoaires des côtes de France. *Arch. Zool. Exp. Gén.* **6**, 193–304.

Lacourt, A. W. (1968). A monograph of the freshwater Bryozoa–Phylactolaemata. *Zool. Verh. (Leiden)* **93**, 1–155.

Lacourt, A. W., and Willighagen, D. D. (1966). Histochemical investigation of *Pectinatella magnifica* (Leidy, 1851). (Bryozoa: Phylactolaemata). *Proc. K. Ned. Akad. Wet. Ser. C* **69**, 22–23.

Levinsen, G. M. R. (1909). "Morphological and systematic studies on the Cheilostomatous Bryozoa," Nat. Forfatt. Forlag, Copenhagen.

Lutaud, G. (1961). Contribution à l'étude du bourgeonnement et de la croissance des colonies chez *Membranipora membranacea* (Linné). Bryozoaire Chilostome. *Ann. Soc. R. Zool. Belg.* **91**, 157–300.

Müller, F. (1860). On the common nervous system of the Bryozoa. *Q. J. Microsc. Sci. [N.S.]* **1**, 1–300.

Nitsche, H. (1871a). Beiträge zur Kenntniss der Bryozoen. III. Über die Anatomie und Entwicklungsgechichte von *Flustra membranacea* L. *Z. Wiss. Zool.* **21**, 416–498.

Nitsche, H. (1871b). Beiträge zur Kenntnis der Bryozoen. IV. Ueber die Morphologie der Bryozoen. *Z. Wiss. Zool.* **21**, 92–119.

Prenant, M., and Bobin, G. (1956). "Faune de France. Bryozoaires, lère partie (Entoproctes, Phylactolèmes, Cténostomes)," No. 60, pp. 1–398. Lechevalier, Paris.

Prenant, M., and Bobin, G. (1966). "Faune de France. Bryozoaires, 2ème partie (Chilostomes Anasca)," No. 68, pp. 1–647. Fédération française des Sociétés de Sciences Naturelles, Paris.

Reichert, K. B. (1869). Vergleichende anatomische Untersuchungen über *Zoobothryon pellucidus. Abh. K. Akad. Wiss., Berlin* **2**, 233–338.

Robertson, A. (1903). Embryology and embryonic fission in the genus *Crisia. Univ. Calif. Berkeley, Publ. Zool.* **1**, 115–156.

Ryland, J. S. (1970), "Bryozoans" Hutchinson University Library, London.

Silén, L. (1942). Origin and development of the cheilo-ctenostomatous stem of Bryozoa. *Zool. Bidr. Uppsala* **22**, 1–59.

Silén, L. (1944a). On the formation of the interzoidal communications of the Bryozoa. *Zool. Bidr. Uppsala* **22**, 433–488.

Silén, L. (1944b). The anatomy of *Labiostomella gisleni* Silén. *Handl. K. Sven. Vetensk.* [3] **21**, 1–111.

Waters, A. W. (1896). Interzoecial communication in Flustridae, and notes on *Flustra. J. R. Microsc. Soc.* [2], Part 3, 292–299.

Zirpolo, G. (1933). *Zoobothryon verticillatum* Delle Chiaje. *Mem. Pont. Acad. Sci. Nat. Lincei, Roma* [2] **17**, 1–300.

10

The Aging Process in Bryozoans

DENNIS P. GORDON

I. INTRODUCTION

"Ces corps bruns ont bien intrigué les observateurs." So wrote Joliet in 1877 (p. 195) after reviewing the opinions of previous authors. The cyclic phenomenon of polypide regression and renewal is a characteristic feature of bryozoans which is to an extent paralleled by some other sessile, colonial metazoans. The products of regression (one or two ovoid or spherical brown bodies) puzzled many early workers as regards their nature, because it was not always understood that they were regression products. They were thought to be ova or ovaries (Thompson, 1830; Farre, 1837; Van Beneden, 1845), developing embryos, the equivalent of phylactolaemate statoblasts (Allman, 1856; Redfern, 1858), germ capsules (Smitt, 1863, 1865; Hincks, 1871, 1873), and an endocystal secretion (Claparède, 1870). It is amusing that free rein was given to such speculation since the origin of brown bodies from withered polypides had already been recognized by Ellis in 1755 and Grant in 1827. "These black spots are nothing but the dead polypes, or remains of the animals once inhabiting these cells. Of which I had evident proof in my last journey to the seacoast. For after I had examined this coralline, with its polypes alive in seawater I laid this specimen aside; and upon examining it again sometime after, I found the lifeless contracted animals exhibited the appearance above-mentioned." Although the latter part of Ellis' explanation is slightly ambiguous it is clear that he recognized the true nature of brown bodies, which he illustrated in some gymnolaemates (e.g., *Menipea ternata, Bugula plumosa*) and a stenolaemate (*Crisia eburnea*).

Part of the subsequent difficulty in recognizing the true nature of the brown body was owing to the observation of a new polypide bud often in intimate contact with it. Hincks (1871, 1873), supporting Smitt's (1863) germ capsule theory, held that the brown body was not derived from the withered polypide but was a special formation produced at its expense by sudden condensation of the stomach cecum which became detached. The body thus formed served as the origin of a new polypide. Nitsche (1869, 1871) seems to have been the first author after Grant (1827) to have realized that the brown body was derived simply as a regression product of the withered polypide.

Next, Ehlers (1876) described histolysis in *Alcyonidium* and Repiachoff (1876) discovered the incorporation of the brown body into the stomach cecum of the backwardly growing new polypide. Subsequently, it was recognized that the brown body may have alternative fates; see next paragraph. Hincks (1880) was persuaded of the correctness of Repiachoff's observations after reading Joliet's "Histoire" (1877), which was supported by Waters (1878). Authors of this and subsequent periods who accepted the polypide origin of the brown body held varied opinions as to its nutritive value to the developing polypide. There were those who held that the residue was useful (Repiachoff, 1876; Hincks, 1880; Zschiesche, 1909; Waters, 1913; Gerwerzhagen, 1913) and those who held it to be inert (Nitsche, 1869, 1871; Joliet, 1877; Marcus, 1926a), although Joliet felt that mesenchyme cells incorporated with the brown body into the new gut could be broken down, and Marcus observed that the incorporated brown body became smaller. Waters (1913), seeing mesenchymal strands between a brown body and ovicell in a zooid, thought that the one supported the development of the other.

Incorporation of the brown body into the new gut is known in *Cryptosula pallasiana* (Repiachoff, 1876), *Scruparia chelata* (Joliet, 1877), *Carbasea papyrea*

(Harmer, 1891), *Carbasea carbasea* (Gerwerzhagen, 1913), *Flustra foliacea* and *Electra pilosa* (Marcus, 1926a), *Chartella papyracea* and *Callopora lineata* (Rey, 1927), *Watersipora cucullata* (Mawatari, 1952), and *Crassimarginatella papulifera* and *Fenestrulina malusii* (Gordon, 1968). It may, however, remain in the zooecial cavity as it is said to occur in *Bugula neritina* and *B. avicularia* (Harmer, 1891), *B. simplex* and *Scrupocellaria scruposa* (Calvet, 1900), *Bugula turbinata* (Römer, 1906), *Flustrellidra hispida* (Rey, 1927), *Alcyonidium gelatinosum* (Bobin and Prenant, 1957), and *Victorella argilla* (Banta, 1967). Alternatively, the brown body was said to be totally destroyed by phagocytosis (Korschelt and Heider, 1912) or eliminated through the intertentacular organ of some species in the reproductive season (Hincks, 1880, p. 89; Prouho, 1892, p. 589).

The nature and origin of the brown body having been established, causal factors in its formation and reasons for its existence were put forth. Brown body formation has been attributed to unfavorable environmental conditions [lack of oxygen; low food supply; extremes of temperature, pH, and salinity; mechanical (e.g., sediment) and chemical inhibition of evagination]; detachment of colonies (Marcus, 1926a; Rey, 1927); and reproductive activity including oogenesis and embryo maturation (Calvet, 1900; Römer, 1906; Borg, 1947; Mawatari, 1952; Bobin and Prenant, 1957; Braem, 1951; Chrétien, 1958; Matricon, 1963; Banta, 1968). Accumulation of residual material in stomach cells has also been thought to be a prime cause, and this has been tied in with the lack of excretory organs in bryozoans.

Brown body formation has been regarded, therefore, as having an excretory function, an idea first expressed by Ostroumoff (1866), and acknowledged by Harmer (1891, 1896, 1931), Delage and Hérouard (1897), Prouho (1892), Calvet (1900), Marcus (1926b), Rey (1927), Bronstein (1939) and Cori (1941). Calvet, however, felt that, as the brown body is not eliminated in some bryozoans, excretion may not be the prime function. Marcus (1926b) and Gordon (1970) noted that polypide regression (with brown body formation) tides zooids over periods when adverse conditions temporarily prevail, and Prouho (1892) felt that the "bryozoite bénéficie d'un rajeunissement partiel périodique."

In the literature polypides are generally referred to as "young" when newly feeding and "old" when stomach walls are laden with orange-brown inclusions, but how and whether natural senescence is related to accumulation of this material has not been clear. Thus, knowing that brown bodies are regression products and that their formation is induced by both extrinsic and intrinsic factors, can brown body formation be regarded as an excretory or an aging phenomenon? What is the nature of the granular material that comprises the regression product and is it likely that it can be further digested in the gut of the new polypide? This review considers these questions and attempts to explain the significance of cyclic regression in the life of bryozoans.

II. GENERAL FEATURES

A. Longevity

Data on rates of development, longevity, and regression are few. Colony longevity is known to range from a few months in *Bugula simplex* (Grave, 1930) to twelve years in *Flustra foliacea* (Stebbing, 1971). Longevity of individual zooids of a colony is difficult to determine, however, for the duration of life of a zooid can be known only by observation of individual zooids over periods of months or even years because zooidal longevity is a function of that of its components–the polypide (alimentary tract and lophophore) and cystid (body wall). It is theoretically possible for small coelomic spaces derived from the metacoel or the derivative hypostegal coelom to remain alive, albeit dormant, for long periods of time providing they are in organic continuity with the remainder of a colony; the body wall remains alive, of course, during polypide regression.

The duration of life of polypides is easy to determine but published data are few (Table I). The figures in the table are absolute ones which need to be correlated with environmental parameters, but they are indicative of the limits that may be attained by polypides.

B. Regression at the Level of Light Microscopy

In the ascophoran *Cryptosula pallasiana* polypides survive for 2–10 weeks in aquarium conditions. After the final emergence of the lophophore for feeding the polypide is retracted and undergoes tissue regression (Fig. 1). Observations of paraffin sections and live colonies reveal that the lophophore starts degenerating from the distal ends of its tentacles, meanwhile condensing, until it has become a shapeless orange mass. The alimentary tract shrinks in size as the gut lumen diminishes and the stomach becomes compacted. The parts of the stomach that contain orange-brown inclusions in life condense as a very dark mass, but the rectum and pharynx are incorporated with the regressing lophophoral mass. Thus the remains of the polypide take on a bipartite appearance, with an anterior reddish part and a posterior dark-brown body (Gordon, 1973b).

From paraffin sections the lophophore base and its basal lamina are seen to be the last parts of the lophophore to regress. Orange-brown inclusions in the stomach wall become a prominent part of the brown body. They autofluoresce under UV, as lipofuscin does, but fluorescence diminishes as regression proceeds, for unknown reasons. If the polypide does not defecate prior to regression then a fecal pellet with skeletal fragments of food organisms becomes incorporated. Sometimes the regressing anterior part (lophophore, pharynx, rectum) remains separated from the regressing stomach portion so that there are two "brown bodies," but more often the reddish-orange part appears as a protuberance or

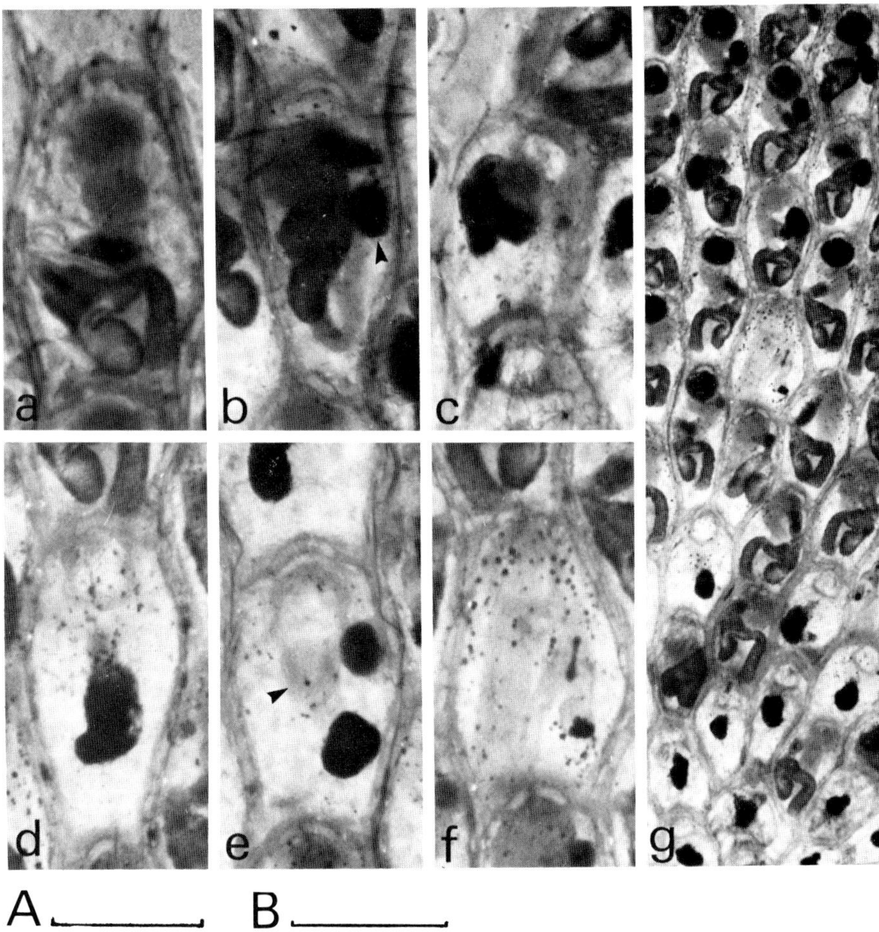

Fig. 1. *Cryptosula pallasiana.* Stages in polypide regression. (a) A normal polypide. (b) Early regression (the arrow indicates an embryo). (c) A later stage in which the lophophore and pharynx have condensed. The different parts of the gut are still recognizable. (d) The lophophore, rectum and pharynx are represented by the lighter mass, the stomach is dense and brown. (e) A polypide primordium (arrow) (adjacent to an embryo) approaches a brown body. (f) The lighter remains of the brown body are seen in the cecum and pylorus of the new polypide. Parts of the regression product are scattered throughout the zooid. (g) Part of a colony showing zooids from the reproductive region to the periancestrular region (all photographs are of the same colony seen in basal view, growing on a glass plate; photographed using a Zeiss Tessovar macrophotography setup). Magnification: (a–f) × 22.5 (scale A, 1 mm); (g) × 55 (scale B, 0.4 mm).

less dark region on the brown body proper. After acridine staining for DNA, fluorescence is confined to the few remaining nuclei (green in UV) in and around the brown body (Gordon, 1973b). Time taken for regression to be completed in *Cryptosula* and some other bryozoans is given in Table I.

In *Cryptosula* the lophophore (and pharynx and rectum) undergoes regression earlier than the stomach, even though the stomach itself is laden with residual material (Gordon, 1973b). This appears to be the general pattern in bryozoans. In *Walkeria cuscuta, Bowerbankia imbricata, Alcyonidium polyoum, Membranipora membranacea*, and "*Lepralia*" *granifera* the lophophore regresses first (Joliet, 1877). Bobin and Prenant (1957) observed that in *Alcyonidium gelatinosum* the tentacles sometimes may be ingested by the pharynx in early regression, and Matricon (1960) observed in *A. polyoum* a similar circumstance, in which the distal parts of the tentacles are introduced into the regressing rectum, although in *Alcyonidium hirsutum* they penetrate as far as the cecum. As the stomach regresses it has been noted that the histological structure disappears (Calvet, 1900), the lumen becomes a receptacle for cytoplasmic debris, and cytoplasm becomes vacuolated (Matricon, 1960).

It seems to be agreed that the brown body is not derived from the entire polypide. Ehlers (1876) recorded separation of the tentacles from the gut during histolysis in *Alcyonidium*. Calvet (1900) indicated regression of the lophophore and pharynx apart from the stomach and rectum, and Gerwerzhagen (1913)

TABLE I
Polypide Longevity and Duration of Regression in Some Bryozoans

Species	Polypide longevity (days)	Duration of regression (days)	Authority
Flustrellidra hispida	21–28[a]	6–7[b]	Rey (1927), Joliet (1877)
Zoobotryon verticillatum	6	–	Bullivant (1967)
Cryptosula pallasiana	15–72	6–17	Gordon (1973b)
Electra pilosa	6–33	3–8	Gordon (1973b)
Eurystomella foraminigera	20–60	5–15	Gordon (1968)
Fenestrulina malusii var. *thyreophora*	35–42	8–10	Gordon (1968)
Bugula flabellata	–	7–9	Joliet (1877)
Bugula neritina	–	2	Harmer (1891)
Carbasea papyrea	–	12	Harmer (1891)
Fredericella sultana	Av. 7.4		Wood (1973)
Plumatella repens	Av. 6.8		Wood (1973)
Plumatella casmiana	Av. 7.8		Wood (1973)

[a]From Rey (1927).
[b]From Joliet (1877).

noted separate regression of the tentacles and foregut. According to Bobin and Prenant (1957) the cardia regresses independently from the rest of the stomach while the rectum may be incorporated into the brown body. Matricon (1960) observed that the sphincters of the stomach delimit those areas that will comprise the brown body proper and those which will not; i.e., lophophore, pharynx, and rectum regress as a unit in *A. polyoum* just as they do in *Cryptosula* (Gordon, 1973b). Rey (1927), on the other hand, recorded for *Flustrellidra hispida* that degeneration starts in the cecum, spreading to other parts of the gut and then the tentacles, and that the entire polypide excluding the muscles contributes to the brown body.

It was mentioned in the introduction that the development of intrazooecial embryos causes polypide regression in some species (Marcus, 1926a,b; Bobin and Prenant, 1972). Observations on *Cryptosula* indicate that this may not always be the case (Gordon, 1973b). In *Cryptosula* colonies are protogynous, ova appearing about two weeks before testes. Some individuals seem to be monoecious but most are dioecious. In any case, the onset of gametogenesis has no apparent effect on the polypide. After presumed fertilization, embryogenesis takes place near the distal wall of the zooid, requiring approximately thirty days, during which time a second embryo may be developing on a side wall. Testes persist for about thirty days. Throughout this reproductivity, including embryonic development, polypides may either continue to feed, regress, or regenerate.

III. ULTRASTRUCTURAL FEATURES OF REGRESSION

The following description is based upon the author's observations on *Cryptosula pallasiana* and *Hippothoa hyalina* (Gordon, 1973a,b).

A. The Regressing Lophophore

Among the first features to be seen in tentacle regression are various changes at the cell surface. The cuticle tends to slough off, carrying with it tips of epidermal microvilli. Abscission of some of the cilia from their basal bodies occurs, leaving them to degenerate extracellularly or to be picked up by macrophagy, apparently by nonciliated cells of the lophophore. In addition, cell junctions weaken and adjacent cells move apart from one another and from the basement membrane; these changes occur both in the outer epithelium and inner peritoneum. At this time, collagen fibrils in the basement membrane still show their usual periodicity of 60 nm. Cytoplasmic features vary from cell to cell as not all cells regress simultaneously. The abfrontal tentacle cells seem to regress more quickly than the lateral and frontal ones and show more autophagic activity. In the case of the ciliated cells, cytoplasm appears almost as in cells prior to regression, but as cells slough off from the basement membrane, changes become apparent. Endoplasmic cisternae become dilated (Fig. 2) and as a result ribosomes are liberated from their surfaces. In detached cells there is some clumping of chromatin as incipient pyknosis occurs (Fig. 3). Pseudopodial processes may be extended and there is occasional intracellular deposition of glycogen. Autophagic vacuoles, multivesicular bodies, and, in some cells, lipoidal pigment granules, are also present. Mitochondria and Golgi bodies appear "normal."

Cells at the lophophoral base and mouth regress later than cells further up the tentacles. The beginning of their regression is marked by the development of myelin figures, residual bodies, and lipid droplets in the cytoplasm and of myelin figures within and around some mitochondria (Fig. 4); later the cells move apart. Peritoneal cells in the tentacles exhibit the same cytoplasmic changes as the outer epidermal cells but show a greater tendency to autolysis.

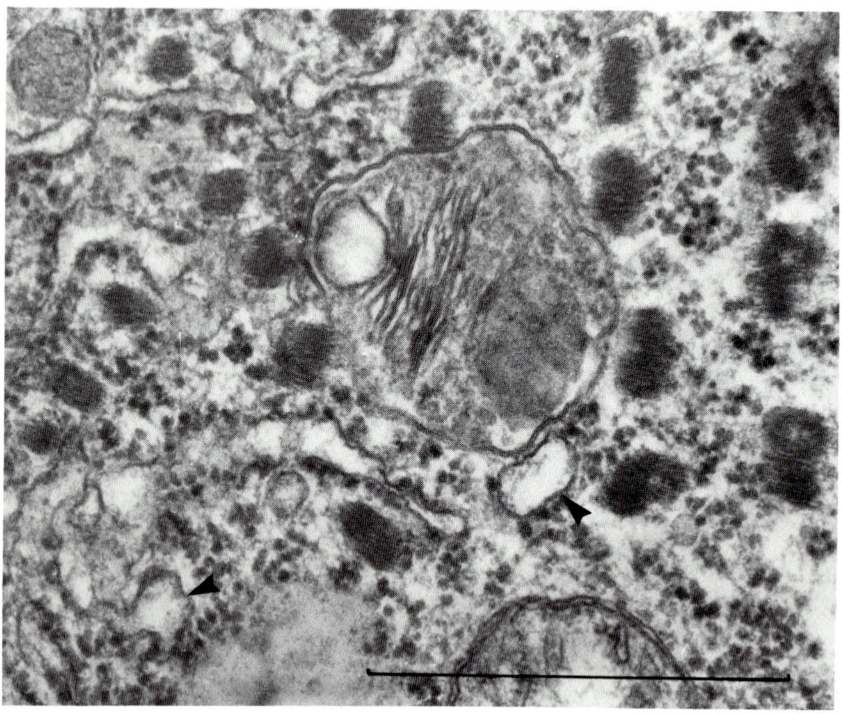

Fig. 2. *Cryptosula pallasiana.* Lophophore regression: Part of a ciliated tentacle cell in early regression with a lytic vacuole containing membrane lamellae. Notice the small vacuoles (arrows) which are derived from endoplasmic cisternae. The fibrous bundles are ciliary rootlets cut obliquely. Magnification: × 57,500 (scale, 1 μm).

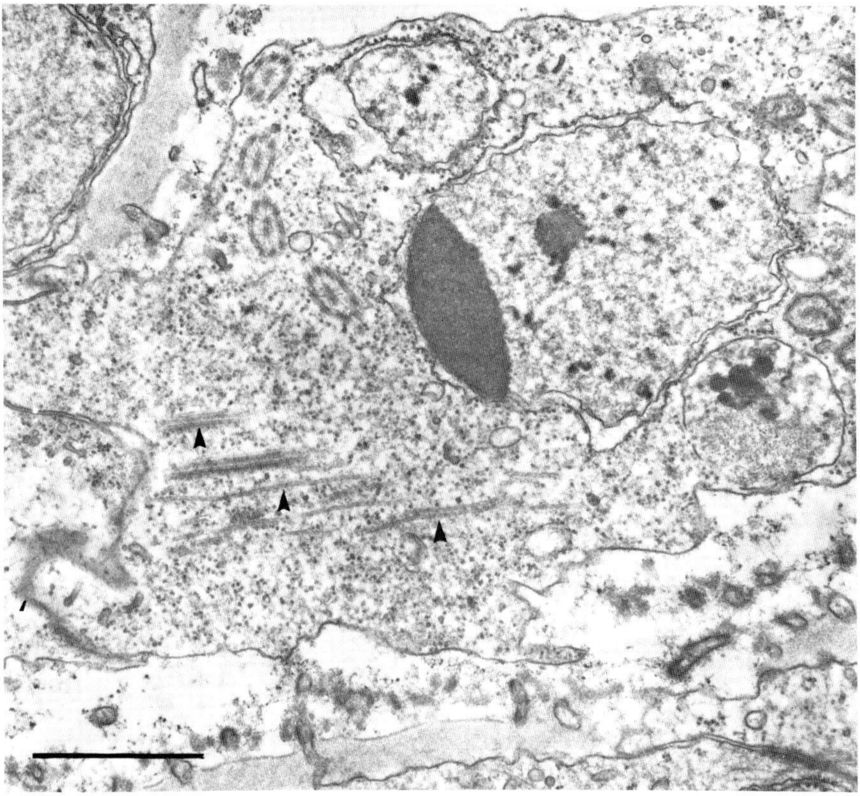

Fig. 3. *Cryptosula pallasiana.* Lophophore regression. A liberated epithelial cell of a tentacle. The nucleus is pyknotic, with dilation of the nuclear envelope. Longitudinal (arrow) and cross sections of ciliary axonemes occur within the cytoplasm Magnification: × 31,000 (scale, 1 μm).

Cross sections of the regressing lophophore consist of seventeen buckled tubes of basement membrane with aggregations of autolyzing peritoneal cells on the inside, attached and detached autolyzing and relatively normal epithelial cells on the outside, as well as organic debris derived from fragmenting cilia and detached cuticle. The tentacles are displaced with respect to one another and in many cases are recognizable only by the irregular profiles of basement membrane that remain.

As these cells regress yet further they develop very irregular outlines, appearing to fragment into smaller pieces; in sections of the regressing lophophore mass many profiles of apparently small cells are encountered. Either they are cell fragments or profiles of irregular cell contours.

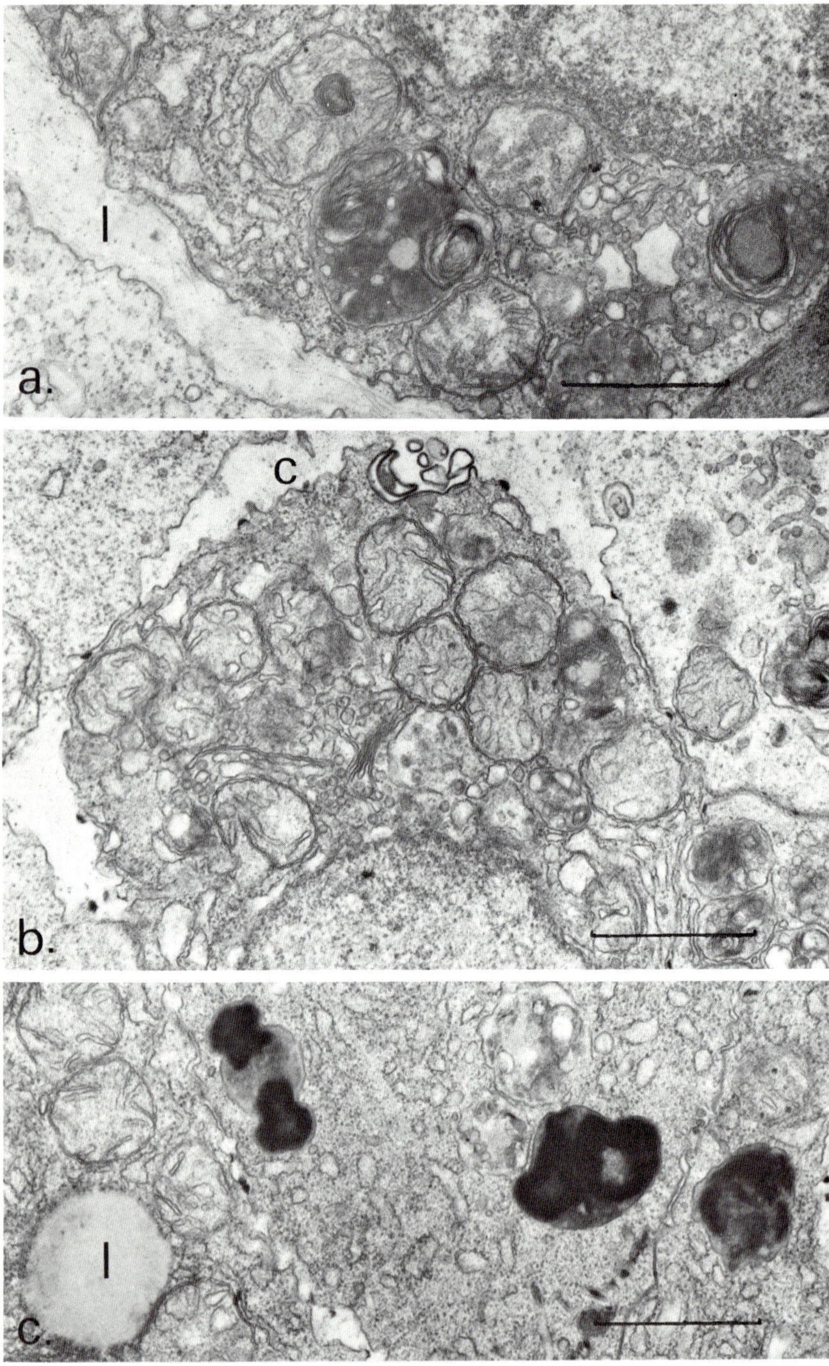
l
a.
c
b.
l
c.

These changes take place over a period of days during which the regressing lophophore condenses by autophagy, heterophagy, and concomitant cell movement. Sections of the condensing mass demonstrate a marked increase in the number of cells that show what Bessis (1964) terms death agony [characterized by the appearance of pseudopods (if liberated) and/or edema of cytoplasm, nucleus, and organelles (Figs. 2, 3, and 5)]. Large numbers are also definitely necrotic (involving nuclear pyknosis and/or karyorrhexis and formation of myelin threads or figures) (Fig. 6). Very few show many signs of normality in late regression and those that do are the former embryonic cells of the lophophoral base (Gordon, 1973b) which are the last to succumb.

Some of the cell debris which occurs in the center of the regressing mass (Fig. 7) is derived from ruptured necrotic cells, although much of the debris becomes incorporated into regressing lophophore cells by heterophagy (Fig. 8a). Some of the ciliated tentacle cells which become liberated behave as amoebocytes, developing pseudopodial processes, but not appearing to be macrophagic. These cells are identifiable by the ciliary rootlets they contain, as well as naked axonemes (Fig. 8b). These axonemes are withdrawn into the cell as though being injected. This phenomenon of ciliary resorption is also known in reorganizing cells of certain protozoans (Paulin, 1973). Within the cells of *Cryptosula* the axonemes break apart, giving random microtubular profiles in the sections. As well as nuclear pyknosis, homogenization and karyorrhexis occur.

The basement membrane maintains its structural integrity for some time. At the beginning of lophophore regression it is buckled and bent but this accordion-like folding normally occurs whenever the lophophore is retracted, in order to accommodate its volume in the cystid. What actually causes the breakdown of the basement membrane during degeneration is not known. Possibly, liberated hydrolytic enzymes accomplish this function as is known to occur during anuran metamorphosis (Eisen and Gross, 1965). At any rate, it does collapse and although some of it becomes reduced to a fine, fibrous form extracellularly, especially in *Hippothoa,* some is incorporated by heterophagy (Fig. 6). At some time during condensation of the lophophore the typical 60-nm nodular periodicity of the collagen is lost and infraperiods of 13–14 nm become apparent.

Fig. 4. *Cryptosula pallasiana.* Lophophore regression. (a) A cell from a lophophore with lytic vacuoles and a mitochondrion containing a small myelin figure (1, basement membrane). Magnification: × 22,400 (scale, 1 μm). (b) Cells from the lophophoral base in varying states of regression. Mitochondria show incipient myelinization and have lost their smooth contours while the nucleus is rather homogeneous (an alternative feature of necrosis) (C, remains of cuticle). Magnification: × 22,800 (scale, 1 μm). (c) Cells from the lophophoral base in early regression, with a lipid droplet (1) and dense residual bodies. Magnification: × 23,000 (scale, 1 μm).

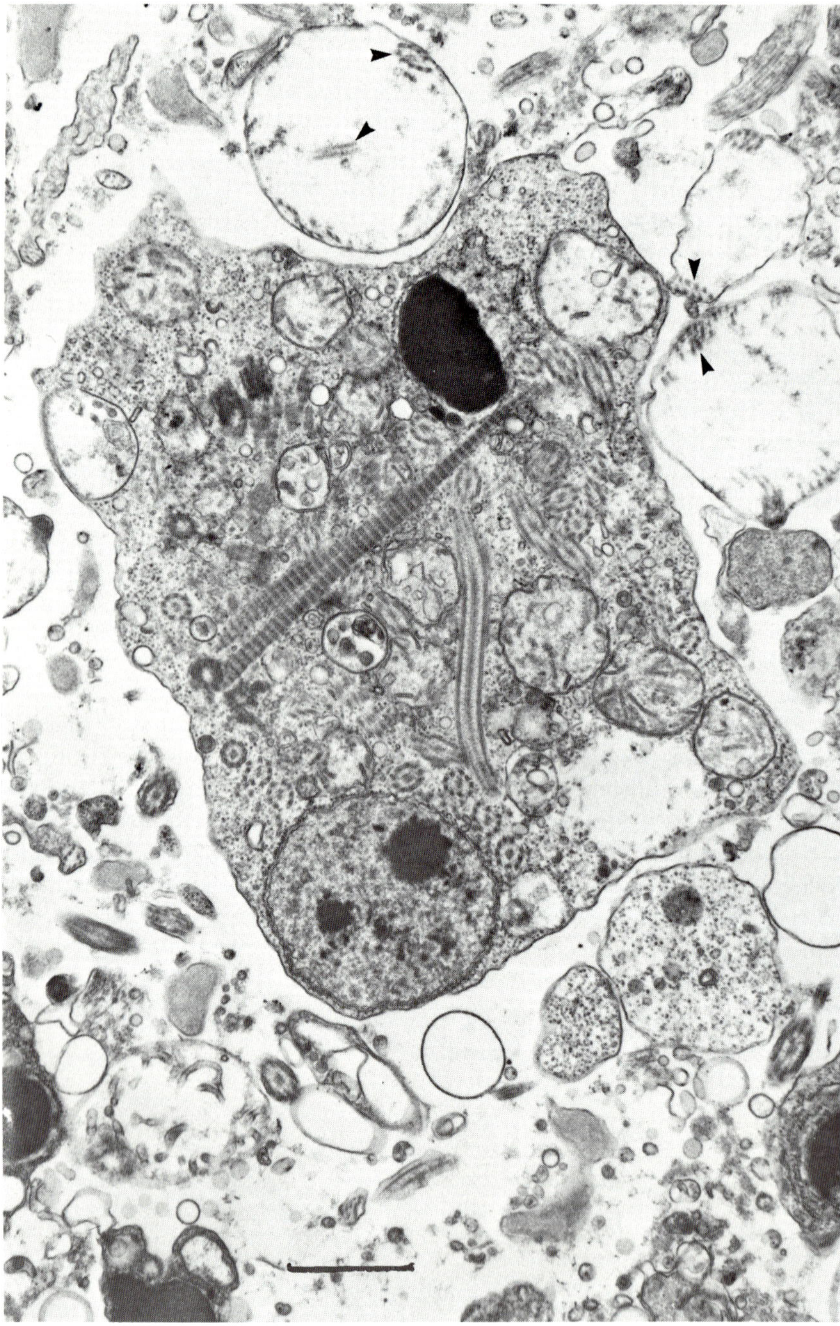

B. Muscle Regression and Phagocytosis

1. *Muscle Regression*

In *Cryptosula* muscle cells in the tentacles and lophophore base shorten and in sections of the regressing mass they are seen to be balled-up with filaments lying in all directions. The balled-up cells may then be incorporated into others by heterophagy or, alternatively, may regress further through autophagy. In muscle cell cytoplasm lipid droplets and vacuoles containing membrane lamellae (Fig. 9), comparable to some described by David (1970, pp. 113, and 149) in other organisms, may appear. Filaments cease to lie in ordered arrays (Gordon, 1973b).

In *Alcyonidium polyoum,* muscles and basement membrane persist for some time while the tentacles are regressing (Matricon, 1960). According to Marcus (1926a) occlusor and parietal muscles remain intact in *Electra pilosa* during regression but are replaced when a new polypide is regenerated, although the lophophore retractors share the fate of the lophophore. In *Watersipora cucullata* the parietals do not degenerate (Mawatari, 1952). Calvet (1900), Rey (1927), and Matricon (1960) agree that regressed muscle does not become part of the brown body proper, but is digested by cells called sarcolytes. According to Matricon, sarcolytes are found only in old zooids, are derived from mesenchyme, and are responsible for phagocytosing musculature. They, in turn, may fuse and become necrotic.

It is not at all certain that sarcolytes are involved in *Cryptosula.* Many of the cells that behave as macrophages during regression appear to be derived from lophophoral cells, where, during some stages of regression, ingestion of cells by neighbors can take place. It is likely that coelomic cells could be found among the regressing cells but it would be difficult to identify them as such as their fine structure has not yet been characterized, and they themselves can undergo regression and necrosis (Matricon, 1960; Bobin and Prenant, 1972). The latter authors have described five types of coelomic cells as seen by light microscopy—vacuolated amoeboid cells, cells with refringent granules, cells with peripheral inclusions, ampoule cells, and glycogen cells in stellate clusters. These cell types were said not to be common to all bryozoans. Sarcolytes are placed in the category of amoeboid cells, and only amoeboid cells are thought to remain after polypide regression. A number of cells within the regressing lophophoral mass contain muscle filaments in varying stages of disarray (Fig. 10). Since the filament aggregates are not enclosed in vacuoles they clearly belong to these cells

---

Fig. 5. *Cryptosula pallasiana.* Lophophore regression. A liberated epithelial cell containing axonemal profiles, rootlets, two nuclei or parts of a former nucleus (one pyknotic), and edemic mitochondria. Outside the cell are membranous formations derived from shed cilia (arrows indicate ciliary tubules). Magnification: × 17,100 (scale, 1 μm).

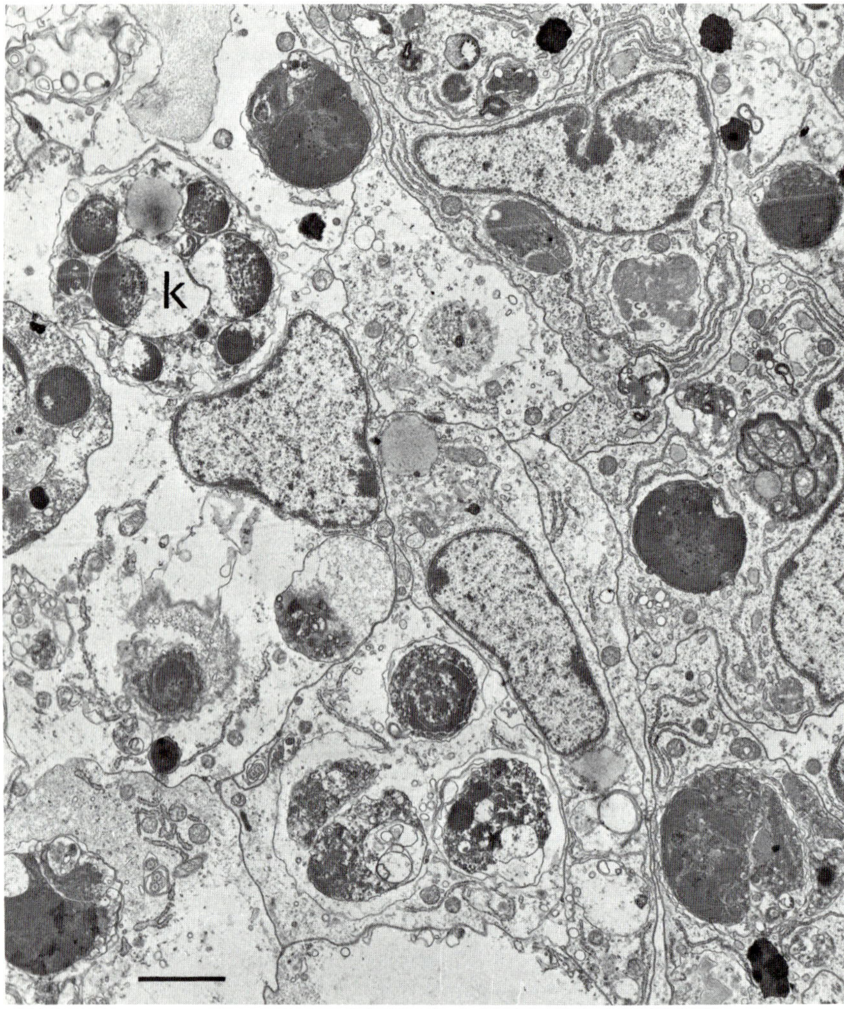

Fig. 6. *Cryptosula pallasiana.* Lophophore regression. Regressing and necrotic cells at a tentacle base, showing a karyorrhexic nuclear cluster (k) in one herniated cell, and islets of basement membrane in other cells (arrows). Magnification: × 5550 (scale, 2 μm).

which must represent former muscle cells, possibly fragmented, as in the case of retractor muscles.

In some of these cells muscle filaments undergo transformation into highly ordered overlapping arrays of helical filaments (Fig. 11a). More compact hexagonal crystalloids of apparently related structure were encountered in the parietal muscle cells which run from the floor of the compensation sac to the basal wall of the cystid. The diameter of the filaments in the loose cells and in the parietal muscle is the same (5–9 nm) and it seems likely that the images seen in the loose cells represent longitudinal profiles of incipient crystalloids, while those seen in the parietal muscle are transverse profiles of more fully formed crystalloids. In transverse sections the rows of filaments are set at angles of 59–64° with respect to one another (Fig. 11b). The tilt angle of the helices in longitudinal section is 34–36°.

The crystalloids are of some interest. Virtually identical crystalloids were found in socket muscle of the entoproct *Barentsia gracilis* (Reger, 1969). In *Barentsia* the filaments comprising the crystalloids were 3–7 nm in diameter with adjacent rows in transverse section set at angles of 58° with respect to each other and with a helical tilt angle of 35–45°. Such crystalloids are known elsewhere; Reger cites (in his text) amphibian oocytes and embryonic cells and human muscle tumor cells. Entoprocts are known to exhibit cyclic degeneration and regeneration of the calyx (Harmer, 1886; Prouho, 1892). Reger did not know the age of his material, but hypothesized that the crystalloids represent a store of muscle proteins prior to their deposition. In his micrographs (Figs. 6 and 7 in his paper), however, early degenerative changes are evident and I am inclined to the view that such crystalloids are a feature of regressing or pathological muscle cells, not of regenerating ones.

2. *Phagocytosis*

Phagocytosis has been said to play an important role in the dissolution of the lophophore, rectum, and pharynx. According to Römer (1906), those parts not entering into the brown body are phagocytosed. Calvet (1900) thought that phagocytes had an excretory function, an idea further supported by Harmer (1931), who observed that certain coelomic cells take up dyes added to seawater. Matricon (1960) stated that the tentacle sheath plays a phagocytic role as it contains sarcolytes. Marcus (1926a), Rey (1927), and Cori (1941) stated that phagocytes laden with residues “disappear” or “degenerate.” According to Rey, phagocytosis of muscle only, by sarcolytes, occurs. Phagocytosis is important in

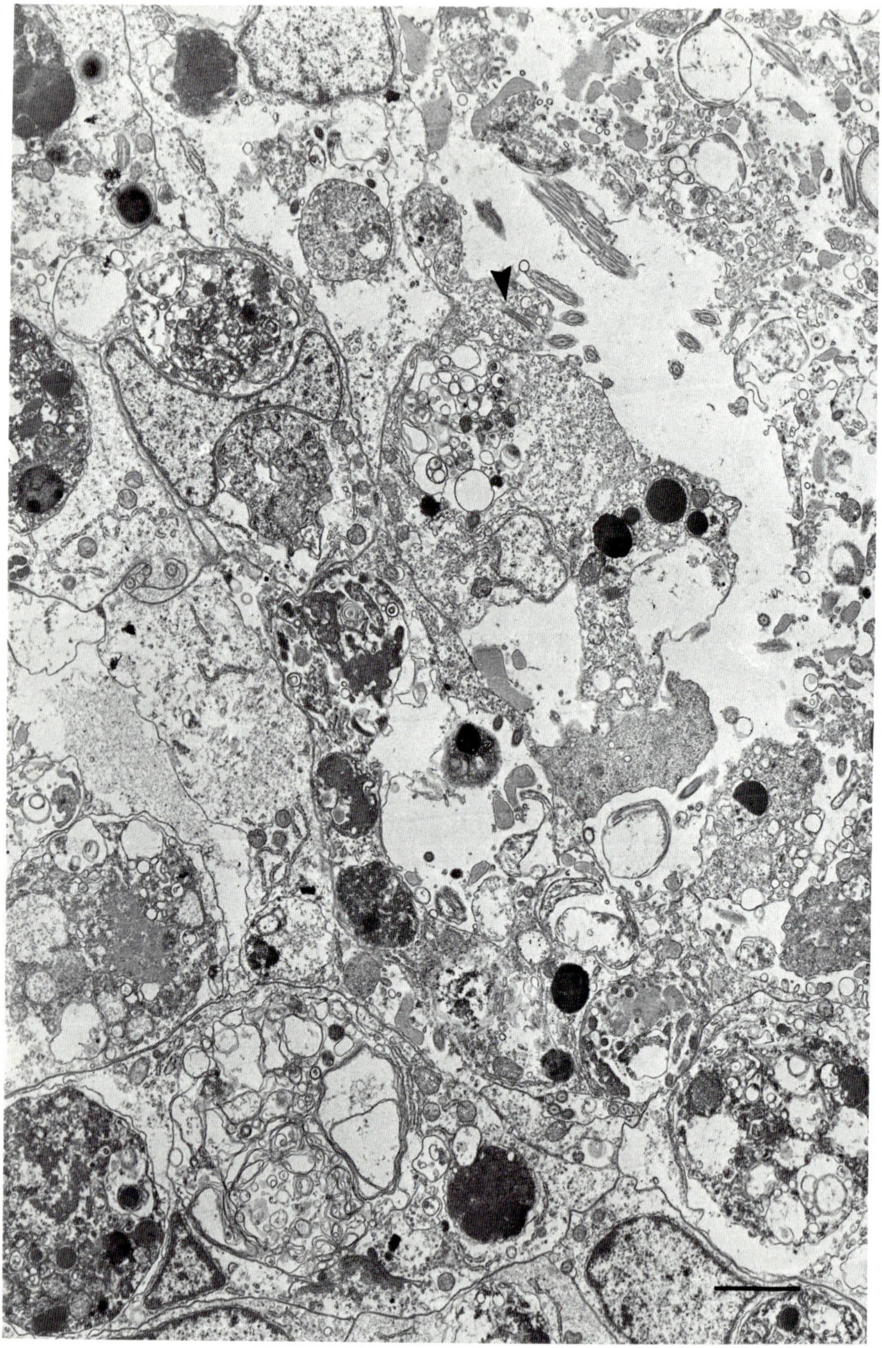

Cryptosula in the condensation of the lophophore and indeed, of the rest of the polypide, but as previously stated, this may not involve special coelomic cells. Cells containing pigmented residue appear throughout the mesenchyme of the degenerating zooid but sometime after a new polypide has been regenerated they are less apparent (Gordon, 1973b).

C. The Regressing Gut

Different parts of the gut regress at different rates. The rectum and pharynx regress concurrently with the lophophore, while the stomach, albeit full of orange-brown inclusions (secondary lysosomes and residual bodies), maintains its integrity slightly longer.

Early regression of the pharynx is marked by increasing density of vacuolar fluid due to the appearance of small granules and membranous structures (Fig. 12a). There is minor blebbing of the nuclear envelope, endoplasmic cisternae become converted to small vacuoles with concomitant liberation of ribosomes, and muscle fibrils become disorganized (Fig. 12b). Myelin figures are formed and the cuticle sloughs off.

In the rectum, products of absorption that accumulate during the life of the polypide are a prominent feature. These comprise vacuoles containing homogeneous contents which may be polysaccharide. Renieri (1970) demonstrated the polysaccharide nature of the matrix between the microvilli of the cecum of *Bugula neritina,* and in *Cryptosula* the matrix is apparently the same from cecum to rectum. Other vacuoles (secondary lysosomes) contain membranous lamellae, dense bodies, and granular material (Fig. 13a). Lipid droplets occur but are not abundant. Regular cytoplasmic changes seen in other regressing cells occur in the rectum. Necrosis leads to the loss of practically all membranes including those around residual bodies.

The entire stomach between the cardiac and pyloric sphincters regresses as a unit. The orange-brown inclusions that have accumulated and enlarged during the feeding life of the polypide come to constitute a major part of the brown body proper and it is these that, through further condensation and digestion, give the dark brown coloration to this body. Even prior to regression myelin figures are commonly encountered at the bases of cells of the cardiac stomach. These may give rise to at least some of the residual bodies that resemble lipofuscin seen near the cell apex in early regression (Fig. 13b). Lipid droplets

---

Fig. 7. *Cryptosula pallasiana.* Lophophore regression. Some cells from the center of the regressing lophophore mass. Here there is much debris from ruptured cells, though some of this is incorporated into others. A former ciliated tentacle cell can be recognized by the resorbed axoneme (arrow). Magnification: × 5150 (scale, 2 μm).

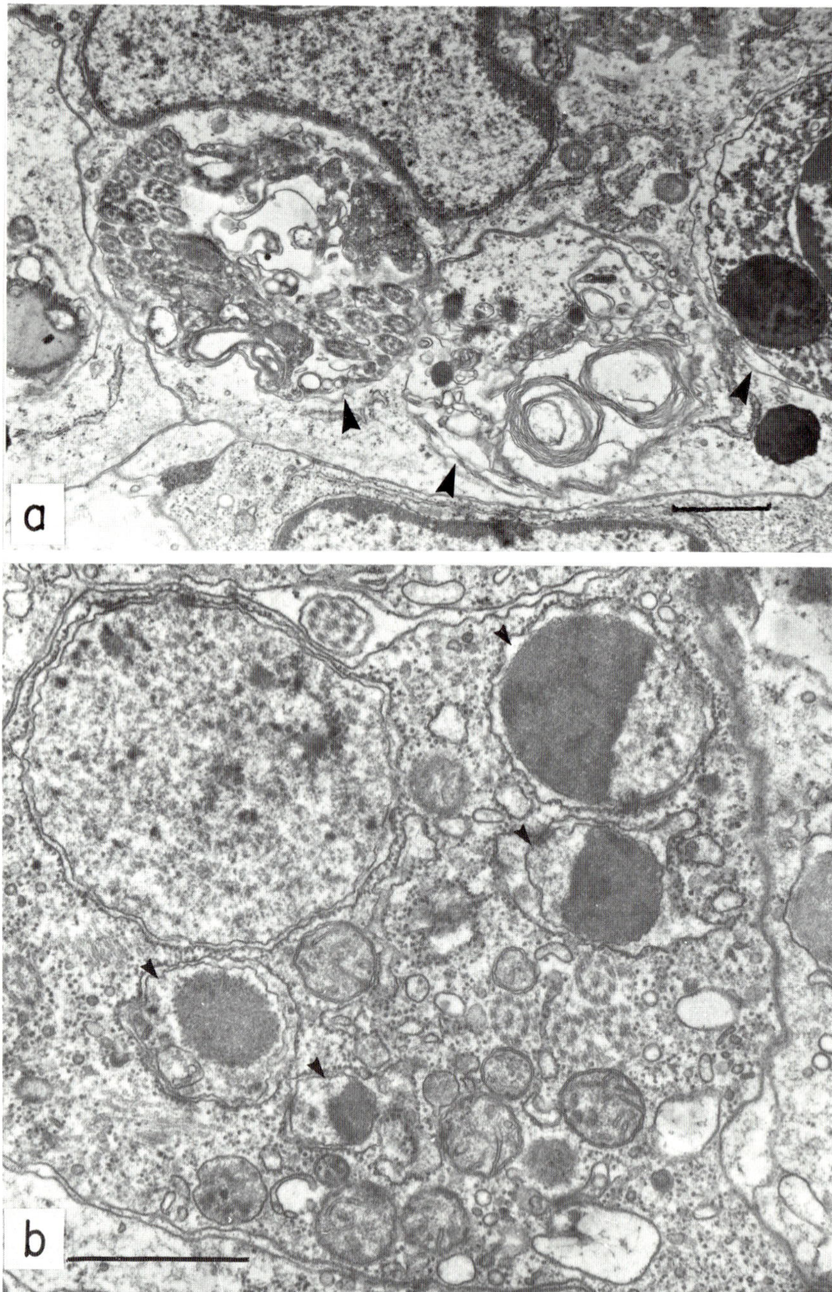
a
b

are also encountered. Endoplasmic cisternae are converted into small vacuoles and mitochondria lose their smooth contours.

Lipid also occurs in regressing cecal cells (Fig. 14). Here the secondary lysosomes that develop during the feeding life of the polypide are huge, up to 8 μm in diameter and twice the size of the nucleus. It is clear from the large numbers of pyknotic and karyorrhexic nuclei in heterophagosomes that condensation of the stomach occurs through heterophagy of some cells by neighbors. Autophagy also takes place. Actively secreting Golgi bodies are seen in regressing cells; most of these produce both small, clear vesicles and macrovesicles containing fibrous elements, but the function of these secretions is unknown (Gordon, 1973a). Mitochondria are irregular or small and rounded with few slender cristae. As noticed in all regressing tissues so far, ribosomes become liberated as endoplasmic membranes are lost. Depending on stomach contents at the time regression was entered into, food fragments such as diatom frustules and unidentifiable skeletal fragments derived from either intracellular material in the cecum or residue in the pylorus may be encountered. Ciliary shafts from the surfaces of the ciliated cells of the stomach may be resorbed in the same manner as those of the epithelial cells of the tentacles.

The muscle cells and peritoneum of the stomach regress but the basement membrane remains intact, not as in the regressing lophophore, and forms a kind of limiting membrane around the resulting regression product (brown body.)

D. Comparisons with Other Organisms

As Gordon (1973a) pointed out, regression of bryozoan polypides resembles regression of aging and pathological cells and tissues of other organisms. Regression of lophophoral cells of *Cryptosula* and *Hippothoa* is comparable to that of leukocytes described by Bessis (1964). During the phase of death agony, polypide cells exhibit incipient vacuolation of the nuclear envelope and vesiculation of endoplasmic cisternae. Mitochondrial membranes frequently become irregular. Loose cells develop short blunt pseudopodia. These features, as well as nuclear pyknosis and/or karyorrhexis and formation of myelin figures, tend to be enhanced during necrosis of cells.

Fig. 8. *Cryptosula pallasiana*. Lophophore regression. (a) Some of the ciliary shafts that are shed are picked up into heterophagosomes, along with other debris, by macrophagic activity of some of the regressing lophophore cells. Arrows indicate the boundaries of three heterophagosomes. Magnification: × 13,200 (scale, 1μm). (b) An epithelial cell with remains of a fragmented nucleus or nuclei. One part is homogeneous in content, four (arrows) contain condensed chromatin. Perinuclear cisternae are typically swollen. Magnification: × 25,00 (scale, 1 μm).

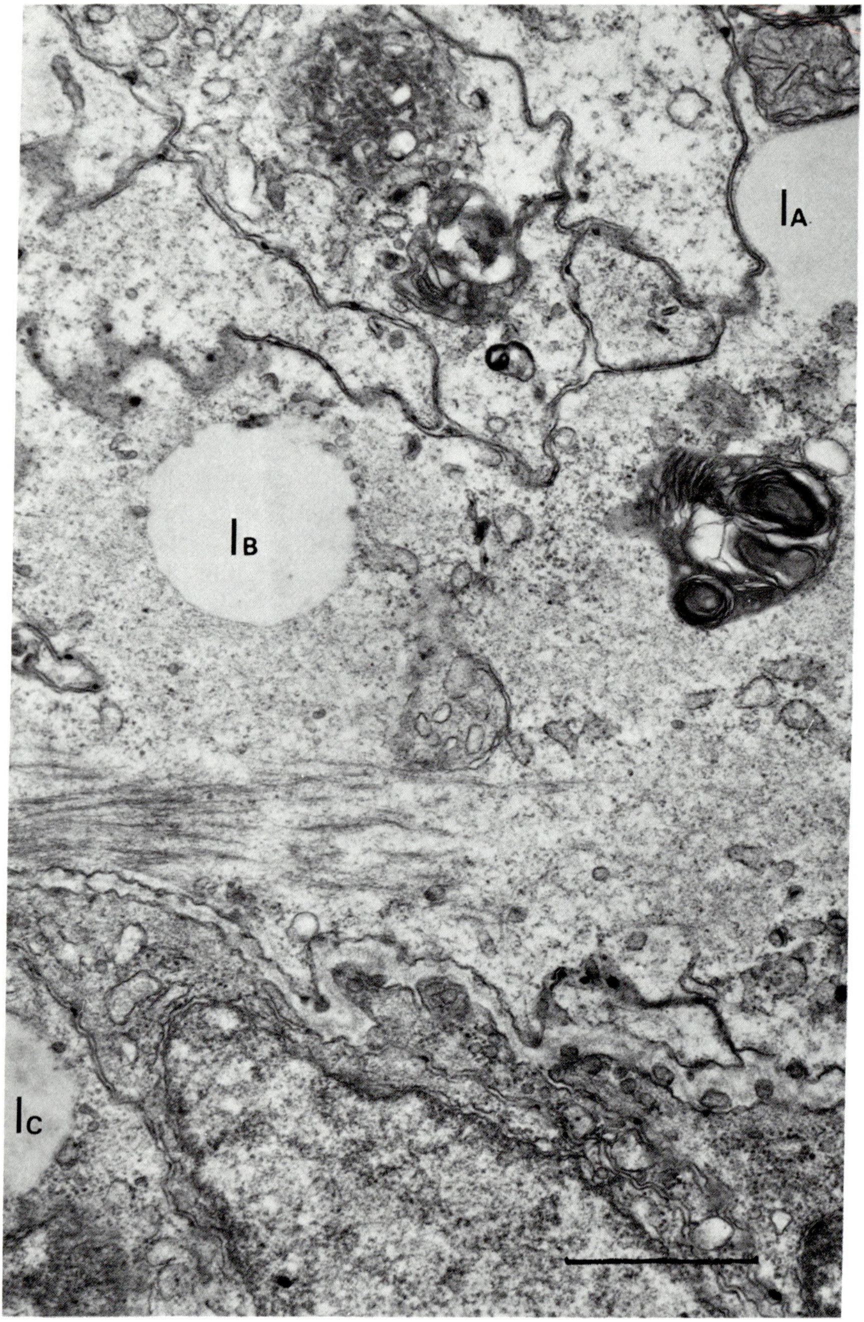

I_A
I_B
I_C

For the death and necrosis of myocardial cells, Manasek (1969) has reported comparable changes, including cells breaking free from neighbors, liberation of ribosomes, occasional occurrence of intracellular glycogen (also seen in *Cryptosula*), enlargement of perinuclear cisternae, and nuclear pyknosis. Some of these features of regression and residual body formation seem to be universal in aging or pathological tissues and have been reported by numerous authors from a variety of organisms including plants under different circumstances (q.v. David, 1970; Berjak and Villiers, 1972).

Mitochondrial swelling has been cited as a feature of cell regression by Bessis (1964), David (1970), and Payen (1972), although Bonneville (1963) reported no marked change during regression of gut cells during bullfrog metamorphosis. Enhancement of swelling can be induced in *Cryptosula* by imprecise fixation, so one must be careful in interpreting regressing cells. Dilation of endoplasmic cisternae is common in regressing or malfunctioning tissues and has been reported by numerous authors (David, 1970, p. 79) and by Geuze (1970) in involuting anuran pancreatic cells.

The behavior of lophophore epidermal cells as macrophages, during regression in *Cryptosula,* is comparable to that of epitheliomuscular cells which may engulf other cells in regressing hydranths of *Campanularia* (Brock, 1970; Fig. 14—Brock's interpretation is different, however). Destruction of a cell in animal tissues is generally intracellular, either by autophagy within its membrane or by heterophagy within neighboring cells should the former become phagocytosed. This is the case in *Cryptosula,* particularly in the stomach and (with the persistence of the basement membrane) accounts for the compactness of the regressing stomach. The lophophore tissues, not bounded by a basement membrane externally, appear more disorganized, especially in *Hippothoa,* and although some cells may lyse in the regressing lophophore, their cellular debris is phagocytosed by others. The chief features of regression are summarized in Fig. 15.

E. The Brown Body (Regression Product)

Continued condensation by autophagy and heterophagy of the regressing polypide results in one or two brown bodies depending on whether the regressed lophophore, pharynx, and rectum detach from the regressed stomach. In

---

Fig. 9. *Cryptosula pallasiana.* Muscle regression. Part of a circular striated cell from around the mouth, showing fibrils no longer organized into regular sarcomeres (the normal condition), and a homogeneous cytoplasm characterized by the paucity of sarcoplasmic membranes. Large lipid droplets ($1_{A,B}$) occur in the muscle cell along with a myelin figure. Another lipid droplet (1_C) is seen in a neuron under the basement membrane. Magnification: × 26,200 (scale, 1 μm).

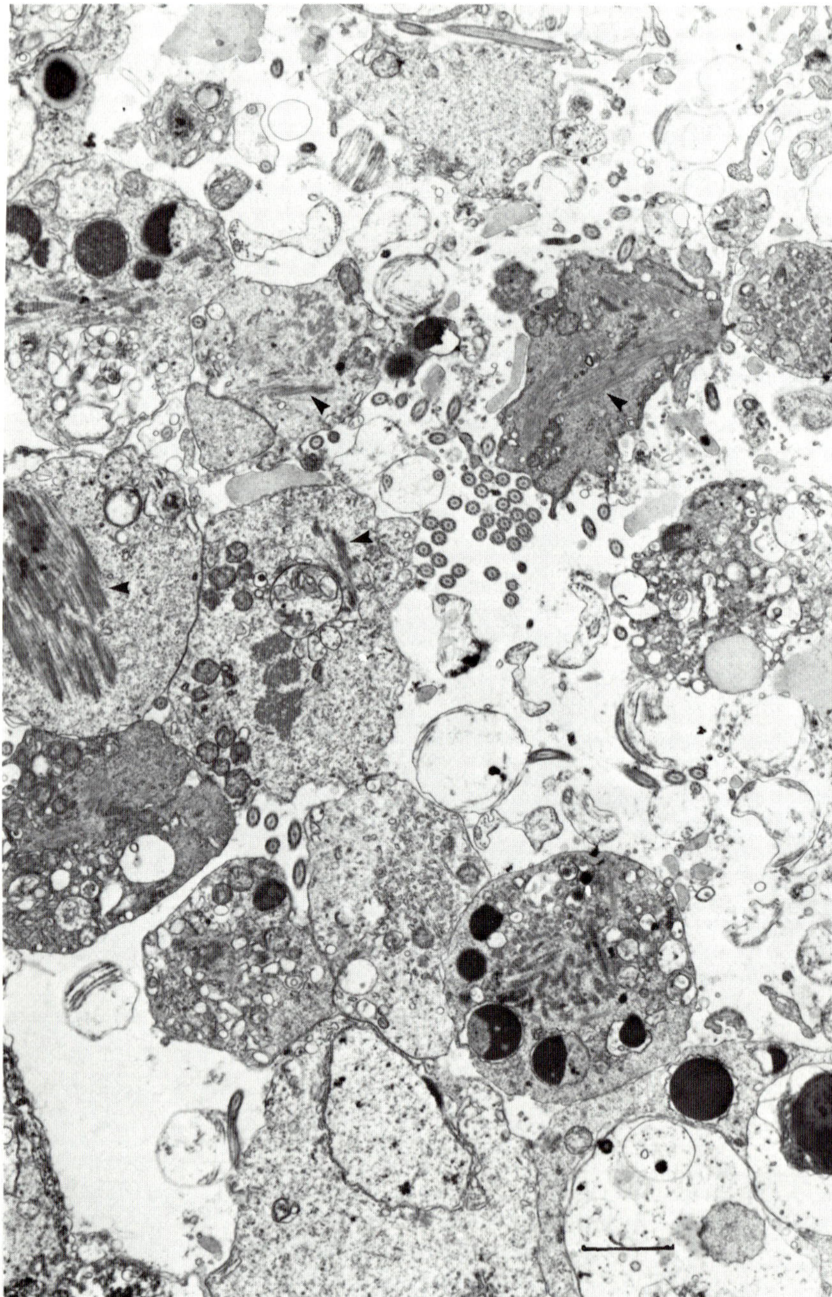

essence, the brown body represents a huge residual body. Actually, it is made up of large cells, each comprising a thin halo of nucleated cytoplasm surrounding a mass of residue. The long dimensions of these cells is comparable to that of normal stomach cells (ca. 15–20 μm) but they are up to three times as wide due to heterophagy, and they sometimes take on a polygonal shape. Their cytoplasm comprises a little rough endoplasmic reticulum, mitochondria, Golgi bodies of two types, and nuclei (Fig. 16). The residual bodies include the remains of several smaller bodies (of membranous elements, opaque granules, myelin figures and amorphous aggregates) no longer membrane-bound, in a more or less homogeneous matrix of fine fibrous material.

Degeneration continues even after the formation of a definitive brown body. What little cytoplasm remains disappears eventually, evidently by autophagy, membranes tend to degrade, and the matrix and residual material become progressively less structured. The part of the brown body derived from the lophophore, pharynx, and rectum is in a most degenerate state. Cells with intact membranes contain residual bodies, myelin figures, lipid, whorls, and bizarre membranous structures quite unlike any in the remainder of the brown body derived from the stomach (Fig. 17).

It has long been recognized that the brown body is chiefly composed of the brownish granules that accumulate in the stomach walls (Joliet, 1877; Harmer, 1891; Prouho, 1892; Marcus, 1926a; Rey, 1927; Cori, 1941; Soule, 1954; Matricon, 1960). Food residues may also occur in the brown body, e.g., foraminiferan, radiolarian, and diatom skeletons (Nitsche, 1871; Joliet, 1877; Rey, 1927), as in *Cryptosula,* and Nitsche (1871) and Rogick (1945) saw gizzard sconces in brown bodies of some stoloniferous ctenostomes.

The membrane around the brown body which was noted by Hincks (1880), Römer (1906), and Gerwerzhagen (1913) is seen in *Cryptosula* to be the former basement membrane of the stomach. In addition, peritoneal or mesenchymal cells may surround the brown body (Joliet, 1877; Calvet, 1900; Bobin and Prenant, 1957). There may be some variation in the nature of the limiting boundary of a brown body according to whether it is destined to remain in a zooid or be eliminated via the digestive tract of a new polypide. Dr. W. C. Banta (American University, Washington, D.C.) sent me photographs of presumed brown bodies from two *Schizoporella* species; these bodies possess a rindlike structure which thickens with age (Fig. 18). According to Joliet (1877) a brown body membrane becomes chitinized. Since this presumably would necessitate an enveloping secretory epithelium such an observation is doubtful, but it is clear

Fig. 10. *Cryptosula pallasiana.* Muscle regression. Many of the cells within the regressing lophophoral mass contain degenerating myofilaments and crystalloids (arrows). The origin of some of these cells is uncertain but some are likely to be fragmented retractor muscle. Magnification: × 6480 (scale, 2 μm).

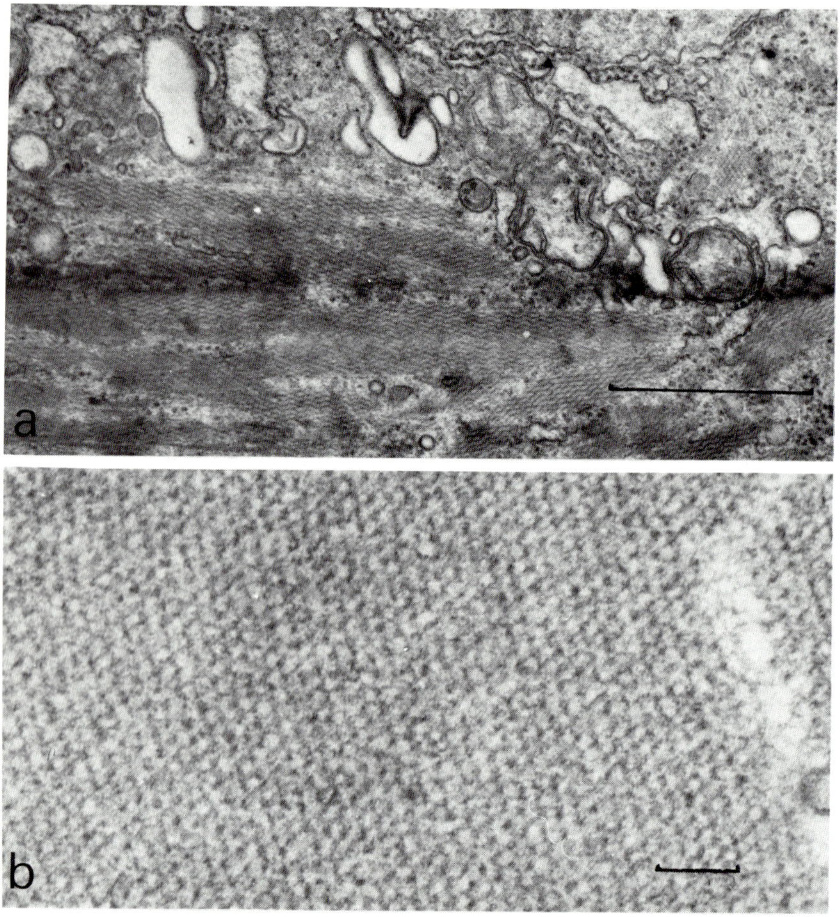

Fig. 11. *Cryptosula pallasiana.* Muscle regression. Muscle filaments are transformed into paracrystalline arrays of filaments with a helicoid disposition, sectioned longitudinally in (a) and transversely in (b). Magnification: (a) × 28,300 (scale, 1 μm); (b) × 110,000 (scale, 100 nm).

Fig. 12. *Cryptosula pallasiana.* Gut regression. (a) The large central vacuoles of the pharyngeal myoepithelial cells become dense and filled with particulate material. Endoplasmic cisternae are transformed into small vacuoles and muscle filaments become dissociated. (b) Lytic vacuoles occur in regressing pharyngeal cells also. Magnification: (a) × 25,300; (b) × 25,900 (both scales, 1 μm).

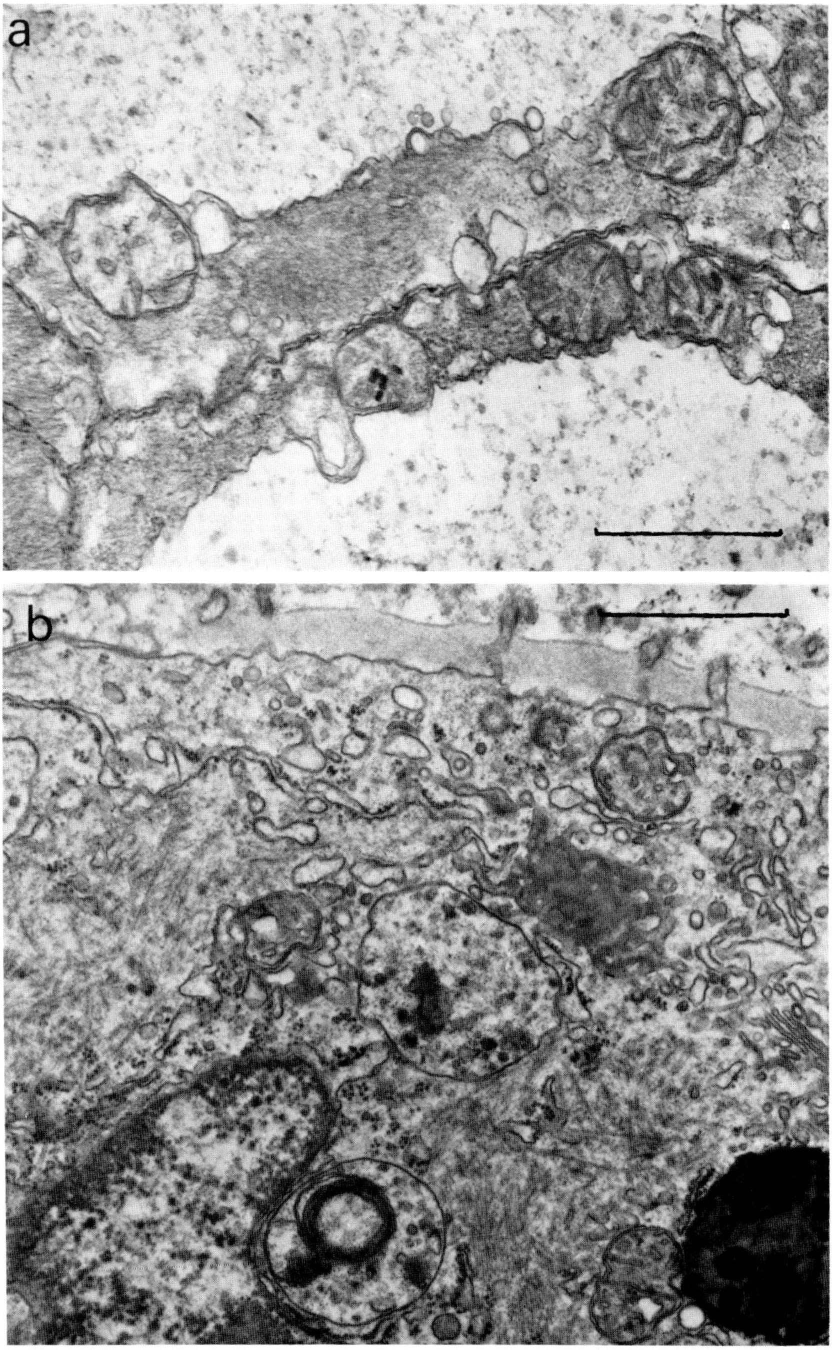
a
b

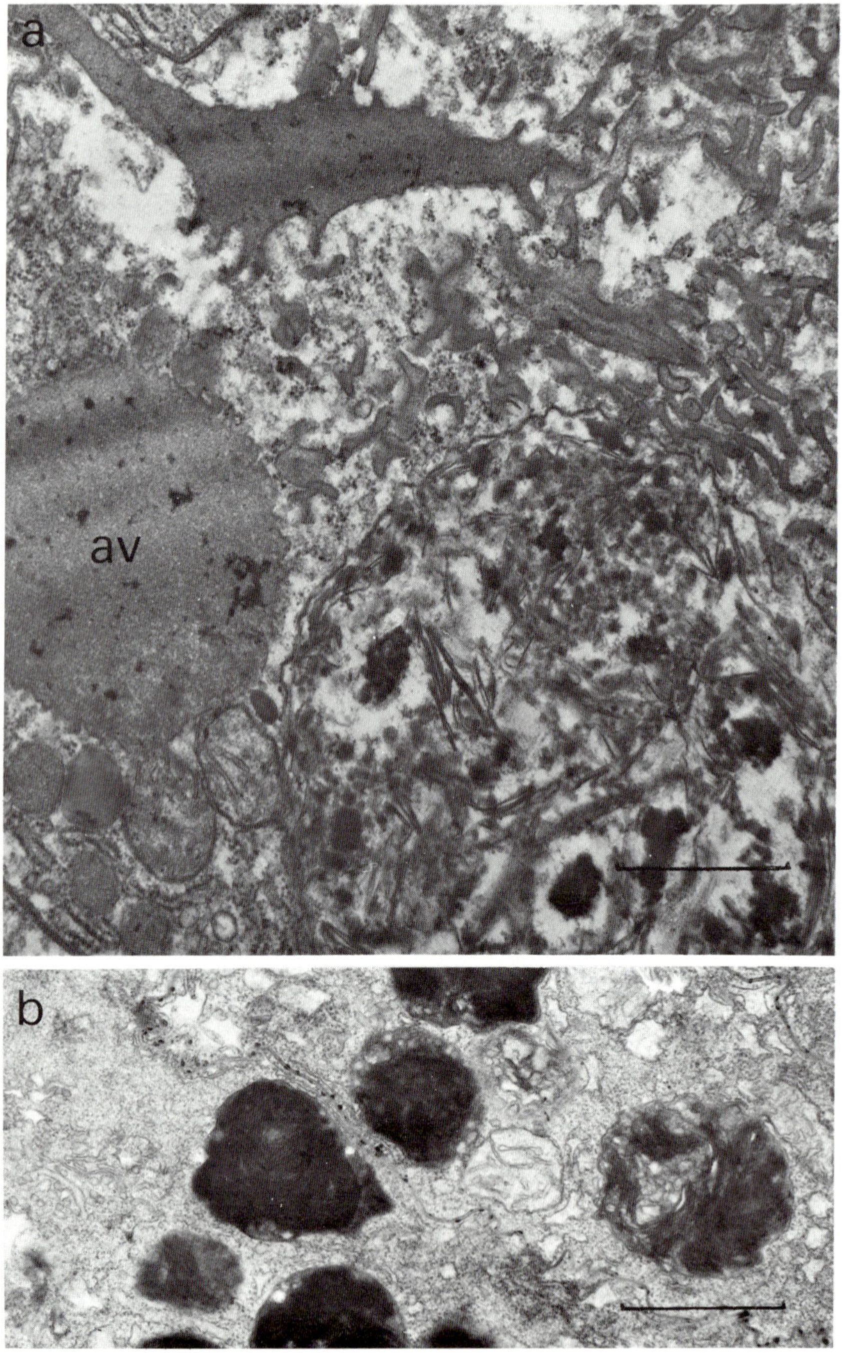
a
av
b

that an electron microscopic investigation of the thickened rinds in *Schizoporella* would be instructive.

F. Glycogen Deposition

During regression, rosettes of α-glycogen may accumulate, sometimes in massive amounts, in cells of the mesenchyme and in the epithelium under the floor of the compensation sac (Fig. 19), but their deposition has not always been encountered in sections of regressing polypides of *Cryptosula* (Gordon, 1973a). This may be due to the tearing-off of mesenchyme cells when a polypide is dissected from a zooid prior to embedment. Some mesenchyme cells always adhere but these do not always contain glycogen. Epon sections of undecalcified zooids stained with Schiff's reagent to locate sites of glycogen deposition in the zooid show PAS-positive sites throughout the mesenchyme, both proximally and distally and under the floor of the compensation sac. Although glycogen occurs in mesenchyme cells, these cells may themselves regress. Some lipid can occur with the glycogen.

None of the cells around the fully formed brown body were ever seen to contain glycogen. Glycogen deposition therefore seems to occur mainly during early regression and then ceases, the deposits being utilized for other aspects of metabolism. Bobin and Prenant (1972) described anastomosing clusters of "glycogen cells" which are common at budding sites and can occur during the life of the polypide but are rare or absent in degenerated zooids. In *Cryptosula* the deposits which are found during regression and which do not persist, could serve as an energy supply for a simultaneously developing polypide (Gordon, 1973a).

IV. THE FATE OF THE BROWN BODY

The historical discoveries of the fate of the brown body have been covered in the introduction. Smitt (1865) first illustrated the uptake of the brown body in *Cryptosula* into the cecum of a new polypide but he interpreted the brown body as a germ capsule from which a new polypide arose. Next, Repiachoff (1876) described this process, again in *Cryptosula,* correctly interpreting what he saw. In *Cryptosula* the following occurs (Gordon, 1973b). After regression begins a polypide primordium is seen developing backward from the body wall. The time

---

Fig. 13. *Cryptosula pallasiana.* Gut regression. (a) In rectal cells large membrane- and granule-containing bodies increase in number and size as regression proceeds. In addition, homogeneous vacuoles containing products of absorption of extrinsic material (av) occur. Magnification: × 23,700 (scale, 1 μm). (b) In cells of the cardiac stomach lipofuscin-like granules are characteristic. Magnification: × 22,100 (scale, 1 μm).

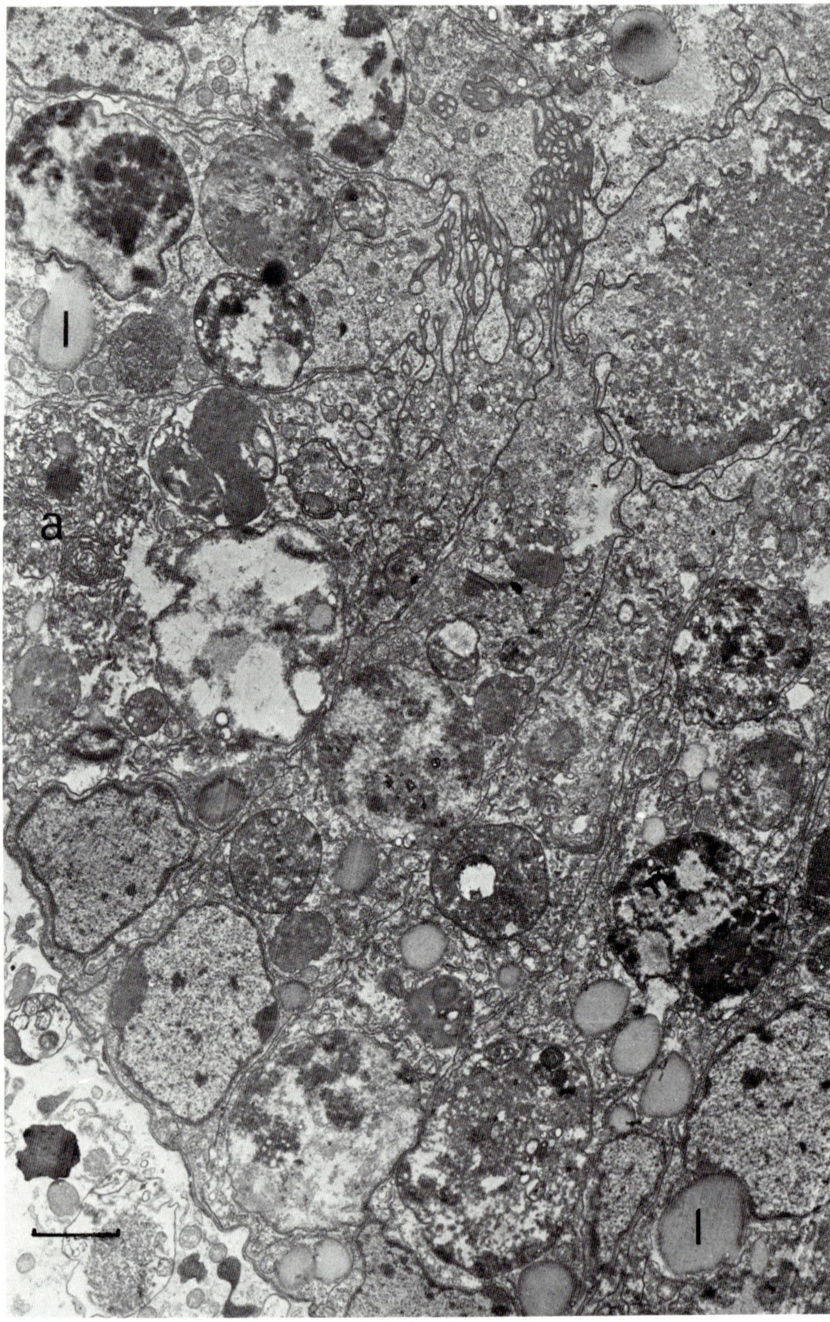
l
a
l

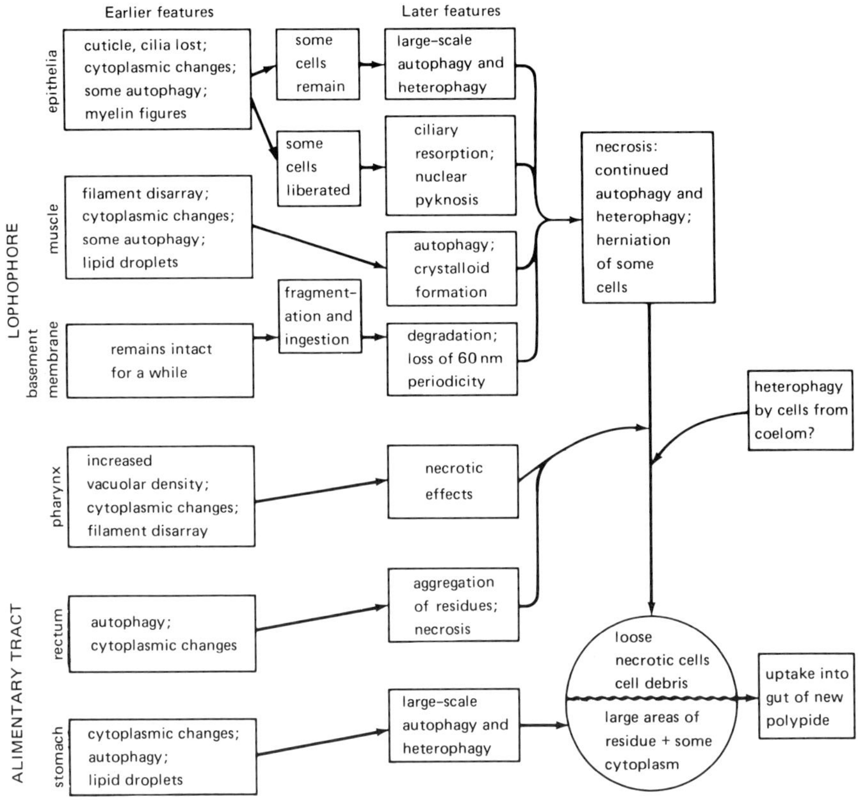

Fig. 15. Summary of events occurring during polypide regression.

of the appearance of the new polypide is related to the rate of regression, such that by the time a brown body is fully formed the backwardly growing primordium encounters the brown body and the gut anlage develops (Fig. 20). If regression is rapid the primordium may appear within 4 days; if regression is slow it may not appear for as long as 15 days. The cecum of the new gut abuts against the brown body and envelopes it by growing over it, and incorporating it into its lumen.

In Fig. 1f the remains of the brown body can be clearly seen in the cecum, pylorus, and rectum. When the polypide evaginates to feed for the first time the remains of the old polypide are defecated.

Fig. 14. *Cryptosula pallasiana.* Gut regression. In the cecum large lytic vacuoles, lipid droplets [two are indicated (1)], and some autophagic vacuoles occur (a). The large vacuoles are secondary lysosomes derived from earlier ingestion of food material. Magnification: × 6080 (scale, 2 μm).

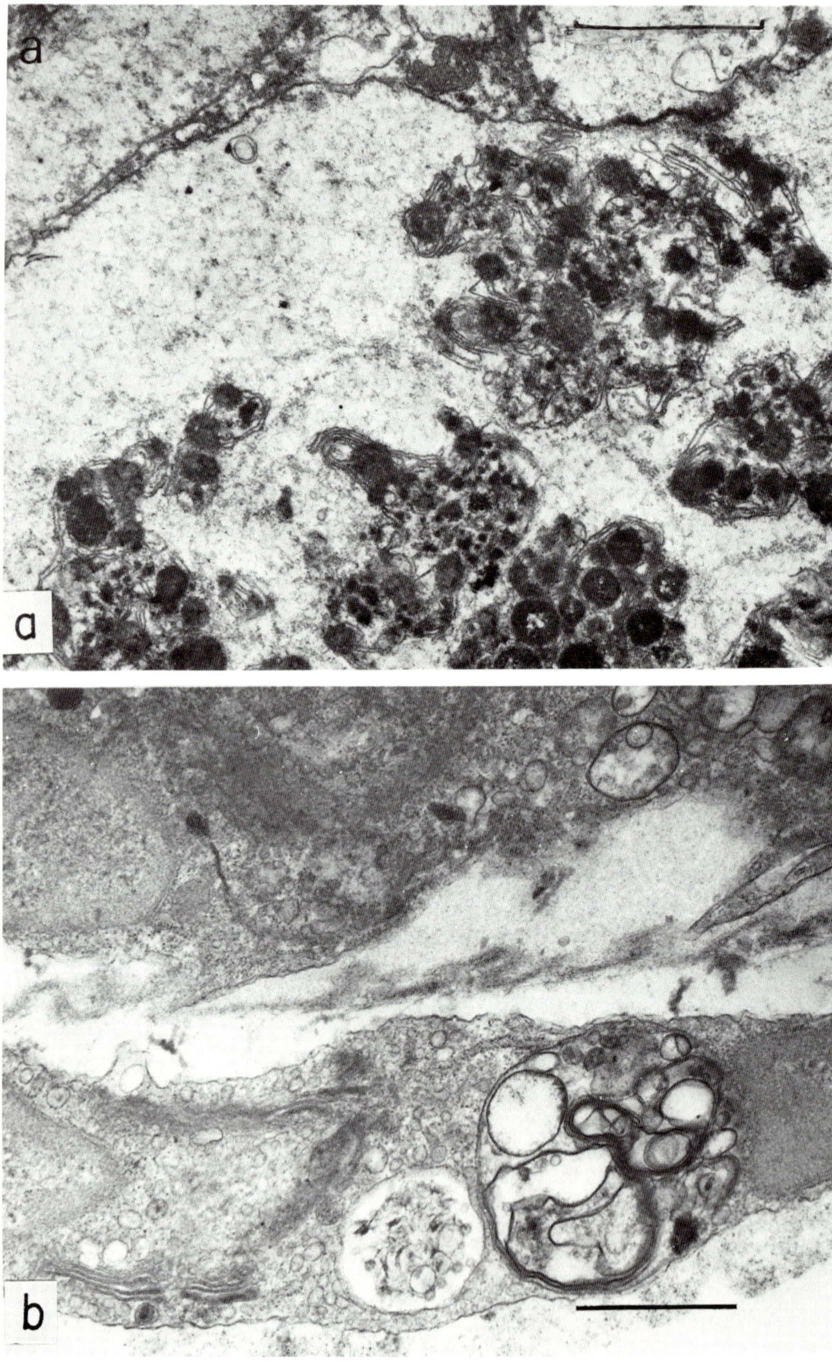
a
a
b

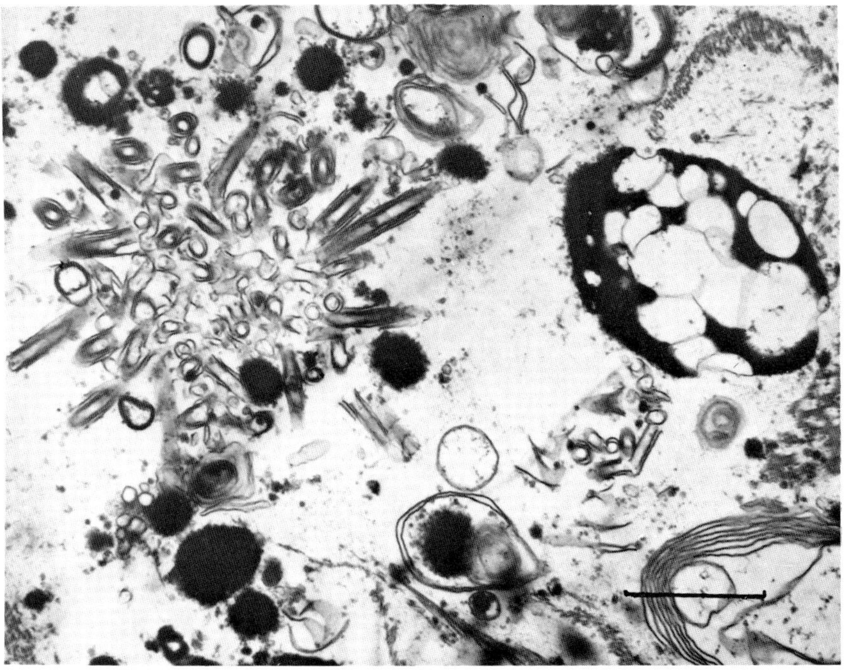

Fig. 17. *Cryptosula pallasiana.* Outside the brown body are seen bizarre formations derived from herniated and regressed cells of the lophophore and peritoneum. Magnification: × 19,400 (scale, 1 μm).

If a new polypide does not grow, the brown body remains in the zooid for some months or until all tissues having disintegrated, the operculum falls off, and ciliates, nematodes, and other organisms enter the dead zoöid (zooecium) and dislodge the brown body.

As mentioned previously, early workers speculated on whether the residue of the old polypide could contribute to the development of the new. Gordon (1973b) has shown that the pigmented inclusions which accumulate in the stomach wall are lipofuscin. As these inclusions accumulate in the old polypide their stainability and fluorescence diminish and disappear. Residual bodies are considered to be totally inert (Cohn and Fedorko, 1969) and it is unlikely that

---

Fig. 16. *Cryptosula pallasiana.* Brown body. (a) Residue comprises membrane lamellae, dense granules, amorphous aggregates and a fine, fibrous material. (b) In some parts of brown bodies are portions of apparently normal organelles [RER, nucleus, Golgi bodies, mitochondria (not shown)] around areas of residue and homogeneous matrix. Magnification: (a) and (b) × 25,400 (scale, 1 μm).

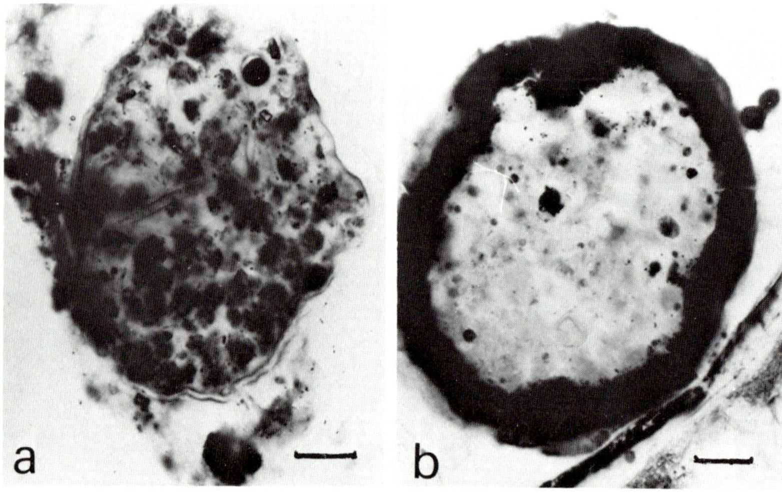

Fig. 18. Epon sections of two apparent brown bodies with thickened "rinds" from *Schizoporella* sp. (photographs courtesy of W. C. Banta, American University, Washington, D.C.). Magnification: × approx. 400 (scale, 20 μm).

they are further digested in the gut of the new polypide. As seen in Fig. 21, however, there may still be some membranous structures in the brown body at the time of incorporation and it is feasible that these and some of the cells from the regressed lophophore are digestible. Revolution of the brown body in the pylorus fragments the residues, allowing greater possibility for absorption of digestible substances.

The new polypide bud in *Cryptosula* does not become evident until some time during advanced regression, but the primordium of the gut has started proliferating by the time regression is more or less complete and envelopment of the brown body by the developing cecum soon takes place. Marcus (1926a) also observed that the polypide bud appears in *Electra pilosa* only after regression is well advanced, but Gerwerzhagen (1913) for *Carbasea carbasea* and Matricon (1960) for *Alcyonidium polyoum* observed that the new bud is evident from the start of polypide regression.

The interaction between dead and living tissues is not clear (Majno, 1964) but in bryozoans it seems that the envelopment of the brown body by the developing cecum is induced by its merely "being in the way," as it were. Bronstein (1938) determined that it is the absence of a polypide from the body cavity that initiates the formation of a new one, rather than a possible "necrohormone" from the residue, suggesting that a polypide exerts an inhibitory effect (hormonal?) on the proliferating tissues of the cystid.

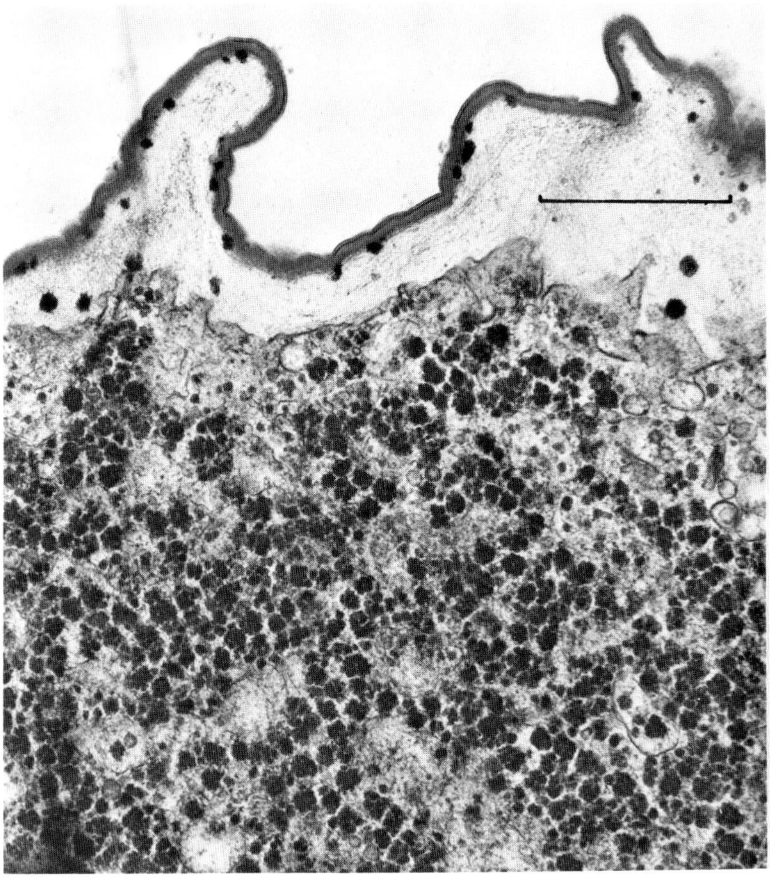

Fig. 19. *Cryptosula pallasiana.* Under the floor of the compensation sac during regression large amounts of α-glycogen are deposited. Magnification: × 26,500 (scale, 1 μm).

V. DISTRIBUTION OF CYCLIC REGRESSION AND REGENERATION

This review has been concerned primarily with the regression phenomenon as it occurs in bryozoans of the class Gymnolaemata. Regression occurs in the two remaining classes Stenolaemata (Ellis, 1755; Harmer, 1898; Borg, 1926) and Phylactolaemata (Marcus, 1934), although brown bodies are definitely known to occur only in the former (but see Harmer, 1931; p. 147). Presumed brown bodies have also been reported from fossil stenolaemates of the order Trepostomata (McKinney, 1969; Boardman, 1971).

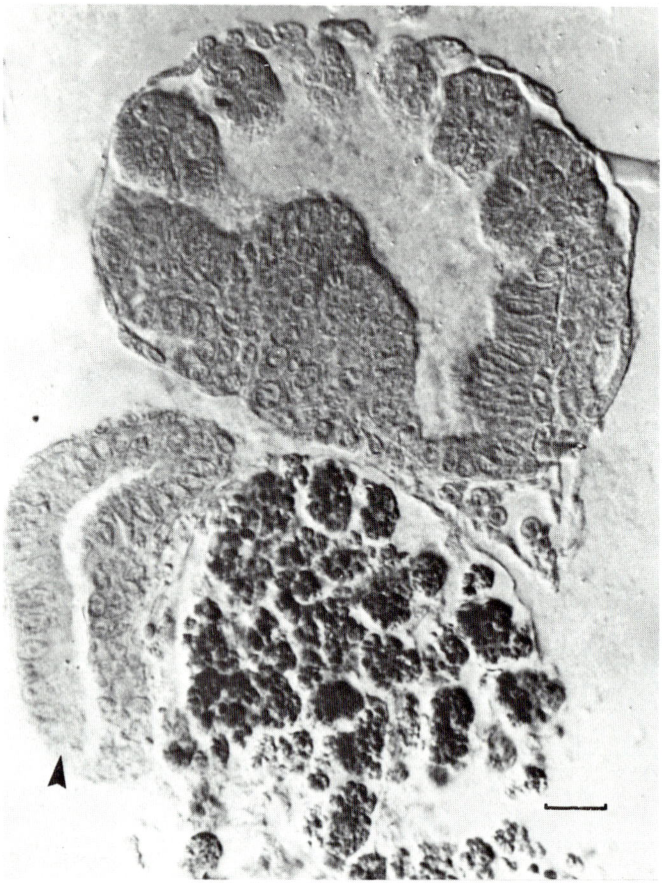

Fig. 20. *Cryptosula pallasiana.* Fate of the brown body. The gut anlage (arrow) of a new polypide is seen beginning to grow alongside a brown body. Above the brown body the lophophore is differentiating (interference micrograph of a paraffin section). Magnification: × 425 (scale, 20 μm).

Among entoprocts the calyx is periodically shed and renewed and Harmer (1886) saw in this a parallel to regression and regeneration in bryozoans. Prouho (1892), however, argued cogently against this point of view, with the facts that the entoproct calyx is not equivalent to a bryozoan zooid, guts are not replaced, and entoprocts possess protonephridia (if lack of excretory organs is a reason for regression). Furthermore, the entoproct larva does not typically lose its gut during metamorphosis (see also Nielsen, 1971 and Chapter 16 of this volume), whereas cyphonautes' gut regresses. The residue of the gut in the cyphonautes

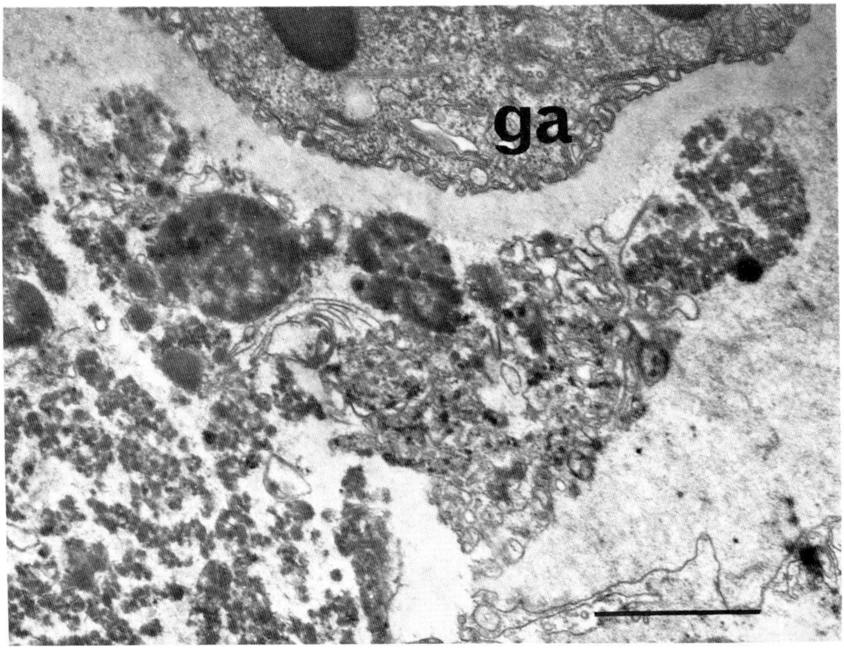

Fig. 21. *Cryptosula pallasiana.* Fate of the brown body. Section of part of a brown body in the process of being enveloped by a new polypide. The interface between a brown body and the differentiating gut anlage (ga) is a structureless matrix. Within the brown body residue some membranes are still present. Magnification: × 23,000 (scale, 1 μm).

larvae of bryozoans is taken into the gut of the ancestrular polypide, foreshadowing later regenerative events (see also Repiachoff, 1875). Prouho is correct in being wary of carrying homologies too far. The great similarities in cyclic regression and regeneration that exist in different phyla is almost certainly related to the colonial way of life, if for no other reason. In some bryozoans entire zooids may drop off after polypide regression, e.g., *Triticella* (Sars, 1874), *Zoobotryon* (Bullivant, 1967), and *Mimosella* (Banta, 1968). Similarly, *Campanularia* hydranths are periodically shed and regenerated (Crowell, 1953). "Brown bodies" have also been described from the hemichordate *Rhabdopleura* (Stebbing, 1970) in which polypide degeneration occurs in response to embryogenesis or suboptimal water conditions. In some ascidians (e.g., *Perophora*) zooids regress into a stolon from which a new zooid subsequently arises (Barth and Barth, 1966). The only common feature of all these animals are that they are sessile and colonial and have small zooids or individuals.

VI. CAUSE AND EFFECT OF BROWN BODY FORMATION

What induces brown body formation? As seen already in the introduction, it may be induced by a range of adverse environmental factors and sometimes embryogenesis, but often occurs in the absence of these factors; in these cases it seems inescapable that the accumulation of residual material in the stomach wall is the prime cause. Bobin (1964) also reached this conclusion ("L'engorgement de l'estomac provoquant la mort du polypide. . . ."), but added, ". . . represént la vraie fonction rénale." Elsewhere, Gordon (1975) has related the form and function of the bryozoan cecum to the molluscan digestive gland and arthropod hepatopancreas. In the former, digestive and secretory cells accumulate residual material as in *Cryptosula* and then regress. The important difference, however, is that as individual digestive gland cells are lost, others replace them. There is no such cell replacement in *Cryptosula.* Thus in bryozoans, when the stomach cells become so full of residual material that they are no longer capable of absorption (even of stains: Harmer, 1931), the whole stomach degenerates. As Marcus (1934) points out, it is interesting that actual regression should start in the lophophore, whose cells are not full of residual material.

Beginning with Ostroumoff (1866), brown body formation was thought to be an excretory phenomenon associated with lack of excretory organs. Marcus (1926a) thought the brown body was toxic and comparable to "uric deposits" in cyclopoids. Cori (1941) attributed degeneration to "toxic nitrogenous substances" that accumulate in the gut. In *Cryptosula,* however, there is no known deposition of uric acid or guanine such as is said to occur in the "hepatic" part of the stomach in entoprocts (Becker, 1938), and the cumulative substances are residual bodies whose contents (in any tissue) are said to be inert (Cohn and Fedorko, 1969) and therefore nontoxic. Furthermore, as Calvet (1900) observed, since brown bodies are not always eliminated from a zooid, excretion cannot be the prime function of the regression–regeneration cycles.

Harmer's (1891) experiments with various biological stains, which were said to indicate an excretory function of the gut walls, "leukocytes" and funiculus, involved suspensions of the dyes in seawater. Since incorporation of particles or molecules into the stomach cells would involve an endocytotic action similar to normal absorption of food material, it is not necessary to attribute this to excretion. The uptake of dyes by "leukocytes" and funicular cells could be regarded as "excretion" perhaps, or a microimmunological response to foreign molecules.

Although lack of cell replacement in the stomach may be one of the reasons for the death of the polypide, it is worth considering that we are dealing with a colonial animal. Polypide regression and renewal may be a manifestation of growth in a colonial unit which cannot increase in overall dimensions. Since there is a maximum size imposed on zooids (both cystids and polypides) it would be unfavorable to the life of the colony as a whole if, their stomachs having accumulated residual material, all polypides were to die with no subsequent regeneration (as some do). This would lead, in the case of encrusting colonies, to a continually thinning band of live zooids at the periphery, relative to an increasingly larger dead central area. It is, therefore, in the interest of the colony to replace polypides after regression.

In this regard, a consideration of regression cycles in other organisms is of some relevance. Crowell (1953), in discussing such cycles in *Campanularia* hydranths, wondered if this was analogous to brown body formation in bryozoans. *Campanularia* hydranths have a mean life span of 7 days. Products of regression are removed into the rest of the colony and the hydrotheca drops off. A new hydranth grows from the old pedicel. The feeding members of a colony are always "youthful" and this phenomenon was therefore thought to be rejuvenatory. As in bryozoans, regression could be induced by suboptimal physical conditions but in their absence the occurrence of regression is nonetheless fixed. In more recent work on *Campanularia,* Toth (1969) determined that repeated damage to, and regeneration of, the pedicel of isolated hydranths extended the polyp's life span to 20 days, and concluded that tissue damage and reorganization function to maintain the hydranth. Results such as these, if carried over to the Bryozoa, would seem to strengthen the idea that repeated regression and regeneration are rejuvenatory for the zooid as a whole, implying that the life-span of a zooid is likely to be extended further through these cycles than if stomach cells were replaced and the polypide were merely capable of living longer. The accumulation of residue in the stomach should merely serve as a trigger for regression in the absence of provocative external stimuli or the onset of embryogenesis. I conclude that brown body formation is not an excretory phenomenon but is rejuvenatory, as Prouho (1892) first suggested, serving to extend the life span of a zooid and therefore the colony.

VII. SUMMARY

A review of the history of our understanding of the phenomenon of polypide regression (brown body formation) is given. The process is described at the levels of light and electron microscopy. Cells of the lophophore move apart from each other and some of the ciliated epithelial cells behave as amoebocytes. Cilia are either sloughed off or completely resorbed. The regressing lophophore condenses by autophagy, heterophagy, and concomitant cell movement. Necrosis of lophophore cells is marked by liberation of ribosomes during vacuolation of endoplasmic cisternae, and nuclei become homogenous in content or there is clumping of chromatin resulting in pyknosis or karyorrhexis. Some cells herniate. Muscle fibrils become balled up or transformed into paracrystalline arrays within the cell. The stomach regresses as a unit and remains bounded by the basement membrane while condensing by autophagy and heterophagy. Brown bodies are formed predominantly from the regressed stomach and comprise large cells containing residue in a homogeneous matrix. Glycogen accumulates under the floor of the compensation sac and in the mesenchyme during regression but is absent from around the fully formed brown body. Reproductive activity has no effect on the state of the polypide in *Cryptosula,* being in no way a contributing factor to regression. Brown body formation is not here regarded as an excretory phenomenon. Its functional significance seems to be that of rejuvenation, serving to extend the life span of the zooid and therefore the colony.

Note added in proof. Cummings (1975) has studied zooid regression in *Schizoporella unicornis floridana,* with particular reference to the body wall and coelom of developing and mature zooids. She usefully considers the superimposed seasonal effects on the response of zooids to normal regression, noting similarities between regenerative pigmented cell masses in dormant zooids and phylactolaemate statoblasts and ctenostome hibernacula. According to Cummings, brown bodies in *S. unicornis floridana* may become calcified, but unfortunately, there are no accompanying micrographs to illustrate this phenomenon.

ACKNOWLEDGMENTS

I gratefully wish to acknowledge the Killam Foundation for financial support and Dr. Eric L. Mills (Institute of Oceanography) for advice and encouragement. Miss Sybilla von Meuleman translated much of the early German literature, for which I am very grateful.

REFERENCES

Allman, G. J. (1856). "A Monograph of the Freshwater Polyzoa." Ray Society, London. (Not seen.)

Banta, W. C. (1967). A new species of *Victorella* from southern California (Bryozoa: Ctenostomata). *Proc. U.S. Nat. Mus.* **122,** 1–8.

Banta, W. C. (1968). *Mimosella cookae,* new species (Bryozoa, Ctenostomata), with a review of the family Mimosellidae. *Bull. S. Calif. Acad. Sci.* **67,** 245–254.

Barth, L. G., and Barth, L. J. (1966). A study of regression and budding in *Perophora viridis. J. Morphol.* **118,** 451–460.

Becker, G. (1938). Untersuchungen über den Darm und die Verdauung von Kamptozoen, Bryozoen und Phoroniden. *Z. Morphol. Oekol. Tiere* **33,** 72–127.

Berjak, P. and Villiers, T. A. (1972). Ageing in plant embryos. V. Lysis of the cytoplasm in non-viable embryos. *New Phytol.* **71,** 1075–1079.

Bessis, M. (1964). Studies on cell agony and death: An attempt at classification. *Cell. Inj., Ciba Found. Symp., 1963* pp. 287–316.

Boardman, R. S. (1971). Mode of growth and functional morphology of autozooids in some Recent and Palaeozoic tubular Bryozoa. *Smithson. Contrib. Paleobiol.* **8,** 1–51.

Bobin, G. (1964). Cytologie des rosettes de *Bowerbankia imbricata* (Adams) (Bryozoaire Cténostome, Vésicularine). Hypothèse sur leur fonctionnement. *Arch. Zool. Exp. Gén.* **104**, 1–44.

Bobin, G. and Prenant, M. (1957). Les cellules cystidiennes et les phénomènes d'histolyse et de phagocytose chez *Alcyonidium gelatinosum* (L.) (Bryozoaire Cténostome). *Bull. Biol. Fr. Belg.* **91**, 203–224.

Bobin, G., and Prenant, M. (1972). Sur les cellules cavitaires de quelques Vésicularines (Bryozoaires Cténostomes). *Cah. Biol. Mar.* **13**, 479–510.

Bonneville, M. A. (1963). Fine structural changes in the intestinal epithelium of the bullfrog during metamorphosis. *J. Cell. Biol.* **18**, 579–597.

Borg, F. (1926). Studies on recent cyclostomatous Bryozoa. *Zool. Bidr. Uppsala* **10**, 180–504.

Borg, F. (1947). Zur Kenntnis der Ökologie und des Lebenszyklus von *Electra crustulenta* (Bryozoa Chilostomata). *Zool. Bidr. Uppsala* **25**, 344–377.

Braem, F. (1951). Über *Victorella* und einige ihrer nächsten Verwandten sowie die Bryozoenfauna des Ryck bei Greifswald. *Zoologica (Stuttgart)* **102**, 1–59. (Not seen.)

Brock, M. A. (1970). Ultrastructural studies on the life cycle of a short-lived metazoan *Campanularia flexuosa.* II. Structure of the old adult. *J. Ultrastruct. Res.* **32**, 118–141.

Bronstein, G. (1938). Mécanisme de la formation du polypide chez *Membranipora membranacea* (L.). *C.R. Hebd. Séances Acad. Sci.* **207**, 506–508.

Bronstein, G. (1939). Sur le tube digestif des Bryozoaires gymnolémides. *C.R. Hebd. Séances Acad. Sci.* **209**, 574–576.

Bullivant, J. S. (1967). Aspects of feeding of the bryozoan *Zoobotryon verticillatum* (delle Chiaje). Ph.D. Dissertation, University of Southern California, Los Angeles. (Not seen.)

Calvet, L. (1900). Contribution à l'histoire naturelle des Bryozoaires ectoproctes marins. *Trav. Inst. Zool. Univ. Montpellier* **8**, 1–488.

Chrétien, M. (1958). Histologie et développement de l'ovaire chez *Alcyonidium gelatinosum* (L.) (Bryozoaire Cténostome). *Bull. Lab. Mar. Dinard* **43**, 25–51. (Not seen.)

Claparède, E. (1870). Beiträge zur Anatomie und Entwicklungsgeschichte der See-Bryozoen. *Z. Wiss. Zool.* **21**, 137–178. (Not seen.)

Cohn, Z. A., and Fedorko, M. E. (1969). The formation and fate of lysosomes. *In* "Lysosomes in Biology and Pathology" (J. T. Dingle and H. B. Fell, eds), Vol. 1, pp. 43–63. North-Holland Publ., Amsterdam.

Cori, C. J. (1941). Bryozoa. *In* "Handbuch der Zoologie" (W. Kükenthal and T. Krumbach eds.), Vol. 3, Part 2, pp. 263–502. de Gruyter, Berlin.

Crowell, S. (1953). The regression-replacement cycle of hydranths of *Obelia* and *Campanularia. Physiol. Zool.* **26**, 319–327.

Cummings, S. G. (1975) Zoid regression in *Schizoporella unicornis floridana* (Bryozoa, Cheilostomata). *Chesapeake Sci.* **16**, 93–103.

David, H. (1970). Zellschadung und Dysfunktion. *Protoplasmatologia* **10**, 1–621.

Delage, Y., and Hérouard, E. (1897). *In* "Traité de Zoologie Concrète." (Schleicher Frères, eds.), Vol. V. Reinwald, Paris.

Ehlers, E. (1876). *Hypophorella expansa.* Ein Beitrag zur Kenntnis der minierenden Bryozoen. *Abh. K. Ges. Wis. Göttingen* **21**, 1–156. (Not seen.)

Eisen, A. Z., and Gross, J. (1965). The role of epithelium and mesenchyme in the production of a collagenolytic enzyme and a hyaluronidase in the anuran tadpole. *Dev. Biol.* **12**, 408–418.

Ellis, J. (1755). "An Essay Towards a Natural History of the Corallines and Other Marine Productions of the Like Kind Commonly found on the Coasts of Great Britain and Ireland. To which is added the Description of a Large Marine Polype Taken near the North Pole, by the Whale-fishers, in the Summer 1753," 27 pp., 15 pl. Front Matter and Co., London.

Farre, A. (1837). Observation on the minute structure of some higher forms of Polypi with views of a more natural arrangement of the class. *Philos. Trans. R. Soc. London,* **1837,** 387–426.

Gerwerzhagen, A. (1913). Untersuchungen an Bryozoen. *Sitzungsber. Heidelb. Akad. Wiss., Math.-Naturwiss. Kl.. 4, Abt. B* **9,** 3–16.

Geuze, J. J. (1970). Light and electron microscope observations on auto- and heterophagy in the exocrine pancreas of the hibernating frog *Rana esculenta. J. Ultrastruct. Res.* **32,** 391–404.

Gordon, D. P. (1968). Growth, regeneration and population biology of cheilostomatous polyzoans. M.Sc. Thesis, University of Auckland.

Gordon, D. P. (1970). Reproductive ecology of some northern New Zealand Bryozoa. *Cah. Biol. Mar.* **11,** 307–323.

Gordon, D. P. (1973a). A fine-structure study of brown bodies in the gymnolaemate *Cryptosula pallasiana* (Moll). *In* "Living and Fossil Bryozoa" (G. P. Larwood, ed.), pp. 275–286. Academic Press, New York.

Gordon, D. P. (1973b). The fine structure of normal and regressing polypides of the fouling bryozoan *Cryptosula pallasiana* (Moll). Ph.D. Thesis, Dalhousie University.

Gordon, D. P. (1975). Ultrastructure and function of the gut of a marine Bryozoan. *Cah. Biol. Mar.* **16,** 367–382.

Grant, R. E. (1827). Observations on the structure and nature of Flustrae. *Edinburgh New Philos. J.* **3,** 107–118 and 337–342.

Grave, B. H. (1930). The natural history of *Bugula flabellata* at Woods Hole, Massachusetts, including the behaviour and attachment of the larva. *J. Morphol.* **49,** 355–383.

Harmer, S. F. (1886). On the life history of *Pedicellina. Q. J. Microsc. Sci [N.S.]* **27,** 239–263.

Harmer, S. F. (1891). On the nature of the excretory processes in marine Polyzoa. *Q. J. Microsc. Sci. [N.S.]* **33,** 123–167.

Harmer, S. F. (1896). Polyzoa. *In* "The Cambridge Natural History" (S. F. Harmer and A. E. Shipley, eds.), Vol. 2, pp. 463–533. Macmillan, New York.

Harmer, S. F. (1898). On the development of *Tubulipora* and on some British and northern species of this genus. *Q. J. Microsc. Sci [N.S.]* **41,** 73–157.

Harmer, S. F. (1931). Recent work on Polyzoa. *Proc. Linn. Soc. London* **143,** 113–168.

Hincks, T. (1871). Note on Dr. Nitsche's paper. *Q. J. Microsc. Sci [N.S.]* **11,** 235–238. (Not seen.)

Hincks, T. (1873). Contribution to the history of the Polyzoa. *Q. J. Microsc. Sci. [N.S.]* **13,** 16–36.

Hincks, T. (1880). "A History of the British Marine Polyzoa," Vols. 1 and 2. Van Voorst, London.

Joliet, L. (1877). Histoire naturelle des Bryozoaires des côtes de France. *Arch. Zool. Exp. Gén.* **6,** 193–304.

Korschelt, E., and Heider, K. (1920). "Lehrbuch der vergleichenden Entwicklungsgeschichte der wirbellosen Thiere." Vierter Abschnitt. Ungeschlechtliche Fortpflanzung und Regeneration 9. "Ungeschlichte Fortpflanzung," pp. 632–694. Fischer, Jena.

McKinney, F. K. (1969). Organic structures in a late Mississippian Trepostomatous ectoproct (Bryozoan). *J. Paleontol.* **43,** 285–288.

Majno, G. (1964). Interactions between dead cells and living tissue. *Cell Inj., Ciba Found. Symp., 1963* pp. 87–98.

Manasek, F. J. (1969). Myocardial cell death in the embryonic chick ventricle. *J. Embryol. Exp. Morphol.* **21**, 271–284.

Marcus, E. (1926a). Beobachtungen und Versuche an lebenden Meeresbryozoen. *Zool. Jahrb., Abt. Syst. Oekol. Geogr. Tiere* **52**, 1–102.

Marcus, E. (1926b). Beobachtungen und Versuche an lebenden Süsswasserbryozoen. *Zool. Jahrb., Abt. Syst. Oekol. Geogr. Tiere* **52**, 279–350.

Marcus, E. (1934). Über *Lophopus crystallinus* (Pall.). *Zool. Jahrb., Abt. Syst. Oekol. Geogr. Tiere* **58**, 501–606.

Matricon, I. (1960). Etude histologique d'*Alcyonidium polyoum* (Hassall): Caractères structuraux spécifiques d'*Alcyonidium mytili* Dalyell (Bryozoaires Cténostomes). *Cah. Biol. Mar.* **1**, 359–395.

Matricon, I. (1963). Dégénérescence du polypide femelle et formation d'une poche incubatrice chez *Alcyonidium polyoum* (Hassall) (Bryozoaire Cténostome). *Arch. Zool. Exp. Gén.* **102**, 79–93.

Mawatari, S. (1952). On *Watersipora cucullata,* a common marine fouling bryozoan. *Misc. Rep. Res. Inst. Nat. Resour. (Tokyo)* **28**, 17–27.

Nielsen, C. (1971). Entoproct life-cycles and the entoproct/ectoproct relationship. *Ophelia* **9**, 209–341.

Nitsche, H. (1869). Beiträge zur Kenntnis der Bryozoen I, II. *Z. Wiss. Zool.* **20**, 1–36.

Nitsche, H. (1871). Beiträge zur Kenntnis der Bryozoen III. Über die Anatomie und Entwicklungsgeschichte von *Flustra membranacea. Z. Wiss. Zool.* **21**, 1–53.

Ostroumoff, A. A. (1866). Contribution à l'étude zoologique et morphologique des Bryozoaires du Golfe de Sebastopol. *Arch. Slaves Biol.* **1**, 557–569; **2**, 184–190, 329, and 355. (Not seen.)

Paulin, J. J. (1973). The resorption of cilia in *Cyathodinium piriforme. J. Protozool.* **20**, 281–285.

Payen, G. (1972). Etude ultrastructurale de la dégéngérescence cellulaire dans la glande androgène du crabe *Ocypode quadrata* (Fabricius). *Z. Zellforsch. Mikrosk. Anat.* **129**, 370–385.

Prouho, H. (1892). Contribution à l'histoire naturelle des Bryozoaires. *Arch. Zool. Exp. Gén.* **10**, 557–656.

Redfern, P. (1858). *Flustrella hispida* and its development. *Q. J. Microsc. Sci [N.S.]* **6**, 96–102. (Not seen.)

Reger, J. F. (1969). Studies on the fine structure of muscle fibers and contained crystalloids in basal socket muscle of the entoproct *Barentsia gracilis. J. Cell Sci.* **4**, 305–325.

Renieri, T. (1970). Submicroscopical features of alimentary canal in Bryozoa. *J. Submicr. Cytol.* **2**, 181–188.

Repiachoff, W. (1875). Zur Entwicklungsgeschichte der *Tendra zostericola. Z. Wiss. Zool.* **25**, 129–142.

Repiachoff, W. (1876). Zur Naturgeschichte der chilostomen Seebryozoen. *Z. Wiss. Zool.* **26**, 139–160.

Rey, P. (1927). Observation sur le corps brun des Bryozoaires ectoproctes. *Bull. Soc. Zool. Fr.* **52**, 367–379.

Rogick, M. D. (1945). Studies on marine Bryozoa I. *Aeverrillia setigera* (Hincks) 1887. *Biol. Bull. (Woods Hole, Mass.)* **89**, 201–214.

Römer, O. (1906). Untersuchungen über die Knospung, Degeneration und Regeneration von einigen marinen ektoprokten Bryozoen. *Z. Wiss. Zool.* **84**, 466–478. (Not seen.)

Sars, G. O. (1874). Om en hidtil lidet kjendt maerkelig Slaegstype af Polyzoa. *Forh. Videnskabselsk. Krist.* pp. 386–400. (Not seen.)

Smitt, F. A. (1863). Bidrag till kännedomen om Hafs-Bryozoernas utvecklung. *Uppsala Univ. Arsskr. (Math.).* (Not seen.)

Smitt, F. A. (1865). Om Hafs-Bryozoernas utvecklung och fettkroppar. *Oefvers. K. Vet.-Akad. Forhandl.* **1**, 5–50.

Soule, J. D. (1954). Post-larval development in relation to the classification of the Bryozoa Ctenostomata. *Bull. S. Calif. Acad. Sci.* **53**, 13–34.

Stebbing, A. R. D. (1970). Aspects of the reproduction and life cycle of *Rhabdopleura compacta* (Hemichordata). *Mar. Biol.* **5**, 205–212.

Stebbing, A. R. D. (1971). Growth of *Flustra foliacea* (Bryozoa). *Mar. Biol.* **9**, 267–272.

Thompson, J. V. (1830). On Polyzoa, a new animal discovered as an inhabitant of some Zoophites. *Zool. Res.* **5**, 89–102. (Not seen.)

Toth, S. E. (1969). Inhibition of hydroid aging in *Campanularia flexuosa. Science* **166**, 619–620.

Van Beneden, P. J. (1845). Recherches sur l'anatomie, la physiologie et le développement des Bryozoaires qui habitent la côte d'Ostende. *Mém. Acad. R. Sci. Lett. Belg.* **18**, 44 pp., pl. I–V.

Waters, A. W. (1878). On Bryozoa. *Proc. Manchester Lib. Philos. Soc.* **17**, 125–138.

Waters, A. W. (1913). The marine fauna of British East Africa and Zanzibar from collections made by Cyril Crossland, M.A., B.Sc., F.Z.S. in the years 1901–1902. Bryozoa-Cheilostomata. *Proc. Zool. Soc.* pp. 458–537, plates 64–73.

Wood, T. S. (1973). Colony development in species of *Plumatella* and *Fredericella* (Ectoprocta: Phylactolaemata). *In* "Animal Colonies Development and Function through Time" (R. S. Boardman, A. H. Cheetham, and W. A. Oliver, Jr., eds.), pp. 395–432. Dowden, Hutchinson & Ross, Inc., Stroudsburg, Pennsylvania.

Zschiesche, A. (1909). Untersuchungen über die Metamorphose von *Alcyonidium mytili. Zool. Jahrb., Abt. Syst. Oekol. Geogr. Tiere* **28**, 1–72. (Not seen.)

11

The Bryozoan Nervous System

GENEVIÈVE LUTAUD

I. INTRODUCTION

The classic works concerning the nervous system of the Ectoprocta, concentrating on the theoretical argument on the significance of the colony as an individual or as a community, advanced the hypothesis of the existence of interzooidal nervous bonds through which a colonial exchange of information and coordination would be possible. This hypothesis originated from an initial misinterpretation by early authors of the continuity of the funicular strands within a colony, but remained an open question in the biology of the Bryozoa when the function of the funiculus in the transportation and storage of metabolites became evident. The first anatomists who described the innervation of the polypide tried to trace the neuronal pathways of colonial information in the cystid wall using methylene blue vital stains or by the usual silver impregnation techniques. Others tried to find evidence of coordinated activities in the rhythms of the protrusion of the polypides, or in the oscillations of mandibles of the sessile avicularia or setae of the vibracula.

The technical obstacle of the hermetic cystid which obscures minute structures on live animals and hinders the penetration of dyes restricted the early observations to transparent species with thin uncalcified walls. This choice led to unequal knowledge of the neuronanatomy among the different classes of the phylum. The nervous system is best known in the Phylactolaemata from the thorough and detailed analysis of Gewerzhagen (1913a) of *Cristatella mucedo,* later corroborated by the concordant results of Marcus (1934) of *Lophopus cristallinus.* Gewerzhagen observed that the wall of the cystid in Phylactolaemata was innervated by a diffuse parietal plexus. A similar but more intricate nerve net was found in the Entoprocta by Hilton (1923) in the body wall and in the stolons of *Barentsia gracilis.* In the Gymnolaemata, Marcus (1926) revealed a similar reticular plexus in the wall of the tubular zooid of the stoloniferan ctenostome, *Farrella repens.* His description of the peripheral and polypidian nerves in the Stolonifera was completed by Graupner (1930), who compared the distribution of the peripheral and lophophoral nerves in *Flustrella hispida* (Carnosa) and *Membranipora membranacea* (Anasca), and studied the structure of the cerebral ganglion in histological sections. Bronstein (1937) generalized the previous findings in a comprehensive survey of the main neuronal pathways and of plexus cells in the interzooidal partitions of the genus *Electra,* representing a ctenostomes and anascan cheilostomes. Hiller (1939) discovered a parietal chain of plexus cells in the interzoidial partitions of the genus *Electra,* representing a most plausible pathway for a colonial coordination. The observation of Hiller was later confirmed by Lutaud (1969) in a general account of parietal innervation in *Electra pilosa.* We do not know how the higher complexity of the calcified wall in the systematic sequence of ascophoran families could modify or complicate the pattern of parietal innervation, or what are the specific characters of the nervous system in cyclostomes.

A thirty-year gap in the literature led to neglect of the very substantial body of information from the pioneer authors and to an underestimate of the real extent of ubiquitous nerve ramifications and the precise organization of the cerebral center in the Bryozoa. The following recapitulation will point out the adaptations of the nervous pathways to the morphological separation of the cystidian and polypidian organs in the autozooid and the functional coordination of the corresponding areas of innervation.

II. GENERAL ANATOMY

A. The Structure of the Cerebral Ganglion

The cerebral center of the Ectoprocta is a small heart-shaped ganglion located in the pharyngeal constriction between the base of the lophophore and the first esophageal subdivision of the digestive tract. The ganglion is situated on the dorsal side of the polypide in retracted position (Plate I, 5). Its size and width/height ratio depend on the specific proportions of the polypide and its average diameter is between 30 μm in small stoloniferan ctenostomes and 60 μm in large encrusting ascophoran cheilostomes. The cerebral ganglion is directly prolonged around the pharynx by a thick perioral belt of the same ganglionic structure lying in a basiepithelial location between the pharyngeal epithelium and the circular lacuna of the basal canal of the lophophore where the internal canals of the tentacles are collected. The peripharyngeal belt, from which the tentacle nerves arise, has a crescentic shape at the base of the horseshoe-shaped lophophore of the Phylactolaemata, but is a complete circle in the Gymnolaemata.

The precise organization of the ganglion has been disregarded, and the cerebral center of the Bryozoa was first believed to be a simple cluster of nerve cells. Gewerzhagen did not consider the cellular arrangement of the cerebral ganglion in the Phylactolaemata, but Marcus introduced the idea of a primitive differentiation of specialized neurons. He identified in the ctenostome *Farrella repens* (a) a pair of "anterior neurons" at the origin of the dorsal tentacle sheath nerves which extend to the parietal plexus and to the parietal muscles and (b) a pair of giant "posterior neurons" at the origin of the ventral recurrent nerves to the retractor muscle and to the intestine. Bronstein developed this idea further and proposed, from his observations on *Alcyonidium*, a basic formula which could be the fundamental pattern of the organization of the cerebral ganglion in all Gymnolaemata, according to which the ganglion includes eight "neurons" or groups of neurons of invariant topographical relationships [see Fig. 2(1)]. These are a central "sagittal piece" with two adjacent "club-shaped neurons," supporting a "transverse piece" with two lateral "distal horns" at the origin of the peripharyngeal belt, and twin "posterior bipolar neurons" at the origin of the common stem of the motor and sensory nerves of the tentacle sheath.

In *Electra pilosa*, nuclear stains of whole mounts, such as hematoxylin, gallocyanine, or the Feulgen reagent, show that the cerebral cells are more numerous than first estimated from methylene blue stains, and reveal betwcen thirty and forty nuclei of different sizes and densities (Lutaud, 1969). Study of ultrathin sections now indicates that the cerebral cells belong to at least three cytological categories including typical neurons with characteristic neurotubules, secretory cells rich in granular reticulum, and investing glial cells. The cerebral ganglion shows an intense neurosecretory activity and dense granules accumulate in the intracerebral nerve bundles.

The cerebral ganglion is not externally subdivided into lobes, but three regions are clearly distinguished in the cellular formations around an excentric neuropil filling the proximal and deep regions of the ganglion: the central cellular complex, the distal margin where the perioral chain of neurons begins, and the proximal groups below the neuropil at the origin of the two tentacle sheath nerves. The three levels are arranged according to a general V-shaped pattern as if the ganglion was derived from a simple initial fold. A tiny central cavity is sometimes observed in ultrathin sections. The median depression of the ganglion is due to the massive exit of the peripheral nerve bundles at this level.

In *E. pilosa,* the central complex is made of three superimposed rows of cells [Fig. 1 (1 and 2)]. The deep row is a crown of five or six cells converging around a large central one supporting two distal "satellites." Below this central cell, the intermediate row includes a range of five cells, and above is another median cell supporting two distal units. The superimposed median cells and the duplicate pairs of distal cells correspond to the "sagittal piece" and to the "distal horns" in Bronstein's interpretation. The superficial row consists of two pairs of dense cells in a transverse array, corresponding to the "club-shaped neurons" described by Bronstein, and from which arise the cerebral connectives of the parietal plexus discovered by Hiller in the partitions of the Electridae. By its situation between the initials of the chains of neurons of the lophophore belt and the bundles of the peripheral nerves in the neuropil, the central cellular aggregate, supporting the presumed pathways of interzooidal information, appears as the center of general coordination.

A median multipolar cell at the distal edge of the ganglion, with prolongations into the central complex, is the first of a chain of periodic multipolar neurons in the perioral ganglion, one at the base of every tentacle. In the alignment of two other marginal cells and of the distal pairs of the central complex, two or three series of nuclei extend laterally in a quincuncial design regularly repeated further in the cellular layer of the perioral ganglion. The fibrillar core of the perioral belt where the roots and endings of the tentacle nerves are entwined is the direct prolongation of the cerebral neuropil; silver stains show that the central bundles, grouping the oral tentacle nerves, join the ganglion beneath the marginal cells, and that the lower bundles, grouping the dorsal tentacles nerves, join the deep region of the cerebral neuropil [Fig. 2 (1)].

Beneath the giant "posterior neurons," the proximal groups include, on either side, at least two other large cells and another more internal cell with a small nucleus. Silver stains show that the transverse commissural bundles of the tentacle sheath nerves in the ganglion constitute most of the external half of the neuropil. Therefore, the polyfibrillar peripheral nerves cannot be considered as

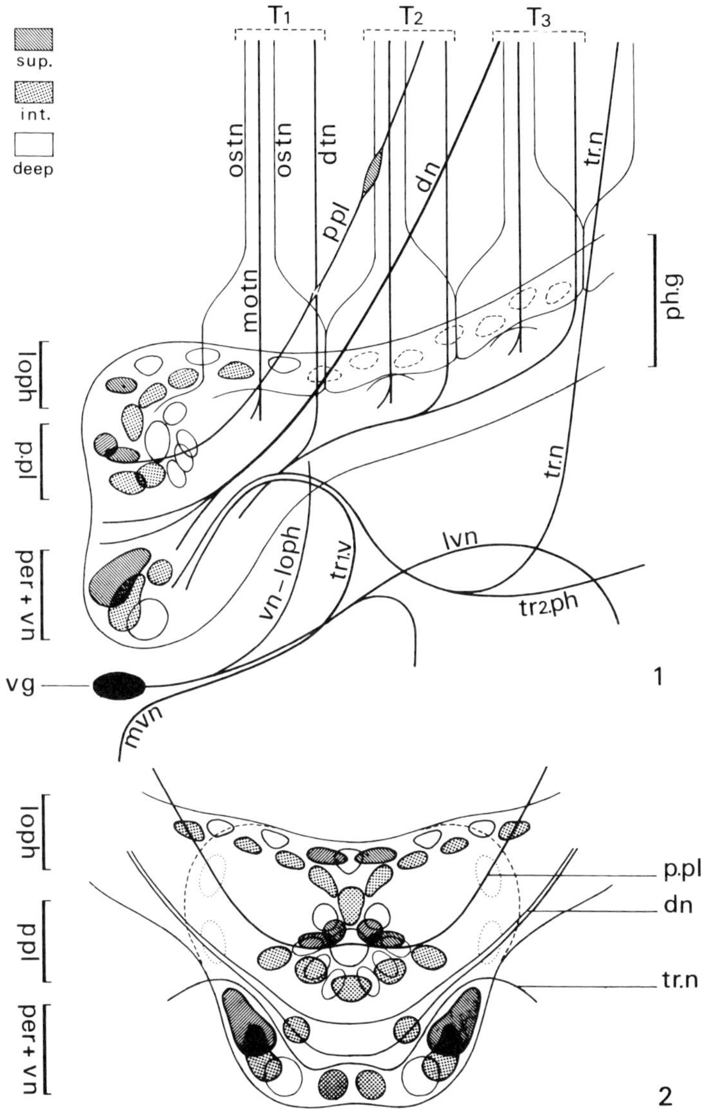

Fig. 1. Interpretation of the structure of the cerebral ganglion in *Electra pilosa* (Linné). (1) In profile; (2) in dorsal view. Abbreviations: loph, lophophoral innervation area; p.pl, parietal plexus; per. + vn, peripheral and visceral innervation area; vg, visceral ganglion; ph.g, peripharyngeal ganglionic belt; T, tentacles; dn, stem of the direct nerve; tr.n, stem of the trifid nerve; $tr._1$ v, visceral anastomose of the trifid nerve; $tr._2$ ph, pharyngeal branch of the trifid nerve; lvn and mvn, lateral and median visceral nerves; vn-loph, connection between the lophophoral and visceral ganglions; motn, motor tentacle nerves; ostn, oral sensory tentacles nerves; dtn, dorsal sensory tentacle nerve.

the simple axonal prolongations of the giant proximal neurons, although the participation of these major cells in the nerve bundles is unquestioned.

The cerebral ganglion and the lophophore belt are wrapped in a thin stratified envelope separating the nervous tissues from the peritoneal lining and from the muscles of the digestive tract wall. Under the external cellular layer, the fibrillar core of the ganglionic formations lies directly against the basal membrane of the epithelial cells in the pharyngeal area.

It would be premature to assume which of the components of this elementary diagram are motor or sensory neurons, relays or neuroglial cells, and what their precise function is in the intracerebral coordinations. In *E. pilosa,* the number and the topographical arrangement of the cerebral cells are rigorously constant. It is probably so in any given species, but specific variance is noticed in the ctenocheilostomes in relation to the systematic position of the considered genus. However, the three levels remain clearly distinct whatever the simplifications or additions are.

B. The Tentacle Sheath Nerves, and the Innervation of the External Muscles, of the Aperture and of the Opesial Area

The two pairs of major nerves, arising from the neuropil and from the proximal cells of the cerebral ganglion and following the tentacle sheath toward the aperture, are the agents of the coordination of the polypide with the external musculature and its cystidean annexes. The term peripheral innervation is used here in a broad meaning to designate these motor and sensory pathways of the coordination of the polypidean and cystidean organs in a functional zooid.

In the Phylactolaemata, Gewerzhagen (1913) first mentioned an "outer ventral nerve" with a ramification to the retractor muscle of the polypide, and an "inner dorsal nerve" to the tentacle sheath wall and the aperture" with a recurrent ramification to the parietal plexus, diverging from a common stem. In the ctenocheilostomes, the two tentacle sheath nerves, together with the cerebral connectives of Hiller's parietal network in the Electridae, emerge on either side through common lateral "buttonholes" within the membranous envelope of the cerebral ganglion and immediately diverge (Fig. 3). The inner nerve is a thick bundle without any autonomous ramification in the early part of its course. The other nerve, called the "trifid nerve" by Bronstein, is a thin, three-branched strand which characteristically loops toward the esophagus before bending up along the sheath and joining the direct fascicle. The two fused nerves remain further associated until the compound axis turns to the frontal side. They break at the top of the sheath on each side into three primary branches of mixed functions, each dividing soon into a motor and a nonmotor secondary ramifica-

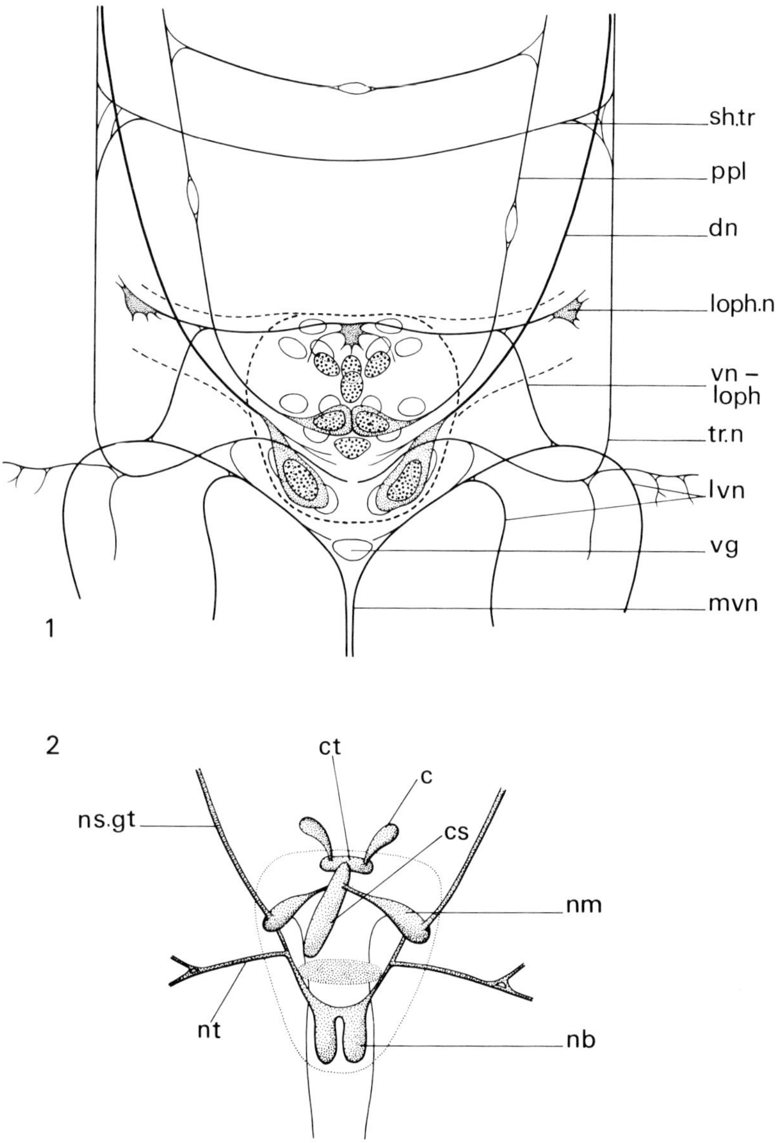

Fig. 2. The basic organization of the cerebral ganglion in Gymnolaemata. (1) Origin of the main nerves in *Electra pilosa;* (2) Structure of the cerebral ganglion in *Alcyonidium polyoum* after Bronstein (1937). Abbreviations: (1) p.pl, parietal plexus; dn, direct nerve of the tentacle sheath; loph.n, peripharyngeal chain of multipolar cells; vn-loph, lophophoral to visceral nerves connectives; tr.n, trifid nerve; sh.tr, annular branch of the trifid nerve in the tentacle sheath; lvn and mvn, lateral and median visceral nerves; vg, visceral ganglion. (2) nb, bipolar posterior neurons; nm, club-shaped neurons; cs, sagittal cell; ct, transverse piece; c, anterior horns; nt, trifid nerve; ns.gt, sensory nerve of the tentacle sheath.

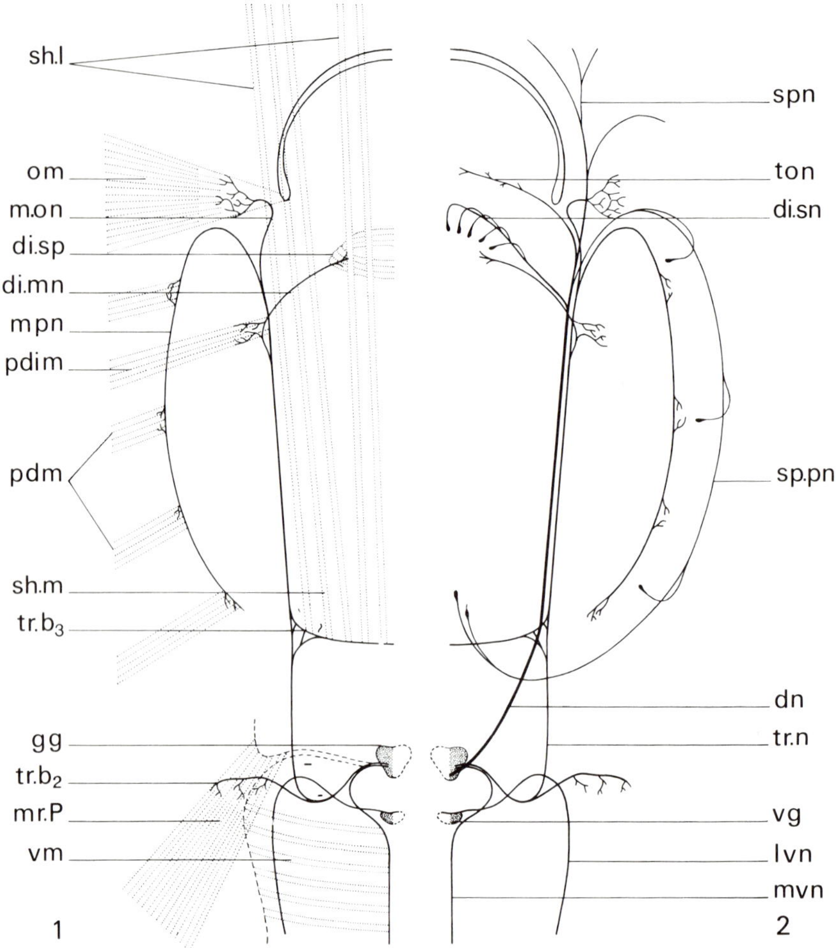

Fig. 3. The motor and sensory pathways of peripheral innervation in *Electra pilosa.* (1) External muscles and motor branches of the tentacle sheath nerves. (2) Motor and sensory branches of the tentacle sheath nerves. Abbreviations: sh.l, ligaments of the tentacle sheath; sh.m, longitudinal muscles of the tentacle sheath; om, abductor muscles of the operculum; m.on, motor opercular branch; di.sp, sphincter of the diaphragm; di.mn, diaphragmatic motor branch; pdim, parietodiaphragmatic muscle; mpn, motor parietal branch; pdm, parietodepressor muscles; gg, cerebral ganglion; $tr.b_3$, annular branch of the trifid nerve; $tr.b_2$, pharyngeal branch of the trifid nerve; mr.P, retractor muscle of the polypide; vm, annular visceral muscles; spn, sensory parietal branch; ton, transopercular branch; di.sn, sensory diaphragmatic branch; sp.p, sensory parietal branch to the spines; dn, direct nerve; tr.n, trifid nerve; vg, visceral ganglion; lvn and mvn, lateral and median visceral nerves.

tion. The first primary branch joins the diaphragm, the second joins the operculum, and the third innervates the frontal wall structures.

1. The Motor Trifid Nerve

According to Marcus in *Farrella* and to Bronstein in *Alcyonidium,* the first ramification of the trifid nerve in ctenostomes is an intestinal branch with dorsal and ventral ramifications; a median dorsal intestinal nerve would proceed from the rear of the cerebral ganglion. This is somewhat different from the observations of Graupner and Lutaud in anascan cheilostomes. In *E. pilosa*, the first ramification of the trifid nerve is a short connective bending back to be absorbed in the thick bundle of the median intestinal nerve. The lateral intestinal nerves do not proceed directly from the trifid stem, but arise from a small accessory ganglion located below and next to the cerebral ganglion, and further cross over the recurrent bend of the trifid nerve [Fig. 2 (1)].

The pharyngeal branch of the trifid nerve, lying between the peritoneal lining and the thickened collagen shield which protects the lophophore canal and supports the insertion of the retractor muscle, expands in ramified arborizations around the muscle fibers. The axial branch then bends up abruptly and, just before joining the direct nerve, develops a third annular ramification around the tentacle sheath. This annular branch occurs at a precise level and links the basal extremities of all the longitudinal muscle fibers of the tentacle sheath wall. These muscle fibers, further collected in the top ligaments of the sheath and inserted on the frontal wall and on the distal transverse partition, pleat the sheath and lift the polypide in the extrusion movement. Thus the autonomous ramifications of the trifid nerve are motor and command the antagonistic great retractor and sheath ligament muscles.

2. The Main Branches of the Mixed Tentacle Sheath Nerves: The Sensory Innervation of the Tentacle Sheath

The diaphragm at the top of the tentacle sheath separates the lophophore atrium from the vestibular space under the operculum. It consists of a sphincter muscle grouping the few annular muscle fibers of the tentacle sheath wall and two opposite fans of more or less permanent epithelial folds enclosing mucocytes. The closing action of the sphincter is compensated by the parietodiaphragmatic muscles which are inserted on the lateral partitions and draw open the diaphragm during protrusion of the tentacles.

In *E. pilosa,* the diaphragmatic branch of the compound tentacle sheath nerve diverges below the parietodiaphragmatic muscle and soon divides: a motor ramification first develops arborizations into the muscle insertion before joining the sphincter in the center of the diaphragm; and a nonmotor ramification reaches the upper folds of the diaphragm and breaks into an ordered fan of single nerve fibers, each bearing a terminal fusiform body which might be a sensory cell

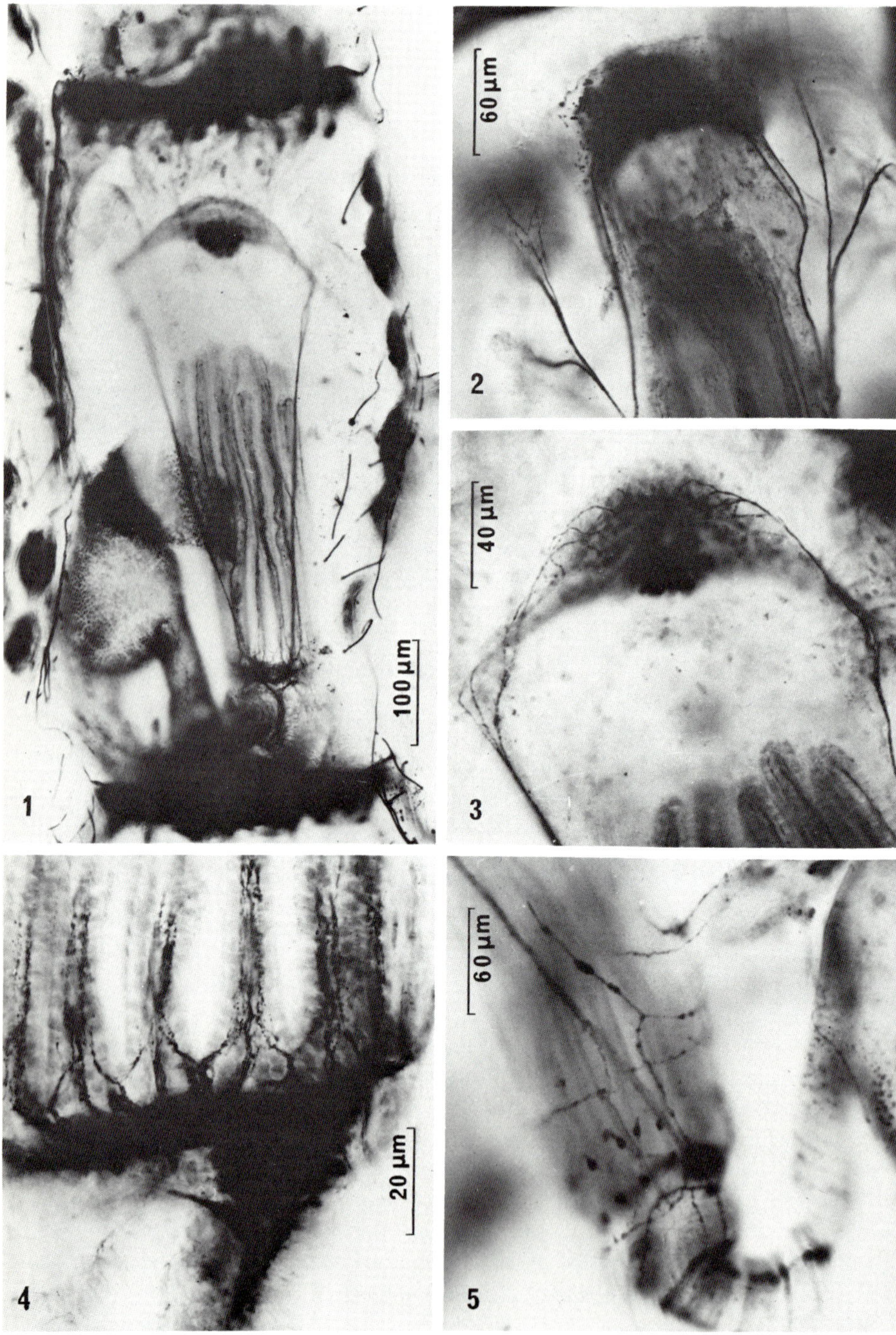

1
100 µm
2
60 µm
3
40 µm
4
20 µm
5
60 µm

(see Plate I,3, p. 380). The nerve fibers in the diaphragm folds are topographically and functionally the homologues of more dispersed nerve fibers with terminal cells found by Marcus (1926) and Graupner (1930) in the vestibular area of the sheath of the Phylactolaemata and the Ctenostomata.

In the Phylactolaemata, the tentacle sheath wall is covered by a nerve net spreading from a ramification of the dorsal tentacle sheath nerves (Gewerzhagen, 1913a; Marcus, 1934). Silver stains do not show such a widespread innervation of the tentacle sheath wall in *E. pilosa.* The base of the sheath in cheilostomes is surrounded by a crown of nerve fibers with terminal cells, in quincuncial order, morphologically similar to the fibers in the diaphragm folds, but proceeding directly from the distal rim of the ganglion (Plate I, 5).

3. The Parietal Branches of the Mixed Tentacle Sheath Nerve. The Innervation of the Frontal Wall Structure in Electridae

The great tentacle sheath nerves, following the frontal ligaments, reach the frontal wall next to the aperture. In *E. pilosa,* two opercular ramifications on each side diverge at the level of the tendon of the abductor muscles of the operculum: a motor branch develops arborizations into the muscles; a nonmotor "transopercular branch" bends in the opposite direction along the hinge of the operculum and fuse medially with the symmetrical strand (Fig. 3).

A recurrent motor branch, called the "parietal nerve" by Marcus, Graupner, and Bronstein, diverges from the axis between the diaphragmatic and the opercular branches and runs around the supple opesia through the insertions of all the parietodepressor muscles which regulate the internal pressure of the body cavity. In *E. pilosa,* this recurrent motor branch is duplicated by another parallel strand running at the edge of the calcified gymnocyst and dispatching a few loose fibers toward the base of every marginal spine (Fig. 3). The long median spine receives the ultimate fibers of both symmetrical ramifications of each side. The innervation of the marginal spines in Electridae, first postulated on the basis of methylene blue stains (Lutaud, 1973a), has since been confirmed by the observation of isolated nerve fibers in electron micrographs of transverse sections of the median spine.

The residual branches of the great tentacle sheath nerves spread into the frontal wall around the aperture and toward the borders of the partitions. The presence of these superficial parietal fibers, although they might seem of minor

---

Plate I. The polypide and peripheral nerves in *Electra pilosa.* (1) The great mixed nerves in the tentacle sheath; (2) the divergence of the primary branches of the mixed nerves at the top of the tentacle sheath; (3) the sensory ramification in the diaphragm; (4) the origin of the oral tentacle nerves at the base of the lophophore; (5) the main nerves diverging from the cerebral ganglion. (1) to (4), Bielschowsky silver stain; (5) methylene blue, preserved specimen.

importance, is significant as they represent, together with the transopercular branch and the fibers to the spines, the means of a widespread nonmotor innervation of the external wall of the zooid. Thus, the cerebral center in the polypide may register external variations at the level of the free wall of the encrusting cystid.

4. The Cytological Features of the Mixed Tentacle Sheath Nerves

The trifid nerve displays motor branches in its autonomous course, and sensory ramifications appear after it joins the direct nerve. It may readily be inferred that the two nerves remain distinct in the common axis. The trifid nerve would thus command all the muscles involved in the protrusion and retraction movements. This interpretation does not take into account the possibilities that the direct nerve could be a mixed bundle or that exchanges may occur between the intermixed fibers of the associated nerves along their common pathway.

Ultrastructural investigation shows that, after the conjunction of the direct and trifid nerves, the great mixed tentacle sheath nerve appears in transverse sections as a unique fascicle of about thirty typical nerve fibers partly or completely surrounded by a single, giant, flattened unit (Plate V, 2) (Lutaud, 1973a). As this strange surrounding element shows the same clear matrix as the central nerve fibers, the possibility that it might be a giant nerve fiber might be considered. However, its nature has not been ascertained, and it could be as well a primitive neuroglial sheath. The nerve runs on the coelomic side of the collagenous lamina, embedded in the peritoneal lining. It differs thus from polypidian nerves which are basiepithelial. We do not know how far this surrounding element persists along the primary and secondary ramifications of the nerve, but the dispersed nerve fibers in the frontal wall are naked.

C. The Innervation of the Polypide

1. The Innervation of the Digestive Tract

The detailed topography of the visceral nerves commanding the annular and longitudinal muscles of the intestinal wall and their precise connections with the different subdivisions of the digestive tract have not been described to date. We have already seen that two pairs of visceral nerves proceed through the relay of a visceral nodule below the cerebral ganglion in anascan cheilostomes (e.g., Section II,B). The median visceral nerve is a thick bundle running in the dorsal groove of the esophagus, and is constituted of fibers issued from the rear of the cerebral ganglion and two confused strands issued from the visceral ganglion. The extent and function of the median nerve are unknown but part of its fibers may be absorbed in the esophageal wall. The lateral visceral nerves arise from the visceral ganglion and their ventral and dorsal ramifications run from the top of the esophagus to the anal pouch. The visceral nerves are periodically linked by annular ramifications in the constrictions between the successive subdivisions of the gut. A complex exchange of nerve fibers occurs at the level of the visceral ganglion. Short constant strands around the cerebral ganglion link the stem of the lateral visceral nerves to the prolongations of the dense multipolar marginal cell of the cerebral ganglion in the peripharyngeal belt. Secondary anastomoses are further established between the lateral visceral nerves and the fibrillar bundles of the peripharyngeal ganglion. Thus the pathways for permanent coordination of the activity of the lophophore and of the peristaltic contractions of the digestive tract are realized. According to Marcus (1926) and to Bronstein (1937), the lateral intestinal nerves diverge directly from the trifid nerves, which would imply the interference of the great motor peripheral nerve in the visceral movements.

We have little information about possible sensory receptors in the wall of the digestive tract. In *L. cristallinus,* Marcus observed a ring of nerve fibers bearing terminal sensory cells in the pharyngeal epithelium, proceeding directly from the peripharyngeal ganglion. Methylene blue-positive cells are found also in the same location in *E. pilosa.* However, these presumed sensory cells around the mouth orifice may be considered as belonging to the base of the lophophore. Ciliated sensory cells have been recently found between the vacuolated cells of the scalariform epithelium of the esophagus in *Zoobotryon verticillatum* (Bullivant and Bils, 1968).

A specific cytological problem in bryozoans is how nerve impulses are conveyed to the intracellular striated myofibrils in the vacuolated cells of the scalariform epithelium of the esophagus. The ultrastructural characters of these cells have been studied by Bullivant and Bils (1968) in *Zoobotryon verticillatum* and by Matricon (1973) in *Alcyonidium polyoum.* In *A. polyoum,* electron micrographs show that nerve bundles are enclosed at the base of the vacuolated cells between the cell membrane and the lamina; groups of vesicles in the nerve fibers suggest the proximity of motor nerve to myoepithelial cell contacts, but the exact mode of such neuroepithelial junctions remains to be elucidated. In *E. pilosa,* similar ultrastructural pictures are observed and silver stains reveal a tight network of intricate nerve fibers all over the esophagus.

2. *The Innervation of the Lophophore*

a. The Tentacle Nerves. In Phylactolaemata, according to Gewerzhagen (1913a), an axillary branch arises from the peripharyngeal ganglion in each interval between adjacent tentacles, and soon divides dichotomously into two ramifications, or "radial nerves," diverging to the two neighboring tentacles [Fig. 4 (1)]. Each tentacle thus receives the opposite ramifications of the two inter-tentacular forks on both sides and is closely connected with the preceeding and

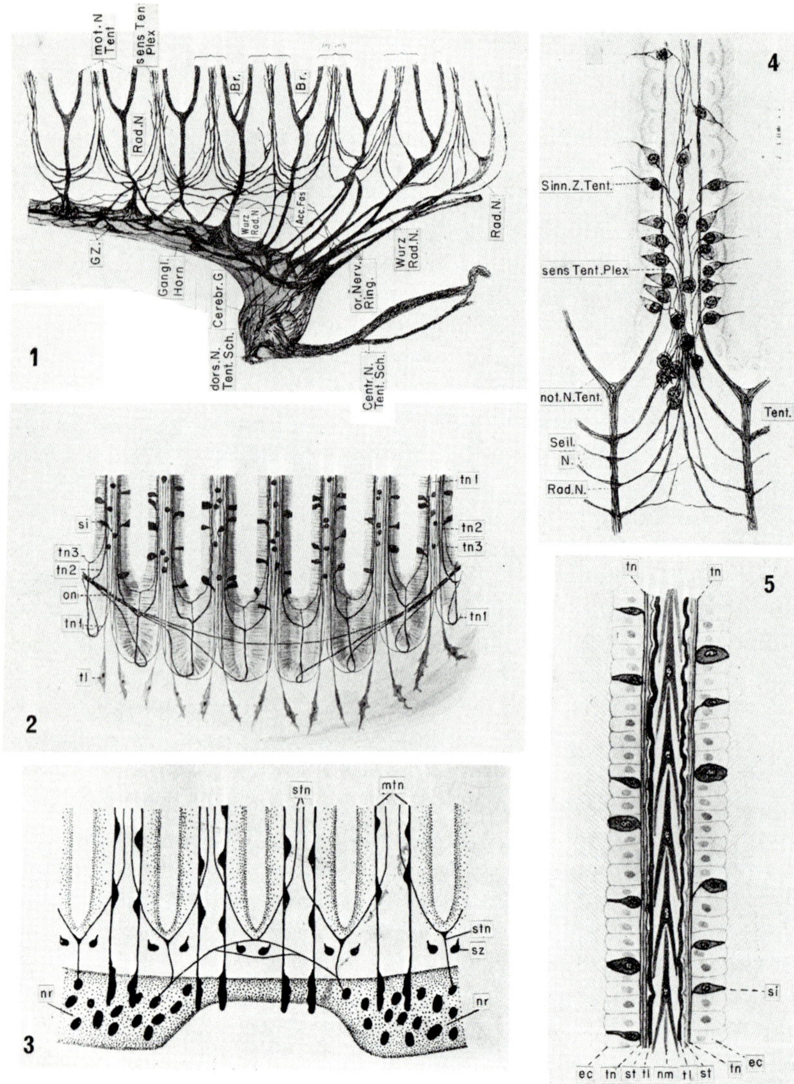

Fig. 4. Innervation of the lophophore. (1) In *Cristatella mucedo* after Gewerzhagen (1913a). (2) In *Lophopus cristallinus* after Marcus (1934). (3) In *Flustrella hispida* after Graupner (1930). (4) Sensory cells in the tentacles of *C. mucedo* after Gerwerzhagen (1913a). (5) Sensory cells in the tentacles of *L. cristallinus* after Marcus (1934). Abbreviations: (1) and (4) GZ, ganglionic cells; Rad.N., radial nerves; Wurz.rad.N, roots of the radial nerves; N.Tent.Sc., nerves of the tentacle sheath; Sinn.Z, sensory cells; Tent.Ple., median sensory strand; mot.N.Tent., motor tentacle nerves; Set.N, roots of the median strand; (3) mtn, motor tentacle nerves; stn, sensory tentacle nerves; sz, sensory cells; nr, initial neurons; (2) and (5) si, interstitial sensory cells; tn_2 and tn_3, radial nerves; tn_1, median strand; tl, motor tentacle nerves; ec, epithelial cells; st, elastic layer of the tentacle; rm, radial muscles.

following ones by the common stem of their "radial nerves," which Gewerzhagen interpreted as motor nerves. Diffuse nerve fibers bearing sensory cells arise from the axillary forks in a loose median strand, or "internal plexus." Marcus (1934), in *L. cristallinus,* considered this median strand as a simple diffuse extension of the "radial nerves'" which he interpreted as subepithelial sensory nerves; along the oral edge of the tentacle, interstitial sensory cells in the epithelium would be sporadically attached on the underlying "radial nerves" [Fig. 4 (3 and 4)]. Another pair of motor nerves would arise from periodic neurons at the base of every tentacle and follow the muscle chevrons inside the tentacle canal. This general pattern is fundamental in all bryozoans. In the Gymnolaemata, Marcus (1926), Graupner (1930), and Bronstein (1937) confirmed that there are two pairs of tentacle nerves which are (a) basiepithelial sensory nerves, arising from the axillary forks and running along the oral edge of the tentacle and (b) oral and dorsal motor nerves in the tentacle canal.

Recent ultrastructural studies of the tentacle by Smith (1971) in a ctenostome (*Flustrellidra*) and by Lutaud (1973) in a cheilostome (*Electra*) confirm that two pairs of typical nerve fascicles, apparently lacking any kind of differentiated envelope, are indeed found in transverse sections of the tentacle. However, all four nerves run between the epithelium and the collagenous substrate (Plate, II, 1). The branches of the axillary forks and the median oral nerve extend under the three cellular rows of the oral edge of the tentacle; the fourth lies midway between the dorsal ridges of the collagen tube. No nerve fascicle of comparable importance is found inside the tentacle canal along the longitudinal tentacle muscles. As muscles must receive nerve impulses, the coinciding position of the median oral and dorsal nerve fascicles and of the subjacent oral and dorsal groups of muscle fibers on both sides of the collagenous layer, is striking enough to presume that these are the motor pathways and that passages pierced through the collagen barrier allow either local contacts between nerves and muscle fibers or the passage of nerve endings.

b. The Sensory Function of the Tentacles. The extended lophophore is, of course, the main agent of the reception of external variations, but we do not know what senses are developed in Bryozoa, or what is the differentiation of potential receptors in the tentacles. On silver-impregnated specimens of *C. mucedo,* Gewerzhagen (1913) observed interstitial fusiform cells with a hairy terminal process, borne by the loose fibers of the median nerve bundle and irregularly distributed in bunches in the epithelial layer of the tentacle [Fig. 4 (4)]. In *L. cristallinus* and *F. repens,* Marcus (1926), using methylene blue, stained two rows of sporadic columnar cells among the epithelial cells, attached on the underlying "radial nerves" [Fig. 4 (2 and 5)]. Graupner (1930) found similar methylene blue-positive cells scattered along the oral edge of the tentacle in *F. hispida* and *M. membranacea* and also indicated that the nerve bundles end with a group of fusiform cells at the tip of the tentacles. Silbermann (1906), in *Alcyonidium* *mytili,* described tiny interstitial cells bearing rigid setae, inserted between the epithelial cells. The existence of such setae has been doubted. However, in all cheilostomes, the cuboidal cells of the oral edge of the tentacle bear two continuous rows of strong cilia with conspicuous basal corpuscles, easily seen with any ordinary dye on histological sections (Lutaud, 1955).

In the Gymnolaemata, the epithelial layer of the tentacle includes nine longitudinal rows of cells (Lutaud, 1973b). The oral and lateral cells, in the prolongation of the intestinal esophageal epithelium, derive from the atrial side of the initial tentacle fold in the polypidean bud; they are all ciliated, but differ in the infrastructure of their ciliature (Smith, 1973a; Lutaud, 1973b). They are separated by the dorsal ridges of the collagen armature from the nonciliated cells of the two dorsal rows which are derived from the dorsal portion of the initial tentacle fold in continuity with the epithelial layer of the tentacle sheath. The median oral cells and the lateral cells bear the oral and lateral fringes of vibratile cilia, with basal corpuscles in kinetal lines. The paired oral cells, next to the median oral cells, have long been noticed for their basophilic cytoplasm and dense nuclei; their ciliary apparatus is reduced to a strong single cilium per cell.

Ultrastructural observation confirms the peculiar differentiation of these monociliated cells (Lutaud, 1973b). They have a pedunculate shape and their narrow convergent extremities are inserted on the collagen substrate between the lateral oral nerves, or "radial nerves," and the median oral nerve running under the median oral cells (Plate II, 1). Their single cilium emerges in a depressed cupula amid extensive and ramified microvilli, and is striking for the reinforcement of its infrastructure. Its large basal corpuscle is supported by three unequal roots with transverse periodic striations. The axial root reaches the basal membrane of the cell, squeezed in the narrow peduncle between the nerve fascicles. Its location is such that any shock or vibration transmitted to the root by the external cilium should be registered by the contiguous nerve fibers. Moreover, the long axial root is surrounded, in the basal region of the cell, by numerous agranular vesicles of 500 Å; concomitant chains of vesicles in the adjacent nerve fibers suggest that the adaptation of these monociliated cells is even more elaborate and that they are connected to the adjacent nerve bundles by synaptic neuroepithelial junctions (Plate II, 2).

Plate II. The sensory innervation of the tentacles in *Electra pilosa.* (1) Basiepithelial location of the oral tentacle nerves and insertion of the monociliated epithelial cells on the collagen (magnification: × 16,000); (2) microvesicles (mv) in the narrowed basal region of the monociliated cells (magnification: × 20,000). Abbreviations: cl Ep, monociliated epithelial cell; m.Ep, median epithelial cell; rcl, ciliary root; l.N, lateral oral tentacle nerve; m.N, medial oral tentacle nerve; C, collagen layer; M, internal muscle fibers.

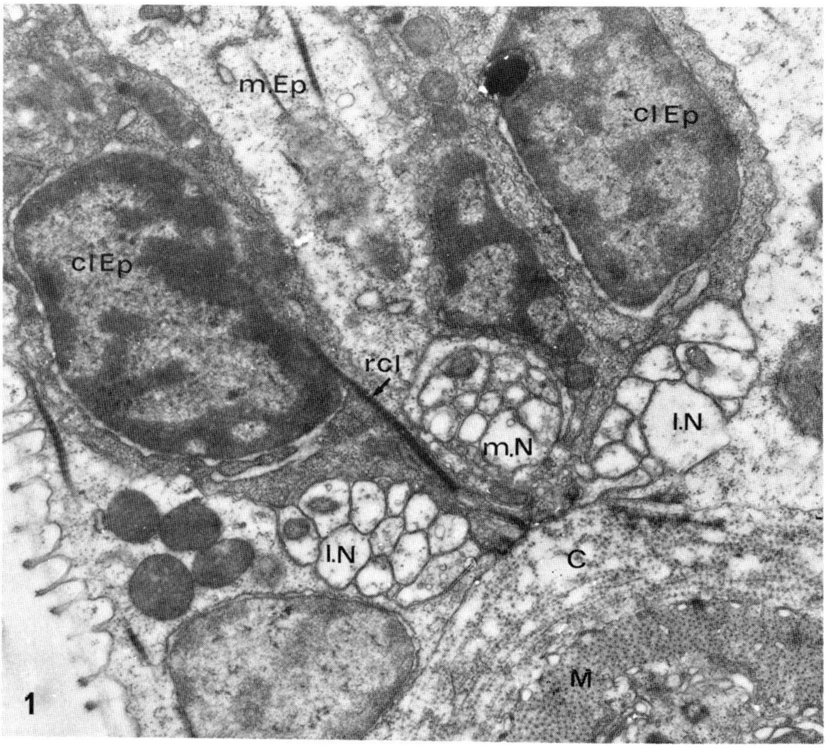

1

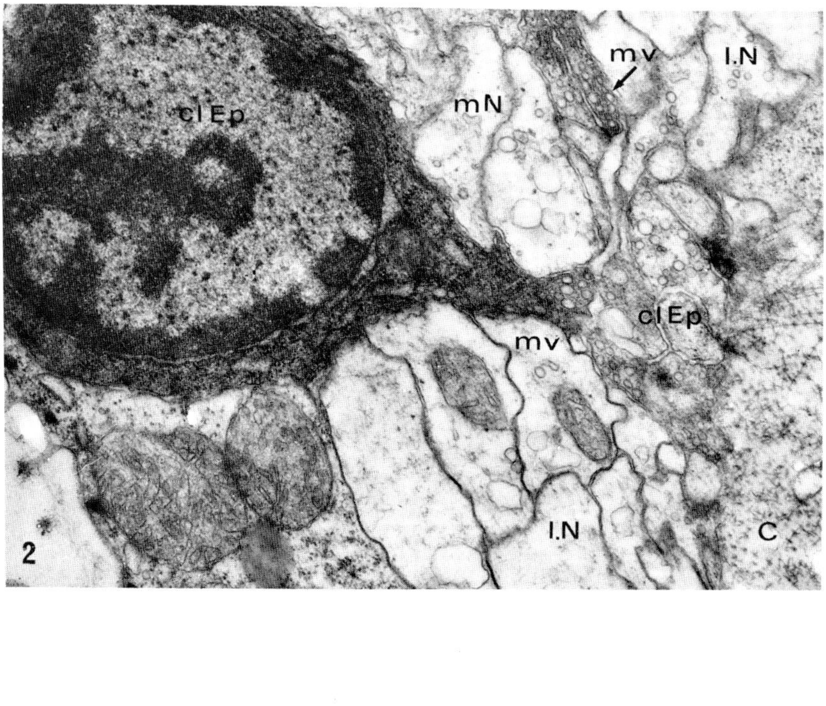

2

Thus, the tentacles of the Gymnolaemata are provided with two continuous rows of mechanoreceptors permitting graded touch possibilities. However, the cytological features and the homology of the different sensory cells mentioned by several authors in different species have not been established, and the specialization of the monociliated epithelial cells does not exclude that other kinds of sensory receptors (either other modified epithelial cells of different adaptation or interstitial nerve endings) able to register light, temperature, or chemical variations, could be found in the tentacles on further investigations.

III. THE INTERZOOIDAL INFORMATION HYPOTHESIS

A. Presumed Manifestations of Colonial Coordination

Two potential functions must be discerned in nonmotor parietal innervation: the individual perception of the environment at the level of the external walls, and the perception of the activity of other zooids within the colony. We may assume that the external wall is not inert and that some parietal sensitivity is possible in the Phylactolaemata and in the ctenocheilostomes either through a reticular parietal plexus or through the diffuse superficial fibers of a sensory parietal ramification of the great mixed tentacle sheath nerves. However, the communication of information or impulses between zooids, inducing coordinated responses, still needs to be proved either by anatomical evidence of the existence of neuronal bonds across the partitions or by a careful demonstration of collective or chain reactions of adjacent or distant zooids to a local experimental excitation.

Simultaneous responses of several polypides or heterozooids in the colony to a natural stimulation have sometimes been noticed in a few behavior observations which do not establish the necessary intervention of nervous pathways. Synchronization in the extrusion periods of groups of polypides has sometimes been alluded to, but may be the simple result of the life cycle and of the alternation of rows of polypides of successive generations within the colony. Collective orientation movements of the extruded lophophores may occur, but individual reactions to incidental stimuli might disturb or counterbalance a collective impulse. Rhythms in the activity of the heterozooids have been recorded. Cook (1963) observed how the activity of the vibracular setae in lunulitiform bryozoans maintains the stability of the conical zoarium on a moving sandy substrate and may restore an overturned colony to its normal position. However, Cook concluded that this phenomenon was the result of the mechanical effect of external pressures on the setae. Marcus (1926) observed the beats and latent periods of the vibracular setae in the Scrupocellariidae by electric recording methods. He tested the influence of light, salinity, and temperature on the rhythms of the oscillations of the pedunculate avicularia of *Bugula flabellata.* He could not detect any unambiguous synchronization or sequence in the activity of the vibracular setae or avicularia along the zoarial branches. Another remarkable manifestation of collective movements in stoloniferous or arborescent colonies is the strange modifications in the orientation of entire zooids or whole zoarial branches. This phenomenon occurs in the phylactolaemates, stoloniferans and cellularine species, and has been studied by Silén (1950) in stoloniferans and the Scrupocellariidae. In *Mimosella* and *Farrella,* the autozooids, under unspecified conditions, can lean collectively toward the stolon, then resume the erect position. In the species where this phenomenon has been observed, the autozooids are articulated and attached on the stolons by intercalary kenozooids. Silén compared this movement to the oscillations of the avicularia in the Bugulidae and reported that the inflection is due, in both cases, to a special parietodepressor muscle at the base of the zooid. In *Kinoteskias,* the fan-shaped arborescent colony is animated by alternating flections and straightenings of the zoarial branches: the contraction of a special muscle at the base bends the autozooids one after another until the whole zoarial branch is curled; with relaxation of the muscle, the strain of the calcified wall restores the straight position. The nature of the initial excitation and the determinant of the propagation of the contraction along the zoarial branches are unknown.

Researchers are now applying modern electrophysiological techniques to bryozoans, trying to detect the intervention of a nervous impulse in the propagation of an electric stimulus from the stimulated zooid to other zooids in the colony. The difficulty of these experiments is to distinguish collective responses to a localized factor from the repercussions of the movements of a polypide in adjacent autozooids through simple hydrostatic regulations. Any change in the position of a polypide causes turbulences in the fluid content of the body cavity and affects the communication organs which are subjected to the tractions of the funicular strands. Taking these indirect effects into account, Thorpe (1973) stimulating electrically the frontal wall of any given zooid in a colony of *Membranipora membranacea* with a silver microelectrode observed that the propagation of the retraction reaction of successive polypides is bidirectional. He measured the latency of the retraction of the polypides and the facilitation of the responses under repeated stimulation, and concluded that the habituation is degressive in space and depends on the interval between repeated stimulations. Recordings of electrical activity show that electric pulses are conducted between zooids as far as 15 successive units. Cuts across the colony stop the propagation of the responses. Thorpe concluded from the shape and the velocity of the responses that the intervention of nervous conduction was necessary and that colonial coordination was a reality. He noticed a sensitivity of the frontal wall and of the spines, thus corroborating the results of anatomical observation.

B. Presumed Pathways of Colonial Coordination

1. The Reticular Parietal Plexus in Phylactolaemata and Stolonifera

In the phylactolaemate *C. mucedo,* Gerwerzhagen (1913a), using methylene-blue injections *in vivo* and the silver impregnation techniques after Cajal or Bielschowsky, revealed a reticular network of large multipolar cells which are united by their delicate processes in the tentacle sheath of *C. mucedo.* This network spreads further into the body wall from the aperture to the common sole of the colony [Fig. 5 (1)] . Marcus (1934) in *L. cristallinus* observed branches of the tentacle sheath nerves spreading into the wall of the tentacle sheath, which he called "Cystidnerven." He mentioned plexus cells in the body wall. Using methylene blue in vital incubations, he had previously stained a similar reticular plexus in the body wall of the erect zooids of the stoloniferan ctenostome *F. repens* [Fig. 5 (3); Marcus, 1926] . He could neither follow the parietal plexus near the obscured insertion of the autozooid on the stolon, nor find nerve cells in the stolon wall. Therefore, he was not convinced that the plexus extended through the basal septum of the zooid and, consequently, that autozooids were connected throughout the colony by the continuity of the nerve net along the common stolons.

The parietal plexus of the Stolonifera has been recently stained again with methylene blue in a vesicularian species, *Bowerbankia gracilis* (Lutaud, 1974). The sudden appearance of the orthochromatic cells, their characteristic bipolar or multipolar shape, and the extent of their thin and branched intercellular prolongations are immediately convincing. In long active zooids, the meshes of the plexus show a tendency to a longitudinal orientation and bipolar cells are prevalent. The parietal plexus persists in regressed zooids with brown bodies.

Methylene blue-positive cells have been found in the wall of the stolons in *B. gracilis* distributed in discontinuous longitudinal chains. The fragmentation of the plexus in stolons is probably a lethal effect of a high concentration of the dye, and the negative result of Marcus in *F. repens* may be attributed to the slow penetration of reagents into the fragile tubes. The observation of the basal insertion of the autozooid, with an active polypide inside, is indeed uncertain and connections between the nerve nets in the autozooid and in the stolon have not been ascertained. However, the very presence of parietal nerve cells in the stolons where they cannot be endings of the peripheral nerves of an absent polypide supports the presumption that the plexus is continuous throughout the colony. Large multipolar cells are stained by methylene blue in precocious autozooidal buds. This embryonic plexus with the other layers of the wall should participate in the formation of partitions. Then interzooidal bonds, if the plexus is not interrupted by the septum, should cross through the central communicative organ.

2. *Hiller's Plexus in the Electridae*

a. General Features. It has been implicitly accepted that the superficial reticular plexus found both in Phylactolaemata and Ctenostomata is a general feature of the wall in all bryozoans. Bronstein (1937) assumed that it could be found also in arborescent or encrusting anascan cheilostomes such as *Bugula* or *Membranipora.* This opinion was soon contradicted by the discovery of another pathway, different in its precise topographical pattern and its internal location. In two species of the genus *Electra, E. crustulenta* and *E. pilosa.* Hiller (1939), using the usual methylene-blue vital technique, revealed a simple continuous filament with nuclear nodes, running at the base of the longitudinal and transverse partitions at the periphery of the encrusting cystid. He interpreted this as a linear chain of plexus cells [Fig. 5 (2)]. The parietal filament of adjacent zooids on both sides of the double-walled partitions are periodically linked by short transverse anastomoses to every poreplate in the longitudinal partitions and through some of the irregular pores in transverse partitions. Twin connections of the same texture, united by a transverse bridge, run along the tentacle sheath between the great compound nerves and farther along the dorsal ligaments of the sheath, linking the cerebral ganglion to the parietal filament at the distal basal corners of the cystid. Thus, the cerebral centers of all the polypides within a colony would be interconnected through interzooidal bonds by the monozooidal meshes of a colonial network.

The network discovered by Hiller in the Electridae has been recently stained in *E. pilosa* by methylene blue as well as by the Bielschowsky silver technique on decalcified whole mounts (Lutaud, 1969). Since then, it has been partly or completely observed in half a dozen other encrusting species of anascan and even ascophoran cheilostomes with little variation in pattern.* In *E. pilosa,* the continuity of the network with the cerebral center is an established fact. The junction of the connections arising from the ganglion along the tentacle sheath with the parietal filament is clearly apparent in preserved silver impregnated specimens as well as in living animals (Plate III, 1 and 2). Silver stains show that the roots of the cerebral connections of the plexus are clearly distinct from the bundles of the two other peripheral nerves in the cerebral ganglion and arise in the prolongation of the superficial pair of dense cells in the central cellular aggregate.

The parietal filament is perennial and persists during the periodic renewal of the polypide in empty cystids with brown bodies. However, the cerebral connections of the network are broken by the destruction of the tentacle sheath and cerebral ganglion of the degenerating polypide. We do not know how new

*These species are *Electra pilosa, Electra posidoniae, Electra monostachys, Tegella tenuis, Membranipora membranacea, Flustra papyracea, Haplopoma impressum, Chorizophora brongnartii.*

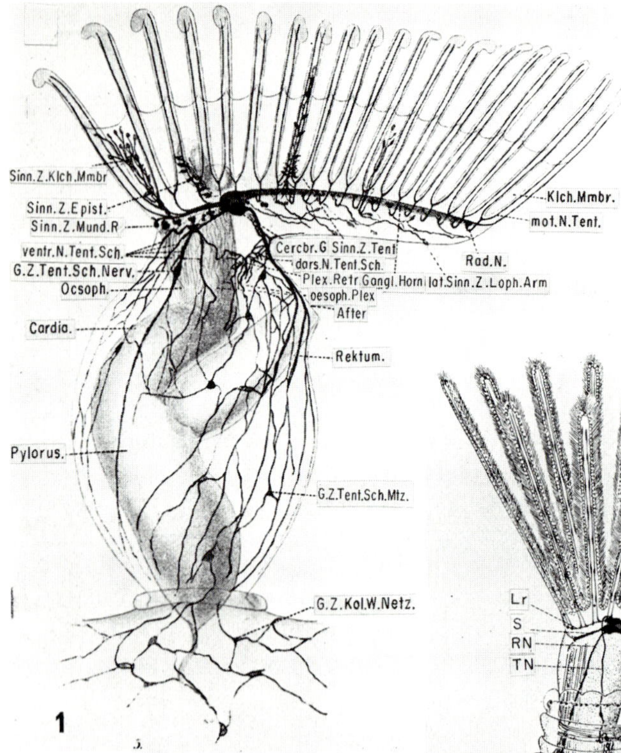

Sinn.Z.Klch.Mmbr
Sinn.Z.Epist.
Sinn.Z.Mund.R
ventr.N.Tent.Sch.
G.Z.Tent.Sch.Nerv.
Ocsoph.
Cardia.
Pylorus.
Cercbr.G Sinn.Z.Tent
dors.N.Tent.Sch.
Plex.Retr Gangl.Horn
oesoph.Plex
After
Rektum.
Klch.Mmbr.
mot.N.Tent.
Rad.N.
lat.Sinn.Z.Loph.Arm
G.Z.Tent.Sch.Mtz.
G.Z.Kol.W.Netz.

1

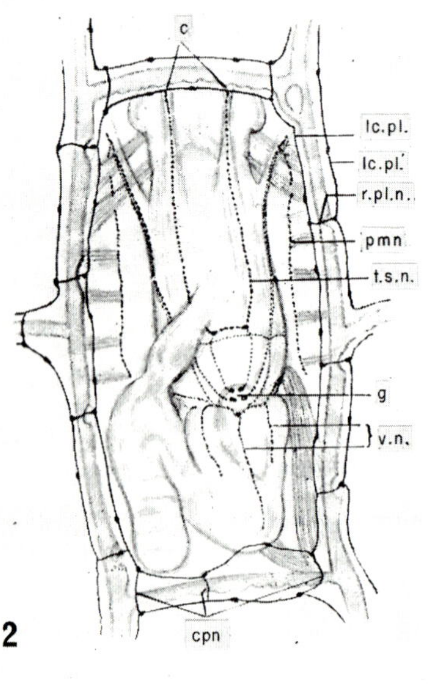

c
lc.pl.
lc.pl.
r.pl.n.
pmn
t.s.n.
g
v.n.
cpn

2

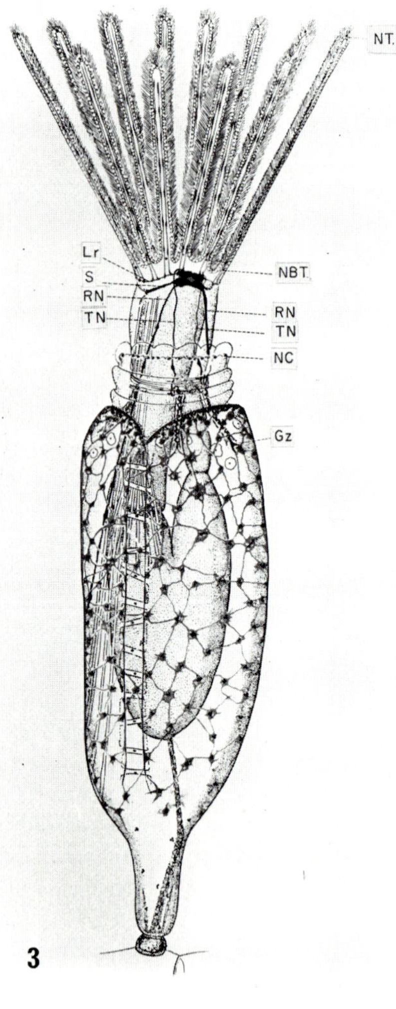

NT.
Lr
S
RN
TN
NBT
RN
TN
NC
Gz

3

connections are reestablished with the cerebral ganglion of regenerated polypides. The parietal filament is always present in the partitions of kenozooids. In *Haplopoma impressum,* it has also been observed around the basal wall of the ovicell.

a. The Presumed Interzooidial Bonds. The transverse bonds which periodically link the opposite parietal filaments on both sides of the partitions as the bars of a scale, are not symmetrical but oriented according to the bulge of the poreplates (Plate IV, 1–5). At the level of every communication chamber on the concave side, a short transverse ramification diverges from the parietal filament, penetrates with the funicular ramifications into the chamber and ends in a terminal knob next to the opposite filament running in the adjacent zooid on the convex side of the pore plate. The pore plate is thin and calcitic on live animals or decalcification of preserved specimens could induce optical distortions. It remains to ascertain whether the transverse anastomoses really cross the pore plate, either through a special pore or between the rosette cells of one of the pores, or whether they are interrupted at the level of the rosettes (Bobin, this volume, Chapter 9). The demonstration of junctions between the plexus chains of adjacent zooids would provide conclusive evidence of the existence of a nervous coordination of the zooidial individuals within a colony. Therefore, the precise structure of the terminals of the transverse ramifications must be very closely examined. However, even, if there is no true junction between the parietal filaments of adjacent zooids across partitions, the mere existence of nerve endings in the interzooidial communication organs is significant. It implies that the cerebral center of any polypide may register perturbations in the position or in the turgidity of the rosettes, and thus be indirectly informed of the activity of the surrounding polypides.

b. The Ultrastructural Characteristics of Hiller's Plexus in Electridae. The neuronal nature of Hiller's network in the partitions of the Electridae is a presumption founded on its affinity for the specific stains for nerves, on its relation with the interzooidal communication organs, on its colonial distribution, and on its continuity with predominant cells in the cerebral ganglion. However, this option has to be corroborated by ultrastructural examination (Lutaud, preliminary account).

Fig. 5. The parietal plexus. (1) In the tentacle sheath and in the body wall of *Cristatella mucedo* after Gewerzhagen (1913a); (2) in the partitions of *Electra pilosa* after Hiller (1939); (3) in the body wall of *Farrella repens* after Marcus (1926). Abbreviations: (2) lc.pl., circular colonial plexus; r.pl.n. and cpn, connections through pore plates and simple pores; pmn, motor parietal nerve; t.s.n., tentacle sheath connectives to the ganglion; g, cerebral ganglion; v.n., visceral nerves; (3) Gz, plexus cells; LR, lophophoral belt; NBT, nerve cells at the base of the tentacles; NC, neurons in the collar; NT., terminal nerve cells in the tentacles; RN, retractor nerve; TN, nerves of the tentacle sheath.

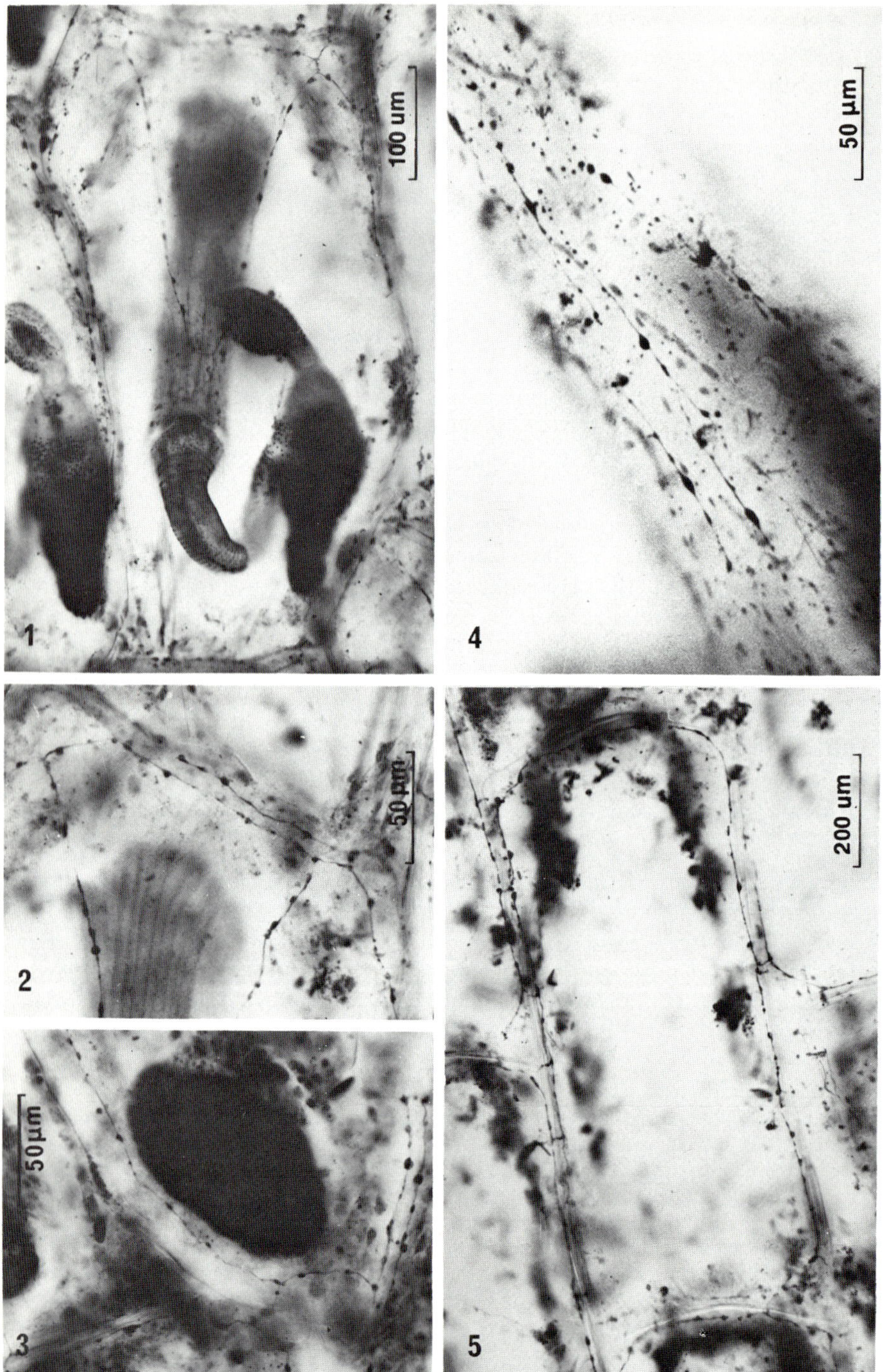
1
100 um
4
50 µm
2
50 µm
3
50µm
5
200 um

The parietal filament and its cerebral connections have been observed in *E. pilosa,* at a direct magnification of 5000–15,000 X on transverse thin sections of the partitions, of the tentacle sheath, and of the sheath ligaments. Electron micrographs show the filament as a single naked fiber 1–2 μm in diameter, lying in the wall between the epithelium and the peritoneal cells, and in the tentacle sheath between the collagen lamina and the peritoneal lining as do the typical fascicles of the peripheral nerves. The clear cytoplasmic matrix encloses a few microtubules, sparse strands of granular reticulum, and characteristic axial mitochondria grouped by threes or fours in the center (Plate V, 3). Except for its comparatively giant size, the section of the parietal filament in internuclear segments is very similar to the section of any ordinary nerve fiber in polypidean or peripheral nerve bundles. The nuclei are enclosed in the filament and do not belong to adventitious cells. This confirms the interpretation of the parietal network as a linear succession of abutting bipolar cells.

*c. **The Homology of the Parietal Plexus in the Stolonifera and Anasca.*** By its superficial distribution in the external wall, the reticular plexus of the tubular zooid in ctenostomes is topographically different from the internal pathway discovered by Hiller in encrusting anascans. However, the plexus of the ctenostomes is cytologically and functionally analogous to the network in the partitions of the Electridae by its cellular structure, by its permanence in the cystidean wall during the phases of the life cycle and in kenozooids or stolons, and by its presumed function in colonial coordination. We do not actually know the pathways of parietal innervation in encrusting ctenostomes (Carnosa). However, we may postulate that there could be a relation between the evolution of the structure of the autozooid from the primitive erect form to the encrusting form and the organization of a more precise pathway with the multiplication of partitions and communication organs in multiserial species, which could lead to the separation of several pathways with different functions in parietal innervation.

Plate III. The parietal plexus in the Stolonifera and in the Anasca. (1) The cerebral connectives of Hiller's plexus in *Electra pilosa*; (2) the parietal plexus along a transverse partition in *E. pilosa*; (3) the parietal plexus in a regressed zooid with a brown body in *E. pilosa;* (4) bipolar cells in the distal region of the tubular autozooid of *Bowerbankia gracilis*; (5) the parietal plexus along the partitions of *Membranipora membranacea* (after excision of the polypide); (1) to (3) and (5), methylene blue, preserved specimens; (4) live specimen, photograph, R. M. Zottoli.

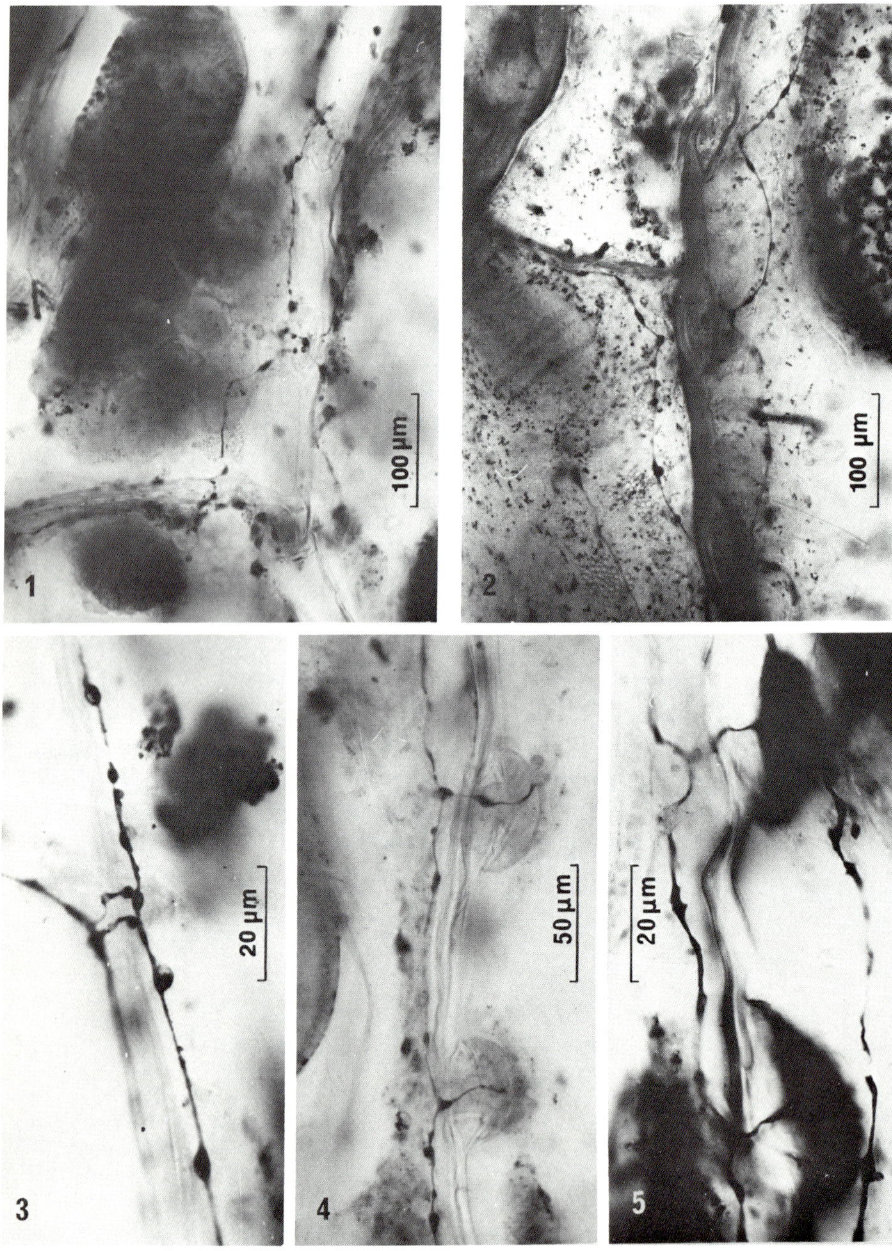
1
100 µm
2
100 µm
3
20 µm
4
50 µm
5
20 µm

IV. DISCUSSION

A. The Adaptations of the Nervous Pathways to Specific Biological Phenomena in Bryozoa

1. Polymorphism and the Innervation of the Heterozooids

The avicularia and vibracula of the cheilostomes are modified zooids produced in certain budding or astogenic conditions. The body cavity of the heterozooids is filled with the predominant muscles which are the homologues of the abductor muscles of the operculum and activate the mandible or seta which correspond to modified opercula (see Silén, 1938). Avicularia and vibracula contain a small organ with the structure and function of a mucous gland, and this is the homologue of an undeveloped polypidean bud. This organ, which has been considered as a sensory organ, is subdivided by a median constriction into a cecal pouch corresponding to the digestive tract and a ciliated anterior atrium corresponding to the lophophore; the long cilia of the anterior subdivision emerge in a bundle through the small aperture.

In the pedunculate avicularium of the Bugulidae, a few nerve cells grouped in the constriction between the lophophoral and intestinal subdivisions of the avicularian organ are stained by methylene blue. The minute avicularian nerves, arising from this cellular cluster that has the topographical location and the significance of a cerebral ganglion, are difficult to follow through the calcified and rounded wall of the avicularium. Preliminary observations lead us to believe that there are a pair of motor nerves joining the tendon of the abductor muscle of the mandible, a pair of nerves joining a few nerve cells at the hinge of the mandible, and a pair joining the internal region of the beak of the avicularium. Interstitial methylene blue-positive cells in the ciliated epithelium of the anterior subdivision of the avicularian organ correspond to the sensory cells of the tentacles in a developed lophophore. It is probable that a close inventory of the avicularian nerves would trace the equivalents of the main pathways of the autozooid. No interzooidal bond with the adjacent autozooid, or link between successive avicularia has been demonstrated so far. The activity of an avicularium is controlled by its own reduced ganglion.

Plate IV. The presumed interzooidal bonds of Hiller's plexus. (1) and (2) Topographical relation between the parietal fiber and the pore chambers in *Electra pilosa;* (3) the transverse branches of the parietal plexus through simple pores of a lateral partition in *Membranipora membranacea;* (4) and (5) the asymmetry of the transverse anastomoses of Hiller's plexus in the pore chambers of a lateral partition in *E. pilosa.* (1) (3) and (4), methylene blue, preserved specimen; (2), methylene blue, live specimen; (5) Bielschowsky silver stain.

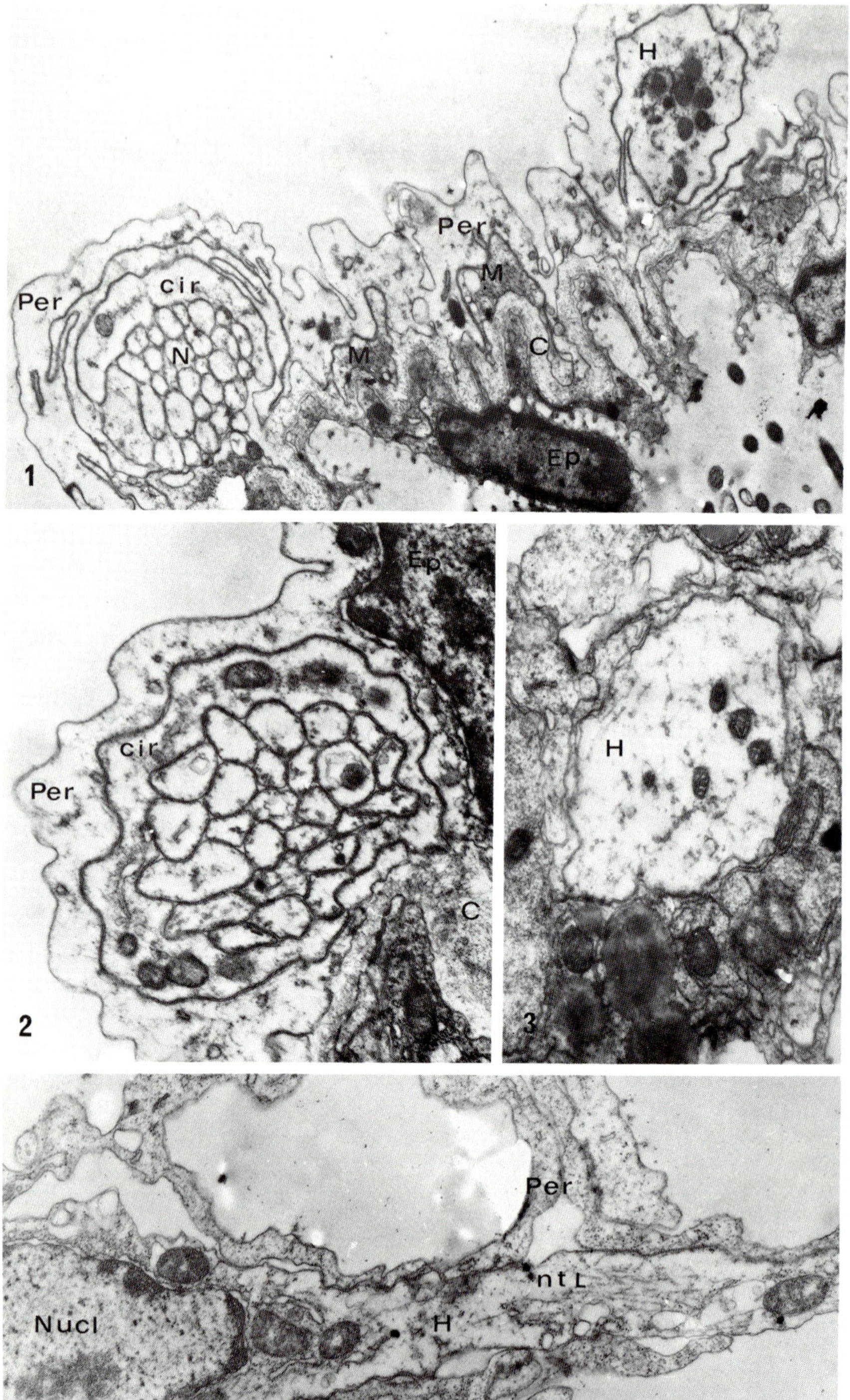
1
H
Per
M
cir
N
C
Ep
2
Ep
cir
Per
C
3
H
4
Per
nt L
Nucl
H

2. *The Periodic Renewals of the Polypide*

The anatomical and functional coordination of the parietal nerves and of the cerebral center are temporarily interrupted when the deciduous polypide degenerates and are reestablished with the polypides of second and following generations. The cerebral ganglion and the peripheral nerves, severed by the retraction of the tentacle sheath around the brown body, are periodically renewed.

In *E. pilosa*, the parietal filament and the presumed interzooidal bonds of Hiller's plexus in the partitions remain morphologically unaffected during the degeneration phases, except for the cerebral connections which are broken by the regression of the tentacle sheath. The sheath ligaments, the sphincter and parietodiaphragmatic muscles, and the great retractor muscle of the polypide are renewed, and there is no reason why the regeneration of the corresponding motor nerves would not follow the ordinary morphogenetic process. In the initial autozooidal bud, the cerebral ganglion is formed precociously, as early as the first subdivision of the polypidean vesicle into a dorsal intestinal pouch and a frontal lophophoral fold with the first indentations of the tentacles. The opercular and diaphragmatic muscles, and the corresponding motor ramifications of the tentacle sheath nerves, are formed simultaneously from the fan of fibrillar strands which join the top of the growing tentacle sheath to the aperture area. During the regeneration phase, the muscles of the parietal series bearing on the frontal wall (opercular abductors and parietodepressors) are not destroyed, and the question is how the motor parietal and opercular branches of the newly formed nerves of the regenerating tentacle sheath grow fibers toward the preexisting muscles. It is not impossible that nonmotor parietal fibers or sensory cells in the wall may persist and eventually perhaps undergo temporary dedifferentiations during the degeneration of the polypide. Contacts between the chains of plexus cells in the partitions and their connections with the new cerebral ganglion may be simply established. The restoration of junctions between the growing axons of new-formed nerves in the wall and preexisting muscle fibers needs a more elaborate explanation.

Plate V. Ultrastructural characters of the mixed peripheral nerve and of Hiller's plexus in *Electra pilosa.* (1) The mixed nerve and the cerebral connectives of the plexus on a transverse section of the tentacle sheath (magnification: × 10,000); (2) transverse section of the mixed nerve in the tentacle sheath (magnification: × 16,000); (3) transverse section of the parietal filament of the plexus at the base of a longitudinal partition (magnification: × 15,000); (4) longitudinal section of a plexus cell in the wall (magnification: × 12,000). Abbreviations: Per, peritoneal layer; N, nerve; cir, circumjacent element of the nerve bundle; H, Hiller's plexus cell or filament; M, muscle fiber; C, collagen; Ep, epithelium; Nucl, nucleus of the plexus cell; ntl, presumed neurotubules.

3. The Innervation of the Larva

A very primitive innervation exists in the larval stages although we do not know if anything remains of the undifferentiated neural strands of the larva after metamorphosis.

In the *Bugula* larva, a central neuromuscular cord of intermixed neurofibers and myofibers joins the apical disc to the roof of the metasomal sac, with ramifications to the group of flagellated cells next to the glandular pyriform organ as well as to an equatorial ring called the "subepithelial plexus" by Calvet (1900). This equatorial ring runs under the ciliated cells of the corona (Woollacott and Zimmer, 1971). A group of ciliated cells, at the insertion of the neuromuscular cord on the apical disc, has been interpreted as a tactile organ by early embryologists. Recent ultrastructural studies by Woollacott and Zimmer (1972) demonstrate that the epithelial cells of the pigmented spots in the corona are connected by simple junctions with underlying neural tracts in the equatorial ring, and that the pigmented spots may be photoreceptor organs.

The larva of *Alcyonidium* differs from the primitive *Bugula* type in the higher differentiation of the neuromuscular cord in which neural strands and myofibers are segregated in distinct bundles (D'Hondt, 1972, 1973a,b). Next to the apical disc lies a group of nerve cells that early embryologists described as a larval ganglion, and that Woollacott and Zimmer do not mention in the *Bugula* larva. The ciliated cells of the apical disc are connected by synapses with the "ganglionic" cells. The evolution of the neuromuscular system would be advanced in the cyphonautes larva of the Flustridae and malacosteges with the differentiation of a coronal muscle, of a pallial muscle, and of two valve adductor muscles which must be innervated by corresponding motor nerves.

B. The Basic Outline of Innervation in Bryozoa

It is difficult to appreciate how the anatomical and cytological features of the nervous system in Bryozoa might be used in a discussion of the phylogenetic affinities of the phylum. From the rudiments of ultrastructural observation that we have begun to assemble, we may simply assume that the differentiation of bryozoan nerves is rather primitive in the invertebrate scale. In the basic organization of the cerebral ganglion with anterior lophophoral arms and the posterior origin of the peripheral and visceral nerves, in the importance of the retractor motor nerve, in the periganglionic connection between the lophophoral and visceral stems, and in the absence of anything assimilable to a neural cord, we may find some similarities with the fundamental structure of the nervous center in other lophophorates which have long been related to the Bryozoa on the basis of other anatomical or ontogenetic characters. The relative topographical autonomy of the peripheral pathways and the divergence of the retractor and parietal motor nerves appear as an immediate consequence of the persistence of the spacious body cavity from the initial bud between the digestive tract and body wall of the adult, and the consequence of the anatomical continuity of the parietal layers of the wall and of the polypide through the tentacle sheath.

The location of the main neurons and commissural strands in the cerebral ganglion provides all coordination potentialities between the initials and terminals of the different areas of innervation which are controlled by clearly delimited anterior, median, and posterior cellular aggregates. Direct extracerebral exchanges occur through permanent connectives or incidental anastomoses between the perioral ring conveying the motor or sensory impulses of the lophophoral and the visceral nerves. The fundamental set of nerves in the polypide includes periodic pairs of motor and sensory tentacle nerves interconnected in the fibrillar bundles of the peripharyngeal ring, and two pairs of visceral nerves of undetermined function. In cheilostomes, the median and lateral visceral nerves are intermixed near their origin in a visceral ganglion. In their first autonomous course, the peripheral motor nerves command the antagonist retractor and tentacle sheath muscles before being absorbed in the great mixed tentacle sheath nerve. The sensory strands of these compound peripheral nerves include fibers to the wall of the tentacle sheath in its upper region and to the zooidal wall. Such fibers bearing terminal cells are particularly developed around the aperture. The evolution of the distribution of nerves within the phylum affects the level of temporary fusions or exchanges between parallel or overcrossing nerves as well as the divergence of the motor and sensory ramifications.

Nerves are simple bundles of coincidental and intermixed axonal and dendritic fibers provided with frequent nerve-to-nerve junctions and free to ramify any moment into the next body structure or associate with overcrossing strands. In the lophophore and the intestine, nerve bundles are basiepithelial and apparently naked, unless some primitive adventitious cell remains inconspicuous among the nerve fibers. However, the possibility of a neuroglial sheath has to be considered in the case of the great mixed peripheral nerves in the tentacle sheath. Neuroepithelial junctions are established between underlying nerve bundles and ciliated epithelial cells or myoepithelial cells which have undergone structural adaptations to some tactile or contractile function.

Two kinds of presumed sensory receptors have been found at present in Bryozoa: modified epithelial cells such as the monociliated cells of the tentacular epithelium or the pigmented "eye spots" in the larva, and fusiform cells stained by methylene blue at the end of the ultimate fibers of nerve bundles in the preapertural folds of the diaphragm, at the base of the tentacle sheath, in the zooidal wall of *Electra* at the base of the marginal spines, in the pharyngeal constriction of the digestive tract, and in the tentacles of the Phylactolaemata. Nerve cells or epithelial receptors are often precisely ordered, but multicellular sensory organs have not been described so far in the adult.

The extraordinary and most primitive character of the nervous system in the

Bryozoa is the incomplete autonomy of the autozooid in the colonial community. Behavioral and anatomical observations now converge toward the conclusion that colonial coordination may be a reality. Interzooidal information is probably conveyed by the parietal plexus which has been demonstrated in the Phylactolaemata and the Gymnolaemata. However, there is no "colonial nerve" as Hiller's plexus has been sometimes misinterpreted, which would imply both a higher degree of cytological differentiation of the nerve fibers and the incomplete individuality of the cerebral ganglion. Interzooidal information can be considered as the simple result of parietal sensitivity and of the colonial continuity of the wall, uninterrupted by the morphogenesis of the transverse or interserial partitions. In the evolution of the eurystomes from the primitive erect type to multiserial growth, the reticular plexus tends toward a more precise distribution, related to its connection with the interzooidal perforations in the partitions. Its connection with the cerebral center becomes separate from the main tentacle sheath nerves. In the present state of investigation, the colonial continuity of the parietal plexus, through simple or elaborate junctions between the networks of adjacent zooids, has not been demonstrated. However, in the Electridae, transverse ramifications of the plexus chain do penetrate into every interzooidal communication chamber, and must be taken into account as the likely means of a permanent direct or indirect perception of the activity of the surrounding zooids, sufficient to induce quick collective responses.

V. SUMMARY

The monographic study of the distribution of the main nervous pathways in the anascan cheilostome *Electra pilosa* (Linné) leads to the conclusion that the innervation is more elaborate in bryozoans than first estimated from early descriptions, although the ultrastructural characters of basiepithelial nerves in the polypide and of presumed sensory endings or epithelial receptors in the wall or in the lophophore show there is a low level of cytological differentiation.

The main nerves serving the different areas of innervation, which are polypidean, parietal, and colonial pathways, arise from distinct cellular aggregates within the cerebral ganglion. Two pairs of sensory and motor nerves arise in every tentacle from the annular prolongation of the cerebral ganglion at the base of the lophophore. Two pairs of visceral nerves, linked to the lophophore belt by permanent connectives, arise from an accessory visceral ganglion. The sensory and motor strands of the mixed peripheral nerves following the tentacle sheath toward the aperture serve the zooidal wall and the external muscles involved in the protrusion and retraction of the polypide.

Two functions are to be discriminated in nonmotor parietal innervation: information about the environment through sensory parietal ramifications of the peripheral nerves and colonial information through a parietal plexus. The hypothesis of interzooidal information leading to colonial responses, through the continuity of the plexus across the partitions, has been confirmed by the demonstration of the existence of a diffuse plexus in the zooidal wall of the Phylactolaemata and of stoloniferous ctenostomes. A chain of plexus cells has been observed in the partitions of encrusting cheilostomes. In *E. pilosa,* the ordered pattern of this plexus is closely related to location of interzooidal communication organs in the transverse or interserial partitions.

REFERENCES

Bronstein, G. (1937). Etude du système nerveux de quelques Bryozoaires Gymnolémides. *Trav. Stn. Biol. Roscoff* **15,** 155–174.

Bullivant, J. S., and Bils, R. F. (1968). The pharyngeal cells of *Zoobotryon verticillatum* (della Chiaje), a gymnolaemate bryozoan. *N. Z. J. Mar. Freshwater Res.* **2,** 438–446.

Bullock, T. H., and Horridge, G. A., eds. (1965). "Structure and Function in the Nervous System of Invertebrates," Vol. 1. Freeman, San Francisco, California.

Cook, P. L. (1963). Observations on live lunulitiform zoaria of Polyzoa. *Cah. Biol. Mar.* **4,** 407–414.

D'Hondt, J. L. (1972). Métamorphose de la larve d'*Alcyonidium polyoum* (Hassall), Bryozoaire Cténostome. *C. R. Hebd. Séances Acad. Sci.* **276,** 767–770.

D'Hondt, J. L. (1973a). La musculature de la larve d'*Alcyonidium polyoum* (Hassall), Bryozoaire Cténostome. Comparaison avec la musculature larvaire d'autres Bryozoaires Eurystomes. *C. R. Hebd. Séances Acad. Sci.* **276,** 1449–1452.

D'Hondt, J. L. (1973b). Etude anatomique, histologique et cytologique de la larve d'*Alcyonidium polyoum* (Hassall), Bryozoaire Cténostome. *Arch. Zool. Exp. Gén.* **114,** 537–602.

Gewerzhagen, A. (1913a), Beiträge zur Kenntniss der Bryozoen. I. *Z. Wiss. Zool.* **107,** 3–345.

Gewerzhagen, A. (1913b). Untersuchungen an Bryozoen. *Sitzungber. Heidelb. Akad. Wiss., Math.-Naturwiss. Kl., Abt.B* **9,** 1–16.

Graupner, H. (1930). Zur Kenntniss der feineren Anatomie der Bryozoen. *Z. Wiss. Zool.* **136,** 38–97.

Hiller, S. (1939). The so-called "colonial nervous system" in Bryozoa. *Nature (London)* **143,** 1069–1070.

Hilton, A. W. (1923). A study of the movements of ectoproctan bryozoans. *Trans. Am. Microsc. Soc.* **42,** 135–143.

Lutaud, G. (1955). Sur la ciliature du tentacule chez les Bryozoaires Chilostomes. *Arch. Zool. Exp. Gén.* **92,** 13–19.

Lutaud, G. (1969). Le "plexus" parietal de Hiller et la coloration du système nerveux par le bleu de méthylène chez quelques Bryozoaires Chilostomes. *Z. Zellforsch. Mikrosk. Anat.* **99,** 302–314.

Lutaud, G. (1973a). The great tentacle sheath nerve as the pathway of an innervation of the frontal wall structures in the cheilostome *Electra pilosa* (Linné). *In* "Living and fossil Bryozoa" (G. P. Larwood, ed.), pp. 317–326. Academic Press, New York.

Lutaud, G. (1973b). L'innervation du lophophore chez le Bryozoaire Chilostome *Electra pilosa* (L.). *Z. Zellforsch. Mikrosk. Anat.* **140,** 217–234.

Lutaud, G. (1974). Le plexus parietal des Ctenostomes chez *Bowerbankia gracilis* Leydi (Vesicularines). *Cah. Biol. Mar.* **15,** 403–408.

Marcus, E. (1926). Beobachtungen und versuche am lebenden Meeresbryozoen. *Zool. Jahrb., Abt. Syst. Oekol. Geogr. Tiere* **52,** 1–102.

Marcus, E. (1934). Über *Lophopus cristallinus* (Pallas). *Zool. Jahrb., Abt. Anat. Ontog. Tiere* **58,** 506–606.

Matricon, I. (1973). Quelques données ultrastructurales sur un myoepithélium: Le pharynx d'un Bryozoaire. *Z. Zellforsch. Mikrosk. Anat.* **136,** 569–578.

Retzius, G. (1905). Das sensible Nervensystem der Bryozoen. *Biol. Unters.* **12,** 49–54.

Silbermann, S. (1906). Untersuchungen über der feineren Bau von *Alcyonidium mytili. Arch. Naturgesch.* **72,** 265–310.

Silén, L. (1938). Zur Kenntniss der Polymorphismus der Bryozoen. Der Avicularien der Cheilostomata Anasca. *Zool. Bidr. Uppsala* **17,** 149–366.

Silén, L. (1950). On the motility of entire zoids in Bryozoa. *Acta. Zool. (Stockholm)* **31,** 349–386.

Smith, L. W. (1973). Ultrastructure of the tentacles of *Flustrellidra hispida* (Fabricius). *In* "Living and Fossil Bryozoa" (G. P. Larwood, ed.), pp. 335–342. Academic Press, New York.

Thorpe, J. P. (1973). An investigation of the colonial responses of the ectoproct *Membranipora membranacea* L. Thesis.

Woollacott, R. M., and Zimmer, R. L. (1971). Attachment and metamorphosis of the cheilo-ctenostome bryozoan *Bugula neritina* (Linné). *J. Morphol.* **134,** 351–382.

Woollacott, R. M., and Zimmer, R. L. (1972). Fine structure of a potential photoreceptor organ in the larva of *Bugula neritina* (Bryozoa). *Z. Zellforsch. Mikrosk. Anat.* **123,** 458–469.

12

Taxes and Tropisms of Bryozoans

J. S. RYLAND

I. INTRODUCTION

Behavioral responses to environmental stimuli, broadly categorized as taxes and tropisms, are exhibited by bryozoans in three phases of their life history: that is, by the free-swimming larva, the metamorphosing larva, and the adult colony. The extent to which larval dispersal is dependent upon currents and diffusion processes in the water mass is unknown, but in laboratory experiments, and presumably in still water, larval reactions to light and gravity apparently determine both the pattern of dispersal and the duration of the free-swimming period. Larval responses appear to be particularly significant with regard to the selection of settlement sites. There may then be further orientation in respect of directional stimuli during the process of metamorphosis. Colonies, with the exception of those of the freshwater *Cristatella mucedo,* have no extensive locomotory powers, but they may exhibit oriented growth responses to light and other stimuli.

The term orientation covers both those reactions which maintain an animal in its normal posture (primary orientation) and those which guide it into one particular habitat or another (secondary orientation). Among the former is included the positional orientation of swimming larvae. Cyphonautes normally swim with the apical organ in advance and the plume cilia directed downward (O'Donoghue, 1926; Cook, 1960), though Silén and Jansson (1972) describe *Electra crustulenta* larvae as adopting an upright swimming posture. Among lecithotrophic larvae it seems that the elongated kinds, found in *Bugula* and other forms with an erect ancestrula, swim with the apical organ directed forward and rotate about their oral–aboral axis; but the flattened larvae of encrusting species maintain the apical organ uppermost and have the pyriform complex in advance (Lynch, 1947). The larval responses to external stimuli considered later in this chapter are all secondary orientations.

Of the descriptive terms applied to orienting movements, tropism is unfortunately still ambiguously used. Originally applied to any directed movement, to most authors its use became restricted to the bending movements of plants and sessile animals (Fraenkel and Gunn, 1940). It is, therefore, a matter of surprise and regret to see some bryozoologists and others still referring to the oriented swimming movements of larvae as tropisms. The usage firmly established by Fraenkel and Gunn will be followed here, so that only the orientations of organisms–usually colonies in the context of the Bryozoa–by growth or bending are referred to as tropisms. The term *taxis* is used for directed orientation reactions involving locomotion. Tropisms and taxes may be positive or negative, depending whether they are directed toward or away from the source of stimulation. A *kinesis* is an undirected locomotory response, in which the speed of movement or frequency of turning depends upon the intensity of the stimulus. Kineses may be high or low depending whether the stimulus in question accelerates or inhibits the behavioral reactions.

Rotational responses of metamorphosing larvae, when the animal is not physically moving, cannot be classified as either taxes or kineses. There appears to be an analogy here with the postural response of locusts in relation to the sun's rays, termed akinesis by Volkonsky (1939); but such a term could easily be misunderstood, and it seems preferable to regard the static orientations of settling larvae as tropisms, as was done by Forbes *et al.* (1971) in their study of barnacle cyprids.

Many of the behavioral responses exhibited by bryozoan larvae relate to the habitats occupied by the adults. Thus there is a firmly established preference by many species for settlement in dark places. The location of lithophilic intertidal species on the undersides of stones will be familiar to anyone who has collected bryozoans on a seashore. It is equally obvious below rock overhangs, where bryozoans are abundant in (and indeed may dominate) what Evans (1949) has called the cryptofaunal community. Here the rock surface may be matted with dense growths of *Bugula, Crisia,* and *Scrupocellaria* in north temperate latitudes, or of *Nellia, Poricellaria, Synnotum,* and *Tetraplaria* in the tropics. Likewise, the prevalence of bryozoans in caves is well established (Riedl, 1966; Harmelin, 1969; Médioni, 1972; Norton *et al.*, 1971). This pattern of occurrence is not confined either to shallow water, for bryozoans occur beneath flat (e.g., agaricid) coral plates at all depths accessible by Scuba diving, or to the sea, for freshwater bryozoans are distributed in an exactly comparable way. However, it must not be assumed that bryozoans are invariably so distributed. *Watersipora arcuata* grows fully exposed on intertidal sandstone reefs in Sydney Harbor, which certainly is a highly unusual situation, especially since this species is darkly pigmented. Erect branching species are sometimes important in the benthic fauna of rock faces near or beyond the continental edge, while lunulitiform species live on sublittoral sands; in neither habitat, presumably, would larvae or adults be skotophilic. In general, if colonies are to survive the effects of insolation and desiccation on the shore, a shaded surface must be selected. If, in the sublittoral euphotic zone, they are to avoid competition from algae or smothering by sediment, they must be established on a more or less downward-facing surface.

Pomerat and Reiner (1942) found that erect *Bugula neritina* and encrusting bryozoans [named in their paper as *Acanthodesia* (now *Membranipora*) *tenuis* and *Electra hastingsae*] not only settled most abundantly on downward-facing panels, but apparently grew faster. In one experiment 165 *M. tenuis* settled below horizontal panels and had an average area after 23 days of 3.29 cm^2, 45 settled below oblique panels and had an average area of 2.71 cm^2, while on the remaining surfaces only 23 colonies were found, their average area being 0.24 cm^2. McDougall (1943), Maturo (1959), and Ryland (1960) obtained essentially similar patterns of settlement, the explanation for which clearly lies in the behavior of the larvae.

There is, of course, no reason for accepting the teleological explanation that larvae settle in certain situations to avoid death from unfavorable physical conditions. Either settlement is random, and ancestrulae established in unsuitable places die; or there are simple behavioral mechanisms which cause aggregation in situations most favorable for survival. There is no evidence in favor of the first proposition; on the contrary, available evidence from invertebrate settlement generally (Meadows and Campbell, 1972) is opposed to the concept of random settlement. Most of this evidence comes from substrate selection experiments, the incentive for which, in bryozoans, arose from observations on substrate specificity. In consequence, the experimental results might give the impression that bryozoans as a whole are markedly stenotopic; but there is really no evidence for any such sweeping generalization. What has been established is, first, that larvae of substrate-specific bryozoans have exploratory behavior patterns that facilitate selective settlement and, second, that the results obtained conform closely with observations made on the shore. Selection of substrata and attachment sites will not be considered any further here, since a separate review is being published elsewhere (Ryland, 1976b). In the following pages the responses of larvae and adults will be considered in turn.

II. ORIENTATION RESPONSES OF FREE-SWIMMING LARVAE AND MOTILE COLONIES

For a comprehensive review, four larval types should be considered: cyphonautes, brooded larvae of gymnolaemates, brooded larvae of stenolaemates, and the "incipient colony" larvae of phylactolaemates. In practice, however, most of the available data refer to the larvae of those gymnolaemates which brood their early developmental stages.

A. Phototaxis

The stimulus triggering the discharge of lecithotrophic larvae from ooecia or other gymnolaemate brood chambers appears always to be light (at least in shallow water species), even when the larvae themselves do not display phototactic responses. Indeed, the technique for obtaining larvae is simply to keep the colonies in the dark until liberation is required: e.g., in *Bugula* (Grave, 1930; Lynch, 1947, 1949; Ryland, 1960), *Watersipora* (Wisely, 1958), *Alcyonidium, Flustrellidra,* and *Scrupocellaria* (Ryland, 1960). Mawatari (1951a) found that *Bugula neritina* larvae were liberated earlier and more abundantly on a sunny day than on a dull one.

On liberation most of these larvae are strongly photopositive but, as stated above, at settlement generally show a preference for surfaces shaded from the light. To investigate this apparent reversal in the response to directional illumination, Ryland (1960) placed larvae in horizontal troughs which were lit from one end by a well-collimated light beam. The distribution of larvae along the length of the troughs was recorded at intervals (Fig. 1). Four behavioral types were distinguished:

A. Initially photopositive, becoming photonegative, e.g., *Hippothoa* (= *Celleporella*) *hyalina.*

B. Initially photopositive, with a partial change to photonegative, e.g., *Alcyonidium hirsutum.* It seems likely that rather than a part only of the population becoming permanently photonegative, the larvae develop changing responses, in which positive and negative periods alternate (see Grave's, 1930, account of *Bugula simplex*).

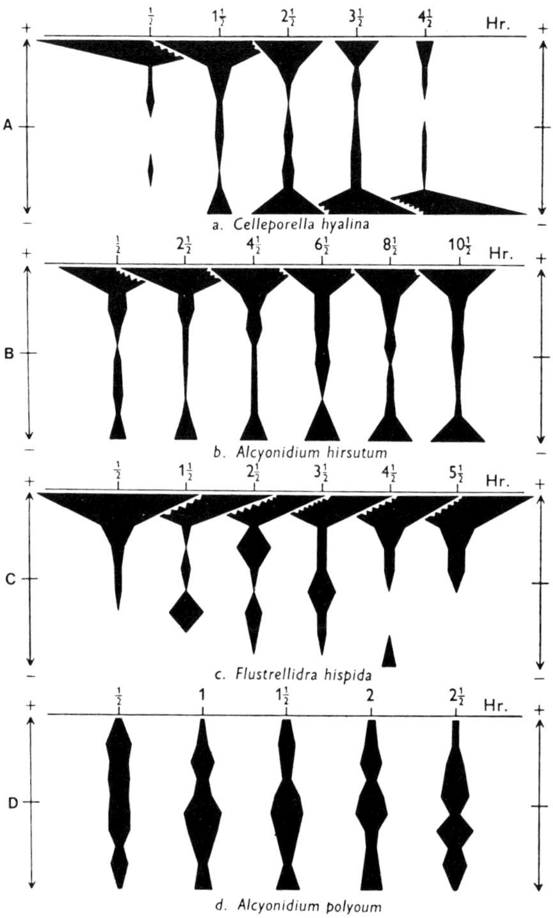

Fig. 1. The responses of bryozoan larvae to directional illumination. The kite diagrams show the distribution at regular time intervals of larvae in troughs illuminated from one end. Four types of behavior were recorded, and one example of each is figured. (A) *Hippothoa hyalina;* (B) *Alcyonidium hirsutum;* (C) *Flustrellidra hispida;* (D) *Alcyonidium polyoum.* From Ryland (1960).

C. Initially photopositive and remaining so, e.g., *Flustrellidra hispida.*

D. Indifferent to light (i.e., larvae randomly distributed along the length of the trough), e.g., *Alcyonidium polyoum.*

Type A seems to be overwhelmingly the commonest in the (exclusively littoral or shallow water) species so far studied, with a smaller number in type B. Species known to show this change in response include: *Bugula fulva* (A), *B. neritina* (B), *B. plumosa* (A), *B. simplex* (as *B. flabellata* in Grave, 1930; Lynch,

1949), *B. turrita, Scrupocellaria reptans, S. scruposa* (A), *Hippothoa* (= *Cellepo-rella*) *hyalina* (A), *Watersipora subtorquata* (B; as *W. cucullata* in Mawatari, 1952), *W. arcuata* (B; as *W. cucullata* in Wisely, 1958), *Cryptosula pallasiana* (B) and *Alcyonidium hirsutum* (B) (Grave, 1930; Lynch, 1947, 1949; Mawatari, 1951a, 1952; Lutaud, 1953; Wisely, 1958; Ryland, 1960, 1962). In addition, larvae of the following species have been recorded as being photopositive on release: *Bugula turbinata, Tricellaria occidentalis, Escharoides coccinea, Micro-porella ciliata, Fenestrulina malusii, Schizobrachiella sanguinea, Bowerbankia imbricata,* and *B. pustulosa* (Joliet, 1877; Hasper, 1913; Mawatari, 1951b; Schneider, 1959; Ryland, 1960, 1976a). The larvae swim along the light beam, either toward or away from the source, rotating by ciliary action as they go. Their response to directional illumination thus appears to depend upon movement of the receptors, which can thus make comparison of intensities at successive points in time; it is, therefore, a photoklinotaxis (Fraenkel and Gunn, 1940).

Many brooded bryozoan larvae are provided with orange-red pigment spots; for example, *Bugula plumosa* has four, *B. simplex* eight, and *B. flabellata* ten [Fig. 2; see also the illustrations in Barrois (1877) and Calvet (1900)]. Calvet described these spots in *B. simplex* (as *B. sabatieri*) as ovoid, red-vermilion, and supporting a cluster of fused cilia. In *B. neritina* larvae, which are rather differently pigmented: "On the side of the larva opposite the median furrow are two, prominent, black, diamond-shaped eye-spots lying almost on the equator and about 90° apart" (Lynch, 1947).

Woollacott and Zimmer (1972) have examined the ultrastructure of the "eyespots" of *B. neritina,* and a diagram based on their findings has been presented elsewhere (Ryland, 1976b, Fig. 29). The pigmented field measures 40–80 μm in length and has at its center a pit 4–6 μm in depth and 1–2 μm in diameter. Cilia forming a refractile granule within this pit project as a tuft, which Lynch evidently perceived as some indeterminate structure, but which corresponds to what Calvet described in other species. The walls of the pit are lined by an epidermis of modified coronal cells, but its basal end consists of a single "basal sensory cell." The cilia of the refractile granule are derived from this basal sensory cell and are twisted into a tightly packed mass 1.0–1.5 μm in diameter. The arrangement is such that the ciliary shafts lie at right angles to the plane of incident light. These cilia possess the usual 9 + 2 construction of the axoneme found in kinocilia, including the occurrence of electron-dense arms associated with the *a*-microtubules of the outer doublets. Subepidermal pigmented cells ensheathe the pit and give rise to the underlying black rhombic field. The basal sensory cell abuts against the equatorial nerve ring, and Woollacott and Zimmer (1972) were able to find morphological evidence of gap junctions between the sensory cell and the neuronal cells, suggesting the pres-

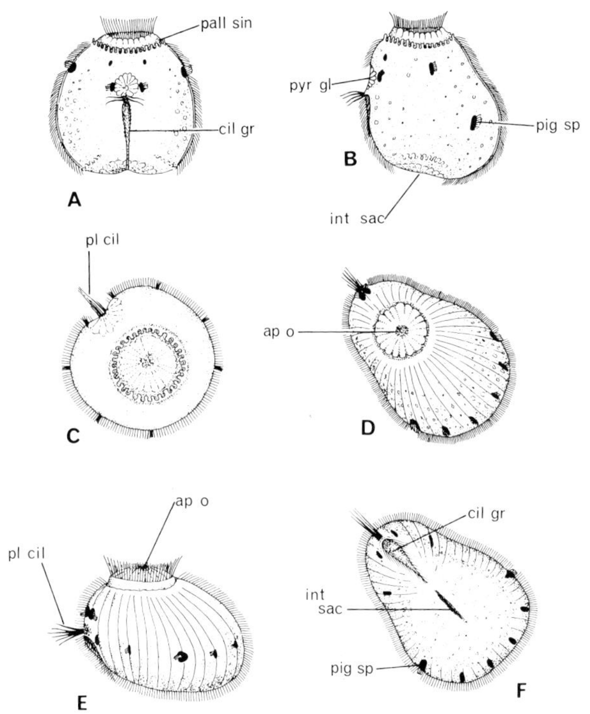

Fig. 2. Lecithotrophic bryozoan larvae. (A) *Bugula simplex,* anterior; (B) *B. simplex,* lateral; (C) *B. turbinata,* aboral; (D) *Chorizopora brongniartii,* aboral; (E) *C. brongniartii,* lateral; (F) *C. brongniartii,* oral. Abbreviations: ap o, apical organ; cil gr, ciliated groove; int sac, internal sac; pall sin, pallial sinus; pig sp, pigmented spot; pl cil, plume cilia; pyr gl, pyriform gland. After Calvet (1900).

ence of a pathway for electrotonic conduction. There is no evidence, however, that these cells are linked by chemical synapses.

This presumed photoreceptor is quite remarkable, apparently being of unique construction among invertebrate ocelli, although it does have the basic deuterostome combination of pigmented cells surrounding a cup containing ciliary shafts (Eakin, 1968). Usually, however, such ocelli have a photosensitive surface provided by the modification of membrane ensheathing the axonemes; such proliferations are absent from the pigmented spots of *B. neritina,* and Woollacott and Zimmer (1972) suggest that the photosensitive surface is provided by the unique arrangement of tightly packed, transversely oriented ciliary shafts. There is no physiological evidence that the larval "eyespots" are photoreceptors, though this seems likely from what is now known of their morphology. However, species such as *Bugula turbinata* (Calvet, 1900, p. 344; Ryland, 1976a) and *Alcyonidium hirsutum* (Ryland, 1960) show orientation to light yet lack pigmented "eyespots."

The transition from photopositive to photonegative appears to be induced by metabolic progress or change, its onset being influenced by temperature but not by light itself. Ryland (1960) showed that *Hippothoa* (= *Celleporella*) *hyalina* larvae kept in the dark, as well as controls subjected continuously to the light beam, had developed a photonegative response after 2 hours. In later experiments on *Cryptosula pallasiana,* Ryland (1962) investigated directly the effect of temperature on the rate of attainment of photonegativity. Perhaps the first surprising feature in his results is that, ignoring the "pipetting effect" (which induces a premature change), the rate of changeover at each temperature is constant (Fig. 2 in Ryland, 1962) and the slopes can be determined by regression analysis (Table I). If the response were due to utilization of some resource, say yolk, the reserves of which would be distributed normally about some mean value, the rate of changeover in a population of larvae should follow a sigmoid curve, with the point of inflection corresponding to the mean. The rates accelerate as temperature rises, but the extent of the increase is about 5.4 times per 10 C°, making it difficult to relate it to the normally expected influence of temperature on metabolism (about 2.3 times per 10 C°). Lynch (1947, 1949) found that increased temperatures accelerated the change to photonegativity in *Bugula simplex* and *B. turrita* also. The influence of temperature on the rate of transition to photonegativity is, of course, a different phenomenon from any change induced by abrupt temperature alterations. Cyphonautes larvae of several species appear generally indifferent to light (Silén and Jansson, 1972) and thus resemble the lecithotrophic larvae in group D.

TABLE I
Effect of Temperature on the Photic Responses (Rate of Attainment of Photonegativity) of *Cryptosula pallasiana* Larvae[a]

Temperature (°C)	Rate (slope)	Time to 50% photonegative: Hours and minutes
10	3.12	11 and 47
15	5.68	7 and 42
17.5	11.72	4 and 01
20	22.90	2 and 17
25	32.62	1 and 35

[a]Data from Ryland (1962).

B. Other Responses to Light and Gravity

The oft-recorded phenomenon of both brooded and planktotrophic bryozoan larvae settling preferentially on lower or shaded surfaces (Grave, 1930; Pomerat and Reiner, 1942; McDougall, 1943; Knight-Jones, 1951; Maturo, 1959; Ryland, 1960; Médioni, 1972; Silén and Jansson, 1972; Humphries, 1973) has already been mentioned. It has sometimes been attributed to skotopositive behavior. Thus Grave (1930) states categorically that: "While the [*Bugula simplex*] larvae are strongly positive to light they quickly seek the lighted part of the field, and when strongly negative to light they as quickly turn toward a shadow." McDougall (1943) thought that negative phototaxis would take larvae into dark places. Ryland (1960) investigated the settlement of *Hippothoa* (= *Celleporella*) *hyalina* on opaque panels in aquaria; here the normal pattern was reversed if the illumination came from below, and the larvae settled on top of the panels. He thought that his results were best explained by what Fraenkel and Gunn (1940) termed high photokinesis, i.e., activity is high in the light, low in dark places, so that larvae once having found a shaded place—by negative phototaxis or other behavior—stay there.

The fact remains, however, that not all settlement of this kind can be explained in terms of light. McDougall's (1943) experiment using horizontal, oblique, and vertical panels fixed in a dark, tunnel-shaped box so constructed "to eliminate any possible effects of the photic factor" produced results (Table II, 5) just like those of Pomerat and Reiner (1942) which were conducted in the open (Table II, 1–4). Maturo (1959) thought that his results (Table II, 6–7) were best explained by the repellent effect of an accumulation of sediment on horizontal panels. McDougall also considered that this was a major factor since the upper horizontal surface "was commonly covered with a layer of fine mud and sand to a thickness of an eighth of an inch or more after two weeks' exposure, and the 135° [upper oblique] plate was also covered, though less thickly." He was, however, bothered by the absence of settlement on the vertical panel, which did not collect sediment. Both Maturo (1959) and Ryland (1960) did obtain quite high settlement on vertical panels in the absence of algal growth (Table II, cf. 8 and 9–14). Silén and Jansson (1972) reported in *Electra crustulenta* quite a different pattern of settlement from that obtained by McDougall. Using a dark tube they found that: "The interior surface of the tube was rapidly covered by evenly distributed colonies," and "colonies were formed over the entire exterior surface" in the absence of algal growth but "exclusively on the downwards turned side" when the upper surface was "covered by initial stages of algae growing to a dense rug." Ryland (1960) also stressed that "those surfaces on which algae grow densely . . . do not receive a high settlement."

It appears that the growth of algae and/or accumulation of sediment prevent settlement, but that in some cases the well-established preference for lower

TABLE II
Records of Settlement on Horizontal, Oblique, and Vertical Surfaces[a]

Species	Number of colonies settled					Reference
	0°	45°	90°	135°	180°	
1. *Membranipora tenuis*	165	125	7	3	6	Pomerat and Reiner (1942), Table I
2. *M. tenuis*	210	123	12	6	2	Pomerat and Reiner (1942), Table II
3. *Electra hastingsae*	82	71	6.5	0	0	Pomerat and Reiner (1942), Table II
4. *Bugula neritina*	299	194	35	18	1	Pomerat and Reiner (1942), Table III
5. *B. neritina*	446	298	34	9	2	McDougall (1943), Table XIX
6. *B. neritina*	181	327	94	12	6	Maturo (1959), June/July, Fig. 4
7. *Schizoporella* sp.	53	25	21	6	4	Maturo (1959), August, Fig. 7
8. *Hippothoa hyalina*	221	–	19	–	5	Ryland (1960), Table XII, Sta. 1, upper panels, outside
9. *H. hyalina*	85	–	41	–	64	Ryland (1960), Table XII, Sta. 1, lower panels, outside
10. *Callopora dumerilii*	260	–	417	–	15	Ryland (1960), Table XII, Sta. 2
11. *Bugula fulva*	120	–	145	–	0	Ryland (1960), Table XII, Sta. 2
12. *Scrupocellaria scruposa*	105	–	100	–	10	Ryland (1960), Table XII, Sta. 2
13. *Bowerbankia gracilis*	95	–	98	–	65	Ryland (1960), Table XII, Sta. 2
14. *Electra pilosa*	195	–	45	–	30	Ryland (1960), Table XII, Sta. 2

[a]Convention: 0°, downward facing horizontal; 45°, downward facing oblique; 90°, vertical; 135°, upward facing oblique; 180°, upward facing horizontal. Experiment 8 differs from 9–14 in that the panels were better lit and an algal sward developed on the 90° and 180° surfaces; for further explanation see text.

surfaces cannot be explained in this way. In *Hippothoa* (= *Celleporella*) *hyalina* light appears to be the controlling factor, while for *Electra crustulenta* Silén and Jansson consider that their results fit best with an assumption that larvae find their settlement site by negative geotaxis.

Gravity responses of larvae have rarely been isolated from photic responses, but there are several references that suggest initial geonegativity, e.g., in *Bugula simplex* (Grave, 1930) and *B. neritina* (Lynch, 1947; Mawatari, 1951a). The change to geopositivity, which might have been expected, has apparently not been recorded [Thorson (1964) seems to have misquoted Grave (1930) in this respect]. Experiments like those of Crisp and Ghobashy (1971) on the ascidian *Diplosoma listerianum* are required, since, as in the bryozoans, larvae of this species remain negatively geotactic until metamorphosis. A tendency to swim upward, combined with appropriate photic responses, may in fact enhance the colonization of lower surfaces, as Silén and Jansson (1972) believe applies to *Electra crustulenta* larvae. A major reinvestigation of the responses of free-swimming bryozoan larvae would be of value.

The studies of Marcus (1926c) on colony motility and larval orientation movements in freshwater bryozoans still provide practically all the information available. It is well known that *Cristatella mucedo* colonies move around slowly on their flat undersurface. Marcus established that they will move from a lighted to a shaded part of an aquarium tank. On the first day of an experiment five 2–3 cm long colonies were situated in the lighted half of an aquarium; one colony had moved into the shaded half by the tenth day, and all five by the fourteenth day. They remained there for two weeks, when observations evidently terminated. This photonegativity in *Cristatella* was described as phototaxis by Marcus, but some other of his experiments make it abundantly clear that the selective mechanism really depends upon high photokinesis. Colonies in illuminated tanks were far more active than those in the dark (Fig. 3); thus colonies moved on

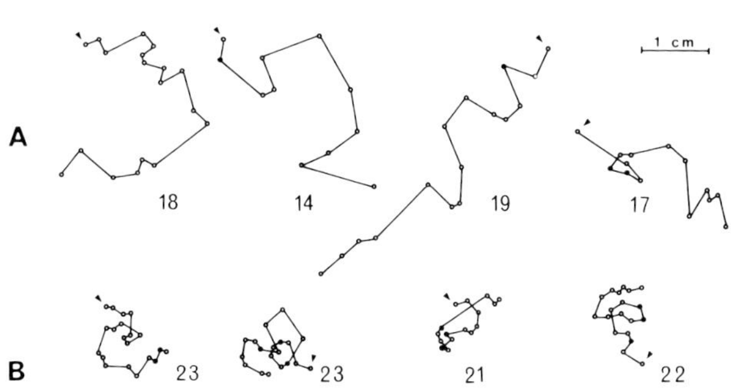

Fig. 3. Daily movements of colonies of *Cristatella mucedo.* (A) In the light, (B) in the dark, for the period of days indicated. At the positions marked by solid circles no movement was recorded for one or more days. From Marcus (1926c).

average 3.4 mm/day in the light and only 1.8 mm/day in the dark. This means that if the wanderings of a colony in the light take it into a dark place, it will tend to stay there.

Marcus (1926c) obtained some rather similar results with *Lophopus crystallinus* in which limited movement is associated with lobulation. Three colonies were established in the lighted half of an aquarium; over a period of two months these divided into nine daughter colonies of which all but one had moved into the shaded half of the tank. Young colonies of *Cristatella,* which hatched from a pile of statoblasts placed on the boundary between the illuminated and shaded halves of a dish, were found predominantly in the darkened side. While the evidence does not wholly rule out the possibility of an oriented response—photonegative or skotopositive—it does seem that the described accumulation of colonies in less well-lit places is best explained in terms of high photokinesis.

Both Williams (1921) in *Pectinatella magnifica* and Marcus (1926c) in *Plumatella fungosa* have described the sexually produced larvae as negatively phototactic. In fact, however, no distinction has been made between responses to the direction of illumination and those to its intensity. That the reaction is again photokinesis seems probable from Marcus' description (1926c, p. 304) of the behavior of the larvae. In a tank divided into lighted and shaded halves, they swam randomly in the light, passed from the lighted to the shaded half of the tank and vice versa, but eventually accumulated and attached in the dark. There is here, however, the alternative or supplementary hypothesis that light directly inhibits larval settlement.

In another paper Marcus (1926a) described a subepithelial nerve net connected with ectodermal sense cells in the larva of *Plumatella fungosa.* This nervous system is presumably involved in the reactions of the larvae to light and other stimuli.

Genuine orientation responses shown by freshwater larvae, according to Marcus' (1926c) studies, are negative geotaxis and positive aerotaxis. The former, he points out, will aid larval dispersal from the parent colony. In general, the behavior of phylactolaemate larvae seems very similar to that of cyphonautes as proposed by Silén and Jansson (1972). In both types of larvae, moreover, it remains unclear how larvae could come to establish colonies at or near the bottom of the water column.

III. ORIENTATION RESPONSES OF SETTLING LARVAE

A. Phototropism

Lynch (1947) described how larvae orient to light during metamorphosis. In *Bugula neritina* the sequence of events is as follows: Having alighted on the substratum, the larvae "rotate counterclockwise on an axis running from a point midway between the eye-spots to the opposite side," where the pyriform organ is in contact with the substratum. Rotation slows and finally stops just before fixation, when the larva appears to be anchored by secretion from the pyriform gland. The substratum is grasped by the ridges delimiting the median groove (Grave, 1930, described this vividly); then the internal sac is suddenly everted and spreads as a rounded mass below the rest of the larva. Simultaneously with the eversion of the internal sac, the median groove releases its grip on the substratum and the larva rotates to bring the apical organ upward, until it faces away from the source of illumination at an angle of 15°–20° from the vertical. By means of squirming, lateral rotations during the brief "hourglass" phase (Fig. 6 in Lynch, 1947), the larva brings the eyespots to the lighted side.

Lynch's account then becomes ambiguous. He recognized that there is a reversal of internal polarity: "The oral end of the larva becomes the aboral end of the ancestrula" and that the direction of incident light is involved, for "the aboral end of the attached larva and later the apical part of the first zooid are oriented so as to point away from the direction of the incoming light." Nowhere is it made clear, however, whether it is the basal or the frontal surface which eventually faces the light; but it appears from general principles that, if the pyriform gland faces away from the source of illumination, so must the frontal surface of the ancestrula and first zooids. According to McDougall (1943) this is indeed what happens: "It was found that larvae invariably orient themselves with reference to light upon becoming attached and in such a way that the first, and therefore all subsequent zooids, face away from the source of illumination." Colonies growing on vertical surfaces were oriented with the zooids facing downward, and McDougall produced the reverse orientation in a "light box" illuminated only from below. Attached colonies were not able to readjust to a changed direction of incident light.

Schneider's (1959) observations on another species of *Bugula,* identified by him as *B. avicularia* (Section IV,B), indicate that it differs from *B. neritina* in several respects. First, the branches can twist during growth, thereby reversing their orientation with respect to light; and second, it is the frontal surface that is brought to the lighted side (see Fig. 7).

However, illumination that is both unidirectional and lateral is obviously artificial. In natural growth the colony is infundibulariform, with the zooids facing into the funnel.

The phototropism in metamorphosing *B. neritina* is interesting from its apparent analogy to the situation described by Forbes *et al.* (1971) in acorn barnacles. Settling cyprids undergo rotation such that the carina of the adult is toward the light and the eyes farthest from it. When the cirral net is spread for feeding, therefore, the light falls on it from behind (as is the case in *B. neritina* zooids). An object approaching against the light will cause a smaller degree of shading than one casting a definite shadow on the better illuminated side. Forbes *et al.* showed that the barnacle is most sensitive to light falling from the rostral end; consequently, the preferred orientation has the effect of increasing the responsiveness in the direction that is likely to receive the least shading from an approaching predator. Their finding suggests that a careful investigation into any shadow response in *Bugula* would be instructive: certainly it seems inconceivable that such a well-developed orientation to light should have no adaptive function (McDougall, 1943). J. P. Thorpe (personal communication) has studied zooidal and colonial responses in *Electra pilosa* and *Membranipora membranacea,* finding that sudden shading of an illuminated zooid does not provoke withdrawal of the tentacles or any other response. *Bugula* might be different: certainly it is established that its zooids contain light-sensitive cells (Schneider, 1960).

Ryland and Stebbing (1971) found that *Scrupocellaria reptans* ancestrulae fixed to *Flustra foliacea* were, almost without exception, oriented toward the periphery of the frond (Fig. 4C). The authors commenting that "... *Flustra* colonies are low, bushy and dense, and live sublittorally, ... light will thus fall rather constantly on the fronds from above," favored light as the orienting stimulus. They overlooked a brief, and possibly confirmatory, statement in a paper by Schneider (1959, p. 258) to the effect that the ancestrular bud in *S. reptans* grows toward the light, thereby determining the initial growth direction of the colony. Unfortunately this was not further clarified, leaving it uncertain how the bud comes to face the light in the first instance. Nevertheless, it seems likely that *S. reptans* provides a further example of positive phototropism at settlement.

B. Rheotropism

Definite orientation of ancestrulae—and therefore of larvae at the time of settlement—in some incrusting bryozoans was reported by Ryland and Stebbing (1971). They established the following descriptive convention. The long axis of an anisomorphic substratum (say an algal frond) was aligned with a perpendicular, upon which was superimposed a circle divided into eight 45° arcs, with the ancestrula positioned at its center. The perpendicular bisects the distal and the proximal arcs, and each octant is labeled in degrees, representing the angle between the perpendicular and the midpoint of each arc. Thus the distal octant is designated 0°, the one to its right 45°, and so on (Fig. 4A). The sector into which the ancestrular orientation falls is determined, preferably by using a protractor eyepiece in a stereomicroscope.

The least ambiguous results obtained were those relating to *Electra pilosa* and *Membranipora membranacea* settled on fucoid and laminarian algae. Both species have cyphonautes larvae. The ancestrulae of *E. pilosa* were not randomly oriented on *Fucus serratus* fronds for, in a substantial majority (375/589), the median longitudinal axis of the ancestrula and of the frond were parallel (Fig. 4B). It was also established that many more ancestrulae faced the frond apices than faced the frond base. Considering orientation based on the 0° and 180° positions alone, the ratio was 257:118. Taking the combined settlement in the 315°, 0°, and 45° positions and that in the 135°, 180°, and 225° sectors, the

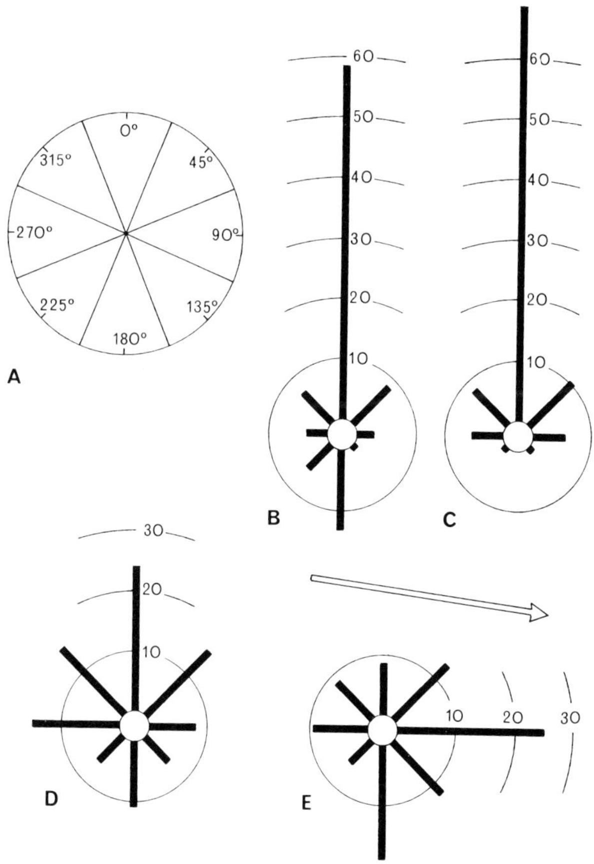

Fig. 4. Oriented settlement in Bryozoa: percentage falling within the range of each octant. (A) Diagram to show the arrangement of octants used for recording; (B) orientation of *Electra pilosa* ancestrulae on fronds of *Fucus serratus;* (C) orientation of *Scrupocellaria reptans* ancestrulae on fronds of *Flustra foliacea;* (D) orientation of *Membranipora tuberculata* ancestrulae on leaflets of *Sargassum* sp.; (E) orientation of *Electra pilosa* ancestrulae on inert panels exposed to directional water flow (arrow). (A–C) from Ryland and Stebbing (1971); (D) after Ryland (1976a).

ratio was 361:176. The trend, however, was much more apparent on some shores than on others, and was seemingly related to the degree of exposure to wave action. *Electra pilosa* ancestrulae thus tend to face the frond apices of *F. serratus*: allowing for the reversal in polarity which occurs during metamorphosis, this means that a majority of the larvae faced the frond base at the moment of fixation.

Ryland and Stebbing assumed that algae living in the turbulent waters of the intertidal zone and just below would be in constant motion. The fronds would be swept to and fro by the waves, always streaming in the direction of water flow, which is from the base toward the extremities of the fronds. The almost sail-like shape of a cyphonautes suggests that an alignment of the larva in the direction of flow is likely to be adopted. Jägersten (1972) has described in an unidentified cyphonautes how locomotion over the substratum is accomplished by the pyriform organ. This structure forms a kind of anteriorly situated foot, upon which the animal creeps. It is easy to see that the only stable orientation in a water current would be facing into it, with the "sail" pivoting rudderlike about the foot.

Although absolute proof of rheotropism at settlement will come only from observation and experiment, some further indirect evidence in its favor has been obtained by Ryland (1976a). Variable, but generally moderate to strong, tidal currents flow through the Menai Strait in North Wales. A freely swiveling panel suspended below a raft, except at brief periods of slack water, will lead downstream. Its two surfaces will thus be swept by a current of varying velocity but constant direction. Young *E. pilosa* colonies on each of these surfaces were examined to determine the orientation of the ancestrulae. The orientation was nonrandom and strongly biased in the direction of water flow (Fig. 4E). As some settlement must have occurred during periods of slack water, the downstream: upstream ratio (45°, 90°, and 135°:225°, 270°, and 315°) at 42:19 is quite impressive.

Nevertheless, it does not seem possible to generalize that all orientations on algal fronds by settling cyphonautes are rheotropic. Ryland and Stebbing (1971) found that ancestrulae of *Membranipora membranacea* tended to be proximally oriented on *Fucus serratus* (the converse of the situation reported for *E. pilosa*). Most of the observations, however, were made on an unusually sheltered shore. Perhaps more perplexing is the statistically significant orientation found by Ryland (1974) in *Membranipora tuberculata* ancestrulae on leaflets of floating gulfweed, *Sargassum fluitans* and *S. natans* (Fig. 4D). In the pleustal habitat occupied by the gulfweed, the *Sargassum* clumps should move with the supporting water mass, and it is difficult to imagine any consistent pattern of water flow over the leaflets. Again, the solution will be found only by observation of and experiment with living larvae.

IV. ORIENTATION RESPONSES OF FIXED COLONIES

A. Rheotropism

Membranipora membranacea is a widely distributed species found on kelps in many parts of the world. In western Europe it is found most commonly on *Laminaria digitata* and *L. hyperborea,* having the same bathymetric range as these algae. Lutaud (1961) commented that the colonies' maximal growth was directed proximally on the frond (i.e., toward the stipe). Ryland (1967) noted that "growth in this direction would be of benefit to the bryozoon since *Laminaria* grows from an intercalary zone at the base of the frond, so that the *Membranipora* spreads onto the youngest part of the thallus." This led to a full investigation (Ryland and Stebbing, 1971). First, Lutaud's observation was fully confirmed. Using the method of recording direction of growth in octants, comparable to that described in Section III,B, it was found that oriented growth on *Laminaria* fronds evolved from randomly oriented settlement (Fig. 5C). A preferentially oriented growth axis emerged early in the astogeny of the colony, often while it consisted of only 6–15 zooids. Indeed, even the first or second rows of daughter zooids sometimes have their long axes skewed toward the algal stipe (Fig. 5D) and by the time the colonies have reached a diameter of 15–35 mm most have acquired a roughly proximal growth orientation (Fig. 5E). When *Membranipora* colonies of all sizes were considered together, about 80% exhibited growth toward the stipe (Fig. 5F). *Membranipora membranacea* is not particularly common on *Fucus serratus,* but a predominantly proximal direction of growth was noted on this alga also.

Ryland and Stebbing considered as possible causative stimuli light, gravity, water current, and polarity of the surface, quickly eliminating the first two despite their known involvement in oriented growth in other bryozoans in other circumstances. The manner in which water probably sweeps over the fronds of algae in the intertidal and euphotic zones has already been explained. Water passing over a *Laminaria* frond follows the physiological age gradient; but water moving over a *Fucus* thallus is flowing contrary to the age gradient. It thus seemed that the only factor in common was water movement, and it was concluded that the observed oriented growth in *M. membranacea* was indicative of positive rheotropism. This has since neatly been confirmed by Norton (1973). *Saccorhiza polyschides* is a kelp in which the stipe is broad and flat. As in *Laminaria* the plant grows from an intercalary meristem where the stipe merges into the frond. If oriented growth were due to rheotropism, *Membranipora* colonies on both frond and stipe would grow toward the holdfast; if, on the other hand, the response were dependent upon the recognition of a physiological age gradient in the alga, colonies on both frond and stipe would grow toward the

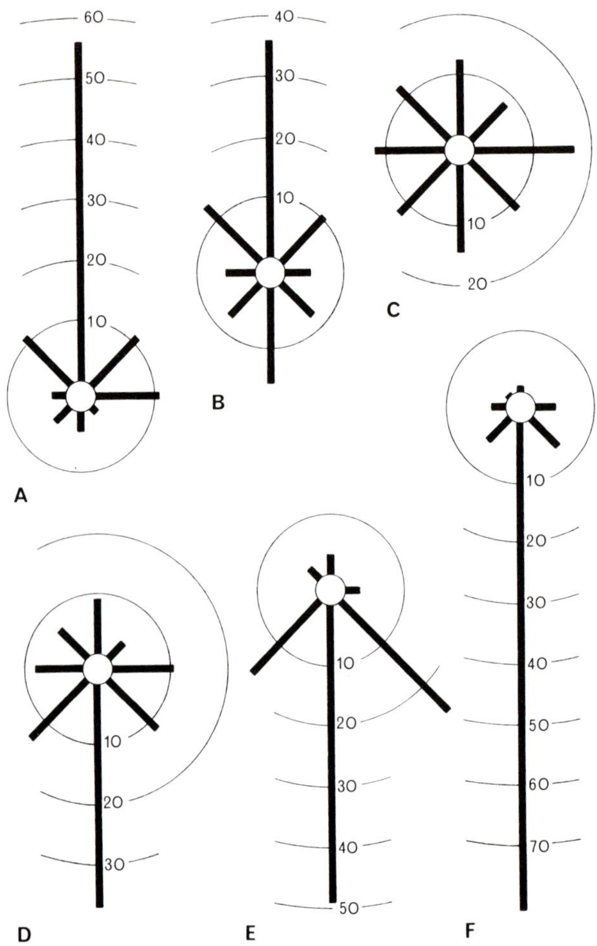

Fig. 5. Oriented growth in Bryozoa: percentage falling within the range of each octant. (A) Direction of growth of *Scrupocellaria reptans* colonies on fronds of *Flustra foliacea;* (B) direction of growth of *Electra pilosa* colonies on fronds of *Fucus serratus;* (C) randomly oriented settlement of *Membranipora membranacea* on fronds of *Laminaria hyperborea;* (D) orientation of growth in *M. membranacea* colonies < 35 mm diameter on *L. hyperborea;* (E) orientation of growth in *M. membranacea* colonies 15–35 mm in diameter; (F) as (E), but using all colonies. Figures or data from Ryland and Stebbing (1971).

meristem. Norton showed quite clearly that growth on the stipe was oriented downward, toward the holdfast, not upward toward the meristem.

Positive rheotropism has also been demonstrated in another—and quite unrelated—bryozoan, the stoloniferous ctenostome *Farrella repens.* Marcus (1926b, pp. 43–44) exposed an established colony to water flowing at about 1 cm/sec. After three days there had been a strongly marked upstream growth of stolons (Fig. 6A and B), whereas growth was symmetrical in still water (Fig. 6C and D). Rather perplexingly, however, Jebram (1970) has been unable to repeat Marcus' results. Using colonies of *F. repens* established on glass slides, no oriented stolonal growth was observed in water flowing at either 3 or 15 cm/sec. No obvious explanation for this contrary result can be established.

Jebram (1970), using a specially developed apparatus, looked for rheotropic responses in other species. None was found in *Conopeum reticulum* in a water current of 3 cm/sec. In *Electra monostachys* no response was detected at a velocity of 15 cm/sec, but at 3 cm/sec the compact colony started to produce monoserial zooid rows predominantly in an upstream direction.

Jebram also conducted experiments with *Electra pilosa.* At current velocities of 3 and 5 cm/sec he found that: "The main parts of these colonies began to grow against the current direction," and concluded: "Both of my experiments show a clearly positive rheotropic growth reaction under the influence of slow water current speeds." The writer of this review is unable to see this in the

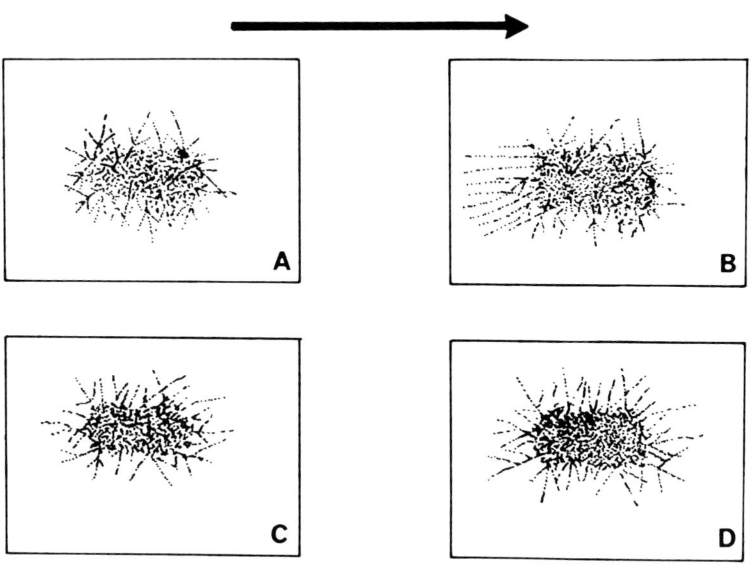

Fig. 6. Rheotropism in *Farrella repens*. (A and B) Colony exposed to water flowing in the direction of the arrow for 3 days; (C and D) control in still water. After Marcus (1926b).

evidence provided (Jebram, 1970, Fig. 11A–D). At 15 cm/sec Jebram's colonies "showed no special reaction." Ryland and Stebbing (1971) found a preponderance of apically directed growth of *E. pilosa* on *Fucus serratus* fronds (Fig. 5B). In view of the characteristic early astogeny of *E. pilosa* (Barrois, 1877, Plate XV, Figs. 5–9; Waters, 1924; Marcus, 1926b), it was considered that this stemmed directly from the orientation of the ancestrula. There was certainly no evidence for the emergence of a proximally directed growth axis, as happens in *Membranipora membranacea* on the same alga. Additionally, when well-grown colonies were found on the swiveling panel used for the study of ancestrular orientation (Ryland, 1976a; and Section III,B), they had the regular stellate form illustrated by Marcus (1926b, Fig. 12). It is concluded that, for the present, there is no satisfactory evidence for rheotropic growth in *E. pilosa.*

Ryland and Stebbing (1971) also described oriented growth in *Scrupocellaria reptans* on *Flustra foliacea* fronds (Fig. 5A), but were fairly certain that this was not a response to water movement. It may have followed inevitably from the orientation of the ancestrulae (Section III); or it might be phototropic, as described for *Bugula* in Section IV,B.

B. Phototropism

The occurrence of phototropism in *Bugula* was briefly first reported by Schneider (1955) and Aymes (1956) independently. Aymes studied *B. neritina*: Schneider, and later Kaissling (1963), worked mainly with a species they named *B. avicularia* but which Ryland (1967) thought was probably *B. stolonifera,* an identification accepted here. These first papers established that colony branches composed of autozooids respond positively, and the attaching rhizoids negatively, to light. While it might be expected *a priori* that the normal growth orientation would be influenced by gravity, if not a direct response to it, this seems not to have been the case. Growing *B. stolonifera* in total darkness, Schneider (1959) found that upward and downward were no longer distinguished and that branches and rhizoids grew into a disoriented tangle. He also found that closely related species may vary in their responses. Thus light is necessary to *B. stolonifera* for typical growth, i.e., for the production of normal-sized zooids in biserial branches; but *B. neritina* will grow normally in the dark. Figure 7 (from Schneider, 1959) shows a branch responding to lateral illumination; it is clear that a genuine growth bending is involved.

The cellular mechanism producing the tropism was described by Schneider (1960). The branches (and rhizoids) grow from an apical bud, in the tip of which lies a plate of dense, spherical cells. These cells secrete the cuticle by intussusception (i.e., by deposition among the elements already present), and move forward with the bud apex at a rate of about 12 μm/hour. The cells display coordinated activity, such that the plate rhythmically pulses and makes other movements. If

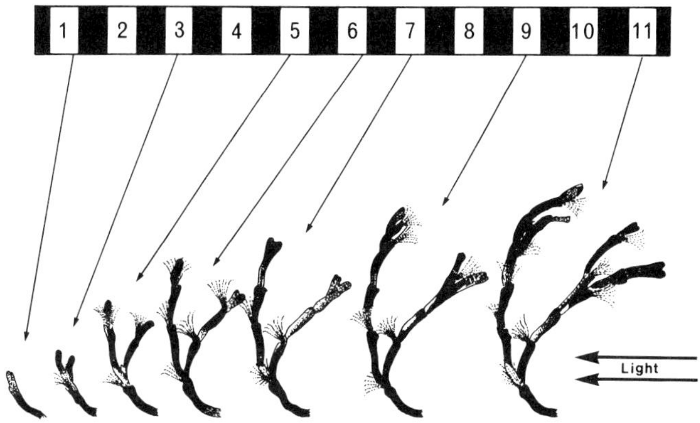

Fig. 7. Phototropism in *Bugula stolonifera*. Branches grow toward the light source and twist so that their frontal surface faces it. The time scale of the experiment is shown above. Redrawn from a photographic illustration in Schneider (1959).

the branch tip is lit from one side, the apical cell plate shifts toward the source of illumination, commencing its movement between 5 and 30 minutes after the onset of the light stimulus. With the apical plate of cells relocated, the bud naturally continues its growth in the new direction. At the growing edge of *Membranipora membranacea* the comparable cell group displays regular side-to-side oscillations (Lutaud, 1961). Presumably a change in growth direction is similarly effected through a lateral shift of these cells.

Later studies in the series (Kaissling and Schneider, 1962; Kaissling, 1963) dealt further with the nature of the stimulus. Using white light it was found that the greatest response was produced by the rather low intensity of about 15 lx (Fig. 8) and, with regard to the action spectrum, blue-green light of wavelength about 500 nm was most effective (Fig. 9). Schneider (1963), in a useful review of his work, considered that the action spectrum indicates a photosensitive system of the retinene type in cells of the apical plate. However, no pigmentation is apparent, so that the amount of photosensitive substance must be very small. How the light is perceived is unknown: "A preliminary electron microscopical study by R. A. Steinbrecht in Schneider's laboratory revealed no membranous or villous structure [of the kind usually associated with photoreception] in the reactive cells" (Eakin, 1968, p. 233).

V. SUMMARY

1. Larvae of gymnolaemates which typically brood their early developmental stages usually display positive phototaxis when first liberated. Under experimen-

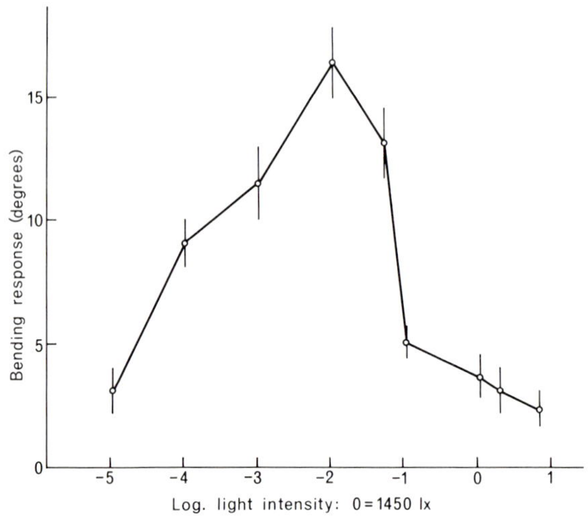

Fig. 8. Phototropism in *Bugula stolonifera.* Amount of bending induced by 15-min exposures of branch tips to unidirectional white light (means ±2 standard deviations). After Kaissling and Schneider (1962).

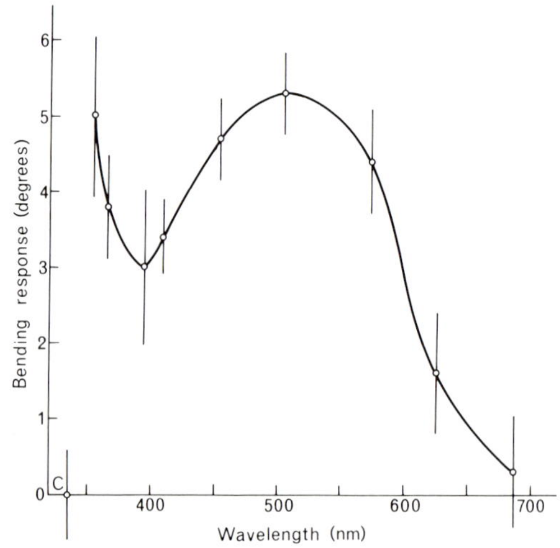

Fig. 9. Phototropism in *Bugula stolonifera.* Amount of bending induced by 15-min exposures of branch tips to unidirectional monochromatic light (means ±2 standard deviations); C, control. After Kaissling and Schneider (1962).

tal conditions at least, except in a few species, the phototaxis becomes completely or partially negative before the larvae settle. The structure and function of larval "eyespots" are discussed. Cyphonautes appear to be indifferent to light; phylactolaemate larvae are at no time photopositive.

2. Larvae of marine and freshwater bryozoans tend to settle on the underside of submerged structures or in other shaded habitats. This may be due to one or more of the following: avoidance of accumulated sediment, competition from algae, aggregation in dark places by high photokinesis, or negatively geotactic swimming from below. Gravity responses in general have been inadequately studied.

3. Colonies of *Cristatella* and some other phylactolaemates move slowly from lighted situations into the shade, but this appears to be more a result of high photokinesis than of any directed locomotory response.

4. *Bugula,* and perhaps *Scrupocellaria,* larvae orient to light during metamorphosis, so that the first zooids in the colony face or grow toward the light irrespective of any subsequent tropism. Circumstantial evidence suggests that cyphonautes larvae of *Electra pilosa* are positively rheotropic during settlement, but no explanation has been found for the oriented settlement of *Membranipora tuberculata* on pelagic *Sargassum* leaflets.

5. Growing colonies of *Membranipora membranacea* on *Laminaria* fronds display positive rheotropism, and there is evidence for this also in *Electra monostachys* and in the stoloniferan *Farrella repens.*

6. In two species of *Bugula* the branches are positively and the rhizoids negatively rheotropic. Light of about 15 lx intensity and 500 nm wavelength has most effect. The sensory cells are those that secrete the cuticle in the branch apices and migrate toward the lighted side of the bud, thus altering the direction of growth. The mechanism of photoreception is obscure.

REFERENCES

Aymes, Y. (1956). Croissance phototropique chez les Bryozoaires du genre *Bugula. C.R. Hebd. Séances Acad. Sci.* **242,** 1237–1238.

Barrois, J. (1877). Mémoire sur l'embryologie des Bryozoaires. Thesis, University of Lille.

Calvet, L. (1900). Contributions à l'histoire naturelle des Bryozoaires ectoproctes marins. *Trav. Inst. Zool. Univ. Montpellier* **8,** 1–488.

Cook, P. L. (1960). The development of *Electra crustulenta* (Pallas) (Polyzoa, Ectoprocta). *Essex Nat.* **30,** 258–266.

Crisp, D. J., and Ghobashy, A. F. A. A. (1971). Responses of the larvae of *Diplosoma listerianum* to light and gravity. *In* "Fourth European Marine Biology Symposium" (D. J. Crisp, ed.), pp. 443–465. Cambridge Univ. Press, London and New York.

Eakin, R. M. (1968). Evolution of photoreceptors. *Evol. Biol.* **2,** 194–242.

Evans, R. G. (1949). The intertidal ecology of rocky shores in south Pembrokeshire. *J. Ecol.* **37,** 120–139.

Forbes, L., Seward, M. J. B., and Crisp, D. J. (1971). Orientation to light and the shading response in barnacles. *In* "Fourth European Marine Biology Symposium" (D. J. Crisp, ed.), pp. 539–558, Cambridge Univ. Press, London.

Fraenkel, G. S., and Gunn, D. L. (1940). "The Orientation of Animals." Oxford Univ. Press, London and New York.

Grave, B. H. (1930). The natural history of *Bugula flabellata* at Woods Hole, Massachusetts, including the behavior and attachment of the larva. *J. Morphol.* **49**, 355–383.

Harmelin, J. G. (1969). Bryozoaires des grottes sous-marines obscures de la région. Marseillaise: Faunistique et écologie. *Tethys* **1**, 793–806.

Hasper, M. (1913). On a method of rearing larvae of Polyzoa. *J. Mar. Biol. Assoc. U.K.* **9**, 435–436.

Humphries, E. M. (1973). Seasonal settlement of bryozoans in Rehoboth Bay, Delaware, U.S.A. *In* "Living and Fossil Bryozoa" (G. P. Larwood, ed.), pp. 115–128. Academic Press, New York.

Jägersten, G. (1972). "Evolution of the Metazoan Life Cycle." Academic Press, New York.

Jebram, D. (1970). Preliminary experiments with Bryozoa in simple apparatus for producing continuous water currents. *Helgol. Wiss. Meeresunters.* **20**, 278–292.

Joliet, L. (1877). Contributions à l'histoire naturelle des bryozoaires des côtes de France. *Arch. Zool. Exp. Gén.* **6**, 193–304.

Kaissling, K.-E. (1963). Die phototropische Reaktion der Zoide von *Bugula avicularia* L. *Z. Vergl. Physiol.* **46**, 541–594.

Kaissling, K.-E., and Schneider, D. (1962). Aktionsspektrum und Intensitätsabhängigkeit des phototropischen Wachstums von *Bugula avicularia* (Bryozoa). *Verh. Dtsch. Zool. Ges.* **1962**, 286–296.

Knight-Jones, E. W. (1951). Aspects of the setting behavior of larvae of *Ostrea edulis* on Essex oyster beds. *Rapp. P.-V. Réun., Cons. Perm. Int. Explor. Mer.* **128**, 30–34.

Lutaud, G. (1953). Premiers stades de la croissance zoariale chez le Bryozoaire Chilostome *Scrupocellaria reptans* Thompson. *Arch. Zool. Exp. Gén.* **90**, 42–58.

Lutaud, G. (1961). Contribution à l'étude du bourgeonnement et de la croissance des colonies chez *Membranipora membranacea* (Linné), Bryozoaire Chilostome. *Ann. Soc. Zool. Belg.* **91**, 157–300.

Lynch, W. F. (1947). The behavior and metamorphosis of the larvae of *Bugula neritina* (Linnaeus): Experimental modification of the length of the free-swimming period and the responses of the larvae to light and gravity. *Biol. Bull. (Woods Hole, Mass.)* **92**, 115–150.

Lynch, W. F. (1949). Modification of the responses of two species of *Bugula* larvae from Woods Hole to light and gravity: Ecological aspects of the behavior of *Bugula* larvae. *Biol. Bull. (Woods Hole, Mass.)* **97**, 302–310.

McDougall, K. D. (1943). Sessile marine invertebrates at Beaufort, North Carolina. *Ecol. Monogr.* **13**, 321–374.

Marcus, E. (1926a). Sinnesphysiologie und Nervensystem der Larve von *Plumatella fungosa* (Pall.). *Verh. Dtsch. Zool. Ges.* **31**, 86–90.

Marcus, E. (1926b). Beobachtungen und Versuche an lebenden Meeresbryozoen. *Zool. Jahrb., Syst. Oekol. Geogr. Tiere* **52**, 1–102.

Marcus, E. (1926c). Beobachtungen und Versuche an lebenden Süsswasserbryozoen. *Zool. Jahrb., Syst. Oekol. Geogr. Tiere* **52**, 279–350.

Maturo, F. J. S. (1959). Seasonal distribution and settling rates of estuarine Bryozoa. *Ecology* **40**, 116–127.

Mawatari, S. (1951a). The natural history of a common fouling bryozoan, *Bugula neritina* (Linnaeus). *Misc. Rep. Res. Inst. Nat. Resour. (Tokyo)* **20**, 47–54.

Mawatari, S. (1951b). On *Tricellaria occidentalis* (Trask), one of the fouling bryozoans in Japan. *Misc. Rep. Res. Inst. Nat. Resour. (Tokyo)* **22,** 9–16.
Mawatari, S. (1952). On *Watersipora cucullata* (Busk). II. *Misc. Rep. Res. Inst. Nat. Resour. (Tokyo)* **28,** 17–27.
Meadows, P. S., and Campbell, J. I. (1972). Habitat selection by aquatic invertebrates. *Adv. Mar. Biol.* **10,** 271–382.
Médioni, A. (1972). Les peuplements sessiles des fonds rocheux de la région de Banyuls-sur-Mer: Ascidies–Bryozoaires. *Vie Milieu* **213,** 591–656.
Norton, T. A. (1973). Oriented growth of *Membranipora membranacea* (L.) on the thallus of *Saccorhiza polyschides* (Lightf.) Batt. *J. Exp. Mar. Biol. Ecol.* **13,** 91–95.
Norton, T. A., Ebling, F. J., and Kitching, J. A. (1971). Light and the distribution of organisms in a sea cave. *In* "Fourth European Marine Biology Symposium" (D. J. Crisp, ed.), pp. 409–432. Cambridge Univ. Press, London and New York.
O'Donoghue, C. H. (1926). Observations on the early development of *Membranipora villosa* Hincks. *Contrib. Can. Biol. Fish. [N.S.]* **3,** 249–63.
Pomerat, C. M., and Reiner, E. R. (1942). The influence of surface angle and of light on the attachment of barnacles and other sedentary organisms. *Biol. Bull. (Woods Hole, Mass.)* **82,** 14–25.
Riedl, R. (1966). "Biologie der Meereshölen." Parey, Berlin.
Ryland, J. S. (1960). Experiments on the influence of light on the behavior of polyzoan larvae. *J. Exp. Biol.* **37,** 783–800.
Ryland, J. S. (1962). The effect of temperature on the photic responses of polyzoan larvae. *Sarsia* **6,** 41–48.
Ryland, J. S. (1967). Polyzoa. *Oceanogr. Mar. Biol.* **5,** 343–369.
Ryland, J. S. (1974). Observations on some epibionts of gulf-weed, *Sargassum natans* (L.) Meyen. *J. Exp. Mar. Biol. Ecol.* **14,** 17–25.
Ryland, J. S. (1976a). Behavior, settlement and metamorphosis of bryozoan larvae: A review. *Thalassia Jugosl.* **10,** 239–262.
Ryland, J. S. (1976b). Physiology and ecology of marine bryozoans. *Adv. Mar. Biol.* **14,** 285–443.
Ryland, J. S., and Stebbing, A. R. D. (1971). Settlement and orientated growth in epiphytic and epizoic bryozoans. *In* "Fourth European Marine Biology Symposium" (D. J. Crisp, ed.), pp. 105–123. Cambridge Univ. Press, London and New York.
Schneider, D. (1955). Phototropische Wachstum der Zoide und Rhizoide von *Bugula avicularia. Naturwissenschaften* **42,** 48–49.
Schneider, D. (1959). Der Aufbau der *Bugula*-Tierstöcke und Seine Beeinflussung durch Aussenfaktoren. *Biol. Zentralbl.* **78,** 250–283.
Schneider, D. (1960). Über den Mechanismus des phototropischen Knospenwachstums bei marinen Bryozoen. *Verh. Dtsch. Zool. Ges.* **1959,** 238–247.
Schneider, D. (1963). Normal and phototropic growth reactions in the marine bryozoan *Bugula avicularia. In* "The Lower Metazoa: Comparative Biology and Physiology" (E. C. Dougherty ed.), pp. 357–371. Univ. of California Press, Berkeley and Los Angeles.
Silén, L., and Jansson, B.-O. (1972). Occurrence of *Electra crustulenta* (Bryozoa) in relation to light. *Oikos* **23,** 59–62.
Thorson, G. (1964). Light as an ecological factor in the dispersal and settlement of larvae of marine bottom invertebrates. *Ophelia* **1,** 167–208.
Volkonsky, M. (1939). Sur la photo-akinèse des acridiens. *Arch. Inst. Pasteur Algér.* **17,** 194–220.
Waters, A. W. (1924). The ancestrula of *Membranipora pilosa* L., and of other cheilostomatous Bryozoa. *Ann. Mag. Nat. Hist.* [9] **14,** 594–612.

Williams, S. R. (1921). Concerning "larval" colonies of *Pectinatella. Ohio J. Sci.* **21,** 123–127.

Wisely, B. (1958). The settling and some experimental reactions of a bryozoan larva, *Watersipora cucullata* (Busk). *Aust. J. Mar. Freshwater Res.* **9,** 362–371.

Woollacott, R. M., and Zimmer, R. L. (1971). Attachment and metamorphosis of the cheilo-ctenostome bryozoan *Bugula neritina* (Linné). *J. Morphol.* **134,** 351–382.

Woollacott, R. M., and Zimmer, R. L. (1972). Fine structure of a potential photoreceptor organ in the larva of *Bugula neritina* (Bryozoa). *Z. Zellforsch. Mikrosk. Anat.* **123,** 458–469.

13

Fouling and Bioadhesion: Life Strategies of Bryozoans

JOHN D. SOULE and DOROTHY F. SOULE

I. INTRODUCTION

Fouling organisms represent the marine equivalent of terrestrial organisms called weeds; plants become weeds when they occur where man does not want them to occur, and sessile marine invertebrates become fouling organisms when they come in conflict with man's investments in the sea. There are innumerable habitats and microhabitats in the marine environment, but many of these represent surfaces that man would prefer uninhabited for reasons of his economy or convenience. As in the case of weeds, the battle is unending, since these organisms are remarkably persistent.

Some authors have extended the use of the word fouling to virtually all sessile organisms that are not benthic or permanently planktonic. There is a vast difference, however, between organisms capable of surviving on wood, metal, or plastic hulls of boats and ships that move through the waters, encountering varied temperatures, salinities, and other parameters, and those organisms that settle on fixed pilings, anchored boats, bulkheads, or even artificial reefs.

Bibliographies of bryozoan fouling organisms have at times included many of the major faunal inventories of various geographical areas of the world in which substrates are identified. Hutchins (1952) listed the citations which specifically indicated colonization of ships, lightships, wrecks, floats, buoys, cables, pipes, gates, ropes, and test surfaces. If all man-made installations are to be included in fouling, then a majority of bryozoans are probably capable of behaving as fouling organisms and only a relatively few species exclusively inhabit natural surfaces such as rocks, coral, or algae. Fouling studies done with plates or racks suspended in areas where ships do not or cannot anchor thus are really studies of the local fauna carried out by offering an additional artificial substrate choice. Probably fouling colonization occurs when ships are anchored or berthed for some time (hours or days) in harbors; certain colonies then are able to survive or grow during transport. It seems unlikely that settlement occurs when a moving ship encounters planktonic larvae at the exact moment when attachment is to take place. Rather ship fouling may require focal areas on docks or wharves for the export of larvae.

Several important factors are involved in the fouling mode of existence. The first are the reproductive strategies such as brooding, transport, and selection of substrate; these are followed by adult strategies such as mode of attachment, colony form, resistance to predation, and tolerance to environmental stress.

II. REPRODUCTIVE STRATEGIES

Bryozoans have representatives of those sessile invertebrate organisms that do not brood ova and those that do, species with very yolky eggs, those with little yolk, and those with extraembryonic nutrition provided by the parent. There are species with larvae capable of surviving as plankton for long periods of time and those that settle within a few hours after release.

In general, ova with little yolk develop as feeding (planktotrophic) larvae capable of remaining in the plankton for days and sometimes weeks. Both P. Bougis of Villefranche-Sur-Mer, France, and the late G. Thorson of Helsingor, Denmark (personal communications), spoke of having observed populations of bryozoan cyphonautes larvae that remained in a water mass for weeks when conditions for settling were not normal (i.e., low temperature). Thorson (1961) stated that he was able to assemble published data on pelagic larval periods for only about 200 species of invertebrates. The scanty information in existence is widely scattered in the literature of taxonomy, ecology, fisheries biology, reproduction, and more recently in environmental impact research documents. That information is sometimes too general in nature to include data on conditions obtaining at the time of observation as well.

Strathmann (1974) proposed that, among advantages to be gained for organisms by having long planktonic larval periods, there is a possible short-term advantage in competition for space and food by large-scale dispersal of sibling larvae. Because bryozoans are colonial, this could be of great long-term importance; individuals in a single colony would be of identical genetic composition instead of having the adaptive capabilities inherent in populations of single individuals with a wider range of generic variation. Dispersal offers opportunity for needed genetic diversity.

Evolution of reproductive strategies, in general, has proceeded along two different lines in time: by development of complex life cycles, and by short-cutting or condensing of complex cycles or stages in moves toward direct development. The more complex cycles seem to be either those of organisms that have become linked with parasitic modes of existence, or of free-living organisms that have developed one or more metamorphic stages with differing food requirements. Such cycles would seem to be very vulnerable due to the potential for interruption at any point, as in transfers between hosts or due to the scarcity of a specific food or host species. Enormous production of ova generally serves to compensate for such hazards, ensuring some minimal measure of success.

In some bryozoans the shelled planktonic cyphonautes larvae are capable of feeding, but the known species having true feeding cyphonautes larvae are few in number. These are included in the cheilostome genera *Membranipora* and *Electra* and in the ctenostome genera *Alcyonidium, Farella,* and *Hypophorella* (Hyman, 1959). Many cheilostomes have modified, so-called cyphonautes larvae which lack feeding organs, have no shell, and may be foreshortened in shape in the vertical dimension. Others are elongated vertically and are covered with cilia, as in *Bugula* (Woollacott and Zimmer, 1971). With the loss of feeding organs, larval settling times are greatly shortened and hence the dispersal distances of larvae are greatly reduced. This can be compensated for in part by yolky (lecithotrophic) eggs, or by brooding devices such as ovicells. Fewer ova are released in such cases, presumably with greater security for each.

The ability of bryozoan larvae to test substrates before settling was indicated by Lynch in publications between 1947 and 1959, in connection with a pioneering series of studies at Woods Hole, Massachusetts, on the experimental modification of settling and metamorphosis (Lynch 1957a,b). From 1951 on, at Plymouth Laboratory in England, D. P. Wilson (1951, cf.) published articles on experiments which indicated that other invertebrate larvae have a minimal period of time in which they could test substrata before final attachment and metamorphosis.

There have been suggestions that an initial bacterial film is required for colonization on smooth surfaces such as glass or metal (ZoBell and Allen, 1935;

Daniel, 1960). Scheer (1945) tested the settlement of bryozoans experimentally and found that the larvae selected plates with a diatom film rather than plates with bacterial coating. He determined that there was a consistent sequence of colonization in which diatoms were followed by bryozoans, then followed by *Styela* (ascidian) or by *Mytilus* (mussel) communities; *Mytilus* might follow *Styela* or follow directly from the bryozoan-dominant stage. Crisp and Ryland (1960) tested filmed and cleaned surfaces and found that, contrary to other larvae tested, *Bugula flabellata* preferred a clean surface.

Some larvae apparently are equipped with chemoreceptors that enable them to detect the presence of mature individuals of their own kind. This could lead the larvae to selection of suitable habitats, but could also serve to overpopulate a given area. Some bryozoans will frequently settle on mature colonies of other bryozoan species. A rather specific instance of this was found by Soule and Soule (1968) on boat hulls in a Hawaiian marina. Round patches of young, reddish, encrusting colonies of *Watersipora edmondsoni* showed the central ancestrular zooecium of each colony to be invaded by the ancestrula of an erect colony of *Bugula californica.* Each patch resembled a small, round island with a tree in the center. Seemingly this pattern was more consistent than random settling would allow, indicating some degree of selectivity is possible for the *Bugula* larvae.

Stebbing (1971) discussed the epizoic fauna of *Flustra foliacea.* He plotted the distribution of *Bugula flabellata* and found the larger colonies to be located near the growing edge of the perennial *Flustra* fronds. This distribution he explained as being linked to the annual cycle of growth and reproduction of the *Bugula.* In two species of *Crisia,* he found one species primarily near the growing edge, whereas the other was centered some distance from it, with some overlap. The same was true of *Scrupocellaria* species. It would appear that, in each case, one of the species was better suited to the exposed area than the other.

Pinter (1969) reported that certain algal substrates inhibit settling on growing tips by exuding toxic substances; some brown algae apparently secrete tannins, and it was suggested that red algae may secrete phenols. Crisp and Williams (1960) noted conversely that fucoid algae seemed to attract bryozoan larvae by release of substances into the waters. Stebbing (1972) reviewed possible explanations of site selection for settling on *Laminaria,* including the possible presence of microflora or fauna, and described settlement on the youngest areas of algal fronds. He noted that different populations seemed to have evolved different substrate preferences, however.

One might note that epizoic and epiphytic bryozoans are not considered to be "fouling organisms" by man, but might be so interpreted by the living organisms serving as substrates. Fouling of shellfish by *Bowerbankia gracilis* is a great nuisance in commercial or experimental mariculture. Following the suggestion of J. D. Soule, five species of shellfish reared at the Lamont–Doherty laboratories in St. Croix were successfully defouled by placing them in a vigorously aerated, saturated NaCl dip for 1 minute, after which they were air-dried for 1 hour and returned to their substrate; J. Dupuy suggested a 15-second dip in 5% clorox after drying as a fouling control for *Ostrea edulis* (O. Roels, personal communication).

Mawatari and Kobayashi (1954) tested fouling communities of bryozoans with nine different colors of glass panels for settling preferences, and discovered that orange was the most frequently chosen color, followed in order by black, red, and green. More than twice as many chose orange as chose white, the least preferred color. Among bryozoans, however, *Watersipora cucullata, Bugula neritina,* and *Bugula avicularia* preferred black, although *Tubulipora pulchra* preferred green and yellow, and *Electra angulata* selected white. *Electra angulata* normally forms a whitish colony over growing barnacles, which may account for its color selection.

Bryozoan larvae mostly exhibit positive phototaxis in the early phase of their planktonic larval period, followed by negative or neutral phototaxis prior to settling behavior, as shown by Lynch (1949), Mawatari (1952b), and Ryland (1967) (see Ryland, Chapter 12).

Many chemical substances have been tested as coatings to prevent the settling of fouling organisms. Heavy metal ions are generally toxic to invertebrate larvae but many antifouling paints probably oxidize too rapidly to serve as repellents for lengthy periods. Recently, paints that are designed to flake off, exposing new toxic surfaces, have been used with some success but have come under criticism because they add to the pollutant load of sediments and water.

Lynch (1947 cf.) was among the first to report effects of an array of chemicals on settling bryozoan larvae; he summarized some of his and other investigators' results in 1959 (Lynch, 1959a,b). In testing antifouling compositions on barnacle settling, de Wolf and van Londen (1966) noted that attachment was dispersed in a nonrandom fashion, suggesting variation in release of copper from the coating. The cyprid larvae are apparently insensitive to the copper but metamorphosed barnacles probably are killed by contact, through diffusion of the copper directly into the body rather than into surrounding seawater. Watersiporids are apparently less affected by copper than some other bryozoan species, as indicated by Ryland (1965), and by the evidence of primary colonization mentioned herein earlier, on boats in the Hawaiian marina. The role of microflora such as bacteria and diatoms has not been investigated in conjunction with the watersiporid colonization, however, and cannot be ruled out as a factor.

It is the ionic form of a heavy metal that produces the actual poisoning effect, and sequestering agents (chelators) are able to suppress ionic activity. Chlorophyll is a chelating agent, and may thus have a role in detoxification of surface antifouling coatings. Poisoning by nickel, mercury, lead, copper, vanadium, cadmium, and manganese are known to respond to chelation (Gornitz, 1972).

III. BIOADHESIVES AND ATTACHMENT

The principal studies of larval attachment have been performed with species of *Bugula,* but it seems probable that the mechanism is generally related to the eversion of the internal adhesive sac. Woollacott and Zimmer (1971) reviewed previous studies and by means of transmission electron microscopy revealed the unexpectedly complex morphogenetic movements involved in attachment of *Bugula neritina.*

Although the adhesive of the metamorphosing larvae has not been identified chemically, J. D. Soule (1973) identified as a simple acid mucopolysaccharide the substance attaching the basal wall of colonies of *Cryptosula pallasiana, Hippothoa hyalina, Membranipora tuberculata,* and *M. villosa* to the substrates. Lutaud (1961), referring to the basal layer of ectocyst of *M. membranacea,* indicated that it had stain characteristics of mucins and certain affinities with acid mucopolysaccharides. Banta (1968) identified "a muco- or glycoprotein or a protein–carbohydrate complex" in *Watersipora* cuticle.

The growth patterns of the colony following the attachment and metamorphosis of the ancestrula differ among the species, but the mode of growth has a basic pattern. It is apparent that the growing epithelial tip or bud extrudes the cement (bioadhesive), which flows over the substrate and hardens very rapidly. Tavener-Smith and Williams (1972) showed, in transmission electron microscopy, the secretion droplets in the growing tip, which they said were likely to be composed of polysaccharide–protein complexes. They also mentioned that droplets of mucopolysaccharide are present.

Scanning electron microscopy (Fig. 1) shows that the adhesive film is capable of flowing over the substrate, conforming closely to any irregularities, covering diatoms or debris, and filling in depressions or gaps in coral colonies, which effectively act as anchors for the colony when the film hardens. In Figs. 1 and 2, colonies of the cheilostome ascophorans of the family Smittinidae are shown bonded to glass and to coral. Soule and Soule (1974) discussed the importance of the bioadhesive attachment in the stabilization of surfaces of older reef corals.

Although bryozoans do not appear to settle close to living polyps of corals, they are apparently among the first to colonize in areas of dying or dead polyps. Thus they would tend to slow the clastic effects leading to breakdown of the surfaces of the skeletal remains.

Deposition of calcium carbonate by the basal epithelium apparently occurs very rapidly after the matrix component of the growing edge of the colony is secreted. Figure 3 shows the single calcareous crystals, which are actually

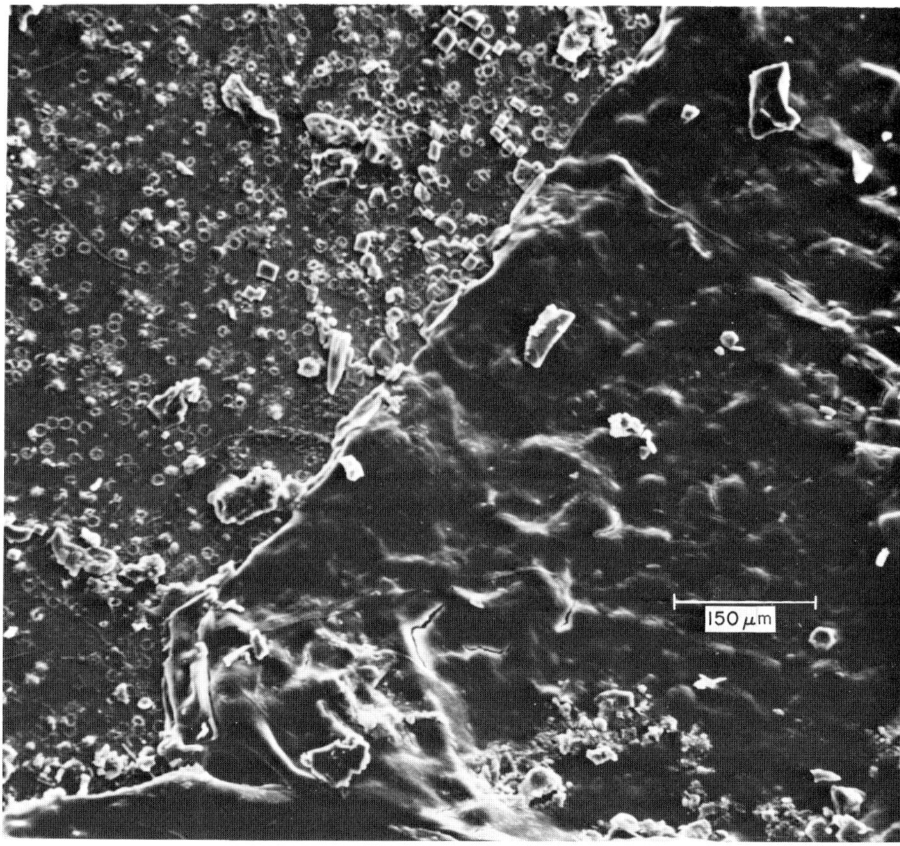

Fig. 1. Growing edge of *Smittoidea* colony, showing where secretions covered diatoms on glass slide. Photographed at 440 ×.

deposited around the curves of the basal layer as it conforms to the irregular substrate. There are apparently differences in the extent of the basal area capable of forming the attachment in the zooecium. Some anascans are attached only by a small basal area, while others are almost completely attached. Species of *Hippothoa,* although relatively easy to remove from some substrates, are apparently also able to erode mollusk shells sufficiently to form distinct shallow depressions for the individual zooecia (Marcus, 1938; Morris, 1975). Hincks (1880), Waters (1904), and others of the early bryozoan workers attempted to relate the burrowing *Terebripora,* known only at that time by their tracings in shells, to the eroding *Hippothoa.*

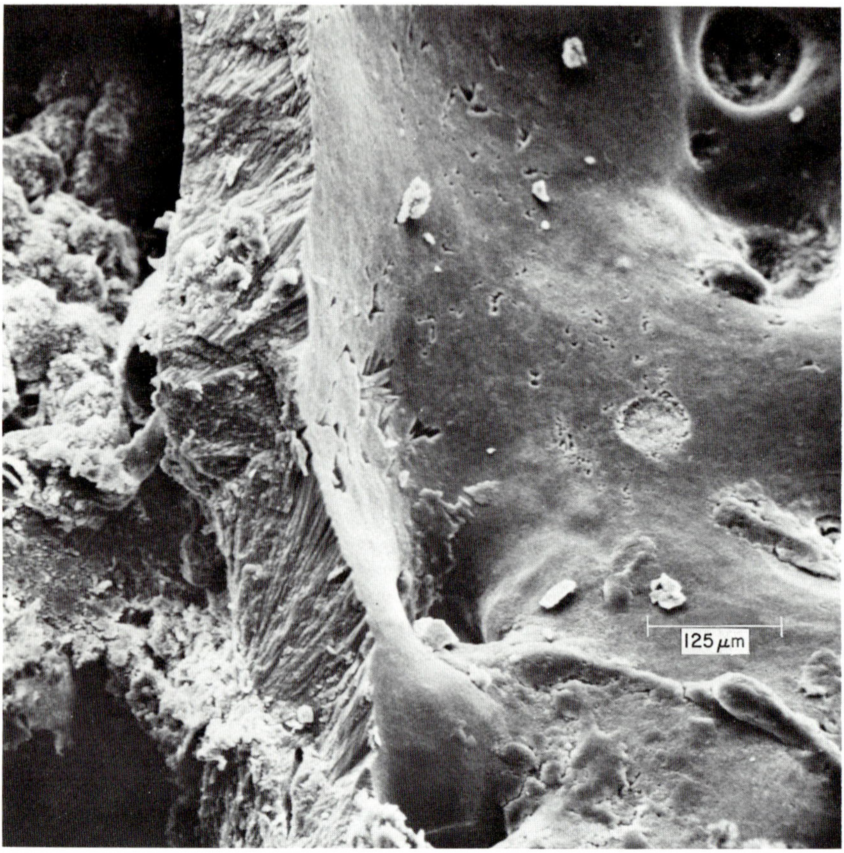

Fig. 2. Broken edge of basal wall (center) of *Parasmittina* as it has flowed over coral (left) and calcified. Lateral wall communication pore is at top right. Photographed at 540 ×.

Investigations of animal secretory products in search of a bioadhesive that is acceptable for clinical use by the dental and medical professions have been underway for many years. Much of this effort has been expended on studies of "barnacle cement." "Adhesion in Biological Systems" (1970) is a comprehensive multiauthored review of the subject up to that date. It contains articles by Barnes, by Hillman and Nace, by Liguori, and by Cook, all dealing with barnacle cement glands, secretory products, and composition. These authors have shown that the barnacle cement is mainly protein in composition, with a minor lipid component.

In the same volume, W. A. Corpe (1970) discusses the attachment of marine

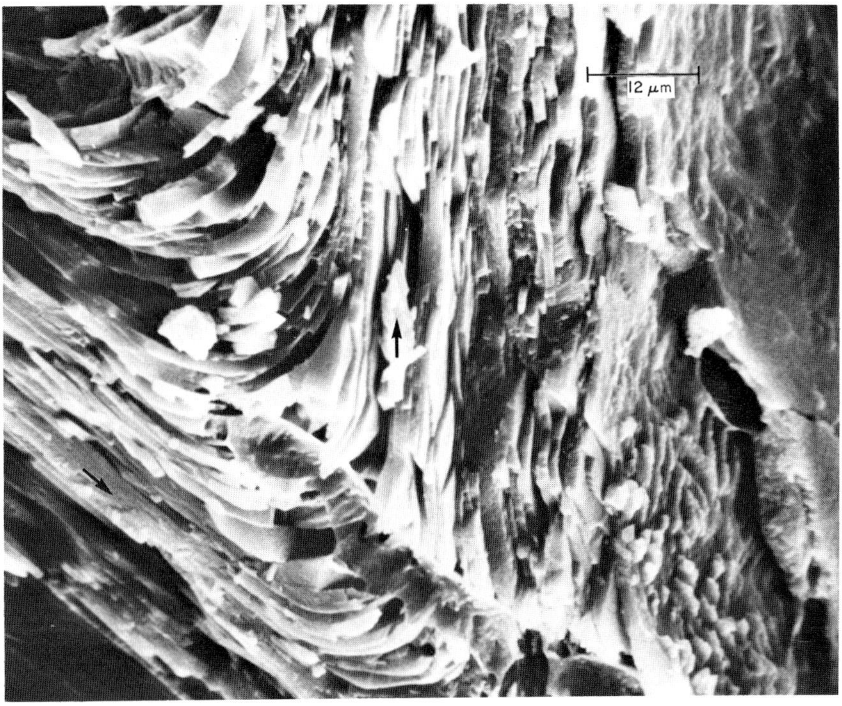

Fig. 3. Broken edge of basal wall (left arrow) and vertical wall (center arrow) showing alignment of calcium crystals. Interior wall is on right. Colony of *Parasmittina* had been riddled by minute, round filaments, possibly algae, shown by central tubular formation. Photographed at 5500 ×.

bacteria to solid substrates. He cites several papers that show that the intercellular substance cementing certain marine bacteria is a polysaccharide, which in one case was determined to contain glucose and amino sugars.

A recent paper (Knight-Jones *et al.,* 1972) shows that the adhesive agent by which the embryos of the polychaete *Romanchella,* (Spirorbinae) adhere is a mucopolysaccharide. In a series of papers from 1960 to 1966, Chaet and Philpott (1960, 1961; Philpott and Chaet, 1962; Chaet, 1965; Philpott *et al.,* 1966) found a mucopolysaccharide acting as the medium for the attachment of the coelenterate *Hydra pirardi.*

The presence of acid mucopolysaccharides have been linked to the phenomenon of calcification. Decapod crustacea have been shown by Travis (1957, 1960, 1963) to have acid mucopolysaccharides in the cuticle. Sulfated acid mucopolysaccharides are secreted by the molluscan mantle, associated with the mollusk shell, and may be involved in the calcification of the shell (Horiguchi, 1956, 1959).

Hunt (1970) brought together the literature on the polysaccharide–protein complexes found in a wide variety of marine invertebrates. He pointed out that the significance of sulfated mucopolysaccharides in the calcification process is in some dispute. The presence of this complex could be interpreted as a consequence of the trapping of the molluscan mantle mucin secretions. As an alternative, it may be possible that the polysaccharides serve as ion-exchange agents binding calcium ions and thereby initiating or inhibiting the deposition of these ions as inorganic calcium carbonate crystals. Hunt cites Simkiss (1964, 1965), reporting that calcite does not form readily in seawater, the inhibition possibly being due to the presence of a variety of interfering cations and anions. Therefore, the presence of natural ion-exchange polymers may be necessary to remove or act as a barrier to these ions in order that the insoluble calcite, rather than the less stable or more soluble form of calcium carbonate, aragonite, may be deposited.

The presence of a simple acid mucopolysaccharide, shown by histochemical methods to be associated with the bryozoan cuticle (Lutaud, 1961; Soule, 1972, 1973), may thus be of some significance in calcification as well as in adhesion. Although this work is in the initial stages, it would be of interest in the future to discover whether there is a chemical difference between the adhesive in bryozoans capable of ship transport and those that do not remain attached to hulls, between those settling on soft substrates (algae) as compared to those on firm (metal, rock), or between those that adhere tenaciously and those that are easily removed.

In primary calcification, calcitic lateral walls are produced by folding of the basal epithelium and in some cases there is evidence of folding to form at least the lower portion of transverse walls (Soule and Soule, 1972). Mixed mineralogy, that is, the presence of both calcite and aragonite (Sandberg, 1971) may possibly be due to secondary calcification and to the further intussusceptive growth of transverse walls after the secretion of mucopolysaccharide by the body wall epithelium has decreased or ceased.

IV. FOULING COMMUNITIES

It has generally been said that a relatively few species comprise an almost world-wide fouling community. The list of fouling organisms compiled in Table I tends to show a greater diversity than might then be expected. Of the 129 species listed (with some duplications in identity) only a very few have been reported from widely diverse localities. The first four columns of localities shown are representatives of some 500 miles of California coastline, and the next five are representatives of other Pacific Basin localities. These may be compared with reports from India, the Mediterranean, and the British Isles, respectively.

The surveys presented are not exhaustive in listing all fouling studies, and species lists that are not fouling but merely shallower water fauna have been omitted. A few nominal species are found in both Pacific and European (British and Mediterranean) waters. These include: *Bowerbankia gracilis, Bugula avicularia* (?), *Bugula neritina, Cryptosula pallasiana, Savignyella lafonti, Scruparia ambigua, Scrupocellaria bertholleti, Scrupocellaria scruposa, Watersipora cucullata, Watersipora subovoidea,* and *Zoobotryon verticillatum*. A number of these and other species are found on both sides of the Pacific and in Hawaii.

Several questions remain unanswered. These species may be the few that are easily and commonly recognized in fouling communities by specialists and nonspecialists alike. Also, some of these species bear few distinguishing characteristics, and subtle differences in size, colony form, and zooecial anatomy may not have been recognized. The *Schizoporella unicornis* is probably several species (Ryland, 1967). For example, this species had been identified by Osburn (1952) from the Baja California area, but when comparative material was examined, it became clear that a separate new species, *Schizoporella occidentalia* (Soule and Soule, 1964) existed there. Similarly, *Watersipora cucullata (sensu lato)* which has also been identified as *W. subovoidea* in cosmopolitan distributions, has sometimes included *W. subtorquata* from the Caribbean, *W. atrofusca* Busk, and *W. arcuata* Banta (1969) from California and Baja California, and *W. edmondsoni* (Soule and Soule, 1968) from Hawaii (Soule and Soule, 1975). Mawatari (1952a,b) earlier noted differences suggesting the presence of two forms of *W. cucullata.* Lack of good comparative material available for examination may result in local acceptance of an identity in one part of the world and another name commonly in use in another; conversely, a species regarded as cosmopolitan may in fact represent several.

Ryland (1965, 1967, 1971) has presented excellent reports or summaries of a number of fouling studies, two of which are utilized in the present table to show British and Mediterranean fouling. Unpublished records by J. D. Soule from Corona del Mar (Newport Bay), California were gathered at the Kerckhoff Marine Laboratory, California Institute of Technology, during settling and development studies in 1957–1958. Other unpublished records include those assembled for sea-level Panama Canal feasibility studies. The Los Angeles–Long Beach Harbor unpublished records by J. D. Soule are from a continuing 5-year environmental baseline inventory project in which settling racks are being used at 24 stations in the harbor. Reish's (1964) bryozoan records (identified by J. D. Soule) were from a new marina in Alamitos Bay, immediately southeast of Los Angeles–Long Beach harbor.

TABLE I
Distribution of Fouling Organisms[a]

Bryozoan	Monterey	Los Angeles area	Corona del Mar	La Jolla	Panama	Hawaii	Japan	Australia	New Zealand	India	Mediterranean	British Isles
1. *Aetea sica*												(13)
2. *Aetea truncata*						(18, 8, 4)						
3. *Aeverrillia setigera*										(10)		
4. *Aimulosia uvulifera*					(21)							
5. *Alderina arabianensis*										(10)		
6. *Amathia* sp.						(8, 4)					(14)	
7. *Amathia distans*						(19)						
8. *Anguinella palmata*					(21)			(1)				
9. *Antropora tincta*					(11)							
10. *Barentsia gracilis*	(5)		(17)			(19)						
11. *Bowerbankia gracilis*	(5)		(17)			(19)				(10)		(14)
12. *Bowerbankia imbricata*		(22)				(19)						
13. *Bugula avicularia*							(9)		(16)		(14)	
14. *Bugula californica*	(5)	(22)				(18)	(6, 7)					
15. *Bugula cucullata*										(10)		
16. *Bugula flabellata*								(1)	(16)			
17. *Bugula fulva*												(14, 13)
18. *Bugula neritina*	(5)	(22, 12)	(15)	(3)	(11)	(4, 8 18)	(6, 7, 9)	(23)	(16)		(14)	(14)
19. *Bugula plumosa*											(14)	
20. *Bugula simplex*												(14, 13)
21. *Bugula stolonifera*					(11)						(14)	(13, 14)
22. *Caberea zelandica*									(16)			
23. *Catenaria lafonti* (=#88)						(4)						

24. *Cellaria diffusa*				(3)						
25. *Cellaria mandibulata*				(3)						
26. *Cellepora dumerilli*										(14, 13)
27. *Celleporaria aperta*						(7)				
28. *Celleporaria brunnea*	(5)	(22)			(11)					
29. *Celleporella hyalina* (=#59)	(5)									(14, 13)
30. *Cleidochasma porcellanum*					(11)					
31. *Cedonellina montferrandi*						(7)				
32. *Conopeum reticulum*		(22)								
33. *Conopeum tubigerum*							(1)			
34. *Crepidacantha longiseta*					(11)					
35. *Crepidacantha solea*					(11)					
36. *Crisia* (sp.)	(5)				(21)					
37. *Crisia eburnea*				(3)						
38. *Crisia geniculata*				(3)						
39. *Crisulipora occidentalis*	(5)		(15)							
40. *Cryptosula pallasiana*	(5)	(12, 22)	(17, 15)			(7)		(16)		(14)
41. *Dendrobeania lichenoides*	(5)									
42. *Disporella californica*					(21)					
43. *Electra angulata*						(9)				
44. *Electra bengalensis*					(11)				(10)	
45. *Electra crustulenta*									(10)	
46. *Electra monstachys*					(11)					
47. *Electra pilosa*										(14)
48. *Electra tenella*						(7)				
49. *Escharoides praestans*						(7)				
50. *Eucratea clavata*			(15)	(3)						
51. *Fenestrulina malusi*						(7)				
52. *Filicrisia* (sp.)					(21)					
53. *Filicrisia franciscana*	(5)									
54. *Gemilliporidra multilaminosa*					(11)					
55. *Hippopodina feegeensis*					(11)	(7)				

TABLE I (*continued*)

Bryozoan	Monterey	Los Angeles area	Corona del Mar	La Jolla	Panama	Hawaii	Japan	Australia	New Zealand	India	Mediterranean	British Isles
56. *Hippoporella gorgonensis*					(11)							
57. *Hippoporella rimata*					(11)							
58. *Hippoporina verrilli*					(11)							
59. *Hippothoa hyalina*			(17)	(3)								
60. *Holoporella aperta* (=#27)			(15)									
61. *Holoporella brunnea* (=#28)		(12)										
62. *Lichenopora novozelandiae*					(21)							
63. *Lichenopora intricata?*					(11)							
64. *Lichenopora radiata*				(3)			(9)					
65. *Membranipora* (sp.)			(17)									
66. *Membranipora annae*					(11, 21)							
67. *Membranipora arborescens*					(11)							
68. *Membranipora membranacea*		(22)		(3)								
69. *Membranipora savarti*					(11)					(10)		
70. *Membranipora tehuelcha* (=#71)				(3)								
71. *Membranipora tuberculata*		(22)			(21)	(18)						
72. *Membranipora villosa*		(22)										
73. *Menipea occidentalis* (=#117)				(3)								
74. *Microporella ciliata*			(17)				(9)					
75. *Microporella umbracula*					(11)							
76. *Nolella blakei*			(17)									
77. *Onychocella alula*					(11)							
78. *Parasmittina* (sp.)					(21)							
79. *Parasmittina collifera*				(3)								
80. *Parasmittina crosslandi*					(11, 21)							

Species												
81. *Parasmittina trispinosa*					(11)							
82. *Parellisina curvirostris*					(11, 21)	(18)						
83. *Reptadeonella violacea*					(11)							
84. *Reteporellina denticulata* var. *gracilis*					(21)	(18, 8)						
85. *Rhynchozoon nudum*						(4)						
86. *Rhynchozoon rostratum*					(11)	(18)						
87. *Rhynchozoon tumulosum* (=#86)			(15)									
88. *Savignyella lafontii*					(21)	(18, 19)				(10)		
89. *Schizoporella cochinensis*										(10)		
90. *Schizoporella errata*											(14)	(13)
91. *Schizoporella linearis* var. *inarmata*										(10)		
92. *Schizoporella unicornis*		(12, 22)	(15)			(4, 18, 19)	(9, 6, 7)	(1)				
93. *Scruparia ambigua*	(5)	(22)	(17)									(13)
94. *Scruparia chelata*												(13)
95. *Scrupocellaria* (sp.)			(17)									
96. *Scrupocellaria bertholleti*		(12)			(11)						(14)	
97. *Scrupocellaria californica*	(5)	(22)										
98. *Scrupocellaria diegensis*			(15)	(3)								
99. *Scrupocellaria scruposa*					(21)							(14)
100. *Scrupocellaria sinuosa*						(19)						
101. *Sineopelta costazi*							(9)					
102. *Steginoporella magnilabris*					(11)		(7)					
103. *Stephanosella cornuta*					(11)							
104. *Smittina trispinosa* (=#81)							(9)					
105. *Smittoidea prolifica*							(7)					
106. *Smittia collifera* (=#79)				(3)								
107. *Smittia reticulata* (=#105)				(3)								

TABLE I *(continued)*

Bryozoan	Monterey	Los Angeles area	Corona del Mar	La Jolla	Panama	Hawaii	Japan	Australia	New Zealand	India	Mediterranean	British Isles
108. *Tegela unicornis*												(13)
109. *Thalamoporella californica*			(17)				(7)					
110. *Thalamoporella delicata*						(20)						
111. *Thalamoporella hawaiiana*						(8, 20)						
112. *Thalamoporella rozieri,* var. *californiensis* (=#109)				(3)								
113. *Thalamoporella verrilli*						(20)						
114. *Trematooecia aviculifera*					(11)							
115. *Trematooecia turrita*					(11)							
116. *Trypostega venusta*						(18)						
117. *Tricellaria occidentalis*							(9)					
118. *Tubulipora pacifica*	(5)				(21)							
119. *Tubulipora pulchra*							(9)					
120. *Tubulipora tuba*			(17)									
121. *Victorella pavida*										(10)		
122. *Vittaticella* (sp.)					(21)	(8)						
123. *Vittaticella contei*					(21)							
124. *Watersipora arcuata*				(2)				(23)	(2)			
125. *Watersipora cucullata* (=#127?)							(6, 9)	(1)	(16)			
126. *Watersipora edmondsoni*						(8, 19)						
127. *Watersipora subovoidea*							(7)			(10)		(13)
128. *Zoobotryon pellucidus* (=129)				(3)		(4)					(14)	
129. *Zoobotryon verticillatum*						(18, 19)	(9)					

[a]Key to references: (1) Allen (1953); (2) Banta (1969); (3) Coe and Allen (1937); (4) Edmondson and Ingram (1939); (5) Haderlie (1969); (6) Kawahara (1965, 1969); (7) Long and Rucker (1969); (8) Long (1970); (9) Mawatari (1953); (10) Menon and Nair (1971); (11) Powell (1971); (12) Reish (1964); (13) Ryland (1965); (14) Ryland (1967); (15) Scheer (1945); (16) Skerman (1960); (17) J. D. Soule (1957, unpublished); (18) Soule and Soule (1967); (19) Soule and Soule (1968); (20) Soule and Soule (1970); (21) D. F. Soule (1972, unpublished); (22) J. D. Soule (1971–1974,

V. FOULING COLONIAL STRATEGIES

Among the various principal species mentioned previously in fouling communities, several adaptive strategies can be seen. On boat hulls, docks, and pilings, the flat, encrusting circular colonies of *Cryptosula, Hippopodina, Schizoporella,* and *Watersipora* are noticeable. Where these colonies are dense, growth patterns may be disrupted and competition may result in an upward curling pattern or one species may overgrow another. Certainly the thin encrusting mode offers the least resistance to waterflow on a ship underway.

Another colony form is the flaccid, branching mode seen in the ctenostomes *Amathia, Bowerbankia,* and *Zoobotryon.* These species drift in gentle waters from their attachments to docks and anchored ship bottoms, but large colonies cannot withstand water turbulence, since longer strands would be torn loose. Other branching colonies include the anascan species, *Bugula neritina* which is mostly chitinous, and *B. californica* and *Scrupocellaria* species, which are more calcified and hence more erect in growth mode.

The small, delicate, jointed ascophoran colonies of *Savignyella* and *Vitaticella* would seem to be unimportant as foulers. Yet on docks in Hawaii it was observed that these species grew between the flaccid ctenostome colonies and tended to stiffen the entire mass into an almost abrasive padding.

ACKNOWLEDGMENTS

Field observations and collections made by the authors in Hawaii were supported in part by National Science Foundation grants GB 5208 and GB 7723. Observations at Corona del Mar were made possible by a postdoctoral fellowship at California Institute of Technology for the senior author. Los Angeles area collections were funded in part by the USC Sea Grant Program and Harbors Environmental Projects of the Allan Hancock Foundation, University of Southern California. Penny Morris assisted greatly in making a survey of fouling literature.

REFERENCES

Abbott, B. C., Soule, D. F., and Oguri, M., and Soule, J. D. (1973). *In situ* studies of the interface of natural and man-made systems in a metropolitan harbor. *Helgo. Wiss. Meeresunters.* **24**, 455–464.

Allen, F. E. (1953). Distribution of marine invertebrates by ships. *Aust. J. Mar. Freshwater Res.* **4**, 307–316.

Banta, W. C. (1968). The body wall of the encrusting cheilostome *Watersipora nigra* (Canu and Bassler). *Atti Soc. St. Nat. e Museo Civ. St. Nat. Milano* **108**, 93–96.

Banta, W. C. (1969). The recent introduction of *Watersipora arcuata* Banta (Bryozoa, Cheilostomata) as a fouling pest in southern California. *Bull. S. Calif. Acad. Sci.* **68**, 248–251.

Barnes, H. (1970). A review of some factors affecting settlement and adhesion in the cyprids of some barnacles. *In* "Adhesion in Biological Systems" (R. S. Manly, ed.), pp. 89–111. Academic Press, New York.

Chaet, A. B. (1965). Invertebrate adhering surfaces: Secretions of the starfish, *Asterias forbesi,* and the coelenterate, *Hydra pirardi. Ann. N. Y. Acad. Sci.* **118,** 921–929.

Chaet, A. B., and Philpott, D. E. (1960). Secretory structures in the tube foot of starfish. *Biol. Bull. (Woods Hole, Mass.)* **119,** 308 (abstr.).

Chaet, A. B., and Philpott, D. E. (1961). Formation and possible function of the "secretory packets" of the starfish tube foot. *Biol. Bull. (Woods Hole, Mass.)* **121,** 373 (abstr.).

Coe, W. R., and Allen, W. E. (1937). Growth of sedentary marine organisms on experimental blocks and plates for nine successive years (at the pier of the Scripps Institution of Oceanography). *Bull. Scripps Inst. Oceanogr.* **4,** 101–136.

Cook, M. (1970). Composition of mussel and barnacle deposits at the attachment interface. *In* "Adhesion in Biological Systems" (R. S. Manly, ed.), pp. 139–150. Academic Press, New York.

Corpe, W. A. (1970). Attachment of marine bacteria to solid surfaces. *In* "Adhesion in Biological Systems" (R. S. Manly, ed.), pp. 73–87. Academic Press, New York.

Crisp, D. J., and Ryland, J. S. (1960). Influence of filming and of surface texture on the settlement of marine organisms. *Nature (London)* **185,** 119.

Crisp, D. J., and Williams, G. B. (1960). Effects of extracts from fucoids in promoting settlement of epiphytic Polyzoa. *Nature (London)* **188,** 1206–1207.

Daniel, A. (1960). The primary film as a factor in the settlement of marine foulers. *J. Madras Univ., Sect. B* **30,** 189–200.

de Wolf, P., and van Londen, A. M. (1966). Anti-fouling compositions. *Nature (London)* **209,** 272–274.

Edmondson, C. H., and Ingram, W. M. (1939). Fouling organisms in Hawaii. *Occas. Pap. Bishop Mus.* **14,** 251–300.

Gornitz, V. (1972). Chelation. *In* "Encyclopedia of Geochemistry and Environmental Sciences" (R. W. Fairbridge, ed.), pp. 149–152. Van Nostrand-Reinhold, Princeton, New Jersey.

Haderlie, E. C. (1969). Marine fouling and boring organisms in Monterey Harbor. II. Second year of investigation. *Veliger* **12,** 182–192.

Hillman, R. E., and Nace, P. F. (1970). Histochemistry of barnacle cyprid adhesive formation. *In* "Adhesion in Biological Systems" (R. S. Manly, ed.), pp. 113–121. Academic Press, New York.

Hincks, T. (1880). "A History of the British Marine Polyzoa," Vol. 1. van Voorst, London.

Horiguchi, Y. (1956). Biochemical studies on *Pteria (Pinctada) martensii* (Dunker) and *Hyriopsis schlegelii* (v. Martens). II. Separation of crude sulfomuco-polysaccharides from various tissues of *Pteria (Pinctada) martensii* (Dunker) and *Hyriopsis schlegelii* (v. Martens). *Bull. Jpn. Soc. Sci. Fish.* **22,** 463–466.

Horiguchi, Y. (1959). Biochemical studies on *Pteria (Pinctada) martensii* (Dunker) and *Hyriopsis schlegelii* (v. Martens). VII. On the separation and purification of sulfomucopolysaccharide and detection of its component sugars. *Rep. Fac. Fish., Prefect. Univ. Mie* **3,** 399–406.

Hunt, S. (1970). "Polysaccharide-Protein Complexes in Invertebrates." Academic Press, New York.

Hutchins, L. W. (1952). The Bryozoa. Marine fouling and its prevention, U. S. Naval Institute. *Woods Hole Oceanogr. Inst. Contrib.* **580,** 141–143, 163, and 190–192.

Hyman, L. H. (1959). "The Invertebrates: Smaller Coelomate Groups," Vol. 5, pp. 273–515. McGraw-Hill, New York.

Kawahara, T. (1965). Studies on the marine fouling communities. III. Seasonal changes in the initial development of test block communities. *Rep. Fac. Fish., Prefect. Univ. Mie* **5**, 319–364.

Kawahara, T. (1969). Studies on the marine fouling communities. IV. Difference in the constitution of fouling communities according to localities. a. Nagasaki Harbor. *Rep. Fac. Fish., Prefect. Univ. Mie* **6**, 109–125.

Knight-Jones, E. W., Knight-Jones, P., and Vine, P. J. (1972). Anchorage of embryos in Spirorbinae (Polychaeta). *Mar. Biol.* **12**, 289–294.

Liguori, V. R. (1970). Histological and histoenzymological characteristics of the cement apparatus in barnacles. *In* "Adhesion in Biological Systems" (R. S. Manly, ed.), pp. 123–138. Academic Press, New York.

Long, E. R. (1970). Second year of marine biofouling studies off Oahu, Hawaii. *In* Oceanographic cruise summary. Informal report. No. 70-48, pp. 1–12. Naval Oceanographic Office, Washington, D. C.

Long, E. R., and Rucker, J. B. (1969). A comparative study of cheilostome Bryozoa at Yokosuka, Maizuru and Sasebo, Japan. *Pac. Sci.* **23**, 56–69.

Lutaud, G. (1961). Contribution à l'étude du bourgeonnement et de la croissance des colonies chez *Membranipora membranacea* (Linné), bryozoaire, chilostome. *Ann. Soc. R. Zool. Belg.* **91**, 157–300.

Lynch, W. F. (1947). The behavior and metamorphosis of the larva of *Bugula neritina* (Linnaeus): Experimental modification of the length of the free-swimming period and the responses of the larvae to light and gravity. *Biol. Bull. (Woods Hole, Mass.)* **92**, 115–150.

Lynch, W. F. (1949). Modification of the responses of two species of *Bugula* larvae from Woods Hole to light and gravity: Ecological aspects of the behavior of *Bugula* larvae. *Biol. Bull. (Woods Hole, Mass.)* **97**, 302–310.

Lynch, W. F. (1959a). Factors inhibiting metamorphosis in *Bugula* and *Amaroecium* larvae. *Wilhelm Roux' Arch. Entwicklungsmech. Org.* **151**, 164–180.

Lynch, W. F. (1959b). Factors influencing metamorphosis of larvae of some of the sessile organisms. *Proc. Int. Congr. Zool. 15th, 1958* Sect. 3, Pap. No. 9, pp. 1–3.

Marcus, E. (1938). Bryozoarios perfuradores de conchas: Shell burrowing Polyzoa. *Arq. Inst. Biol., Sao Paulo* **9**, 273–296.

Mawatari, S. (1952a). On *Watersipora cucullata* (Busk). I. *Misc. Rep. Res. Inst. Nat. Resour. (Tokyo)* **25**, 14–17.

Mawatari, S. (1952b). On *Watersipora cucullata* (Busk). II. *Misc. Rep. Res. Inst. Nat. Resour. (Tokyo)* **28**, 17–27.

Mawatari, S. (1953). On *Electra angulata* Levinsen, one of the fouling bryozoans in Japan. *Misc. Rep. Res. Inst. Nat. Resour. (Tokyo)* **32**, 5–10.

Mawatari, S., and Kobayashi, S. (1954). Seasonal settlement of animal fouling organisms in Ago Bay, Middle Part of Japan. I. *Misc. Rep. Res. Inst. Nat. Resour. (Tokyo)* **35**, 37–47.

Menon, N. R., and Nair, N. B. (1971). Ecology of fouling bryozoans in Cochin waters. *Mar. Biol.* **8**, 280–307.

Morris, P. A. (1975). A comparative study of decalcification of Mollusca shells by various bryozoans. *In* "Bryozoa 1974" (S. Pouyet, ed.), Doc. Lab. Géol. Fac. Sci. Lyon (France). H.S. 3 (fasc. 1), pp. 109–113, 2 pls, Univ. Claude Bernard, Lyon.

Osburn, R. C. (1952). Bryozoa of the Pacific Coast of America. Part 2, Cheilostomata-

Ascophora. Allan Hancock Found. Pacific Exped., Univ. Southern California (Los Angeles), **14,** 271–612, pls. 30–64.

Philpott, D. E., and Chaet, A. B. (1962). Electron microscope observations of secretory granules in the adhesive surface of *Hydra pirardi. Biol. Bull. (Woods Hole, Mass.)* **123,** 509 (abstr.).

Philpott, D. E., Chaet, A. B., and Burnett, A. L. (1966). A study of the secretory granules of the basal disk of *Hydra. J. Ultrastruct. Res.* **14,** 74–84.

Pinter, P. (1969). Bryozoan-algal associations in southern California waters. *Bull. S. Calif. Acad. Sci.* **68,** 199–218.

Powell, N. A. (1971). The marine Bryozoa near the Panama Canal. *Bull. Mar. Sci.* **21,** 766–778.

Reish, D. J. (1964). Studies on the *Mytilis edulis* community in Alamitos Bay, California. II. Population variations and discussion of the associated organisms. *Veliger* **6,** 202–207.

Ryland, J. S. (1965). Catalogue of main marine fouling organisms (found on ships coming into European waters). 2. "Polyzoa," pp. 1–83. Organization for Economic Cooperation and Development, Paris.

Ryland, J. S. (1967). Polyzoa. *Oceanogr. Mar. Biol.* **5,** 343–369.

Ryland, J. S. (1971). Bryozoa (Polyzoa) and marine fouling. *In* "Marine Borers, Fungi and Fouling Organisms of Wood" (E. B. Gareth Jones and S. K. Eltringham, eds.), pp. 137–154. Organization for Economic Cooperation and Development, Paris.

Sandberg, P. A. (1971). Scanning electron microscopy of cheilostome bryozoan skeletons: Techniques and preliminary observations. *Micropaleontology* **17,** 129–151.

Scheer, B. T. (1945). The development of marine fouling communities. *Biol. Bull. (Woods Hole, Mass.)* **89,** 102–121.

Simkiss, K. (1964). Phosphates as crystal poisons of calcification *Biol. Rev., Cambridge Philos. Soc.* **39,** 487–505.

Simkiss, K. (1965). The organic matrix of the oyster shell. *Comp. Biochem. Physiol.* **16,** 427–435.

Skerman, T. M. (1960). Ship-fouling in New Zealand waters: A survey of marine fouling organisms from vessels of the coastal and overseas trades. *N. Z. J. Sci. Technol.* **3,** 620–648.

Soule, D. F., and Soule, J. D. (1964). The Ectoprocta (Bryozoa) of Scammon's Lagoon, Baja, California. *Am. Mus. Novit.* **2199,** 1–56.

Soule, D. F., and Soule, J. D. (1967). Faunal affinities of some Hawaiian Bryozoa (Ectoprocta). *Proc. Calif. Acad. Sci.* **35,** 265–272.

Soule, D. F., and Soule, J. D. (1968). Bryozoan fouling organisms from Oahu, Hawaii, with a new species of *Watersipora. Bull. S. Calif. Acad. Sci.* **67,** 203–218.

Soule, D. F., and Soule, J. D. (1972). Ancestrulae and body wall morphogenesis of some Hawaiian and eastern Pacific Smittinidae (Bryozoa, Ectoprocta). *Trans. Am. Microsc. Soc.* **91,** 251–260.

Soule, D. F., and Soule, J. D. (1974). The bryozoan-coral interface on corals and coral reefs. *In* "Proceedings of the International Coral Reef Symposium," Vol. I, pp. 335–340. Great Barrier Reef Committee, Brisbane, Australia.

Soule, D. F. and Soule, J. D. (1975). Species groups Watersiporidae. *In* "Bryozoa 1974" (S. Pouyet, ed.), Soc. Lab. Geol. Fac. Lyon (France). H. S. 3 (fasc. 1), pp. 299–309. Univ. Claude Bernard, Lyon.

Soule, J. D. (1972). Investigation into the nature of the bioadhesive from marine bryozoans. *J. Dent. Res.* **51,** Suppl. 147 (abstr.).

Soule, J. D. (1973). Histological and histochemical studies of the bryozoan-substrate interface. *In* "Living and Fossil Bryozoa" (G. P. Larwood, ed.), pp. 343–347. Academic Press, New York.

Soule, J. D., and Soule, D. F. (1970). New species of *Thalamoporella* (Ectoprocta) from Hawaii, examined by scanning electron microscopy. *Am. Mus. Novit.* **2417**, 1–18.

Stebbing, A. R. D. (1971). Growth of *Flustra foliacea* (Bryozoa). *Mar. Biol.* **9**, 267–272.

Stebbing, A. R. D. (1972). Preferential settlement of a bryozoan and serpulid larvae on the younger parts of *Laminaria* fronds. *J. Mar. Biol. Assoc. U. K.* **52**, 765–772.

Strathmann, R. (1974). The spread of sibling larvae of sedentary marine invertebrates. *Am. Nat.* **108**, 29–44.

Tavener-Smith, R., and Williams, A. (1972). The secretion and structure of the skeleton of living and fossil Bryozoa. *Philos. Trans. R. Soc. London Ser. B* **264**, 97–159.

Thorson, G. (1961). Length of pelagic larval life in marine bottom invertebrates as related to larval transport by ocean currents. *In* "Oceanography," Publ. No. 67, pp. 455–474. Am. Assoc. Adv. Sci., Washington, D.C.

Travis, D. F. (1957). The molting cycle of the spiny lobster, *Panulirus argus* Latreille. IV. Post-ecdysial histological and histochemical changes in the hepato-pancreas and integumental tissues. *Biol. Bull. (Woods Hole, Mass.)* **113**, 451–479.

Travis, D. F. (1960). The gastroliths of the crayfish as convenient sites for the study of mechanisms of matrix and mineral deposition in calcified structures. *Fed. Proc., Fed. Am. Soc. Exp. Biol.* **19**, 248 (abstr.).

Travis, D. F. (1963). Structural features of mineralization from tissue to macromolecular levels of organization in the decapod Crustacea. *Ann. N.Y. Acad. Sci.* **109**, 177–235.

Waters, A. W. (1904). Bryozoa. *Exped. Antarct. Belgo-Neerl.* **10**, 1–114.

Wilson, D. P. (1951). Larval metamorphosis and the substratum. *Ann. Biol.* **27**, 491–501.

Wisely, B. (1963). Effects of antifouling paints on settling larvae of the bryozoan *Bugula neritina* L. *Aust. J. Mar. Freshwater Res.* **14**, 44–59.

Woollacott, R. M., and Zimmer, R. L. (1971). Attachment and metamorphosis of the cheilo-ctenostome bryozoan *Bugula neritina* (Linné). *J. Morphol.* **134**, 351–382.

ZoBell, C. E., and Allen, E. C. (1935). The significance of marine bacteria in the fouling of submerged surfaces. *J. Bacteriol.* **29**, 239–251.

14

Population Genetics of Bryozoans

THOMAS J. M. SCHOPF

Population genetics is a branch of genetics that deals with the breeding structure and genetic composition of populations.

Bruce Wallace, 1968

I. INTRODUCTION

The initial step in achieving an understanding of the population genetics of bryozoans is to be able to recognize the genotype of an individual organism (i.e., a colony). External morphological features are perhaps under the control of many gene loci. In any event, no morphological feature of bryozoans has yet been found to vary in discrete classes as is typical of single gene characters. In addition, no controlled laboratory breeding experiments of bryozoans have yet been attempted. Accordingly, the discovery through electrophoresis of variations in electrical charge in enzymes and other proteins that can be attributed to different states (alleles) of a single gene locus in many plants and animals, including bryozoans (Gooch and Schopf, 1969), has permitted many previously unanswered questions to be examined.

Several electrophoretic studies have been conducted since the significance of the technique was broadly demonstrated in 1966 (Hubby and Lewontin, 1966; Lewontin and Hubby, 1966), and these studies have used a wide variety of animals [largely summarized by Gottlieb (1971) and by Selander and Kaufman (1973)]. Taken together these investigations showed that while many loci are monomorphic for a single "wild-type" allele, considerable genetic polymorphism is the rule. As of now, the largest number of loci (by a factor of 2) that has been identified by electrophoretic techniques is in *Homo sapiens.* Of 71 loci, each of which determines the structure of an enzyme or other protein, 28% are polymorphic, and the average heterozygosity per locus is 6.7% (Harris and Hopkinson, 1972). Many species have 20–50% of their loci polymorphic, and 6–15% heterozygosity. For *H. sapiens* one significant result of the analysis of variation (1) within populations, (2) within races, and (3) between races is that approximately 85% of human genetic variation is accounted for by variation within local populations, and only 6% of human genetic variation may result from differences between races (Lewontin, 1972). A considerable portion of this variation may be due to natural selection (Lewontin and Krakauer, 1973).

In other species, the description of geographic variation of gene frequencies has shown that some taxa are strongly differentiated into local populations but that other taxa appear to have similar alleles over a wide geographic range. Indeed, within a single species, genetic differentiation is seen in some polymorphic loci and not in others. Whether the degree of genetic differentiation is closely related to the degree of dispersal in most cases, and indeed whether the observed genetic variability is explicable in terms of natural selection, are questions currently under intense study, and for which a considerable literature exists (Manwell and Baker, 1970; Lewontin, 1974; Gooch, 1974, for marine species).

The purpose of this chapter is to present what has been discovered by using the electrophoretic technique in studies of the breeding structure and genetic composition of bryozoan populations. The data refer to population size, outbreeding and inbreeding, linkage, geographic differentiation, selection vs. migration, and genetic variability. The chief experimental species has been the marine bryozoan *Schizoporella errata* (= *S. unicornis* of older Woods Hole literature; see Gooch and Schopf, 1971). This species is among the most abundant coastal-zone, truly marine species of any phylum. My original motivation for investigating the genetic characteristics of bryozoans was to establish guidelines for the mechanism of speciation in these animals. The implications of the present data for models of speciation are presented in the final section of the paper. Since some students and colleagues have been interested in the techniques utilized for bryozoans, these are summarized in the Appendix (Section VIII).

II. POPULATION SIZE

My field observations of *Schizoporella errata* at some 25 localities indicate a census population of approximately 10 to 300 colonies per local population. This and other bryozoan species are substrate limited (for *S. errata,* chiefly attached to pilings and rocks) and the distribution of local populations is therefore patchy. If the assumption of the same breeding structure can be applied to each local population, then the effective (i.e., the breeding) population (N) can be calculated if $\bar{q}$ and its variance ($\sigma^2_{\delta q}$) are known for the sampled populations according to the expression shown in Eq. (1) (Li, 1955, p. 315).

$$N = \frac{\bar{q}\,(1 - \bar{q})}{2\sigma^2_{\delta q}} \qquad (1)$$

In order to have as large and yet as homogeneous samples as possible in calculating N, I used the grouped data which later serve, in this chapter, as the basis for the construction of fitness surfaces. Gene frequencies are for the enzymes leucine aminopeptidase (hereafter referred to as LAP) and glutamate-oxalate transaminase (GOT) from warm water, intermediate temperature, and cool water areas (Tables V–VII). As summarized in Table I, the most striking result is that N is small (mean of 75.3).

The relationship expressed in Eq. (1) is based on the assumption that the variance in $\bar{q}$ is due to random effects. Since N and the variance of $\bar{q}$ are inversely proportional in Eq. (1), any of the observed variance due to nonrandom effects (for example, selection) would have the result of decreasing N from its true value. The data were chosen to minimize such effects. However, the computed effective population size (N) appears to be best considered as an approximate lower limit.

TABLE I
Estimates of Effective Population Size, *N*, in *Schizoporella errata*

	Locus		
Grouping	*LAP*	*GOT*	Mean
Cool water	22.8	119.4	71.1
Intermediate	177.5	33.4	105.4
Warm water	72.1	27.2	49.6
Mean	90.8	60.0	75.3

Unless some factor of differential selection is operating with respect to LAP and GOT loci, data from each locus should yield an identical N for each group of samples. In the present instance, differences in N exist between loci, but there are no consistent ecologic or genetic trends. Accordingly, given the small number of populations on which the calculation is based, I regard the observed variation in N as probably a reflection of the differences between sampled q, and its variance, with respect to the population q and its variance. In any event, the important conclusion supported by all of the data is that the effective breeding population is small. This is consistent with the sessile nature of reproducing colonies of *Schizoporella errata,* and its chiefly annual breeding cycle.

III. OUTBREEDING AND INBREEDING

Many species of bryozoans are hermaphroditic (Hyman, 1959), but at least some possess the *possibility* for outbreeding first demonstrated by Silén (1966). However, the lack of recognition of traits that reassort into the well-known Mendelian ratios prevented any testing of whether outbreeding was in fact carried on. For cheilostome bryozoans, the first agreement of a particulate character with the Mendelian ratios was shown by Gooch and Schopf (1970) for one locus of *Schizoporella errata* at each of three localities, and for three loci of *Bugula stolonifera* from one locality.

Allele frequencies are now known in *S. errata* at a 2-allele locus which is responsible for the enzyme leucine aminopeptidase (LAP), and from a 2-allele locus which determines the enzyme glutamate-oxalate transaminase (GOT). Data on these loci have been obtained from more than 20 local populations, 16 of which have been followed for 2 successive years, and 7 of which have been followed (for LAP) for 3 to 5 years (data in Schopf, 1974a). [See Addendum.] The range in allele frequencies for each locus over a 120-km geographic transect is substantial. In all 32 of the populations sampled for GOT, and in 30 of the 32 populations sampled for LAP (data of 1972 and 1973), the observed genotype frequencies are within the expectations of the Hardy–Weinberg binomial expansion.

Because allele frequencies generally remain uniform within a population from year to year, routine outbreeding appears to be likely, though very large samples are usually required to demonstrate deviation from the binomial expansion, whether due to inbreeding or selection (Lewontin and Cockerham, 1959; Ward and Sing, 1970). There is little reason to believe that these loci would yield results compatible with Hardy–Weinberg in many populations and with widely different allele frequencies, unless these loci were all reflecting random mating in each local population (additional data in Schopf, 1973).

Data on genotype frequencies for 1972 and 1973 are summarized here in de Finetti diagrams (Li, 1955; Cannings and Edwards, 1968). Each datum point in the triangle (Figs. 1 and 2) represents a population such that the perpendicular from one side is the proportion of one homozygote class (D), the perpendicular from another side the proportion of heterozygotes (H), and the perpendicular from the third side the proportion of the other homozygote class (R). The sum of the three proportions (D + H + R) equals one. Allele frequencies for each locus can be read off the horizontal axis with p (frequency of "slow" allele, "S") read from the right and q (frequency of "fast" allele, "F") read from the left. For example, for LAP, the highest point to the right in the triangle represents a population with these characteristics: 2 slow homozygotes (8.3%), 14 heterozygotes (58.3%), and 8 fast homozygotes (33.4%); $p = 0.37$; $q = 0.63$; $n = 24$).

The solid-line parabola indicated in each diagram is a continuous set of populations, each of which is in perfect Hardy–Weinberg equilibrium. The deviation of a point from that parabola is one measure of heterozygote excess or deficiency. An approximate line of significance for the 0.5% confidence limit is indicated for the heterozygote deficiency case (n = 24) since it appears that some populations may deviate considerably in that direction. Because the usual

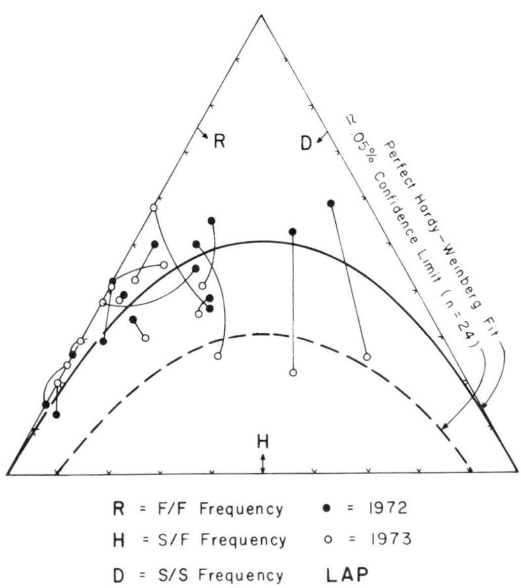

Fig. 1. de Finetti diagram for leucine aminopeptidase (LAP). For 1972(●), the pooled data are p = 0.677, q = 0.323 (n = 496), and for 1973(○), p = 0.690, q = 0.310 (n = 407); p = frequency of "slow" (S) allele, q = frequency of "fast" (F) allele. Lines connect data points from the same locality.

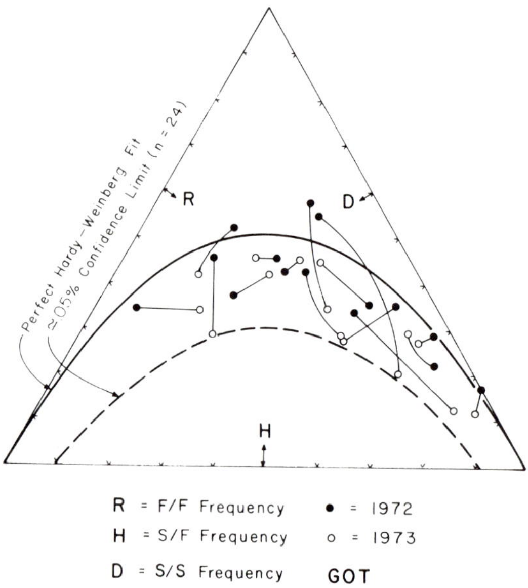

Fig. 2. de Finetti diagram for glutamate-oxalate transaminase (GOT). For 1972, the pooled data are $p = 0.392$, $q = 0.608$ ($n = 378$), and for 1973, $p = 0.373$, $q = 0.626$ ($n = 407$); p = frequency of "slow" (S) allele, q = frequency of "fast" (F) allele. Lines connect data points from the same locality.

field sample was 24 individuals, and since individual populations are further partitioned into three separate classes, a change in the genotype of a very few individuals could result in large changes in ratios of genotypes. Thus the first general conclusion is that the usual small change in populations from year to year indicates that most populations are genotypically stable.

The second general aspect of these de Finetti diagrams is that individual populations remain fairly close to Hardy–Weinberg equilibrium over a wide range of allele frequencies. Only the two populations for LAP (Fig. 1, at p = 0.44 and 0.58) may be significantly different from the Hardy–Weinberg equilibrium ($\chi^2_{(1)}$ = 4.2 and 6.3). However, the chance collection of one more heterozygote and one fewer rare homozygote in each case would make these nonsignificant (with recalculated expected distributions). In addition, approximately one of 20 populations would be expected to deviate from Hardy–Weinberg at the 0.05 level. Accordingly, no great weight can be attached to these particular deviations from the apparent general pattern of random reassortment of alleles during breeding. However, it must be emphasized that sample sizes per population are small, and that Hardy–Weinberg is a weak procedure.

A third general conclusion from these data is that the allele frequencies of *S. errata* show a bias toward an apparent shortage of heterozygotes when all populations are grouped together (the Wahlund effect, see Wallace, 1968, p. 100; Li, 1968, p. 297). The pooled data for LAP (1973) and GOT (1972 and 1973) show a strong deficiency of heterozygotes compared with the binomial expectation ($\chi^2_{(1)}$ = 10.07, 13.46, and 46.35 respectively; $P < 0.005$ in each case). Since it is inevitable that pooled populations of sharply different allele frequencies, each in Hardy–Weinberg equilibrium, will show the Wahlund effect, this conclusion is in agreement with expectations.

Even though the general pattern is of outbreeding in each local population, nevertheless, a small amount of inbreeding is also consistent with the data although it may not be statistically significant in each generation. First, considering all populations together, inbreeding certainly exists in local populations in the sense that at fertilization there is a greater likelihood of two gametes being brought together from more closely related members of a local population than from distantly related members of widely separated populations. This "isolation by distance" develops from two factors. (1) Sperm are transported through the water to eggs of other colonies, and the probability of a given fertilization would seem to decrease exponentially with distance between parental colonies. (2) Larval transport in *S. errata* is limited to a few hours which probably results in a net distance traveled of less than a kilometer, on the average.

For the inbreeding coefficient F, the calculation of genotype proportions at equilibrium is $D = p^2 + Fpq$, $H = 2pq\,(1-F)$, and $R = q^2 + Fpq$, where D, H, and R are proportions of genotypes SS, SF, and FF, respectively. If F were as much as 0.15, the theoretical curve on the GOT de Finetti diagram (Fig. 2) would have its high point, of course, at p = 0.5, for which H = 42.5% heterozygotes. This curve would approximately bisect the observed GOT (1973) data points (Fig. 2).

An additional comparison can be made of the percentage of heterozygotes in each population relative to the proportion expected from the binomial expansion for each population. Considering separately the heterozygotes in each population, GOT is deficient in heterozygotes, on the average, by 3.9% (1972) and 9.3% (1973). For LAP, the comparable data are 1.1% (1972) and 1.5% (1973) excess.

Heterozygote deviation from estimated expected values can also be calculated using the statistic as in Eq. (2) (Selander, 1970).

$$D = \frac{\text{observed} - \text{expected}}{\text{expected}} \tag{2}$$

whose mean $\bar{D} \cong 0$ and variance $\cong 1/n$, where n is the number of observations in each population, in this case nearly always 24 (D is plotted in Fig. 3). The use of

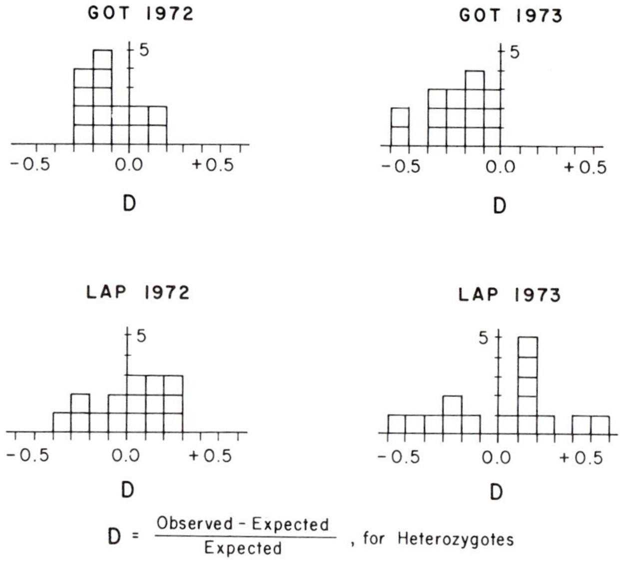

Fig. 3. Histogram of values of D for glutamate-oxalate transaminase (GOT) and leucine aminopeptidase (LAP) to indicate the distribution of heterozygotes relative to the expected ($\bar{D} \simeq 0$).

D is equivalent to the $\chi^2_{(1)}$ statistic. Under the null hypothesis, one would expect that $\bar{D}$ would fall within the range of Eq. (3).

$$-1.96[(15)(24)]^{-1/2} \leqslant \bar{D} \leqslant +1.96[(15)(24)]^{-1/2}$$

or (3)

$$-0.098 \leqslant \bar{D} \leqslant +0.098$$

For GOT (1972 and 1973), $\bar{D} = -0.12$ and -0.25, each of which is clearly highly significant. For LAP (1972 and 1973), $\bar{D} = +0.03$ and $+0.004$, neither of which is significant.

If the apparent greater deficiency in heterozygotes for GOT than for LAP is real, then a mechanism must be sought to explain this differential result. One mechanism would be significant heterosis affecting the *LAP* locus, but not the *GOT* locus.

Since $\bar{D}$ for GOT is significant, this indicates some inbreeding. But $\bar{D}$ for LAP is not significant, perhaps because heterosis could be compensating for potential heterozygote deficiency. Alternatively, negative heterosis could yield the same result. A better understanding of this situation might be obtained from biochemical data on the activity of the various enzyme phenotypes.

Considering both the probable existence of inbreeding in local populations, and yet the general conformity with the Hardy–Weinberg equilibrium, the situation in *S. errata* seems comparable to that of some social or religious groups compared with *Homo sapiens* in general. The change that would be expected in such small groups relative to their more widely distributed counterparts would be a gradual loss of genetic variance as homozygosity gradually increased. This should be reflected in a reduction in both the proportion of loci polymorphic within a population, and in a lower heterozygosity per individual.

IV. LINKAGE

One case of probable linkage has been discovered in the cheilostome bryozoan *Bugula stolonifera*, for two esterase loci (Gooch and Schopf, 1970). Whereas *esterase-5* obeys the binomial distribution, *esterase-3* does not; in addition, genotypes of *esterase-3* are predictable from those of *esterase-5* (Table II).

V. GEOGRAPHIC VARIATION

Data are available on gene and genotype frequencies in *Schizoporella errata* over areas of square meters, to tens of square meters and along geographic transects of 30 to 100 km. The previously observed patterns are summarized in Tables III and IV.

The most comprehensive picture of geographic variation exists for changes in allele frequency at the two biallelic loci *LAP* and *GOT* from each of 15 localities of a 120-km transect along the southern shore of Cape Cod and the adjacent

TABLE II
Possible Linkage between Two Esterase Loci in *Bugula stolonifera*[a]

	Esterase-3 genotypes		
Esterase-5 genotypes	SS	SF	FF
SS	19	0	0
SF	0	13	8
FF	0	0	3

[a] *Esterase-5* obeys Hardy–Weinberg, *esterase-3* does not.

TABLE III

Comparisons of Gene and Genotype Distributions from Bryozoans[a,b]

Locality	Substrate	Species	n	Locus	Conclusions	Reference
1. Woods Hole	Floating drum, 0.4 m^2	*B. stolonifera*	36	*E-5*	Random distribution	(1)
2. Green Pond	Floating dock, 9.0 m^2	*S. errata*	43	*LAP*	One side with heterozyogte excess in one year	(1,2)
3. Woods Hole	6 Pilings; ca. 2 m between each	*S. errata*	50	*LAP*	No differences among pilings	(1)
4. Woods Hole–Sheep Pen Harbor	Pilings: exposed vs. unexposed sides	*S. errata*	38 on exposed 36 on unexposed	*LAP*	Gene frequency not significant; heterozyogte excess on unexposed sides	(2)
5. Robinsons Hole–Cuttyhunk	As above	*S. errata*	58 on exposed 14 on unexposed	*LAP*	No significant differences in gene or genotype frequency	(2)
6. Woods Hole–Sheep Pen Harbor	Pilings: protected vs. unprotected sides	*S. errata*	20 on protected 18 on unprotected	*LAP*	No significant differences in gene or genotype frequency	(2)
7. Robinsons Hole–Cuttyhunk	As above	*S. errata*	31 on protected 21 on unprotected	*LAP*	Heterozyogte deficiency on unprotected sides	(2)
8. Woods Hole-1	Pilings: shallow vs. deep samples	*S. errata*	24 at 0–2 m 23 at 3–7 m	*LAP* *GOT*	Each homogeneous	(3)
9. Woods Hole-2	Pilings: shallow vs. deep samples	*S. errata*	24 at 0–2 m 24 at 6–7 m	*LAP* *GOT*	LAP significant difference; GOT homogeneous	(3)

[a] Abbreviations: *B.*, *Bugula; S.*, *Schizoporella; n*, number of colonies; *LAP, LAP-3* locus; *E-5*, esterase-5 locus.
[b] Key to references: (1) Gooch and Schopf (1970); (2) Gooch and Schopf (1971); (3) Schopf (1974a).

TABLE IV
Regional Differentiation of Gene Frequencies in the Bryozoan *Schizoporella errata*[a]

Distance in km over which allele frequency changed	K	Locus	$\|p_1 - p_2\|$	Reference
30	3	*LAP*	0.24	Gooch and Schopf (1970)
35	5	*LAP*	0.45	Schopf and Gooch (1971)
120 (1972 data)	15	*LAP*	0.55	Schopf (1974)
120 (1972 data)	15	*GOT*	0.54	Schopf (1974)
120 (1973 data)	15	*LAP*	0.61	Schopf (1974)
120 (1973 data)	15	*GOT*	0.65	Schopf (1974)
ca. 1000	9	*LAP*	0.69	Gooch and Schopf (1971)

[a]Abbreviations: *K*, number of populations along transect (nearly all of the stations are along the southern shore of Cape Cod and the adjacent offshore islands); *LAP*, leucine aminopeptidase locus; *GOT*, glutamate-oxalate transminase locus; $|p_1 - p_2|$, maximum absolute difference in allele frequencies.

offshore islands. The transect parallels a regional temperature gradient and the correlation of allele frequency with temperature yields a virtually identical slope for both loci. R^2, the proportion of the variance in allele frequency accounted for by temperature, is 64% in LAP and 68% in GOT (data and figures substantiating these points are in Schopf, 1974a).

In the present paper, I have utilized these data in order to compute general fitness surfaces for cool-water, intermediate-temperature, and warm-water localities (Fig. 4). The fitness surfaces enable us to visualize the changing fitness of particular genotypes, and to see the way that the composition of a local population changes relative to these genotypes.

The data on which these surfaces are based are neither as comprehensive nor as internally consistent as is desirable (Tables V–VII). For the cool-water data, LAP of Cuttyhunk Channel (1973) is not quite in Hardy–Weinberg equilibrium (but see above), although the allele frequency (but not the genotype frequency) is identical to the previous year which was in close agreement with Hardy–Weinberg. LAP of Robinsons Hole (1973) is close to being out of Hardy–Weinberg equilibrium, and approaches being different from the 1972 and 1973 Cuttyhunk Channel data (test of homogeneity, $\chi^2_{(1)} = 2.14$ and 2.57, $0.75 < P < 0.90$ for both $\chi^2_{(1)}$s). For the intermediate-temperature data, LAP of New Seabury (1973) is not quite in Hardy–Weinberg equilibrium (but see above), although the allele frequency is close to that of the previous year. These difficulties are partially averaged out by the remaining data which seem good, and by using data for two or three localities, each for two successive years. The results can be considered a first approximation toward the true fitness surfaces.

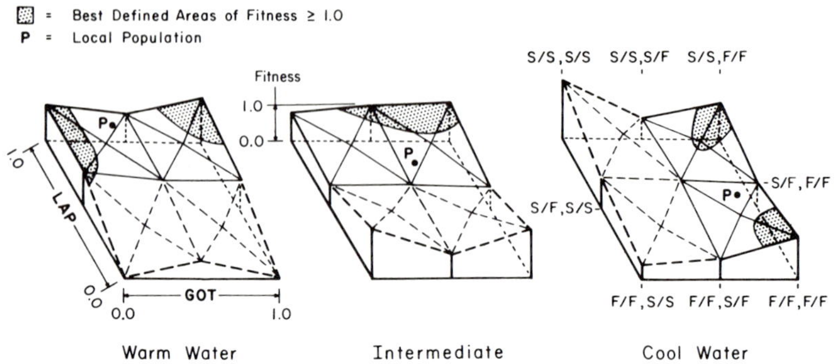

Fig. 4. Fitness surfaces for leucine aminopeptidase (LAP) and glutamate-oxalate transaminase (GOT) genotypes along a temperature gradient. *P* represents the position of the local populations from which the data for each diagram were derived.

TABLE V
Data from Localities Used to Compute Cool-Water Fitness Surface[a,b]

	LAP						GOT					
Locality	*N*	AA	Aa	aa	*p*	*q*	*n*	AA	Aa	aa	*p*	*q*
1972												
Cuttyhunk Channel	27	5	14	8	0.44	0.56	27	2	6	19	0.19	0.81
Robinsons Hole	24	2	14	8	0.37	0.63	24	1	6	17	0.17	0.83
Sum	51	7	28	16	0.412	0.588	51	3	12	36	0.176	0.824
1973												
Cuttyhunk Channel	24	8	5	11	0.44	0.56	24	2	7	15	0.23	0.77
Robinsons Hole	24	4	6	14	0.29	0.71	24	2	6	16	0.21	0.79
Sum	48	12	11	25	0.365	0.635	48	4	13	31	0.219	0.781
Total	99	19	39	41	0.388	0.611	99	7	25	67	0.197	0.803

[a]From Schopf (1974a).

[b]Abbreviations: *n*, number of colonies; AA, Aa, and aa, numbers of "slow," heterozygote, and "fast" genotypes; *p*, frequency of A; *q*, frequency of a; LAP, leucine aminopeptidase; GOT, glutamate-oxalate transaminase.

TABLE VI
Data from Localities Used to Compute Intermediate-Temperature Fitness Surface[a,b]

	LAP						GOT					
Locality	n	AA	Aa	aa	p	q	n	AA	Aa	aa	p	q
1972												
New Seabury	24	9	12	3	0.63	0.37	24	6	11	7	0.48	0.52
Quissett Harbor	24	10	9	5	0.60	0.40	24	3	14	7	0.42	0.58
Sum	48	19	21	8	0.614	0.385	48	9	25	14	0.448	0.552
1973												
New Seabury	24	11	6	7	0.58	0.42	24	7	11	6	0.52	0.48
Quissett Harbor	24	11	8	5	0.62	0.38	24	5	8	11	0.38	0.62
Sum	48	22	14	12	0.604	0.396	48	12	19	17	0.448	0.552
Total	96	41	35	20	0.609	0.391	96	21	44	31	0.448	0.552

[a]From Schopf (1974a).
[b]Abbreviations: *n*, number of colonies; AA, Aa, and aa, numbers of "slow," heterozygote, and "fast" genotypes; *p*, frequency of A; *q*, frequency of a; LAP, leucine aminopeptidase; GOT, glutamate-oxalate transaminase.

The method utilized in the present paper follows closely that presented by Lewontin and White (1960). The fitness of each of the nine observed genotypes was estimated by the ratio of the number of individuals found to those expected by the binomial expansion, using the correction given by Haldane (1956) to remove bias. The genotypic composition of the cold-, intermediate-, and warm-water populations, and their deviations from the expected composition as calculated by the binomial expansion using the observed gene frequencies are given in Tables VIII–X). Based on the nine fitness values thus obtained, an additional four values were computed for each surface for the four genotype frequency permutations of 0.75/0.25, 0.75/0.25. The values are plotted (Fig. 4) as computed to take into account Turner's (1972) objection to referring fitnesses to an arbitrary standard. Since the number of individuals in rarer genotypes is very few (14 of the 27 cells have fewer than 10 individuals each), the half of each fitness surface which suffers the most from this sampling problem is dotted in each fitness surface.

The fitness values are based on expected genotype frequencies calculated from observed single locus values, and assuming no association between loci. However, it is possible that these would differ from the fitness as estimated from

TABLE VII
Data from Localities Used to Compute Warm-Water Fitness Surface[a,b]

Locality	LAP						GOT					
	n	AA	Aa	aa	*p*	*q*	*n*	AA	Aa	aa	*p*	*q*
1972												
Meetinghouse Pond	24	20	3	1	0.90	0.10	24	9	11	4	0.60	0.40
Bass River	23	17	6	0	0.87	0.13	23	7	12	4	0.56	0.44
Uncle Roberts Cove	24	20	4	0	0.92	0.08	24	11	8	5	0.63	0.37
Sum	71	57	13	1	0.894	0.106	71	27	31	13	0.598	0.402
1973												
Meetinghouse Pond	24	19	5	0	0.90	0.10	24	11	7	6	0.60	0.40
Bass River	24	17	7	0	0.85	0.15	24	10	10	4	0.63	0.37
Uncle Roberts Cove	24	18	6	0	0.88	0.12	24	14	8	2	0.75	0.25
Sum	72	54	18	0	0.875	0.125	72	35	25	12	0.659	0.341
Total	143	111	31	1	0.884	0.115	143	62	56	25	0.629	0.370

[a]From Schopf (1974a).
[b]Abbreviations: *n*, number of colonies; AA, Aa, and aa, numbers of "slow," heterozygote, and "fast" genotypes; *p*, frequency of A; *q*, frequency of a; LAP, leucine aminopeptidase; GOT, glutamate-oxalate transaminase.

TABLE VIII
Genotypic Composition of the Cool-Water Population[a]

LAP	GOT			Totals
	SS	SF	FF	
SS	3 (0.6)	4 (4.7)	12 (9.5)	19
SF	3 (1.8)	13 (15.0)	23 (30.0)	39
FF	1 (1.4)	8 (11.8)	32 (23.6)	41
Total	7	25	67	

[a]Numbers in Table are *observed* number of individuals with SS (slow homozygote), SF (slow–fast heterozygote), and FF (fast homozygote) genotypes. *Expected* genotype numbers in parentheses.

TABLE IX
Genotypic Composition of the Intermediate-Temperature Population[a]

LAP	GOT SS	GOT SF	GOT FF	Totals
SS	7 (7.1)	20 (17.6)	14 (10.8)	41
SF	8 (9.2)	18 (22.6)	9 (13.9)	35
FF	6 (2.9)	6 (7.3)	8 (4.5)	20
Total	21	44	31	

[a]Numbers in Table are *observed* number of individuals with SS (slow homozygote), SF (slow–fast heterozygote), and FF (fast homozygote) genotypes. *Expected* genotype numbers in parentheses.

gametic frequencies, as Turner (1972) had found in analyzing the chromosome frequencies utilized by Lewontin and White (1960). Accordingly, fitness values were recomputed so that the predicted genotype numbers were derived from the observed gametic frequencies (Table XI), assuming random assortment for each combination. Fitness values obtained in this way are virtually identical to those obtained from single locus frequencies, differing in only 1 of the 27 cells by as much as 0.04 for fitness values that ranged from 0.62 to 1.87. Thus there seems to be no significant epistatic interaction between loci.

The chief point from the fitness surfaces (Fig. 4) is that the fitness of particular genotypes changes markedly along an ecological gradient. For *Schizo-*

TABLE X
Genotypic Composition of the Warm-Water Population[a]

LAP	GOT SS	GOT SF	GOT FF	Totals
SS	48 (44.2)	42 (52.0)	21 (15.3)	111
SF	14 (11.5)	13 (13.5)	4 (3.9)	31
FF	0 (0.7)	1 (0.9)	0 (0.3)	1
Total	62	56	25	

[a]Numbers in Table are *observed* number of individuals with SS (slow homozygote), SF (slow–fast heterozygote), and FF (fast homozygote) genotypes. *Expected* genotype numbers in parentheses.

TABLE XI
Gamete Frequencies in the Populations of *Schizoporella errata*

Sample	S/S	S/F	F/S	F/F
Warm water	158.5	21.5	94.5	11.5
Intermediate	51	35	66	40
Cool water	19.5	19.5	57.5	101.5

porella errata, there is clearly no single adaptive peak but rather a series of adaptive mounds. This general situation has been claimed to be customary (Kojima, 1971), and the *Schizoporella* data may therefore be of interest in seeking a basis for the maintenance of the polymorphism of these loci. The importance of selection of different genotypes in different habitats was emphasized by Levene (1953), and placed in a more general context by Smith (1970). He showed that "a stable genetic polymorphism in a varied environment does not require that the heterozygote is the fittest genotype in any environment, and is in fact possible if one allele is dominant in all environments."

An additional, but possibly spurious feature of the fitness surfaces is the apparent lower fitness of heterozygotes relative to homozygotes in the better defined values. An increase in heterozygotes of, say, 10% and a concomitant decrease in each of the two homozygote classes by 5% (proportioned according to the number of individuals in each class) raises heterozygote fitness and decreases homozygote fitness. If subsequent field data for *Schizoporella errata* warrant supporting the general picture thus far obtained, then an additional correction for slight inbreeding may be applied. Such a correction was applied to the data of Lewontin and White (1960) by Allard and Wehrhahn (1964) and this resulted in the alteration of the fitness surface so that fitness points of actual populations were no longer at saddles between peaks but were on peaks.

It is possible to evaluate the relative importance of migration and selection in accounting for the patterns of geographic variation in LAP and GOT allele frequencies. Let migration per generation (m) and the selection for or against an allele (s) be of the same order of magnitude. The absolute difference in allelic frequency between two populations selected in opposite directions then will be very close to that shown in Eq. (4) (Prakash *et al.*, 1969),

$$|p_1 - p_2| = |1 - (1 - \bar{p})^{1/2} - (\bar{p})^{1/2}| \tag{4}$$

where $\bar{p}$ is the average frequency of the allele over both types of populations. For 1972 and 1973 pooled data, $\bar{p} = 0.68$ for LAP, and $\bar{p} = 0.38$ for GOT. The absolute difference of $|p_1 - p_2|$ is 0.39 for LAP, and 0.40 for GOT. These values are about two-thirds of the largest observed differences between populations, i.e., 0.55 (1972) and 0.61 (1973) for LAP, and 0.54 (1972) and 0.65

(1973) for GOT. Thus, considering the cline as a whole, the migration rate tending to equalize allele frequencies is smaller than the intensity of the selection leading to differentiation.

If $s > m$, the difference in allelic frequencies between two populations is approximately that shown in Eq. (5) (Li, 1955 p. 304; see also Prakash *et al.* 1969).

$$|p_1 - p_2| = 1 - (m/s) \quad (5)$$

For the observed largest difference between populations of about 0.6 in both LAP and GOT, $m = 0.4s$. In other words, for the extreme case of greatest differences, the adjustment in allelic frequencies leading to the observed LAP and GOT clines could be developed if changes in allele frequencies due to selection are only 2.5 times the change due to migration.

Adjacent populations are much more likely to be influenced by migration than are widely spaced populations. And it is, therefore, of interest to compare localities which are ecologically different but geographically similar. Three pairs of adjacent localities have the same parent body of water, and differ only in their degree of water circulation and hence temperature (ecological data in Schopf, 1974a). There are (1) Cuttyhunk Channel compared with Cuttyhunk Harbor (1.5 km distant), (2) Lewis Bay (Hyannis) compared with Uncle Roberts' Cove (2.9 km removed from the open channel), and (3) Stage Harbor (Chatham) compared with its counterpart Meetinghouse Pond (East Orleans), about 22 km "upstream" at the end of a long marine passageway (see Fig. 5 for localities).

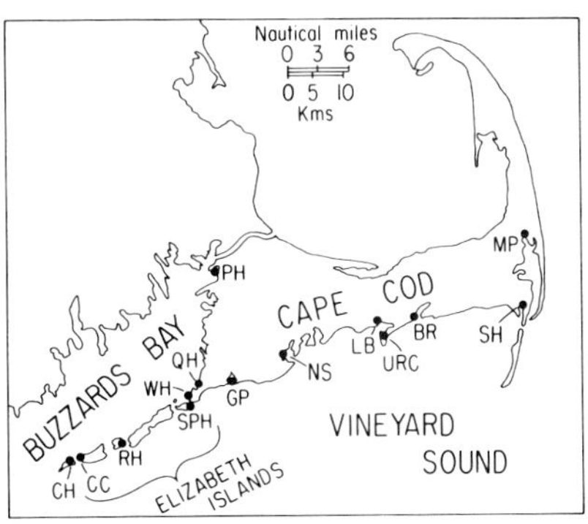

Fig. 5. Places discussed in text. From southwest to northeast, CH (Cuttyhunk Harbor), CC (Cuttyhunk Channel), RH (Robinsons' Hole), SPH (Sheep Pen Harbor), WH (Woods Hole), QH (Quissett Harbor), PH (Phinney's Harbor; at Monument Beach), GP (Green Pond), NS (New Seabury), LB (Lewis Bay; at Hyannis), URC (Uncle Roberts' Cove), BR (Bass River), SH (Stage Harbor), and MP (Meetinghouse Pond; at East Orleans).

Using LAP and GOT data from the first of these pairs in Eq. (5), $|p_1 - p_2|$ = 0.33 and 0.41, respectively; corresponding data from the second and third pairs of localities are 0.31 and 0.41, and 0.41 and 0.38. These values, on the average, are about twice the observed differences between populations. Thus, for these local cases the influence of migration might seem to be much greater than the influence of local selection.

If $m > s$, the difference in allelic frequencies between two populations is approximately that shown in Eq. (6) (Prakash *et al.*, 1969).

$$|p_1 - p_2| = 2\frac{s}{m}\bar{p}(1 - \bar{p}) \tag{6}$$

For LAP and GOT for the first pair of localities listed above, $s = 0.533m$, and $0.301m$; for the second pair, $s = 0.392m$, and $0.701m$; and for the third pair, $s = 0.827m$, and $0.896m$. In other words, the observed differences could be developed if changes in allele frequencies due to selection are a minimum of 30 to 90% of the changes due to migration. However, this does not necessarily mean that migration must also be significant since similar ecological settings may yield similar allele frequencies, independent of migration. This is shown by populations that are widely separated, for example, those at Woods Hole and Phinney's Harbor (see Fig. 5) which have similar allele frequencies but certainly do not share extensive gene flow by migration. Accordingly, I conclude, on the one hand, that data from nearby localities permit but do not demand extensive migration, but that on the other hand these data require at least some selection.

If my estimate of migration of less than 1 km per generation is approximately correct, then there is very little migration among the 15 sampled localities, most of which are kilometers apart, over a total distance of 120 km. As a consequence, the selection intensity necessary to develop the observed clines appears to be very small.

The same result may be illustrated another way. Several coefficients of selection in natural populations have been measured between 0.05 and 0.5 (summarized by Antonovics, 1971). If these are applicable to marine bryozoans, then (using the relationship $m = 0.4s$), m ranges from 0.02 to 0.20, or from 2 to 20% of the individual colonies in a population are exchanged each generation. If these migration rates are too high, as seems very likely, then selection coefficients are smaller than 0.05. An intuitively more reasonable migration rate for these populations is less than 0.1% of the individuals exchanged per generation, and for this, s is less than 0.0025.

In a model developed by Slatkin (1973) the change in allele frequencies divided by the distance over which it occurs is proportional to the square root of the selection pressure divided by the average migration distance per generation. Using 1 km as the average migration distance, s is of the order of 0.02.

If selection pressures are as small as seems indicated from these three approaches, then the question arises how these can be effective in natural populations. Franklin and Lewontin (1970) demonstrated that although extremely small fitness effects per single locus may exist, the fitness effects may yet be effective since the selection of the chromosome as a whole may be "the overriding determiner of allelic frequencies." Stated differently, "a population held in a stable polymorphic equilibrium by natural selection has a mean fitness at equilibrium that generally increases with tighter linkage" (Lewontin, 1971). If *LAP* and *GOT*, and perhaps other loci are each slightly subject to selection, then the cumulative effect may be what is being selected for. Unfortunately we do not know whether these two loci are on the same chromosome, and in fact we do not know the number of chromosomes in *S. errata*.

My general conclusion from these considerations of geographic variation is that the clinal pattern permits us to disregard any effects of random drift, and hence that we can concentrate on selection vs. migration in accounting for the observed data. From several lines of evidence, selection appears to be a more satisfactory explanation for these values from field populations than does migration.

VI. GENETIC VARIABILITY

Only in *Schizoporella errata* have more than 11 loci been designated in a bryozoan, and thus this species provides the best current estimate of genetic polymorphism. Of 27 scorable loci, 3 (11.1%) are polymorphic, and of 34 tentatively defined loci, 7 (20.6%) are polymorphic (Table XII). These figures are on the low side for animals in general, but at present we do not know if this difference is significant.

VII. IMPLICATIONS FOR SPECIATION

If the data presented and summarized in this paper can be taken as characteristic for at least several species of the phylum, then these studies have a direct bearing on the process of speciation. To begin with, although many or most bryozoan species are hermaphroditic, outbreeding would be the general rule. In addition, bryozoans would be approximately as genetically variable as other diploid species. Models of speciation based on outbreeding species of normal genetic variability may then apply to bryozoans instead of having to consider constraints imposed by a unique breeding structure.

The studies show that considerable genetic differentiation over distances of a few kilometers is to be expected, even though the geographic range of a species may extend over hundreds of kilometers. Indeed, over a distance of about 100

TABLE XII
Presumptive Gene Loci for *Schizoporella errata*[a,b,c,d]

Locus	Polymorphism	*n*	Comments
1. *DIP* - 1	M	43	
(*DIP* - 2 M)		43	Redundant; same band position as *LAP* - 2
2. *DIP* - 3	M	43	
(*CBP* - 1 M)		5	Redundant; same band position as *LAP* - 2
3. *LAP* - 1	M	~400	Reported in Gooch and Schopf (1970)
4. *LAP* - 2	M	~200	Reported in Gooch and Schopf (1970)
5. *LAP* - 3	P	~900	Reported in Gooch and Schopf (1970)
6. *E* - 1	M	23	Reported in Gooch and Schopf (1970)
7. *E* - 2	M	23	Reported in Gooch and Schopf (1970)
8. *E* - 3	M	23	Reported as P in Gooch and Schopf (1970)
9. *E* - 4	M	23	Reported in Gooch and Schopf (1970)
10. *E* - 5	M	23	
11. *GP* - 1	M	12	
12. *GP* - 2	M	20	
13. *GP* - 3	M	18	
14. *GP* - 4	M	12	
15. *AKP* - 1	M?	29	
(*AP* - 1 M?)		48	Reported in Gooch and Schopf (1971) Redundant; same position as *AKP*-1
16. *AP* - 2	M	24	
17. *AP* - 3	M	24	
18. *AP* - 4	M	24	
19. *GOT* - 1	M	~800	
20. *GOT* - 2	P	~800	
21. *GDH* - 1	P?	50	
22. *MDH* - 1*	P?	~100	
23. *MDH* - 1	P	~300	Reported as M in Gooch and Schopf (1970)
24. *ODH* - 1	M	33	
25. *NDH* - 1	M	~150	Found in 13 enzyme systems
26. *TO* - 1	M	>100	Variously developed in ~25 enzyme systems
27. *TO* - 2	M	>100	
28. *TO* - 3	M	>100	Reported in Gooch and Schopf (1971)
29. *TO* - 4	M	>100	
30. *TO* - 5	M	>100	
31. *PGI* - 1	P?	~100	
32. *PGI* - 2	M?	~100	
33. *PGI* - 3	M?	~100	
34. *PGI* - 4	P?	~100	
Totals	3 24 4 3		

Scorable loci: 3/27 = 11.1% polymorphic
All loci: 7/34 = 20.6% polymorphic

[a]Based on samples collected in 1972 and 1973 from the southern coast of Cape Cod, chiefly at Woods Hole, Massachusetts.

[b]Abbreviations for enzyme systems given in Appendix, Section VIII.

[c]Abbreviations: M, monomorphic; P, polymorphic; M?, provisionally monomorphic; P?, provisionally polymorphic; *n*, number of individuals.

[d]Within a protein system, loci are listed in increasing mobility from the origin.

km, selection appears to be a much stronger influence on gene frequencies of local populations than does the influence of migration. The fitness surfaces indicate that the same genotype has considerably different fitness under different ecological conditions. This geographic variation appears attributable to ecological trends, particularly to environmental temperature or some aspect of the environment which is strongly correlated with temperature (such as phytoplankton abundance and diversity).

Since *Schizoporella errata* and other bryozoans species have ranges of several hundred kilometers, and since migration is very restricted, selection leading to a particular set of gene frequencies in a local population would seem to be the rule. Given (1) a certain probability of the isolation of a local population (or group of populations) from migration, and sufficient enhancement of local selection pressures together with the possibilities for adaption, and (2) a given probability of persisting, then isolating mechanisms should evolve resulting in speciation. Put in different terms, "In regions of temporal stability, there will be greater geographic differentiation and speciation per unit spatial heterogeneity" (Levins, 1965). Several examples have now been examined in other taxa in which sibling species have been shown to share a very high proportion of their genes [see Schopf and Murphy (1973) for a marine example]. Thus the process of speciation may be carried out with only a small part of the genome affected, and indeed this may be the typical pattern.

In order to understand quantitatively the rates of speciation, we need to estimate the probability of local disturbances resulting in isolation of local populations, together with the probabilities of persistence. Other things being equal (especially normalizing for dispersal ability), there should be a higher probability for geographic isolation of local populations the greater the species range. And there should be a greater probability of persistence the more uniform the environment. For bryozoans, the most extensive areas permitting wide species ranges are in the marine tropics and in the deep sea, and these regions also have environments less likely to be physically limiting to local populations. Accordingly, I suspect that simply by stochastic processes that high species diversities would occur in both areas.

This way of thinking about the development of bryozoans diversity (developed in graphic form by Schopf, 1972) does not depend upon any particular ecological or biological factor per se, such as temperature, trophic resources, etc., and appears qualitatively to explain the general patterns of latitudinal species diversity in bryozoans (Schopf, 1970). In the deep-sea environment (Schopf, 1969), the number of bryozoan species per station decreases, but this only indicates that for bryozoans the carrying capacity of the deep-sea environment is lower than the continental shelf, and not that the species diversity (a measure of number and relative abundance of species obtained per standard number of specimens sampled) decreases.

VIII. APPENDIX: SUMMARY OF TECHNIQUES

Two hundred to eight hundred individuals of a colony (an area about the size of a quarter to a half-dollar) were examined to remove attached barnacles, detritus, etc., drained of seawater, and ground in 1–5 drops of seawater in centrifuge tubes by means of a plexiglass rod driven by an electric eraser. The ground-up colony was centrifuged at about 20,000 rpm for 3 minutes (tubes and centrifuge from Microchemical Speciality Corporation, Berkeley, California). Ten to twenty μl of the supernatant was transferred to each gel pocket. Enzyme separation was achieved on acrylamide gels by electrophoresis (Aardvark Instruments), using as buffer Tris–borate, pH 8.9. Runs were usually 2–3 hours at constant voltage of 300–400 V, and initial milliamperage of 70–100. A wide slot former and spacer were sometimes used and the gel then sliced in two, analogous to starch gel techniques. Several criteria are useful for determining if the bands on the gel reflect activity attributable to single gene loci (summarized by Gooch and Schopf, 1970). In addition to the 14 protein systems whose loci are reported here, I have tried an additional 28 systems whose activity was usually absent in the laboratory conditions, mostly using formulas of Shaw and Prasad (1970) and Brewer (1970).

Dipeptidase, DIP (personal communication from W. H. Johnson)

DL-Leucyl-DL-alanine	80 mg
Bothrops snake venom	16 mg
Peroxidase	40 mg
0.25 *M* $MnCl_2$	0.5 ml
0.1 *N* Tris–HCl buffer, pH 7.5	150 ml
O-Dianisidine	20 mg (add at end)

Carboxypeptidase CBP (personal communication from K. W. Wishner)

N-Carbo-β-naphthoxy-DL-phenylalanine (dissolve in acetone)	60 mg
O-Dianisidine, tetrazotized	45 mg
25% $CaCl_2$	3 drops
10% $ZnCl_2$	2 drops
Tris–HCl buffer, pH 7.0	150 ml

Leucine aminopeptidase, LAP (modified from Hubby and Lewontin, 1966) (Incubate gel in 0.5 *M* boric acid in cold 1/2 hour before staining to lower pH)

L-Leucyl-β-naphthylamide HCl (dissolve in acetone)	30 mg

Fast Black K (dissolve in buffer)	75 mg
$MnCl_2$	3 drops
0.2 *M* Tris–maleate buffer, pH 5.2	150 ml

Esterase, E (modified from Hubby and Lewontin, 1966) (Incubate gel in 0.5 *M* boric acid in cold 1/2 hour before staining to lower pH)

Fast Red TR salt	75 mg
α-Naphthyl acetate (dissolve in acetone)	40 mg
0.1 *M* Phosphate Buffer, pH 6.5	150 ml

Alkaline phosphatase, *AKP* (from Hubby and Lewontin, 1966)

α-Naphthyl acid phosphate (Na salt)	75 mg
Polyvinylpyrolidone	750 mg
Fast Blue RR salt	75 mg
NaCl 20% solution	15 ml
$MgCL_2$ 0.1 *M* solution	7.5 ml
0.1 *M* Tris–HCl buffer, pH 8.5	150 ml

Acid phosphatase, AP (from Shaw and Prasad, 1970) (Incubate gel in 0.5 M boric acid in cold 1/2 hour before staining to lower pH)

α-Naphthyl acid phosphate (Na salt)	150 mg
Black K salt	150 mg
0.05 *M* Acetate buffer, pH 5.0	150 ml

Glutamate-oxalate transaminase, GOT (from Selander *et al.*, 1971)

Pyridoxal 5′-phosphate	5 mg
α-Aspartic acid	600 mg
α-Ketoglutaric acid	300 mg
Fast Blue BB salt	450 mg
0.2 *M* Tris–HCl buffer, pH 8.0	150 ml
(method given by Shaw and Prasad, 1970, also works)	

Glutamate dehydrogenase, GD (modified from Shaw and Prasad, 1970)

TPN	90 mg
1 *M* Na glutamate, pH 7.0	7.5 ml

PMS	3 mg
NBT	45 mg
0.5 *M* phosphate buffer, pH 7.0	150 ml
Incubate in dark	

Malate dehydrogenase, MDH (modified from Hubby and Lewontin, 1966)

DPN	50 mg
NBT	50 mg
PMS	2 mg (add last)
L-Malic acid (Na salt)	100 mg
0.1 *M* Tris–HCl buffer, pH 7.6	150 ml
Incubate in dark; patterns enhanced by addition of a few mg KCN	

Octanol dehydrogenase, ODH (from Hubby and Lewontin Laboratory)

DPN	45 mg
NBT	39 mg
Octyl alcohol	1.5 ml
PMS	3 mg
0.1 *M* Tris–HCl buffer, pH 7.5	150 ml
Incubate in dark	

Phosphoglucose isomerase, PGI (from Shaw and Prasad, 1970)

TPN	15 mg
$MgCl_2$	120 mg
PMS	5 mg
MTT	30 mg
Glucose-6-phosphate dehydrogenase	0.05 ml
Fructose 6-phosphate	150 mg
0.1 *M* Tris–HCl buffer, pH 8.0	150 ml
Incubate in dark; bands fade fast	

"Nothing" dehydrogenase, NDH

Many dehydrogenase and other systems yield bands in the absence of the particular substrate required for specific enzyme activity.

Tetrazolium oxidase, TO

Many of the dehydrogenase systems and others develop clear areas on gels (Brewer, 1967). For *Schizoporella errata,* there occur 14 such zones, operationally identified as tetrazolium oxidase. These zones do not stain equally well on the gels of the 25 different enzyme systems on which they occur. The five most prominent band systems that have been identified in the largest number of individuals are listed in the table of loci.

General protein, GP (From Hubby and Lewontin, 1966)

Methyl alcohol, water, and glacial acetic acid in 5:5:1 mixture	150 ml
Amido black or coomassie blue	75 mg
Destain with the same solution, minus the stain	

IX. ADDENDUM

Since this chapter was accepted in October, 1973, additional work pertaining to bryozoans has appeared and is briefly summarized here.

Data on genetic variability in all animals were gathered together by Powell (1975) (in his tabulation, *Schizoporella unicornis* and *S. errata* are erroneously listed as distinct species).

Schopf's laboratory at the Marine Biological Laboratory, Woods Hole, has been the sole source of data using marker genes in population studies of bryozoans. Clines in genetic differentiation (as revealed by *LAP* and *GOT* loci) are now known along the southern shore of Cape Cod to be stable over periods of 4 to 6 years, with one locality having been followed for 7 successive years (1969–1975) (Schopf, 1974b; Schopf and Dutton, 1976). These clines in gene frequencies were discovered in 1975 to be parallel to a cline in the length of the defensive polymorph (the avicularium) (Dutton and Schopf, 1975; Schopf and Dutton, 1976). This conclusion is based on genotype determinations of more than 1500 animals, and on some thousands of morphologic measurements from colonies at 13 localities over the 102-km transect (Schopf and Dutton, 1976).

Since bryozoans are colonial, it is possible to partition morphologic variation into individual (colony) and population components. A comparison was made between the deep-sea genus *Euginoma* and the shallow-water species *Schizoporella errata* (Schopf and Dutton, 1976) in order to assess the influence of environmental variability on morphologic variability.

> The shallow water species in general appears to have a larger proportion of its total variability within a colony than between colonies . . . in the more variable shallow water environment, the environmental contribution to each individual colony may be expressed to a greater degree than in the more stable deep-sea regime. This is as though each shallow water colony is more of a morphologic 'generalist" and each deep-sea colony more of a morphologic 'specialist' (Schopf and Dutton, 1976, p. 262–263).

The implications of this further work on models and patterns of speciation are further considered by Schopf (1977). These and related data on marine species (Schopf and Murphy, 1973) seem most easily interpreted as indicating a pattern of gradual and continual genetic and morphologic differentiation as the environment changes both laterally, and through time. Most genetic and morphologic change would appear to result from continual change spread over a long period of time rather than being largely limited to the "instant" in which reproductive isolation is achieved.

ACKNOWLEDGMENTS

I gratefully acknowledge the assistance of Dr. Joe Whittaker, University of Chicago, for advice on statistics, and of Dr. Eliot Spies, University of Illinois, Chicago Circle, Dr. Roger Milkman, University of Iowa, Dr. Richard Lewontin, Harvard University, and Dr. James L. Gooch, Juniata College, for criticism of the genetic aspects of the paper. Data discussed here were obtained through support by NSF Grant GB-30870, and by the Block Fund, University of Chicago.

REFERENCES

Allard, R. W., and Wehrhahn, C. (1964). A theory which predicts stable equilibrium for inversion polymorphisms in the grasshopper *Moraba scurra*. *Evolution* **18**, 129–130.

Antonovics, J. (1971). The effects of a heterogeneous environment on the genetics of natural populations. *Am. Sci.* **59**, 593–599.

Brewer, G. J. (1967). Achromatic regions of tetrazolium stained starch gells: inherited electrophoretic variation. *Am. J. Hum. Genet.* **19**, 674–680.

Brewer, G. J. (1970). "An Introduction to Isozyme Techniques." Academic Press, New York.

Cannings, C., and Edwards, A. W. F. (1968). Natural selection and the de Finetti diagram. *Ann. Hum. Genet.* **31**, 421–428.

Dutton, A. R., and Schopf, T. J. M. (1975). Morphologic differentiation can approximately parallel genetic differentiation in the ectoproct *Schizoporella errata* in coastal populations from Woods Hole to Orleans. *Biol. Bull. (Woods Hole, Mass.)* **149**, 425–426.

Franklin, I., and Lewontin, R. C. (1970). Is the gene the unit of selection? *Genetics* **65**, 707–734.

Gooch, J. L. (1975). Mechanisms of evolution and population genetics. *In* "Marine Ecology" (O. Kinne, ed.), Vol. 2, Pt. 1, pp. 349–409. Wiley, New York.

Gooch, J. L., and Schopf, T. J. M. (1969). Genetic studies on marine species of the Phylum Ectoprocta. *Biol. Bull. (Woods Hole, Mass.)* **137**, 400–401.

Gooch, J. L., and Schopf, T. J. M. (1970). Population genetics of marine species of the Phylum Ectoprocta. *Biol. Bull. (Woods Hole, Mass.)* **138**, 138–156.

Gooch, J. L., and Schopf, T. J. M. (1971). Genetic variation in the marine ectoproct *Schizoporella errata*. *Biol. Bull. (Woods Hole, Mass.)* **141**, 235–246.

Gottlieb, L. D. (1971). Gel electrophoresis: New approach to the study of evolution. *BioScience* **21**, 939–944.

Haldane, J. B. S. (1956). The estimation of viabilities. *J. Genet.* **54**, 294–296.
Harris, H., and Hopkinson, D. A. (1972). Average heterozygosity per locus in man: An estimate based on the incidence of enzyme polymorphisms. *Ann. Hum. Genet.* **36**, 9–20.
Hubby, J. L., and Lewontin, R. C. (1966). A molecular approach to the study of genic heterozygosity in natural populations. I. The number of alleles at different loci in *Drosophila pseudoobscura. Genetics* **54**, 577–594.
Hyman, L. H. (1959). "The Invertebrates: Smaller Coelomate Groups." McGraw-Hill, New York.
Kojima, K. (1971). Is there a constant fitness value for a given genotype? No! *Evolution* **25**, 281–285.
Levene, H. (1953). Genetic equilibrium when more than one ecological niche is available. *Am. Nat.* **87**, 331–333.
Levins, R. (1965). The theory of fitness in a heterogeneous environment. IV. The adaptive significance of gene flow. *Evolution* **18**, 635–638.
Lewontin, R. C. (1971). The effect of genetic linkage on the mean fitness of a population. *Proc. Natl. Acad. Sci. U.S.A.* **68**, 984–986.
Lewontin, R. C. (1972). The apportionment of human diversity. *Evol. Biol.* **6**, 381–398.
Lewontin, R. C. (1974). "Genetic Basis of Evolutionary Change." Columbia Univ. Press, New York.
Lewontin, R. C., and Cockerham, C. C. (1959). The goodness-of-fit test for detecting selection in random mating populations. *Evolution* **13**, 561–564.
Lewontin, R. C., and Hubby, J. L. (1966). A molecular approach to the study of genic heterozygosity in natural populations. II. Amount of variation and degree of heterozygosity in natural populations of *Drosophila psuedoobscura. Genetics* **54**, 595–609.
Lewontin, R. C., and Krakauer, J. (1973). Distribution of gene frequency as a test of the theory of the selective neutrality of polymorphisms. *Genetics* **74**, 175–195.
Lewontin, R. C., and White, M. J. D. (1960). Interaction between inversion polymorphisms of two chromosome pairs in the grasshopper, *Moraba scurra. Evolution* **14**, 116–129.
Li, C. C. (1955). "Population Genetics." Univ. of Chicago Press, Chicago, Illinois.
Manwell, C., and Baker, C. M. A. (1970). "Molecular Biology and the Origin of Species." Univ. of Washington Press, Seattle, Washington.
Powell, J. R. (1975). Protein variation in natural populations of animals. *Evol. Biol.* **8**, 79–119.
Prakash, S., Lewontin, R. C., and Hubby, J. L. (1969). A molecular approach to the study of genic heterozygosity in natural populations. IV. Patterns of genic variation in central, marginal and isolated populations of *Drosophila pseudoobscura. Genetics* **61**, 841–858.
Schopf, T. J. M. (1969). Geographic and depth distribution of the Phylum Ectoprocta from 200 to 6,000 meters. *Proc. Am. Philos. Soc.* **113**, 464–474.
Schopf, T. J. M. (1970). Taxonomic diversity gradients of ectoprocts and bivalves and their geologic implications. *Geol. Soc. Am. Bull.* **81**, 3765–3768.
Schopf, T. J. M. (1972). Varieties of paleobiologic experience. *In* "Models in Paleobiology" (T. J. M. Schopf, ed.), pp. 8–25. Freeman, Cooper & Co., San Francisco, California.
Schopf, T. J. M. (1973). Population genetics of ectoprocts: Status as of January, 1972. *In* "Living and Fossil Bryozoa" (G. P. Larwood, ed.), pp. 585–592. Academic Press, New York.
Schopf, T. J. M. (1974a). Survey of genetic differentiation in a coastal zone invertebrate: The ectoproct *Schizoporella errata. Biol. Bull. (Woods Hole, Mass.)* **145**, 78–87.
Schopf, T. J. M. (1974b). Long-term (3 to 5 year) records of gene frequencies in natural

populations of an abundant, sub-tidal species (the bryozoan *Schizoporella errata*). *Biol. Bull. (Woods Hole, Mass.)* **147**, 498.

Schopf, T. J. M. (1976). Environmental versus genetic causes of morphologic variability in bryozoan colonies from the deep sea. *Paleobiology*. **2**, 156–165.

Schopf, T. J. M. (1977). Patterns and themes of evolution among the Bryozoa. *In* "Patterns of Evolution" (A. H. Hallam, ed.). Elsevier, Amsterdam.

Schopf, T. J. M., and Dutton, A. R. (1976). Parallel clines in morphologic and genetic differentiation in a coastal zone marine invertebrate: the bryozoan *Schizoporella errata*. *Paleobiology*. **2**, 255–264.

Schopf, T. J. M., and Gooch, J. L. (1971). Gene frequencies in a marine ectoproct: A cline in natural populations related to sea temperature. *Evolution* **25**, 286–289.

Schopf, T. J. M., and Murphy, L. S. (1973). Protein polymorphism of the hybridizing seastars *Asterias forbesi* and *A. vulgaris* and implications for their evolution. *Biol. Bull. (Woods Hole, Mass.)* **145**, 589–597.

Selander, R. K. (1970). Behavior and genetic variation in natural populations. *Am. Zool.* **10**, 53–66.

Selander, R. K., and Kaufman, D. W. (1973). Genic variability and strategies of adaptation in animals. *Proc. Natl. Acad. Sci. U.S.A.* **70**, 1875–1877.

Selander, R. K., Smith, M. H., Yang, S. Y., Johnson, W. E., and Gentry, J. B. (1971). Biochemical polymorphism and systematics in the genus *Peromyscus*. I. Variation in the old-field mouse (*Peromyscus polionotus*). Studies in Genetics. VI. *Univ. Tex. Publ.* **7103**, 50–90.

Shaw, C. R., and Prasad, R. (1970). Starch gel electrophoresis of enzymes–a compilation of recipes. *Biochem. Genet.* **4**, 297–320.

Silén, L. (1966). On the fertilization problem in the gymnolaematous Bryozoa. *Ophelia* **3**, 113–140.

Slatkin, M. W. (1973). Gene flow and selection in a cline. *Genetics* **61**, 733–756.

Smith, J. M. (1970). Genetic polymorphism in a varied environment. *Am. Nat.* **104**, 487–490.

Turner, J. R. G. (1972). Selection and stability in the complex polymorphism of *Moraba scurra*. *Evolution* **26**, 334–343.

Wallace, B. (1968). "Topics in Population Genetics." Norton, New York.

Ward, R. H., and Sing, C. F. (1970). A consideration of the power of the test to detect inbreeding effects in natural populations. *Am. Nat.* **104**, 355–365.

15

An Adaptive Model for the Evolution of the Ectoproct Life Cycle

JACK D. FARMER

I. INTRODUCTION

The phylogenetic relationships among higher taxa pose some of the most intriguing and challenging problems of evolutionary biology. The lophophorate phyla Ectoprocta, Phoronida, and Brachiopoda are of special interest in this regard, for although a majority of authors consider them to form a close-knit clade (e.g., Marcus, 1958; Hyman, 1959; Brien, 1960; Zimmer, 1964, 1973; Jägersten, 1972), the exact relationship of these phyla to other metazoan groups is unclear. This confusion stems in part from the fact that, as a group, the lophophorates share features common to both the Protostomia and Deuterostomia and appear to lie close to the divergence of those two major evolutionary lines. Nielsen (1971; Chapter 16, this volume) considers bryozoans to be protostomes closely related to entoprocts and isolated from the brachiopods and phoronids. However, other recent work (Zimmer, 1964, 1973; Emig, 1973, 1974a) suggests that the lophophorates in actuality share a larger proportion of features with the Deuterostomia, of which they may represent an early evolutionary offshoot.

Interestingly, the phylogenetic relationships among the lophophorate phyla have also been difficult to resolve, as evidenced by the diversity of opinion that currently exists in the literature. This derives in part from the fact that comparisons based on early stages of development are often in opposition to those based only on a comparison of adult features. Consequently, most authors have emphasized only one stage of the life cycle in framing hypotheses of lophophorate phylogeny. Considering the developmental patterns among the lophophorates (Zimmer, 1964, p. 267), such inconsistency is not entirely unexpected. Despite the strong similarities in cleavage (but see Rattenbury, 1954), regulative potential, and in the adult coelomic architecture, few similarities appear to exist among these phyla with regard to mesoderm formation and its early differentiation during larval development.

During evolutionary change, natural selection may operate independently and in quite dissimilar ways at different stages of the life cycle. Therefore, a desired approach in reconstructing phylogeny is one in which the entire life cycle is considered (Jägersten, 1972, p. 250). Hypotheses that involve all aspects of the life cycle, although they are difficult to construct, possess greater explanatory power than those based on only one stage. This should be especially true for benthic marine invertebrate groups, such as the lophophorates, which have a planktonic larval stage. Prior to metamorphosis in these groups, larval development and behavior are subject to a totally different regime of natural selection than is the adult stage (see Vance, 1973). Thus, evolution in such groups may have followed adaptive pathways in the larval stage quite independent of those in the adult. For groups where morphological divergence between stages of the life cycle has been extreme, a total reorganization at metamorphosis (such as that observed in the Ectoprocta) is not unexpected.

Most taxonomy is based upon morphological features of the adult, and these are the features that are commonly preserved in the fossil record. In addition, phylogenetic speculation ideally proceeds from a knowledge of the functional basis of morphology—a knowledge that at present appears to be far more complete for the adult stage than for earlier phases of the life cycle of most groups. An incomplete knowledge of the adaptive basis of larval morphology and behavior in many groups makes it difficult to identify the features that may be important to considerations of phylogeny. For higher metazoan taxa, phylogenies based on strictly embryological or larval features may only be tested in a strict historical sense where they are borne out independently by patterns of morphological change in the fossil record of the adult. For these reasons, a comparison of adult morphologies is believed to be the most reasonable starting point for considerations of the origin of higher categories.

II. THE GROUND PLAN CONCEPT AND ITS BEARING ON THE ORIGIN OF PHYLA

At progressively higher levels in the taxonomic hierarchy, organisms share a more inclusive set of characters and the basic body plan shared by all individuals is increasingly more generalized (Ghiselin, 1972). At the phylum level, the definition of similarity is based on a suite of characters shared by a large number of species and which group particular subsets of species (genera, families, etc.) into that phylum. These features are embodied in what has been called the "ground plan" of the phylum (e.g., see Valentine, 1973a, p. 51). Ground plans need not include only aspects of adult morphology, and in fact each phylum is characterized by larval or embryological features that also participate in the definition of the phylum. These features are fundamental in a taxonomic sense and equally, therefore, in the reconstruction of phylogeny. But more than that, they may reflect a suite of adaptations that were present in the ancestral stocks that originally gave rise to the phylum.

The goal of most systematists is a usable classification that also reflects phylogeny and an important assumption is that the morphological similarities used in constructing a taxonomic framework more often reflect common ancestry than convergence. The characteristics comprising the ground plan are important in the detection of convergence in that all subsequent adaptations must develop within the constraints imposed by earlier adaptations, representing variations on the major morphological theme imposed by the ground plan.

Although morphological similarities are fundamental in determining phylogeny, differences in morphology play an equally important role, and one that goes beyond providing a basis for the detection of convergence. Because morphological differences are assumed to be underpinned by differences in biological function, they also possess paleoecological value in providing the basis for reconstructing and understanding the possible evolutionary pathways that groups may have followed since their divergence.

III. THE POSITION OF THE ECTOPROCTA IN LOPHOPHORATE PHYLOGENY

Recent proposals that concern the relationship of the Ectoprocta to the other lophophorate phyla may be considered in light of the dichotomy of approach mentioned earlier. Most authors who have made comparisons of adult morphology regard the Phoronida as possessing the greatest number of primitive lophophorate characters, thus placing them nearest to the ancestral lophophorate stock (e.g., Silén, 1944; Marcus, 1958; Hyman, 1959). An interesting argument for this is found in the work of Clark (1964, pp. 235–238), who traces the evolutionary trend among the lophophorates from the infaunal, tubicolous habit of the Phoronida to the sessile, epibenthic habit of living Ectoprocta and most Brachiopoda. This model is based on Clark's belief that the primitive function of the coelom was as a hydrostatic skeleton in burrowing, and that elaboration of the role of the coelom in locomotion among early metazoan precursors was the basis for the adaptive radiation and early differentiation of these phyla. It seems likely that this radiation occurred in late Precambrian time (Valentine, 1973b). Among the lophophorates, the hydrostatic system of the Phoronida has probably been modified least from this assumed primitive condition, serving in burrowing, but primarily in moving the animal within its tube. In the ectoproct adult, the hydrostatic system has undergone changes to accompany a completely sessile existence.

Cowen (1974) has pointed out recently that respiration may have played an equally important role in early metazoan evolution in association with the locomotory functions stressed by Clark (1964). Among deeper burrowing, metamerically segmented coelomates, respiratory exchange would have been accomplished by currents set up by peristalsis associated with active burrowing. However, among the more sedentary oligomerous groups (such as the Phoronida), incapable of prolonged active burrowing, alternative adaptations for effective respiratory exchange may have been required. One possible adaptive solution to this problem could have been the development of a specialized tentaculate organ which functioned in both respiration and suspension feeding.

Nielsen (1971; Chapter 16, this volume) derives the ectoprocts from a solitary, entoproct-like ancestor based upon a consideration of both larval and adult features. Although this work represents a noteworthy contribution toward a phylogeny based on entire life cycles, it dismisses the significance of several differences in the adult ground plans of ectoprocts and entoprocts that have led others (e.g., Hyman, 1951, 1959; Hadzi, 1958; Brien, 1970; Jägersten, 1972) to place the Entoprocta in a separate phylum, much nearer the annelid line of evolution. The view of these later authors is adopted here because of the marked difference in the body cavity architectures of adult ectoprocts and entoprocts. It is by no means clear that the similarities between ectoprocts and entoprocts identified by Nielsen (1971, pp. 315–331) cannot be accounted for by convergent evolution in light of the strong similarity in the life styles of the two groups.

Jägersten (1972, pp. 25–64) derived a model for lophophorate phylogeny, based primarily on larval ontogenies, that traces the entire group to a sessile, "protentaculate" ancestor that possessed a cyphonautes-type larva. In this scheme, the Brachiopoda arose first, with the Phoronida and Ectoprocta appearing successively from later evolutionary branches. This model fails to consider the functional origin of important adult features, such as the coelom. Furthermore, it rests in large part on the assumption that the cyphonautes larva of the ectoprocts is primitive among all lophophorates. The cyphonautes is viewed by Jägersten (1972, p. 40) as being poorly adapted to a pelagic existence owing to the presence of features in its ground plan (notably, a bivalved shell and ciliated oral creeping "foot"–the pyriform organ) that are alleged to have evolved in the benthonic adult stage of a remote ancestor, and were gradually shifted to the larval stage by "adult pressure" to speed metamorphosis. The work of Atkins (1955b) and Strathmann (1973) suggest, however, that the cyphonautes is, in fact, highly adapted to a planktotrophic existence. In view of the high estimates of larval mortality rates (>99%) that characterize many benthic invertebrate species with planktotrophic larvae (Thorson, 1946; Mileikovsky, 1971), there seems to be little reason to suspect that selection prior to metamorphosis in these groups should be less effective in eliminating poorly adapted traits than selection in the adult stage. Considering the types and intensities of selection that appear to operate during the pelagic larval stage of the life cycle (see Vance, 1973), it seems unreasonable to suggest, as Jägersten does, that ancient adult characters which lack functions beyond the acceleration of metamorphosis should persist there at all, particularly where they are no longer present in the adult stage.

Accepting in principle Clark's (1964) proposal for lophophorate phylogeny and remaining within the theoretical framework developed previously, the following model is an extension of recent work by Farmer *et al.* (1973) that derives the ectoprocts from a phoronid ancestor by adaptation leading to coloniality and small size. In developing this hypothesis further, the implications of the proposed phylogeny for evolutionary trends at the larval stage will also be considered.

IV. GENERAL FEATURES OF THE PHORONIDA AND ECTOPROCTA

A. General Statement

In order to postulate a phoronid ancestry for the ectoprocts, all major morphological differences in the ground plans of the two groups must be accounted for in a reasonable and efficient way by adaptation along evolutionary pathways that at every step contain viable intermediate forms operating in appropriate modes of life. As a first step in the development of this model, a general comparison of living Phoronida and Ectoprocta is necessary in order to elucidate the important differences in the adult ground plans of the two groups. The account that follows is based largely on the reviews of Hyman (1959), Ryland (1970), and Emig (1971).

B. Comparative Analysis

The Phoronida are small (most adult species range from 80–250 mm in length), vermiform coelomates possessing a tentaculate horseshoe-shaped lophophore that is nontractable (Fig. 1B). The phylum is small, consisting of approximately a dozen species distributed among two genera (Emig, 1974b). The Ectoprocta are a much larger phylum consisting of an estimated 4000 living

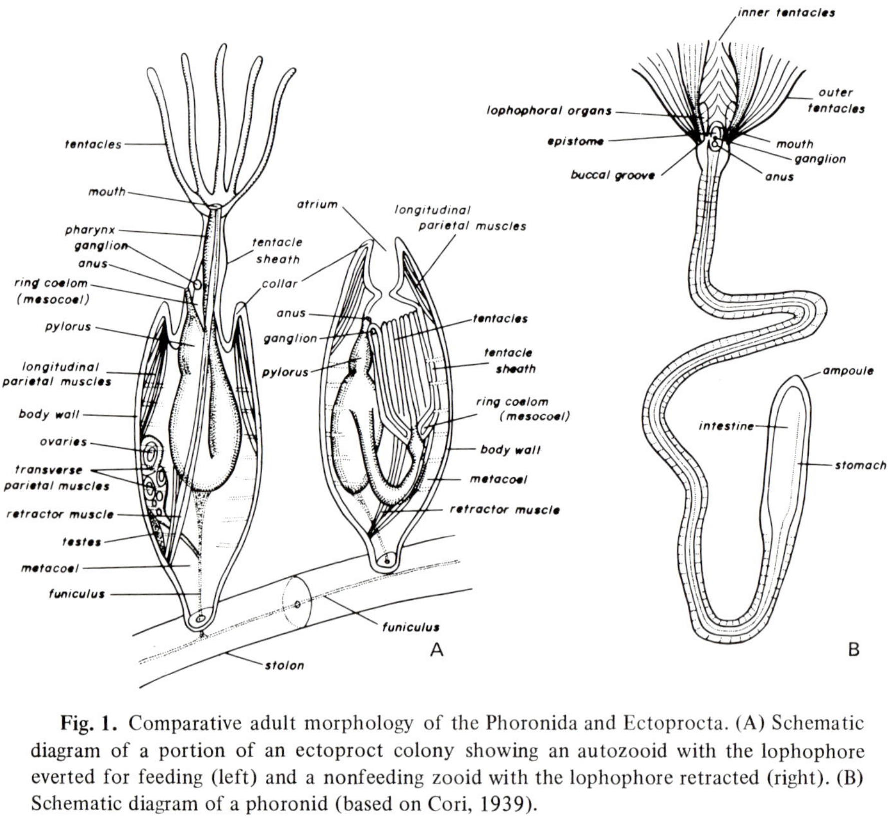

Fig. 1. Comparative adult morphology of the Phoronida and Ectoprocta. (A) Schematic diagram of a portion of an ectoproct colony showing an autozooid with the lophophore everted for feeding (left) and a nonfeeding zooid with the lophophore retracted (right). (B) Schematic diagram of a phoronid (based on Cori, 1939).

species and perhaps four times as many fossil species. They are smaller in size (usually <0.5 mm), are coelomate, and possess a crescentic or circular lophophore that is retractable (Fig. 1A). In both groups the digestive tract is distinctly regionated and U-shaped with the mouth being located within the circlet of tentacles and the anus "mid-dorsally," outside the lophophore.

In the phoronid lophophore, ciliated tentacles arise in single rows from two ridgelike extensions of the body wall which are separated along their length by a ciliated buccal groove. The paired arms of the lophophore coil spirally away from the mouth, which consists of an oval or crescentic aperture that is located between the two lophophoral ridges and which is continuous with the buccal grooves on either side. The mouth is covered dorsally by a crescentic flap of body wall called the epistome.

Although differing in detail, the general constructional features of the ectoproct lophophore are remarkably similar to those seen in the Phoronida. In the stenolaemate and gymnolaemate ectoprocts, the lophophore is formed from a circular ridge of the body wall, bearing a single row of tentacles which encircle the centrally located mouth. In the phylactolaemate (freshwater) ectoprocts, the lophophore is generally crescentic or horseshoe-shaped and the mouth is covered dorsally by a crescentic epistomal flap of body wall, similar to that of phoronids.

Although controversy exists concerning the details of food particle capture and transport, several authors (Bullivant, 1968; Strathmann, 1973; Winston, this volume) have noted strong similarities in the feeding methods of all the lophophorate phyla. In phoronids, ciliary currents enter between the double rows of tentacles, passing outward at their base. Food particles impinging on the frontal surfaces of the tentacles are captured and transported by ciliary action into the buccal grooves, and eventually carried to the mouth. In the ectoprocts, currents enter at the top of the tentacle crown, exiting radially between the tentacle bases. Apparently mucus has not been demonstrated to play a role in the trapping or transport of food particles in the lophophorates. Bullivant (1968) coined the term "impingement feeding" to describe the mechanism of food particle capture in the lophophorate phyla, by which particles entering the tentacle crown are carried across flow lines by their own momentum and essentially "thrown" at the mouth. But Strathmann (1971) doubts the effectiveness of the impingement mechanism in food capture because most particles are only slightly denser than water and possess negligible momentum of their own. As an alternative explanation, Strathmann *et al.* (1972; Strathmann, 1973) suggest that most food particle capture and transport in these groups is accomplished by the intermittent reversal of the beat of lateral cilia which directs the particles toward the frontal surface of the tentacle.

In both groups a spacious coelom is divided into two distinct regions, an anterior mesocoel and posterior metacoel, by a septum located at the base of the lophophore; this septum is complete in phoronids, but incomplete in bryozoans.

The presence of a third division of the coelom, corresponding to the protocoel of other oligomerous groups, is believed by some authors to be represented by the epistome present in the phoronids and phylactolaemate ectoprocts.

In both the phoronids and ectoprocts, the nervous system consists of a plexus throughout the body wall, with a main ganglionic mass situated between the mouth and anus that is continuous with a nerve ring that encircles the base of the lophophore. In most phoronid species, there is generally a giant nerve fiber that passes from the mid-dorsal region of the nerve ring to the left side, where it descends within the epidermis at the attachment site of the left-lateral mesentery. In a few species, a second giant nerve is present in a complementary position on the right side. Each of the giant fibers gives off throughout its length transverse branches that enter the basement membrane of the muscular region of the trunk.

Phoronids possess a circulatory system of the closed type and a pair of metanephridia that also function as gonoducts. In contrast, circulatory and nephridial systems are absent in ectoprocts.

Phoronids secrete a cylindrical, chitinous tube within which they can move freely. The secretion is initially soft and sticky and in appropriate sediments picks up grains of sand or shell fragments that adhere to it, forming a rigid, agglutinated structure. Most species live infaunally in soft, sandy to muddy substrates of the littoral zone, where the tubes stand vertically within the sediment. However, in some species aggregations of tubes form entangled masses attached to some hard substrate, such as rocks or pilings. An ususual habit of boring into hard calcareous substrates, such as mollusk shells or limestone, is known for three species, *Phoronis hippocrepia, P. ijimai,* and *P. ovalis.*

The Ectoprocta are strictly colonial, the colonies originating by asexual budding of a single founding individual or ancestrula, derived from the metamorphosis of a sexually produced larva. The colony is composed of individual zooids, each enclosed in an exoskeletal case (zooecium) composed of chitinous cuticle, or more commonly, cuticle and calcite. Functional differentiation and specialization among the individuals of the colony is a common feature of ectoprocts. In living cheilostomes, a high degree of polymorphism has been achieved through the specialization of individuals for feeding, reproduction, protection, etc. (see Silén, this volume, Chapter 6).

Phoronids may be hermaphroditic or dioecious. In many species, embryos are held within the concavity of the lophophore of the parent and in one species, within the parental tube, where they are brooded to larval stage. In a few species, no brooding is observed and all development occurs externally in the water. Embryos of all but one species develop into the characteristic actinotroch larva. Phoronids tend to be gregarious, occurring in aggregations that in some species are propagated in part by asexual reproduction (see Harmer, 1917; Marcus, 1949; Emig, 1972).

The majority of ectoproct species are hermaphroditic (monoecious), but the dioecious condition is also known. The hermaphrodotism is protandrous and this led to confusion among early workers regarding the dominant mode of fertilization in the group. However, observations of sperm release in a number of gymnolaemate and stenolaemate ectoprocts (Silén, 1966, 1972; Bullivant, 1967; Ström, 1969) and the detection of high genetic variability in populations of other species (Gooch and Schopf, 1970) suggest that outbreeding is probably the general rule. Fertilized eggs are shed directly into seawater where they rapidly develop into the distinctive cyphonautes larval form or, in most marine forms, are brooded to the larval stage within the maternal coelom, or externally attached to the parent in specialized reproductive structures.

The Phoronida exhibit remarkable powers of regeneration of the anterior end in particular, and they regularly cast off their lophophores and regenerate new ones following the onset of unfavorable conditions. The Ectoprocta show similar powers of regeneration as polypides undergo regular cycles of degeneration and regeneration due to natural senescence, the onset of unfavorable conditions, or in association with sexual reproduction. The powers of regeneration generally exhibit a gradient within the colony, being greater in the younger, more distal regions. Marcus (1926) explained degeneration not associated with sexual reproduction as a method of excretion resulting from the accumulation of substances in the stomach epithelium. Capabilities for regeneration are not confined to the polypide alone, and the repair of structural damage to the colony or replacement of feeding zooids by specialized polymorphs has also been observed.

V. THE ADAPTIVE BASIS OF MORPHOLOGICAL DIFFERENCES IN THE ADULT STAGE OF PHORONIDS AND ECTOPROCTS

Based on the preceding comparison, it is apparent that the major differences in adult morphology that separate living ectoprocts from living phoronids include a colonial habit, small size, differences in lophophore complexity and feeding, the absence of nephridial and closed circulatory systems, and the hydrostatic function of the coelom. In developing an evolutionary model linking phoronids and ectoprocts, the possible adaptive basis of each of these differences in adult morphology will now be considered.

Colonial organisms have been defined as associations in which the interests of the individual are subordinated to those of the group (Barrington, 1967, p. 443). The advantages of coloniality are well understood and are generally considered to include the regulation of competition among individuals (Knight-Jones and Moyse, 1961); coordinated manipulation of the environment, as in the generation of colonial feeding currents to function in feeding or waste disposal (Cowen and Rider, 1972; Banta *et al.,* 1974); protection of the colony by specialized sensory (see Thorpe *et al.,* 1975) or protective devices; and the potential for differentiation and specialization among individuals to perform such functions as colony support, reproduction, protection, etc., resulting in increased efficiency (Mackie, 1963).

The adaptive basis of small size is more difficult to assess, although a possible advantage in small colonial organisms like ectoprocts is that the individuals may not grow large enough during ontogeny to encounter the problems associated with allometric size/area/volume increases in such functions as feeding and respiration. Small individual body size may hold disadvantages for some other colony functions, however, where late astogenetic developments in many erect colonies include the "overcalcification" of older individuals in proximal regions of the colony in order to strengthen the colony against mechanical stresses during growth (see Cheetham, 1971, p. 12).

The small size of ectoprocts is also correlated with other facets of their biology. Hyman's (1959, p. 500) suggestion that the absence of nephridial and closed circulatory systems in ectoprocts may be associated with their small size is appealing, considering that the efficiency of closed systems for the transport of nutrients and wastes must decline in relation to that achieved by direct exchange, as size decreases.

The implication of Clark's (1964) work is that differences in the hydrostatic systems of phoronids and ectoprocts can be understood in terms of differences in individual mobility. Phoronids are mobile creatures that can burrow or move freely within their tube, while ectoprocts lack this ability. The highly integrated colonial habit of ectoprocts, in which organic contact between zooids is maintained implies a loss of mobility for the individual. The body wall of each zooid is stationary, being attached to the substrate or to rigid skeletal walls of the zoarium and the lophophore is everted relative to this fixed surface. The function of the hydrostatic system presumably has been modified to accompany this sessile mode of life. It seems reasonable, therefore, that the retractable lophophore and difference in hydrostatic function of the coelom in ectoprocts may be interpreted as coadaptations to colonial life.

VI. A MODEL FOR THE ORIGIN OF THE ADULT FEATURES OF THE ECTOPROCT GROUND PLAN

It seems apparent that the differences between phoronids and ectoprocts represent a suite of characters, all associated in a general way with two contrasting modes of life. Phoronids are relatively large, somewhat mobile creatures which live an individualized existence, while ectoprocts are small, sessile, and colonial. Based upon the differences that characterize the ground plans of these two groups, a model is proposed that accounts for the observed differences within a framework of adaptive evolution.

The common lophophorate ancestor in the model is envisioned as a tubicolous, infaunal, vermiform coelomate very similar in adult morphology and behavior to living Phoronida. At least by late Cambrian–early Ordovician time one or more lineages of this primitive stock became epifaunal, some propagating asexually on hard substrates to form entangled masses of tubes. This trend toward an epifaunal existence may have been correlated with the general expansion and diversification of other higher taxa of epifaunal filter feeders that occurred in the lower Ordovician, perhaps the result of an overall amelioration of global marine climates (Valentine and Moores, 1970). The disadvantages of intraspecific competition associated with crowding in the early gregarious lineages were gradually overcome by the regulation of spacing and dispersion patterns among individuals with the adoption of a colonial habit. As a preadaptation to colonial life, the hydrostatic system underwent change to function in the eversion of the lophophore. This implies a loss of individual mobility prior to the development of coloniality, perhaps also associated with crowding or the development of a boring habit similar to that observed in some living species of *Phoronis* (see Section VIII). Associated with the change toward a sessile, colonial existence was a trend toward decreasing size, perhaps related to trophic specialization in the early epifaunal lineages. Decreasing size was accompanied by the simplification of the lophophore and feeding system and the loss of nephridial and closed circulatory systems.

VII. HISTORICAL EVIDENCE BEARING ON THE MODEL

The Ectoprocta first appear in the fossil record of the early Ordovician and rapidly diversified into the major taxa which dominated the remainder of the Paleozoic (Larwood *et al.,* 1967; Ross, 1964). Due to the almost simultaneous appearance of early ectoproct lineages, phylogenetic relationships are unclear. Among the earliest reported genera is a stoloniferous ctenostome, *Marcusodictyon,* preserved as excavations in brachiopod shells from the early Ordovician of Estonia (Bassler, 1953). The explosive appearance of several well-delineated groups in the lower Ordovician may reflect the contemporaneous appearance of skeletonization in several soft-bodied lineages that were well-differentiated by at least early Ordovician time.

As might be expected in the absence of a hard skeleton, body fossils of phoronids are unknown. However, trace fossils that may be attributable to phoronid activities have been referred to the ichnogenus *Skolithos* (Fenton and Fenton, 1934; Alpert, 1974), some members of which have been reported from rocks in Australia dated at 790 million years (Glaessner, 1969). Anvimeleck

(1955) reported trace fossils attributed to the genus *Phoronopsis* from Maestrichtian limestones, as well as a boring species—possibly *Phoronis hippocrepia*—in hard, calcareous substrates in the late Cenomanian, early Turonian, and early Pliocene of Israel.

Evidence for the presumed existence of phoronids in the Precambrian or early Paleozoic awaits further study. Results so far suggest an approach emphasizing a comparative analysis of burrow morphology in living phoronids and their potential fossil representatives to be a fruitful line of research. The smaller boring species of *Phoronis* may well have older fossil records that have heretofore gone unnoticed. This is certainly deserving of further study.

Fossil remains of temporally intermediate forms would provide the strongest evidence for the model proposed. However, such "missing links" are notoriously absent in the fossil record of most groups and the model suffers for a lack of testability in this regard. The model is, nonetheless, potentially testable and, hopefully, will serve to direct attention to the fossil record for evidence that may permit a more critical evaluation of the proposed relationships.

VIII. THE VIABILITY OF INTERMEDIATE MODES OF LIFE

Even in the absence of intermediate fossil forms linking phoronids and ectoprocts, it is still possible to evaluate the viability of intermediate modes of life predicted by the model in the light of some living phoronid species. Such modes of life are believed to be represented in the aberrant phoronid species, *Phoronis ovalis.* The following account of the biology of this interesting species is based on the observations of Harmer (1917), Marcus (1949), Silén (1954), and Emig (1970, 1974b, personal communication).

The smallest known living phoronid (usually <6 mm in length), *P. ovalis,* bores into hard, calcareous substrates where it reproduces extensively by asexual fission, achieving densities of up to 150 individuals/cm^2. The lophophore of *P. ovalis* is highly simplified in relation to other species of the genus, consisting of a single indented ring of ciliated tentacles, much reduced in number (15–28) which arise from a single oval ridge of body wall that surrounds a centrally located, oval mouth. An epistome is present but is much reduced in size, originating as a slight fold in the body wall at the base of the tentacles and projecting toward the mouth. The nephridial system is also simplified, but the closed circulatory system is apparently more complex, as it includes a second accessory vessel not found in other phoronid species. The giant fibers of the nervous system that characterize larger phoronid species are absent in *P. ovalis,* the muscular part of the body wall being supplied with only a diffuse nerve plexus. In *P. ovalis* bundles of strong longitudinal muscles are restricted to a relatively short proximal part of the metasome. The distal end of the metasome, bearing the lophophore, possesses an extremely thin body wall which lacks the strong longitudinal muscles found in the proximal part of the trunk. *Phoronis ovalis* is capable of completely retracting its lophophore and tentacles into a shallow introvert formed by the invagination of the thin distal portion of the body wall into the thickened posterior muscular region of the trunk (Harmer, 1917, pp. 139–141). Apparently, no retractor muscles comparable to those in ectoprocts have been documented in *P. ovalis* and the exact method of lophophore retraction and eversion needs further study. It is envisioned that lophophore retraction probably involves shortening of longitudinal muscles in the proximal region of the trunk, with eversion being accomplished by muscular deformation of the coelomic cavity to create a positive internal hydrostatic pressure similar to the mechanism found in ectoprocts. Individuals of *P. ovalis* excavate shallow burrows more or less parallel to the surface of the substrate, which is often the shell of a mollusk. During feeding only the tentacles are protruded from the tube. It is conceivable that the adaptations for lophophoral retraction in *P. ovalis* have developed as a consequence of a lack of individual mobility associated with the boring habit and of crowding due to the high population densities commonly achieved by asexual reproduction. *Phoronis ovalis* also reproduces sexually, and the large, yolky embryos are retained in the distal end of the tube where they are brooded to the larval stage. Immediately prior to sexual reproduction, the lophophore is cast off and fertilized eggs apparently exit from the temporary opening created at the anterior end of the body. The parent then retracts to the proximal end of its tube, entering a stage of dormancy until the larvae escape, at which time a new lophophore is generated and feeding resumes. This behavior is similar to the cycles of degeneration and regeneration exhibited by some ectoprocts during sexual reproduction.

The intermediate morphological and behavioral patterns exhibited by *P. ovalis*—many of them apparently associated with coadaptations for small size and the opportunistic use of hard substrates—correlate nicely with those predicted by the model and provide an indication of the nature and viability of possible intermediate adult forms linking phoronids and ectoprocts.

IX. EVOLUTION OF THE LARVAL STAGE OF THE LIFE CYCLE

In benthic marine invertebrates that possess a pelagic larval stage, the life cycle may be subdivided into larval and adult phases that are linked to each other by only those processes that operate at metamorphosis. The adaptive demands placed on each of these stages of development clearly differ and evolutionary divergence in the adult stage may follow adaptive pathways quite independent of those in the larval stage. In theory, this observation offers a potential for approaching phylogenetic hypotheses along two lines of inquiry and improves our chances of arriving at a model with the highest possible explanatory power.

In developing the previous evolutionary model linking ectoprocts and phoronids, only morphological features of the adult stage of the life cycle have been emphasized, for reasons discussed earlier (see Section I). In the absence of supporting evidence from the fossil record, the problem of directionality in evolution has been approached indirectly on theoretical grounds, based on the assumed primitive function of the coelom in the metazoa (Section III), and the advantages of a colonial life style (Section VI). But, for the larval stage of the life cycle, where the relative adaptive advantages of morphological differences are less clear, it is more difficult to visualize a theoretical framework that will independently impose directionality to the course of evolution. However, an attempt can be made to evaluate the consistency of the interpretations of adult morphology with morphological and behavioral patterns in the larval stage. To accomplish this, it is necessary to delineate the salient features of the larval ground plans of phoronids and ectoprocts and to evaluate the probable adaptive basis of morphological differences in these groups. Much of the account of larval biology which follows is based on reviews by Hyman (1959) and Zimmer (1964).

X. THE ACTINOTROCH LARVA OF THE PHORONIDA

The fully differentiated actinotroch larva of the phoronids is characterized by an elongate (1–4 mm) tripartite body plan (Fig. 2A). At the anterior end, a large hoodlike feature–the preoral lobe–projects downward over the ventral surface of the larva, covering the mouth. At the apical pole of the larva, in the central region of the hood, is a thickened plate of ectoderm bearing long cilia. This apical plate functions as a sensory organ and constitutes the main nervous center of the larva. Posterior to the preoral lobe, slanting ventrally beneath it, is the collar region of the larva. Along the posterior margin of this collar, a girdle of tentacles arises from a ridge of thickened ectoderm that encircles the vestibule formed beneath the overarching hood. The vestibular cavity changes size and shape with movements of the hood and is confluent with the mouth, located anteriorly (and dorsally). The tentacles, and the ridge bearing them, are heavily ciliated and in advanced larvae apparently function primarily to generate feeding currents that enter the vestibule (Lebour, 1922; p. 666; Strathmann, 1973; p. 133). Proximal to the tentacular girdle is an elongate trunk region. The trunk terminates in a heavily ciliated ectodermal thickening that surrounds the anus and functions as the principal locomotory organ of advanced larvae.

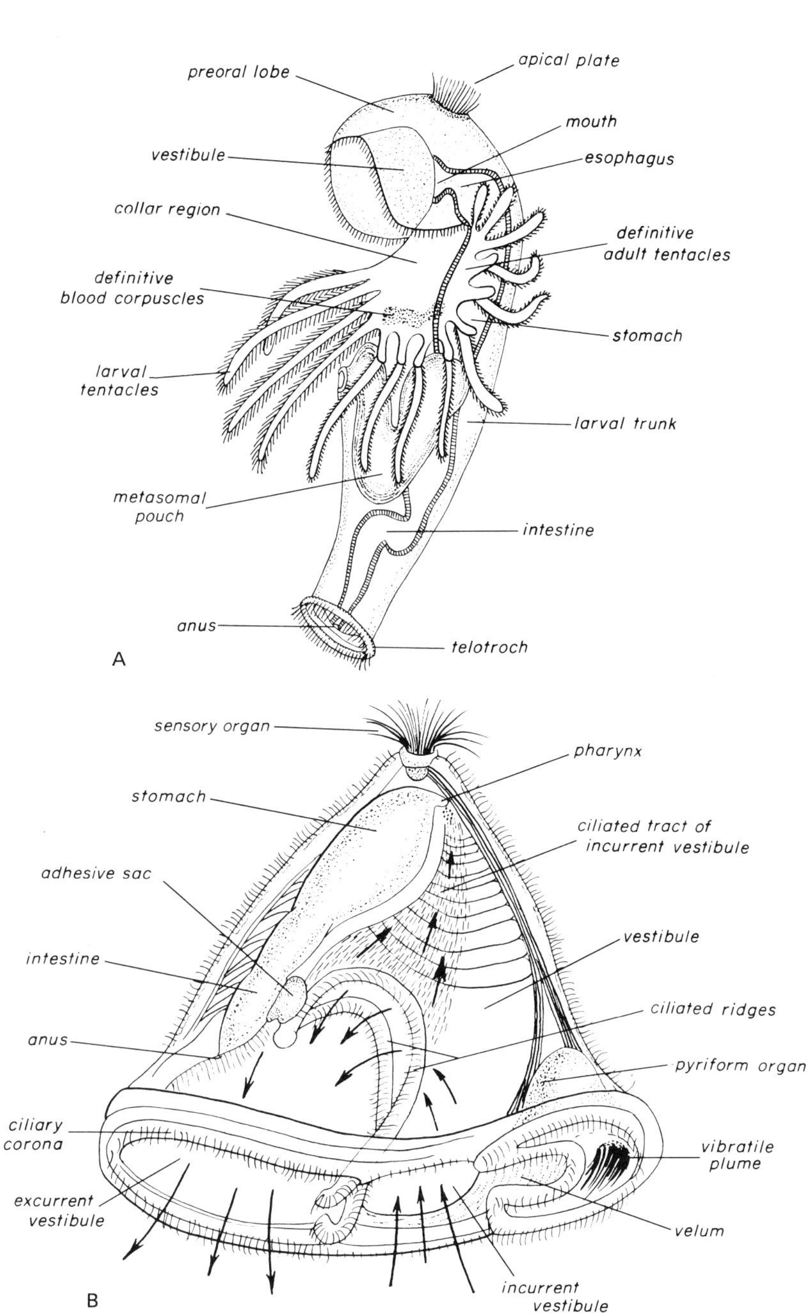

Fig. 2. Comparative morphology of planktotrophic larval forms in the Phoronida and Ectoprocta. (A) Schematic showing the major features of the actinotroch larva of the Phoronida (based on Hyman, 1959; and Jägersten, 1972). (B) Schematic showing the major features of the cyphonautes larva of the Ectoprocta (based on Atkins, 1955b).

The larval coelom also exhibits a tripartite construction, consisting of a preoral protocoel and postoral mesocoel and metacoel (see Zimmer, 1964; pp. 275–278 for discussion). The trunk coelom or metacoel is spacious, and is separated from the space of the collar region by a septum at the base of the tentacular girdle.

The metasome is occupied largely by the digestive tract, which is essentially straight and regionated into an esophagus, elongate stomach, and intestine. Located midventrally immediately below the tentacular girdle is an epidermal invagination called the metasomal pouch. This invagination elongates throughout larval development, eventually forming a large, tubular structure infolded into the coelomic space surrounding the gut.

A larval protocoel occupies much of the preoral lobe of the young actinotroch. During growth of the larva there is a gradual flattening of the hood, the protocoel being reduced (commonly to a small vesicle lying beneath the apical plate of the larva) or even obliterated (see Zimmer, 1964, pp. 225–278). Most of the hood cavity of the advanced larva is blastocoelic space that in some species, at least, becomes filled with a gel-like substance probably secreted by amoeboid mesenchymal cells. The significance of this substance is debatable, but it may in part function in supporting the preoral lobe (Zimmer, 1964, p. 277). The remaining preseptal space is occupied by a hollow blastocoelic cavity in the collar region. This collar blastocoel persists throughout larval life, but prior to metamorphosis it is partially replaced by a horseshoe-shaped collar coelom, the larval mesocoel. This coelom forms at the level of the tentacular girdle, just anterior to the collar–trunk septum.

The protonephridia of the actinotroch larva originate as a single midventral ectodermal invagination just posterior to the tentacle ridge. This primordium soon bifurcates and the paired protonephridia eventually come to lie on either side of the stomach. The nephridiopores remain just below the tentacular girdle and most of the solenocytes project into the preseptal blastocoelic space of the collar region.

The actinotroch larva is provided with a complex musculature, the details of which vary considerably between species. Major elements that are often present include a pair of longitudinal muscles spanning the trunk coelom on the ventral and/or dorsal sides from anterior attachments in the collar region to posterior attachments near the anus; hood and apical organ retractors, arising beneath the apical plate and traversing the preseptal space to attachments in the dorsal region of the collar; hood levators arising on the dorsal surface of the hood and attaching to the dorsal wall of the collar and trunk; tentacle levators and depressors; radial and concentric muscles of the preoral lobe; and outer circular and inner longitudinal muscles of the trunk and metasomal sac.

The principal nervous center of the actinotroch larva is a ganglion situated beneath the apical plate in the center of the dorsal surface of the preoral lobe. In addition, the larva is provided with an intraepidermal nervous plexus that is continuous with the ganglion through several discrete nerve tracts. The latter are subject to considerable interspecies variation with regard to their location and design, but commonly present are longitudinal tracts arising from the ganglion, traversing the ventral and dorsal margins of the hood, and connecting to a nerve ring at the base of the tentacular girdle; a sensory element located on the midventral surface of the hood, associated with thickenings of the ventral longitudinal tracts; longitudinal dorsal or ventral nerves of the trunk; and a nerve ring encircling the anus.

The actinotroch larva possesses no functional circulatory system, but the major elements of the adult system are anticipated at the larval stage. The details of the larval system vary between species, but one or two longitudinal vessels—actually channels between the stomach wall and lining of the metacoel which are continuous with the preseptal collar blastocoel—are generally present in dorsal (and dorsolateral) positions along the stomach. The lophophoral ring vessels of the adult originate from the larval collar blastocoel during the transformations at metamorphosis. Definitive blood corpuscles occur as discrete aggregates, usually attached to the body wall of the collar region or in the preoral lobe. The circulatory elements of the larva and adult are considered to be blastocoelic in origin, but become lined by mesenchymal cells after metamorphosis (Zimmer, 1964, p. 322).

Late in the larval development of some species, the definitive tentacles of the adult appear as buds at the base of the larval tentacles. These tentacle buds are associated with evaginations of the horseshoe-shaped mesocoel that develops in the collar cavity late in larval development. In other species, the larval tentacles are retained in the adult, but also become invaded by tubular extensions of the mesocoel prior to metamorphosis. Metamorphosis of the actinotroch larva is usually accomplished very rapidly(15–25 minutes), due to the high degree of preformation of the adult (Hyman, 1959, p. 259; see also Silén, 1954, p. 251). Following the selection of a suitable substrate, the larva everts its metasomal pouch through convulsive contractions of the larval trunk. Shortening of the longitudinal retractor muscles of the trunk presumably generates high hydrostatic pressures within the larval metacoel that assist in forcing the metasomal pouch to evert. With eversion of the metasomal pouch, the digestive tract is carried out and thrown into a U-shaped configuration; shortening of the larval trunk brings the anus and mouth into proximity at the future distal end of the adult. As a consequence, the everted metasomal pouch becomes the definitive trunk of the adult, while the preoral lobe shrinks and is cast off down to near the level of the definitive lophophoral coelom. The larval tentacles may be retained in the adult, but are sometimes cast off with the preoral lobe, the tentacle buds of the adult growing and rapidly increasing in number. The origin of the adult ganglion is unclear; it is generally believed to arise later as an ectodermal thickening, but in some species the larval ganglion may be retained in the adult (Zimmer, 1964; p. 313). The cavities which form the larval circulatory system are retained in the adult and become functional as the associated corpuscular masses break down during the contractions that occur with eversion of the metasomal pouch. The protocoel—or epistome—is generally believed to form later as a fold in the dorsal body wall. However, it has been suggested that in some species, part of the larval protosome and protocoel may be retained through metamorphosis to form the epistome and its corresponding coelomic division in the adult (Zimmer, 1964; pp. 279–280).

XI. LARVAL DEVELOPMENT IN THE ECTOPROCTA

In the gymnolaemate ectoprocts, larval forms are quite varied, but may be grouped into two main types: the cyphonautes, a planktotrophic form that develops from nonbrooded eggs that are shed directly into the sea, and a lecithotrophic form, quite unlike the cyphonautes, that develops from brooded eggs. The vast majority of living gymnolaemate ectoprocts brood their eggs and lecithotrophic forms are therefore the dominant larval type. Silén (1944) believes such brooded forms to be primitive in the Ectoprocta. However, some authors (Atkins, 1955a; Brien, 1960; Jägersten, 1972; Zimmer and Wollacott, this volume, Chapter 3) have viewed the cyphonautes as an archaic, and presumably primitive form within the Ectoprocta, lecithotrophic types having been derived secondarily with specializations for the brooding of the young. But the strength of any assumption about larval phylogeny is difficult to evaluate in view of the strong dependency of larval biology on reproductive "strategy" in living species. The observed decrease in the number of benthic marine invertebrate species possessing pelagic larvae, from equator to poles, and with increasing depths in modern oceans underlines the important role of the environment in patterning the optimal reproductive "strategies" of living species (Thorson, 1946; Miliekovsky, 1971). Arguing by homology, a case can perhaps be made for the cyphonautes being a primitive larval form within the Cheilostomata. Although known for several species of the Ctenostomata, typical planktotrophic cyphonautes are essentially restricted to certain malacostegoid genera believed to be primitive to Cheilostomata, based on functional criteria and supported by evidence from the fossil record (see Cheetham, 1971, pp. 11–15 for discussion; Pohowsky, 1973). However, we have no evidence for extending this same argument to the entire phylum, which is today dominated by lecithotrophic forms. The implication is that historically, larval biology may have depended more on the geographic site of evolution than on phylogeny. For example, if ancestral ectoproct lineages first appeared in seasonal environments such as those that occur at high latitudes today, evidence from living systems would suggest brooded lecithotrophic forms to be the primitive larval type. Alternatively, if ancestral lineages appeared in stable environments such as those that characterize low latitudes today, the same reasoning would suggest planktotrophic forms to be primitive to the phylum. It is of interest in this context that the earliest reported record of the Bryozoa (the ctenostome *Marcusodictyon*), is from the lower Ordovician of Estonia, which during that period in geologic history is believed from paleomagnetic and geologic evidence to have been positioned within 10°–30° of the paleoequator (see Briden *et al.*, 1974). Unfortunately, our present knowledge of paleobiogeography and the history of global marine climate precludes any choice between these alternatives on purely empirical grounds.

In expanding the model for ectoproct origins to encompass the early stages of the life cycle, we therefore assume the directionality of evolution arrived at from a consideration of adult morphology. The characteristic larval form of the Phoronida is planktotrophic, and because the purpose here is only to investigate the plausibility of the actinotroch as an ancestral larval form for the ectoprocts, comparison is, of necessity, confined to the planktotrophic form of the ectoprocts—the cyphonautes.

XII. THE CYPHONAUTES LARVA

The described species of cyphonautes larvae exhibit a range of variability in size (usually 0.2–0.8 mm in length) and in certain details of their morphology (see Ryland, 1965). However, they share in common a bilaterally compressed body plan, and are enclosed between a bivalved shell of chitinous composition (Fig. 2B). This distinctive morphology, along with the presence of a functional gut, serves to distinguish the cyphonautes from the more common lecithotrophic larval forms found in the vast majority of living ectoproct species.

The chitinous valves of the cyphonautes are generally triangular in outline and may be transparent or minutely agglutinated in structure. The apex of each valve is truncated and often flared to accommodate a knoblike sensory organ that bears tufts of long cilia. Basally the valves gape, and the mantle along the shell margin is thickened and heavily ciliated. This mantle thickening consititutes the ciliary corona which functions as the principal locomotory organ of the larva. Enclosed between the valves is a large mantle cavity, or vestibule, that is divided into anterior incurrent and posterior excurrent regions by two ciliated ridges that project toward each other across the vestibular space.

The digestive system is U-shaped and regionated into ciliated pharynx, stomach, and intestine. The funnel-shaped inhalent chamber extends from the base of the shell to near its apex. The pharynx is confluent with the heavily ciliated apical region of the incurrent chamber, while the anus opens posteriorly within the excurrent chamber. Ciliation is weak or absent in the excurrent chamber, although in some species a ciliated tract is present, leading from the anus to the mantle edge (Atkins, 1955a, p. 457).

The shell margin is also regionated into excurrent and incurrent areas by modifications of the structure of the corona that encircles it. Coronal cilia, which function primarily as locomotory devices, beat downward, away from the apex. Along the excurrent margin of the vestibule, ciliation of the corona is heavier, and there the ciliary beat assists the current exiting from the exhalent chamber. But along the incurrent margin, where the beat of coronal cilia generates currents opposed to those entering the inhalent chamber, cilia are reduced in size and number (Atkins, 1955a, pp. 455–456). This variation in the size and density of cilia along the coronal margin may have an additional function in governing the attitude of the larva during swimming, as the cyphonautes is observed to carry the apical organ forward during swimming (Ryland, 1965, p. 4). Further differentiation of the mantle margin with regard to trophic function is provided by a delicate, epidermal fold (the velum) that is present on the corona surrounding the inhalent aperture. This epidermal fold is provided with circular muscles that regulate the size of the inhalent aperture and fanlike groups of cilia that probably function as a sieve (Atkins, 1955a, p. 454).

The ciliated ridges that divide the mantle cavity into incurrent and excurrent areas bear three longitudinal tracts of cilia that generate the principal feeding current and function in food collection (see Atkins, 1955a, pp. 458–461; Strathmann, 1973). Food particles entrapped on the medial tract of cilia are conveyed up the ridges by frontal cilia, and eventually pass onto a ciliated tract that crosses the upper part of the incurrent mantle cavity. Some food particles also reach this tract directly, due to the strong beat of the cilia located in this region of the mantle. The food particles are carried along the ciliated tract and accumulate in an unciliated region in front of the mouth before being swallowed.

Situated near the oral surface are two organ systems that are common to all gymnolaemate ectoproct larvae. In an anterior position near the oral surface, immediately in front of the inhalant region of the mantle cavity is the pyriform organ—a complex of glandular cells associated with a ciliated invagination. Near the anterior end of this invagination is the vibratile plume—a tuft of long sensory cilia (or flagella). In a posterior position, situated between the vestibule and anus, is the adhesive sac—another invagination that is lined by columnar cells that also have a secretory function. Both of these organs increase in size during growth, eventually occupying much of the larval volume in later development.

The cyphonautes larva is equipped with a complex musculature that serves chiefly to retract the corona and various organ systems into the protection of the valves and to regulate closure of the valves (Kupelweiser, 1905; Hyman, 1959, p. 349). The muscle system of the cyphonautes is intimately associated with the major elements of the larval nervous system and is commonly united with the latter by a single neuromuscular system. Most major muscles originate from attachments beneath the apical sensory organ and associated ganglion. Spanning a narrow blastocoelic space between the pharynx and outer body wall, an anterior branch descends to the pyriform organ where it forks, sending strands to the vibratile plume and adjacent corona. Posterior dorsal branches ascend to attachments in the apical organ itself, or descend between the body wall and intestine where they supply the posterior region of the corona. Pairs of lateral muscles also extend from medial attachments on the inner surfaces of the valves to insertions in the adhesive sac or the anterior and posterior regions of the corona. Closure of the valves is accomplished by transverse adductor muscles that span the anterior region between the adhesive sac and vestibule, probably assisted by a circular muscle band that is present in the corona. Circular muscles are also present in the pharynx where they function in peristalsis, in the velum where they effect changes in the size of the inhalant aperture, and in the pyriform organ where they assist in its protrusion.

Metamorphosis of the cyphonautes larva occurs following an often extended planktonic existence that may last as long as two months (Marcus, 1940, p. 334). Just before settlement, the larva hovers over the bottom using the vibratile plume in a sensory capacity for the selection of a substrate for attachment. When a suitable substrate is found, convulsive contractions cause the adhesive sac to evert and then flatten out over the substrate, adhering to it by the secretions it produces. Simultaneously, the two valves separate by rupturing of the adductor muscles, and then flatten out, covering the mass (Kupelwieser, 1905; Hyman, 1959, p. 352). The valves are retained for one to several days and would seem to take on a secondary protective function during the early stages of metamorphosis, prior to the secretion of the definitive cuticle of the ancestrula (see Atkins, 1955b). All exclusively larval organs are retracted into the interior of the mass and disintegrate. The result is a flattened mass of loose cells and debris, which is enclosed within an epidermis.

The first definitive adult organs begin to form from cells of ectodermal origin derived from the apical plate of the larva which invaginate into the interior and eventually separate from the exterior to form a hollow vesicle. This vesicle becomes covered by mesenchymal cells which differentiate from ectoderm to form peritoneum. The initial vesicle enlarges rapidly and by constriction forms a smaller secondary vesicle. The distal part of the first vesicle differentiates to form the definitive tentacle sheath, while its proximal part gives rise to the adult tentacles and pharynx. The tentacle buds on the ventral side arise before those on the dorsal side, imparting an early bilateral symmetry to the developing lophophore. The smaller secondary vesicle differentiates to form the rest of the digestive tract. The main divisions of the coelom become defined through the rearrangement of mesenchymal cells which line the larval blastocoel. Muscles of the adult also differentiate from mesenchymal cells and the adult ganglion soon forms from an epidermal invagination. Communication is eventually established between the two developing parts of the polypide, and the tentacle sheath becomes associated with an epidermal thickening that eventually breaks through to form an orifice. At this stage in development, the polypide of the ancestrula is essentially complete and colony formation proceeds by growth of the primary zooid to form one or more buds. The time elapsed between attachment and development to this stage varies, but generally falls within the range of one to six days (Hyman, 1959, p. 354).

XIII. DISCUSSION AND CONCLUSIONS

The strong similarities in coelomic architecture, construction of the lophophore, and other features of the ground plan of adult phoronids and ectoprocts have been previously interpreted as having a phylogenetic basis. In pursuing this suggestion, an evolutionary model was proposed (Farmer *et al.,* 1973) which derives the adult ground plan of the Ectoprocta from a phoronid ancestor by adaptation and evolutionary divergence leading to a colonial habit and small size. Directionality of evolution in the model is based on the assumption that the coelom functioned primitively as a hydrostatic skeleton in burrowing (Clark, 1964). Subsequent functional modifications for respiration and feeding in some early oligomerous stocks eventually led to the development of a primitive, protentaculate ground plan (Cowen, 1974), bearing a close resemblance to living Phoronida. Also inherent in this interpretation are the assumed advantages of a colonial "strategy" and small individual body size. Both of these adaptations have been acquired independently in the evolution of many unrelated phyla and would seem to have universal advantages under an appropriate selective regime. Intermediate modes of life linking phoronids and ectoprocts are regarded as viable in light of the biology of the aberrant phoronid species, *Phoronis ovalis.*

Consideration of the larval biology of phoronids and ectoprocts reveals that similarities in morphology, comparable to those observed in the adult ground plans of these phyla, are not apparent at the larval stage. Consequently, it is difficult to visualize a phylogenetic connection between the Phoronida and Ectoprocta based on larval morphology alone. Even if it were possible to associate the two larval forms on the basis of morphological similarity, we still lack an independent theoretical framework, comparable to that developed for the adult stage, which would impose directionality to the course of evolution. However, it does not seem reasonable to reject a relationship between phoronids and ectoprocts on the basis of negative evidence from the larval stage alone. In weighing the relative importance of the observed dissimilarities in the larval morphology of phoronids and ectoprocts, careful consideration must be given to the adaptive basis of morphological differences in the two groups.

Perhaps the most interesting aspect of the ground plan of the phoronid larva is the high degree of preformation of the adult at the larval stage of the life cycle (see Silén, 1954, p. 251); indeed, many of the important morphological attributes of the actinotroch are best explained by "adulation" (Jägersten, 1972, p. 4) or the shifting of adult characters into the larval stage which has facilitated an accelerated metamorphosis. Although the "package" of adaptations we associate with the actinotroch larva cannot be explained entirely from the viewpoint of the adult, they are believed to represent the best functional compromise for fulfilling the requirements of a successful pelagic life, on the one hand, and rapid metamorphosis linking the larval and adult stages of development, on the other. This suggests that in the Phoronida evolution at the larval stage has, in general, been closely linked to evolution at the adult stage as a consequence of a reproductive "strategy" that emphasizes the advantages of rapid metamorphosis.

In contrast, considerable differences exist between the larval and adult ground plans of the Ectoprocta. The cyphonautes larva is viewed as a highly coadapted planktotroph, capable of a sustained pelagic existence. In the Ectoprocta there seems to have been considerable independence in the evolution of the adult and larval stages of the life cycle in that none of the features of the larva bear any clear relationship to those of the adult. In the case of the cyphonautes larva, the strong morphological divergence between the larval and adult stages is undoubtedly linked to a reproductive "strategy" which emphasizes the advantages of planktotrophy and long-range dispersal. In addition, the complex and prolonged metamorphosis that is characteristic of the Ectoprocta as a whole may be a necessary consequence of this evolutionary divergence of larval and adult stages of development during the early history of the group.

A noteable departure from the general pattern of larval development, behavior, and metamorphosis in the Phoronida has been observed for *Phoronis ovalis* (Silén, 1954). Morphologically, the larva of *P. ovalis* is highly divergent from the typical actinotroch larva of other phoronid species. The trunk region of the larva is considerably reduced in size relative to the preoral hood, and is ventrally flattened (see Silén, 1954, p. 245, Fig. 22). The larval stage of *P. ovalis* consists of two stages: an early pelagic lecithotrophic stage, serving in general dispersal, followed by a benthonic stage, serving in substrate selection and during which time the larva "creeps" over the substrate on the flattened ventral region of the body by ciliary action, assisted by mucous secretion. Compared to other phoronid species, the larva of *P. ovalis* is poorly differentiated upon settlement, sharing few features in common with the adult. Although the details of larval development in *P. ovalis* await study, the observations of Silén (1954, p. 248) suggest that metamorphosis is prolonged, involving a more complicated reorganization of larval structures than is required for other phoronid species, and approaching in many respects the style of metamorphosis observed in the Ectoprocta.

The degree of independence of larval and adult stages of the life cycle in evolution, and consequently the potential of morphological divergence of the two stages of development, is clearly linked to reproductive "strategy" in the Phoronida and Ectoprocta. Because reproduction in benthic marine invertebrates appears to be strongly patterned by spatial and temporal variations in biotic and abiotic factors of the environment, "strategies" of reproduction have certainly varied in the past in response to changing paleogeographies and marine climates. In light of the many complex events that have shaped the evolution of the biosphere and the long time spans involved, dissimilarities in larval biology alone do not seem to provide a strong basis for rejecting a relationship between the Phoronida and Ectoprocta.

In considering possible evolutionary trends at the larval stage, it is useful to identify the selection pressures that can operate to pattern the reproductive "strategy" of a species. Vance (1973; see also Underwood, 1974; Vance, 1974) has examined the common modes of reproduction in benthic marine invertebrates from a theoretical standpoint and arrived at some interesting conclusions regarding the selection of optimal reproductive "strategies." Because reproduction involves a nontrivial portion of the energy budget of a species, it is assumed that natural selection favors the pattern with the greatest efficiency. Of the common reproductive modes, planktotrophy and nonpelagic lecithotrophy may be considered as end members with regard to the relative duration of feeding and nonfeeding stages of larval development. For planktotrophic forms the nonfeeding period is regarded as trivial, while in lecithotrophic forms it encompasses the entire period of larval life. For organisms with comparable larval development times, an important advantage of planktotrophy over nonpelagic lecithotrophy is the capability for greater dispersal, which enhances the utilization of a patchy environment by the adult. An additional advantage of planktotrophy is that a large number of eggs can be produced for a given allocation of reproductive energy in that the larva does not depend on a supply of yolk from the parent for its development. The obvious disadvantages are that larvae must depend solely on a planktotrophic food source throughout development and are subject to higher mortality rates due to planktonic predation. Nonpelagic lecithotrophy involves a greater investment of energy per egg as yolk. Thus, the number of eggs that can be produced for a given allocation of reproductive energy is lower, but the rate of planktonic predation is essentially reduced to zero. An important disadvantage of nonpelagic lecithotrophy is the lowered dispersal capability of brooded larvae. Planktotrophic development is favored when (1) the distribution of habitats in space and/or time is "predictable," (2) when planktotonic predation is low, or (3) when planktonic food is abundant. Nonpelagic lecithotrophy is favored when (1) planktonic predation is excessive, (2) planktonic food is limiting, or (3) when the distribution of habitats is "unpredictable," thereby favoring opportunistic use of habitat by the adult. Pelagic lecithotrophy represents a "strategic" compromise in allowing for independence from a pelagic food source and perhaps shorter exposure to planktonic predation compared to planktotrophy, and greater dispersal potential than nonpelagic lecithotrophy.

An interesting framework for considering the implications of the proposed model for evolution at the larval stage is provided by Fig. 3A–D, in which possible morphological homologies between the actinotroch and cyphonautes larval forms through two hypothetical intermediate forms are presented. The interpretations are based on Jägersten (1972, p. 58), who proposed an evolutionary trend that traces the actinotroch from a cyphonautes ancestor. It seems equally reasonable, however, to interpret the sequence in the opposite sense, based upon the directionality suggested by the model for the adult stage—namely, the derivation of the cyphonautes larval form from an actinotroch ancestor by specialization for planktotrophy. Some major morphological

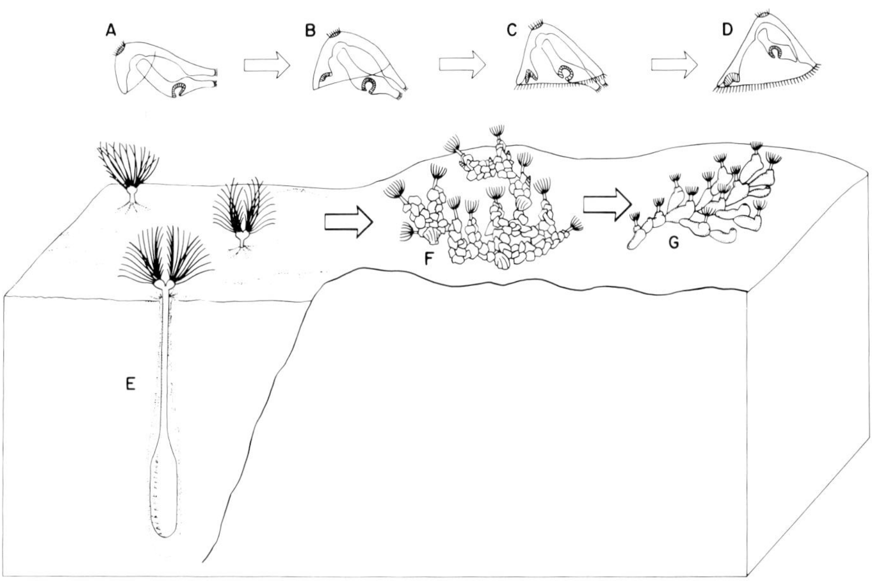

Fig. 3. Diagrammatic representation of the proposed model for evolution of the ectoproct life cycle. (A–D) Evolution at the larval stage of the life cycle tracing the cyphonautes larval form (D) to an actinotroch ancestor (A) through two hypothetical intermediate morphologies (B and C) by specialization for planktotrophy and long-range dispersal ability (diagram based on Jägersten, 1972; p. 58). (E–G) Evolution at the adult stage of the life cycle tracing the Ectoprocta to a phoronid ancestor through an intermediate epifaunal form by specialization for a colonial habit and small size (based on Farmer *et al.*, 1973).

changes required by such specialization include (1) a shortenting of the larval trunk of the actinotroch and modifications in the biological role of the preoral lobe to function not only in feeding, but also as the principal locomotory organ in the cyphonautes; (2) a change in the biological role of the metasomal pouch as the preadult trunk of the actinotroch to function in attachment of the cyphonautes to the substrate prior to metamorphosis; (3) bilateral compression and an associated reduction in the coelomic and blastocoelic body cavities, with acquisition of a bivalved shell in the cyphonautes; (4) the development of the pyriform complex in the cyphonautes to function as the principal organ in substrate selection; and (5) the elaboration of certain neuromuscular systems to function in retraction of various organ systems into the protection of the bivalved shell of the cyphonautes. It is of interest in this context that morphological and developmental divergence within the living Phoronida, as exemplified by *P. ovalis,* is of a nature and magnitude such that the changes suggested by Fig. 3 A–D do not appear unreasonable from an evolutionary standpoint.

Perhaps the most serious objections to an evolutionary model linking phoronids and ectoprocts are those of Nielsen (1971, pp. 309–311), which concern the developments at metamorphosis in the two groups. Assuming a solitary, protentaculate ancestor in which the alimentary canal and larval tentacles are retained in the adult, Nielsen finds difficulty in deriving the typical ectoproct life cycle from that which characterizes living phoronids. Although the general features of the phoronid ground plan fulfill the requirements for a probryozoan ancestor very well, problems arise in achieving a functional phoronid adult through an ectoproct-type of metamorphosis. In the Phoronida, the adult tentacles develop from, or in contact with, the metatroch of the actinotroch larva. However, in the metamorphosis of the ectoproct larva, the comparable region becomes enclosed beneath the episphere of the larva which eventually expands to contact the substratum. Clearly, a functional adult would be impossible to obtain in an ancestor with this type of metamorphosis. But this interpretation relies on the assumption that the ectoproct style of metamorphosis had already evolved in the early probyrozoan lineage. It seems reasonable, however, to envision an intermediate form linking phoronids and ectoprocts which was colonial, but in which the first individual of the colony developed by a phoronid type of metamorphosis, thus yielding a functional adult. Nielsen (1971, pp. 310–311) rejects this notion of an ancestral probryozoan which was already colonial because budding in the Ectoprocta is from the episphere of the cyphonautes larva, a region believed by him to be homologous in position to a narrow region inside the ring of tentacles of the adult phoronid. Clearly, budding from such a position in the adult phoronid seems incompatible (functionally) with general phoronid organization. But the specific location and orientation of budding in the two groups is critical to Nielsen's argument. Nielsen (1971, p. 309) points out the difficulties in assigning an orientation to the polypide in the ectoproct adult. Orientation cannot be based on embryological evidence, as the gut of the polypide has nothing to do with the gut of the cyphonautes larva, the former having an independent origin at metamorphosis. In light of the problems of interpreting the embryological evidence, the traditional orientation of the ectoproct adult is based on comparisons with adult phoronids. The phenomenon of budding in the Phoronida awaits detailed study, but the observations of Marcus (1949, p. 164) for *Phoronis ovalis* suggest that buds originate on the ventral (oral) side of the adult animal in the nonmuscular region of the trunk, and not within the tentacle ring, as suggested by Nielsen (1971, p. 310) on embryological evidence. In the Ectoprocta such oral budding occurs in the Phylactolaemata, while budding in the Gymnolaemata and Stenolaemata is in an anal direction (Marcus, 1949; Jebram, 1973).

The derivation of the ectoproct life cycle from that of a colonial probryozoan with a phoronid-type of metamorphosis seems reasonable using adult criteria for orientation. But we are left with the problem that the shift to an ectoproct type of metamorphosis would have to have been abrupt in that no intermediate style of development has been envisioned that is capable of producing a functional adult. In the evolutionary trends represented in Fig. 3A–D, an ectoproct style of metamorphosis would seem to be necessitated by shortening of the larval trunk of the actinotroch larval form whereby eversion of the metasomal pouch (adhesive sac) would begin to occur beneath the expanded preoral lobe (corona), making it increasingly difficult to achieve a functional adult by a phoronid style of metamorphosis (the details of larval development and metamorphosis in *Phoronis ovalis* may prove enlightening in exploring this aspect of the problem, and are certainly deserving of study). But such an abrupt change in the style of metamorphosis may have held considerable adaptive advantages for those populations in which it first appeared. Under an appropriate selective regime considerable evolutionary potential may have been gained with the appearance of an ectoproct-type of metamorphosis in that the larval ground plan would have been freed from restrictions placed on it by the adult stage for accelerating metamorphosis. The morphological trends represented in Fig. 3A–D, interpreted to represent specialization for planktrophy at the larval stage, actually seem to necessitate increased flexibility in larval development and metamorphosis by a decoupling of the larval and adult stages of the life cycle through a shift away from adultation. The appearance of this novel style of metamorphosis, regardless of ancestry, may in part underlie the rapid evolutionary advances made by the Ectoprocta in the Lower Ordovician.

If we are to accept the proposed model linking phoronids and ectoprocts at the adult stage, we must be willing to accept the possible evolutionary changes implied for the larval stage as well. We also must accept the conclusion that evolution at the larval stage of the life cycle in this system has been less conservative to morphological change than evolution at the adult stage. That the larval stage is a highly adaptable part of the life cycle in living groups is apparent from observations that larval behavior in many aquatic invertebrates varies significantly between closely related species, between populations of single species, and even within single populations (Meadows and Campbell, 1972, p. 351).

Based on these interpretations, an evolutionary model linking the life cycles of phoronids and ectoprocts is derived which emphasizes specialization for planktotrophy and long-range dispersal at the larval stage (Fig. 3A–D), and independent trends in the evolution of the adult stage leading to small size and a colonial habit (Fig. 3E–G). Evolutionary trends at both stages of the life cycle may have been guided by selection for trophic specialization. The general and widespread diversification of filter feeders in the lower Ordovician, suggestive of an abundance of planktonic food, and the widespread stabilization of marine climates is favorable to such an interpretation. An interesting corollary to this interpretation is that the later evolutionary history of the Ectoprocta has witnessed a widespread shift in reproductive mode to the pelagic, lecithotrophic development that characterizes the phylum today.

ACKNOWLEDGMENTS

I am grateful for constructive comments and criticisms of the manuscript by Richard Cowen and James W. Valentine. I also benefited from discussions with Claus Nielsen, William Banta, R. S. Boardman, and other members of the Third Conference of the International Bryozoology Association held at Claude-Bernard University, Lyon, France, in 1974. Attendance of that meeting was made possible by a travel grant from the University of California, Davis Campus. I express special thanks to Drs. James W. Valentine and Richard Cowen who generously permitted the inclusion in this chapter of material that had been published jointed with the writer in 1973. Thanks are also due Ellen Bailey who drafted the illustrations and to Debba Kunk who typed the manuscript.

REFERENCES

Alpert, S. P. (1974). Systematic review of the genus *Skolithos*. *J. Paleontol.* **48,** 661–669.

Anvimeleck, M. (1955). Occurrence of fossil Phoronidea-like tubes in several geological formations in Israel. *Bull. Res. Counc. Isr.* **5,** 174–177.

Atkins, D. (1955a). The cyphonautes larvae of the Plymouth area and the metamorphosis of *Membranipora membranacea* (L.). *J. Mar. Biol. Assoc. U.K.* **34,** 441–449.

Atkins, D. (1955b). The ciliary feeding mechanism of the cyphonautes larva [Polyzoa Ectoprocta]. *J. Mar. Biol. Assoc. U.K.* **34,** 451–466.

Banta, W. C., McKinney, F. K., and Zimmer, R. L. (1974). Bryozoan monticules: Excurrent water outlets? *Science* **185,** 783–784.

Barrington, E. J. W. (1967). "Invertebrate Structure and Function." Nelson, London.

Bassler, R. S. (1953). "Treatise on Invertebrate Paleontology," Part G. Univ. of Kansas Press, Lawrence.

Briden, J. C. and Drewry, G. E., and Smith, A. G. (1974). Phanerozoic equal-area world maps. *J. Geol.* **82,** 555–574.

Brien, P. (1960). Classe des Bryozoaires. *In* "Traité de Zoologie" (P. P. Grassé, ed.), Vol. 5, Part 2, pp. 1053–1355. Masson, Paris.

Brien, P. (1970). Considérations phylogénetiques à-propos des Lophophoriens. *Bull. Cl. Sci., Acad. R. Belg., Cl. Sci. [5]* **56,** 565–579.

Bullivant, J. S. (1968). Release of sperm by Bryozoa. *Ophelia* **4,** 139–142.

Bullivant, J. S. (1968). The method of feeding of lophophorates (Bryozoa, Phoronida, and Brachiopoda). *N.Z. J. Mar. Freshwater Res.* **2,** 135–146.

Cheetham, A. H. (1971). Functional morphology and biofacies distribution of cheilostome Bryozoa in the Danian stage (Paleocene) of southern Scandinavia. *Smithson. Contrib. Paleobiol.* **6,** 1–87.

Clark, R. B. (1964). "Dynamics in Metazoan Evolution." Oxford Univ. Press, London and New York.

Cowen, R. (1974). Respiration in metazoan evolution. *Evolution* **27,** 696–701.

Cowen, R., and Rider, J. (1972). Functional analysis of fenestellid bryozoan colonies. *Lethaia* **5,** 145–164.

Emig, C. C. (1970). Considérations sur la systématique des Phoronidiens. VIII. *Phoronis pallida* (Schneider) Silén, 1952. IX. *Phoronis ovalis* Wright, 1856. *Bull. Mus., Hist. Nat., Paris, 2nd ser.* **41**(6) 1531–1542.

Emig, C. C. (1971). Taxonomie et systématique des Phoronideins. *Bull. Mus. Nat. Hist. Nat. Paris, Zool. 3rd ser.* **8,** 469–568.

Emig, C. C. (1972). Reproduction asexuée chez *Phoronis psammophila. Mar. Biol.* **13,** 247–258.

Emig, C. C. (1973). Les processus de l'ontogénèse, comparés à ceux de la régénération des Phoronida. *Z. Morphol. Tiere* **75,** 329–350.

Emig, C. C. (1974a). Observations et discussions sur le développement embryonnaire des Phoronida. *Z. Morphol. Tiere* **77,** 317–335.

Emig, C. C. (1974b). The systematics and evolution of the phylum Phoronida. *Z. Zool. Syst. Evolutionsforsch.* **12,** 128–151.

Farmer, J. D., Valentine, J. W., and Cowen, R. (1973). Adaptive strategies leading to the ectoproct ground plan. *Syst. Zool.* **22,** 233–239.

Fenton, M. A., and Fenton, C. L. (1934). *Scolithus* as a fossil phoronid. *Am. Geol.* **61,** 341–348.

Ghiselin, M. T. (1972). Models in phylogeny. *In* "Models in Paleobiology" (T. J. M. Schopf, ed.), p. 130–145. Freeman, San Francisco, California.

Glaessner, M. F. (1969). Trace fossils from the Precambrian and basal Cambrian. *Lethaia* **2,** 369–393.

Gooch, J. L., and Schopf, T. J. M. (1970). Population genetics of marine species of the phylum Ectoprocta. *Biol. Bull. (Woods Hole, Mass.)* **138,** 138–156.

Hadzi, J. (1958). Zur abschaffung der zoologisch-systematischen gruppe Bryozoa Ehrenberg 1831. *Razpr. Slov. Akad. Znan. Umet. Hist. Nat.* **4,** 125–147.

Harmer, S. F. (1917). On *Phoronis ovalis,* Strethill-Wright. *Q. J. Microsc. Sci.* **62,** 115–148.

Hyman, L. H. (1951). "The Invertebrates," Vol. 3. McGraw-Hill, New York.

Hyman, L. H. (1959). "The Invertebrates," Vol. 5. McGraw-Hill, New York.

Jägersten, G. (1972). "Evolution of the Metazoan Life Cycle." Academic Press, New York.

Jebram, D. (1973). The importance of different growth directions in the Phylactolaemata and Gymnolaemata for reconstructing the phylogeny of the Bryozoa. *In* "Living and Fossil Bryozoa" (G. P. Larwood, ed.), pp. 565–576. Academic Press, New York.

Knight-Jones, E. W., and Moyse, J. (1961). Intra-specific competition in sedentary marine animals. *Symp. Soc. Exp. Biol.* **15,** 72–95.

Kupelwieser, H. (1905). Untersuchungen über den feineren Bau und die Metamorphoses des Cyphonautes. *Zoologica (Stuttgart)* **19,** Heft 47, p. 1–50.

Larwood, G. P., Medd, A. W., Owen, D. E., and Tavener-Smith, R. (1967). Bryozoa. *In* "The Fossil Record" (W. B. Harland *et al.,* eds.), pp. 379–395. Geol. Soc., London.

Lebour, M. V. (1922). The food of plankton organisms. *J. Mar. Biol. Assoc. U.K.* **12,** 644–677.

Mackie, G. O. (1963). Siphonophores, bud colonies and superorganisms. *In* "The Lower Metazoa" (E. C. Dougherty, ed.), pp. 329–337. Univ. of California Press, Berkeley.

Marcus, E. (1926). Beobachtungen und Versuche an lebenden Meeresbryozoen. *Zool. Jahrb., Abt. Syst. Oekol. Geogr. Tiere* **52,** 1–102.

Marcus, E. (1940). Mosdyr (Bryozóa eller Polyzóa). *Danmarks Fauna* **46.**

Marcus, E. (1949). *Phoronis ovalis* from Brazil. *Univ. Sao Paulo, Fac. Filos., Cienc. Let., Bol. Zool.* **14,** 157–171.

Marcus, E. (1958). On the evolution of animal phyla. *Q. Rev. Biol.* **33,** 24–58.

Meadows, P. S., and Campbell, J. I. (1972). Habitat selection by aquatic invertebrates. *Adv. Mar. Biol.* **10,** 271–382.

Mileikovsky, S. A. (1971). Types of larval development in marine bottom invertebrates, their distribution and ecological significance: A re-evaluation. *Mar. Biol.* **10,** 193–213.

Nielsen, C. (1971). Entoproct life-cycles and the Entoproct/Ectoproct relationship. *Ophelia* **9,** 209–341.

Pohowsky, R. A. (1973). A Jurassic cheilostome from England. *In* "Living and Fossil Bryozoa" (G. P. Larwood, ed.), pp. 447–461. Academic Press, New York.

Rattenbury, J. C. (1954). The embryology of *Phoronopsis viridis. J. Morphol.* **95,** 289–349.

Ross, J. P. (1964). Morphology and phylogeny of early Ectoprocta (Bryozoa). *Geol. Soc. Am. Bull.* **75,** 927–948.

Ryland, J. (1965). Polyzoa (Bryozoa). Zooplankton Sheet 107. Fisches d'identification du zooplancton. Conseil International Pour L'Exploration de la Mer.

Ryland, J. (1970). "Bryozoans." Hutchinson Univ. Library, London.

Silén, L. (1944). The anatomy of *Labiostomella gisleni* (Bryozoa Protocheilostomata). With special regard to the embryo chambers. *K. Sven. Vetenskaps akad. Handl. [3]* **21,** 1–111.

Silén, L. (1954). Developmental biology of Phoronidea of the Gullmar Fiord Area. *Acta Zool. (Stockholm)* **35,** 215–257.

Silén, L. (1966). On the fertilization problem in the gymnolaematous Bryozoa. *Ophelia* **3,** 113–140.

Silén, L. (1972). Fertilization in the Bryozoa. *Ophelia* **10,** 27–34.

Strathmann, R. (1971). The feeding behavior of planktotrophic echinoderm larvae: Mechanisms for regulation and rates of suspension feeding. *J. Exp. Mar. Biol. Ecol.* **6,** 109–160.

Strathmann, R. (1973). Function of lateral cilia in suspension feeding of lophophorates (Brachiopoda, Phoronida, Ectoprocta). *Mar. Biol.* **23,** 129–136.

Strathmann, R. Jahn, T., and Fonseca, J. R. C. (1972). Suspension feeding by marine invertebrate larvae: Clearance of particles from suspension by ciliated bands of a rotifer, pluteus, and trochophore. *Biol. Bull. (Woods Hole, Mass.)* **142,** 505–519.

Ström, R. (1969). Sexual reproduction in a stoloniferous bryozoan, *Triticella koreni* (G. O. Sars). *Zool. Bidr. Uppsala* **38,** 113–127.

Thorpe, J. P., Shelton, G. A. B., and Laverack, M. S. (1975). Colonial nervous control of lophophore retraction in cheilostome Bryozoa. *Science* **189**, 60–61.

Thorson, G. (1946). Reproduction and larval development of Danish marine bottom invertebrates, with special reference to planktonic larvae in the Sound (Øresund). *Medd. Komm. Havunderso., Ser. Plankton* **4**, 1–529.

Underwood, A. J. (1974). On models for reproductive strategy in marine benthic invertebrates. *Am. Nat.* **108**, 874–878.

Valentine, J. W., and Moores, E. (1970). Plate tectonic regulation of faunal diversity and sea level: a model. *Nature* (Lond.) **228**, 657–659.

Valentine, J. W. (1973a). "Evolutionary Paleoecology of the Marine Biosphere." Prentice-Hall, Englewood Cliffs, New Jersey.

Valentine, J. W. (1973b). Coelomate Superphyla. *Syst. Zool.* **22**, 97–102.

Vance, R. R. (1973). On reproductive strategies in marine benthic invertebrates. *Am. Nat.* **107**, 339–352.

Vance, R. R. (1974). Reply to Underwood. *Am. Nat.* **108**, 879–880.

Zimmer, R. L. (1964). Reproductive biology and development of Phoronida. Ph.D. dissertation, University of Washington.

Zimmer, R. L. (1973). Morphological and developmental affinities of the lophophorates. *In* "Living and Fossill Bryozoa" (G. P. Larwood, ed.), pp. 593–599. Academic Press, New York.

16

Phylogenetic Considerations: The Protostomian Relationships

CLAUS NIELSEN

I. INTRODUCTION

Most authors now consider the ectoprocts as related to phoronids, and together with the brachiopods these three groups are usually referred to as tentaculates or lophophorates. The phoronids have certain characteristics relating the group to the deuterostomians. The entoprocts, which earlier were united with the ectoprocts, are considered as protostomians, although their affinities have been much discussed. Newer observations on the life cycles of several species of entoprocts throw light both on the relationship between the entoprocts and the ectoprocts and on their affinities to other groups. The following discussion mainly follows my earlier, more detailed paper (Nielsen, 1971, with full literature references), but new information has been added, e.g., on larval eyes and on the structure of the tentacles.

II. THE RELATIONSHIP BETWEEN ENTOPROCTS AND ECTOPROCTS

The ectoprocts are in principle colonial. In all extant types the polypides arise from the larval body through a budding process. The larval gut is not retained in the adult in any species. In fact, the ectoprocts show an alternation between sexually produced larvae and asexually produced polypides. Consequently, it is not possible on embryological grounds to assign a ventral/dorsal orientation to the polypides.

Some of the entoprocts produce larvae which through a metamorphosis develop into adults which reproduce both sexually and asexually, but in some species of the family Loxosomatidae the larvae produce one or more buds and die; thus, these species have life cycles comparable to those of the ectoprocts. In *Loxosoma jaegersteni* and *Loxosomella vivipara* the budding of the larva takes place in invaginated areas of the episphere and the whole process resembles polypide formation in the ectoprocts. Comparisons between different life cycles of the loxosomatids clearly show how the more complicated type with internal budding in the larvae may have evolved from the primitive type in which the larva develops into an adult (Fig. 1).

The larvae of loxosomatids with a primitive life cycle settle with the frontal organ while most ectoproct larvae settle with the everted adhesive sac. The hitherto known larvae of *Pedicellina* and *Barentsia,* however, settle with the ring of cells above the contracted prototroch and become glued to the substratum by means of secretions from the three pairs of foot glands. Certain variations in the mode of settling occur also in the ectoprocts, and in species of *Victorella* and *Bulbella* the metamorphosing larvae evert the adhesive sac and produce a secretion which after the retraction of the sac attaches the ring of cells above the contracted corona (= prototroch) to the substratum. It is clear that the metamorphosis of these larvae is almost identical to that of the larvae of *Pedicellina* and *Barentsia* (see Fig. 2).

It appears that the ectoprocts with their complicated life cycle can be derived from the entoprocts, and that this theory may contribute substantially to our understanding of their affinities. The following pages summarize our present knowledge of structure and development of entoprocts and ectoprocts from the

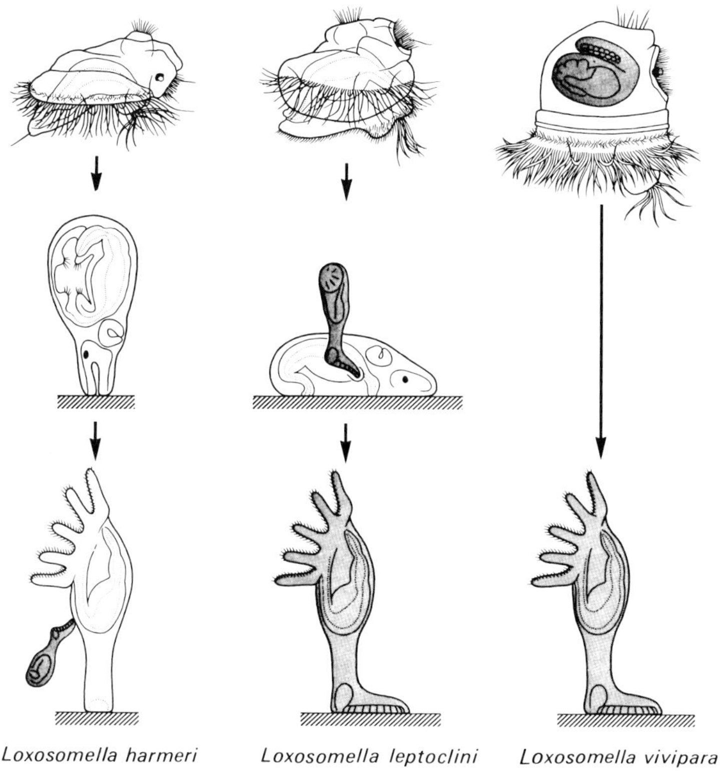

Fig. 1. Developmental types within *Loxosomella* (the buds are dotted). In *Loxosomella harmeri* the larva metamorphoses to form an adult which forms buds. In *L. leptoclini* the larva adheres to the host, forms a pair of buds, and then degenerates. In *L. vivipara* a bud is already formed in the larva retained in the atrium of the mother animal; shortly after the liberation of the larva this bud ruptures the episphere of the larva to get free, and the larval body disintegrates. From Nielsen (1971).

view of my hypothesis: The ectoprocts have evolved from colonial entoprocts which had larvae with internal budding and a settling which resembled that of *Pedicellina.*

A. Embryology

The entoprocts have spiral cleavage with the mesoderm originating from the 4d cell and the entoderm originating from the cells 4a–c and 4A–D. In *Loxosomella vivipara* the spiral pattern cannot be followed beyond the 8-cell stage,

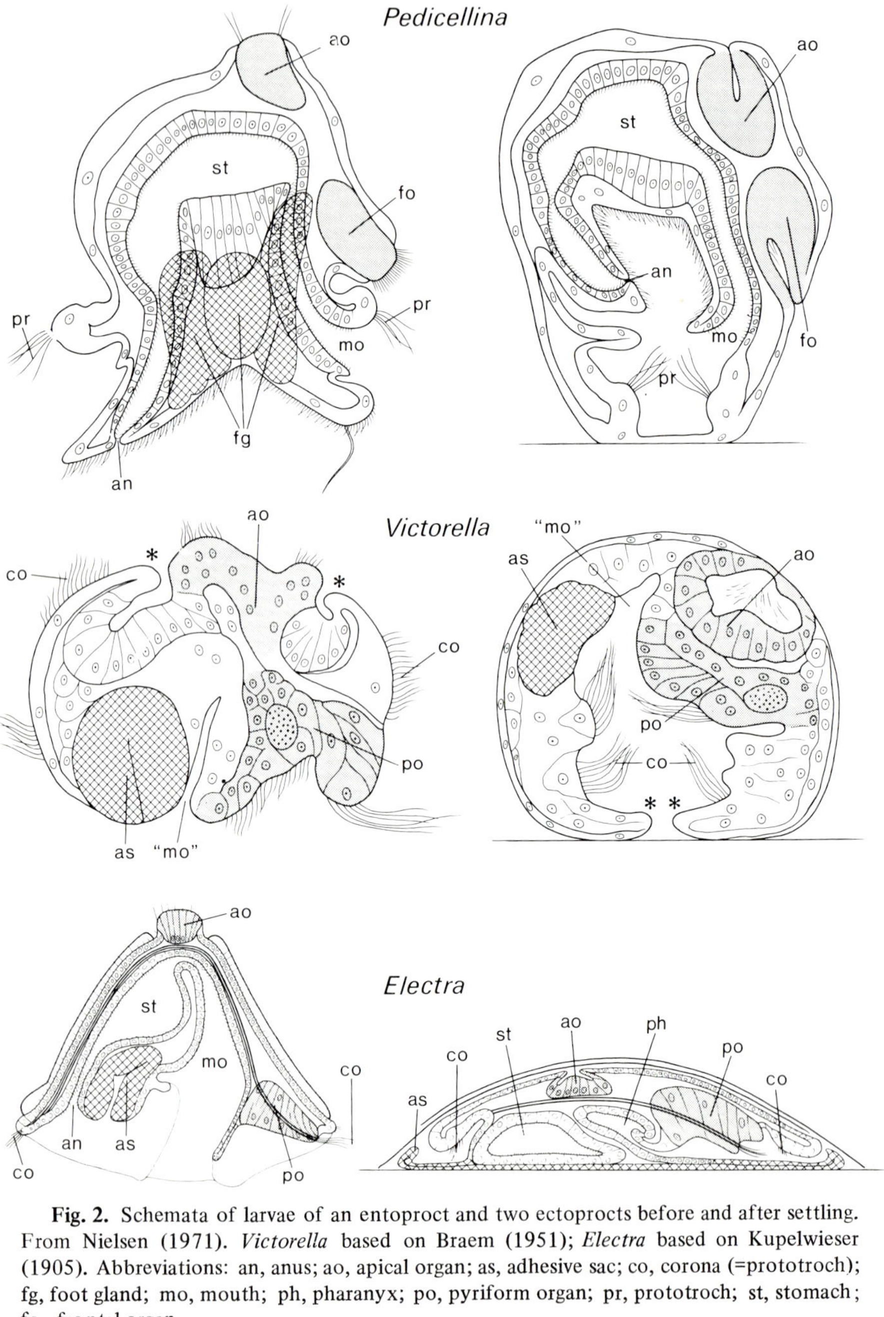

Fig. 2. Schemata of larvae of an entoproct and two ectoprocts before and after settling. From Nielsen (1971). *Victorella* based on Braem (1951); *Electra* based on Kupelwieser (1905). Abbreviations: an, anus; ao, apical organ; as, adhesive sac; co, corona (=prototroch); fg, foot gland; mo, mouth; ph, pharanyx; po, pyriform organ; pr, prototroch; st, stomach; fo, frontal organ.

and entoderm formation has not been observed; the fully grown larva lacks any trace of a gut. The ectoprocts show considerable variation: The cyclostomes have polyembryony, the phylactolaemates have an irregular cleavage, but the eurystomes all have a biradial cleavage. Zimmer (1973) has isolated the blastomeres of the 2-cell stage of *Membranipora isabelleana* and observed full regulation. The formation of the mesoderm has not been described adequately.

B. Larvae

Both planktotrophic larvae with a well-developed gut and lecithotrophic larvae devoid of any trace of a gut occur in the entoprocts as well as in the ectoprocts. The planktotrophic larvae of the two groups are rather similar, resembling polychaete trochophores. The only important difference seems to be that the entoproct larva typically has a frontal sense organ, sometimes provided with a pair of eyes (see Section II,C), whereas at least all the eurystome larvae have a so-called pyriform organ, which appears to be a specialized portion of the corona (= prototroch). A pair of nerves connect the apical organ with the anterior part of the prototroch/pyriform organ, and the frontal organ of the entoproct larva lies in contact with a large ganglion on this pair of nerves. Coelomic cavities are absent in all larvae described.

C. Larval Eyes

Most of the loxosomatid larvae and several of the eurystome larvae are provided with usually red pigment spots that are considered to be connected with photoreception.

The ultrastructure of the eyespots of a loxosomatid larva has been described by Woollacott and Eakin (1973), who found a narrow epithelial invagination with a single sensory cell with a bundle of almost unmodified cilia; behind the sensory cell was found a single pigment cell; a lens cell with large vacuoles is also present. The cilia of the sensory cell have the normal $9 \times 2 + 2$ axoneme, but lack the arms on the *a*-tubules; they are somewhat swollen in their distal portions. The cilia are oriented at right angles to the direction of the incoming light.

The eyespots of the larva of *Bugula neritina* were described by Woollacott and Zimmer (1972). The bottom of a small invagination of the epithelium consists of a cell with a twisted bundle of morphologically unmodified cilia; behind and lateral to this supposed photosensitive cell lie a number of pigment cells which shield the pit so that the cilia are reached only by light perpendicular to their longitudinal axes. Eyespots with almost unmodified cilia oriented at right angles to the light have not been described for any other animal group.

D. Metamorphosis

The transition from the free larval stage to the sessile adult stage involves profound reorganization in both entoprocts and ectoprocts. In loxosomatids with larval budding the larva simply disintegrates after having released the buds. In the other entoprocts and in all ectoprocts the settling involves the enclosing of the prototroch cells in a cavity which soon coalesces with the primary body cavity of the larva. The secretion which glues the larva to the substratum comes from the foot glands in the pedicellinids and from the adhesive sac in the ectoprocts (except the phylactolaemates); the two attachment organs are both formed as invaginations from the ventral surface and are to be considered homologous. In all newly settled larvae, except those of the phylactolaemates, degenerating prototroch cells containing cilia in a large vacuole have been observed in the more or less spacious body cavity. In the entoprocts and in the eurystomes *Victorella* and *Bulbella* the whole outer epithelium of the metamorphosed larva is derived from the episphere of the larva; in the cyclostomes and in the eurystomes *Electra, Flustrellidra,* and *Alcyonidium* the basal wall of the ancestrula is derived from the everted adhesive sac while the frontal wall is derived from the episphere of the larva; in *Bugula* and *Scrupocellaria* the margin of the everted adhesive sac constricts and covers the whole ancestrula.

Just after the metamorphosis the gut of the entoproct larva rotates 90°–180°, the atrium reopens, and the larval gut directly becomes the gut of the adult. In the ectoproct larvae having a gut this degenerates together with other larval organs and the polypides arise through budding (see Section II,E).

The metamorphosis of the larvae of entoprocts and ectoprocts is, thus, similar both in principle and in some details, the most important difference being that the larval gut is retained in some entoprocts while it is not retained in any ectoproct. This type of metamorphosis does not to my knowledge exist in any other group, and the high degree of similarity must be of considerable importance in phylogenetic considerations.

E. Budding

In both groups new individuals arise through a budding process involving only the outer epithelium and the mesoderm (possibly the cyclostomes form an exception). The gut is formed as an invagination which soon differentiates into the various regions; it has no connection with the entoderm of the already existing zooids. The ganglion forms from a small epithelial invagination at the ventral side between mouth and anus in the entoprocts and as a similar invagination from the anal side of the mouth region in the ectoprocts. This difference does not seem to be important, and as a consequence of the similarities in the budding of the two groups I consider the ganglion of the ectoprocts as ventral.

The loxosomatids are solitary and their buds detach from the mother zooid, but the other entoprocts and all ectoprocts are colonial and the new zooids become more or less completely separated by infoldings from the body wall.

The polypides of the ectoprocts degenerate after some time and new ones are formed through the usual budding process. In the ctenostome *Triticella* the whole "head" of the zooid may be cast off and a new one regenerated from the tip of the stalk; the very same process has been observed in the entoproct *Barentsia.*

F. Resting Bodies

Thick-walled cysts have been described in a number of the colonial entoprocts as well as in all groups of ectoprocts. Potts (1891) observed that the single joints of the stalks of *Urnatella* may act as cysts, and cysts with one or more chambers are formed on special stolons in *Barentsia* and *Loxosomatoides.* Among the ectoprocts the phylactolaemates are well known for their statoblasts, but also several eurystomes, such as *Bugula, Victorella,* and *Bowerbankia,* form thick-walled cysts; dormant, closed zooids have been observed in *Conopeum* and in the cyclostome *Crisia.*

The cellular content of the cysts is a mass of cells rich in yolklike material surrounded by an epithelium. At germination new zooids are formed through a normal budding process.

The resting bodies of the two groups are, thus, almost identical, and as similar structures are to my knowledge only known from sponges and hydroids, this similarity must be regarded as an important argument in favor of uniting the two groups.

G. Body Cavity

The entoprocts and the ectoproct larvae have no coelom and it seems highly questionable whether the body cavity of the adult ectoprocts can be considered a "coelom." The origin of the mesoderm after metamorphosis of the larva and in the budding process has not been adequately described; the mesodermic layer is incomplete, for example, in the regions of budding. In my opinion the body cavity of the adult ectoprocts is a specialization of a primary body cavity such as that of the entoproct larva (e.g., *Loxosoma pectinaricola* and *Barentsia gracilis*); in both cases the body cavity functions as a "fluid skeleton" which is used during protrusion of the feeding apparatus: the hyposphere in the entoproct larvae and the polypide in the ectoprocts. It is reasonable to believe that the larger types having a fluid skeleton develop a more or less complete mesodermal covering of the cavities. I find the idea of the development of differentiated body cavities in connection with other morphological changes much more attractive than the various theories which derive all triploblastic groups from an ancestor with coelomic pouches, and which consequently regard all acoelomate and pseudocoelomate groups as secondarily simplified.

H. Ciliation

Both entoprocts and ectoprocts have tentacles with cilia which produce a water current from which food particles are retained.

In the entoprocts the water current flows between the tentacles and away from the atrium. The ciliation is borne on five rows of cells on the frontal side of the tentacle (Fig. 3). The lateral cells have long cilia united in bundles of 6–10 which form a double row almost at a right angle to the longitudinal axis of the tentacle (Figs. 4 and 5); these syncilia beat perpendicularly to the longitudinal axis of the tentacle, and it has not been possible to detect metachronal waves. The rows of laterofrontal cilia likewise move at right angles to the tentacle as indicated by the orientation of the two central microtubules (Nielsen and Rostgaard, 1976). The frontal cilia beat parallel to the longitudinal axis of the tentacle.

The ectoproct tentacle has rows of lateral cilia, which carry a water current in the direction of the mouth and out between the tentacles, i.e., in a direction opposite to that of the entoprocts. These cilia appear single (Lutaud, 1973; Smith, 1973) and the metachronal waves are obvious (Strathmann, 1973). In some species frontal cilia are also present. Strathmann (1973) has observed that the lateral cilia reverse the beat locally when touched by a particle and carry the particle to the frontal surface of the tentacle. Other currents created by the lateral cilia are believed to transport the particle along the tentacle to the mouth. Thus, there appear to be important differences in the ciliary mechanisms of the two groups. The lateral cilia of the adult entoprocts can be compared to the prototroch of the entoproct larva and other trochophores and the frontal cilia correspond to the adoral ciliary band (see Fig. 7). It is not possible to state anything definite about the derivation of the ciliary bands of the adult ectoprocts.

I. Nervous System

The adults of both entoprocts and ectoprocts have a ganglion which is formed as an invagination from the epithelium and which lies in the loop of the alimentary canal. The peripheral systems are only poorly known in most groups, but it seems certain that there are nervous connections between the zooids in both groups.

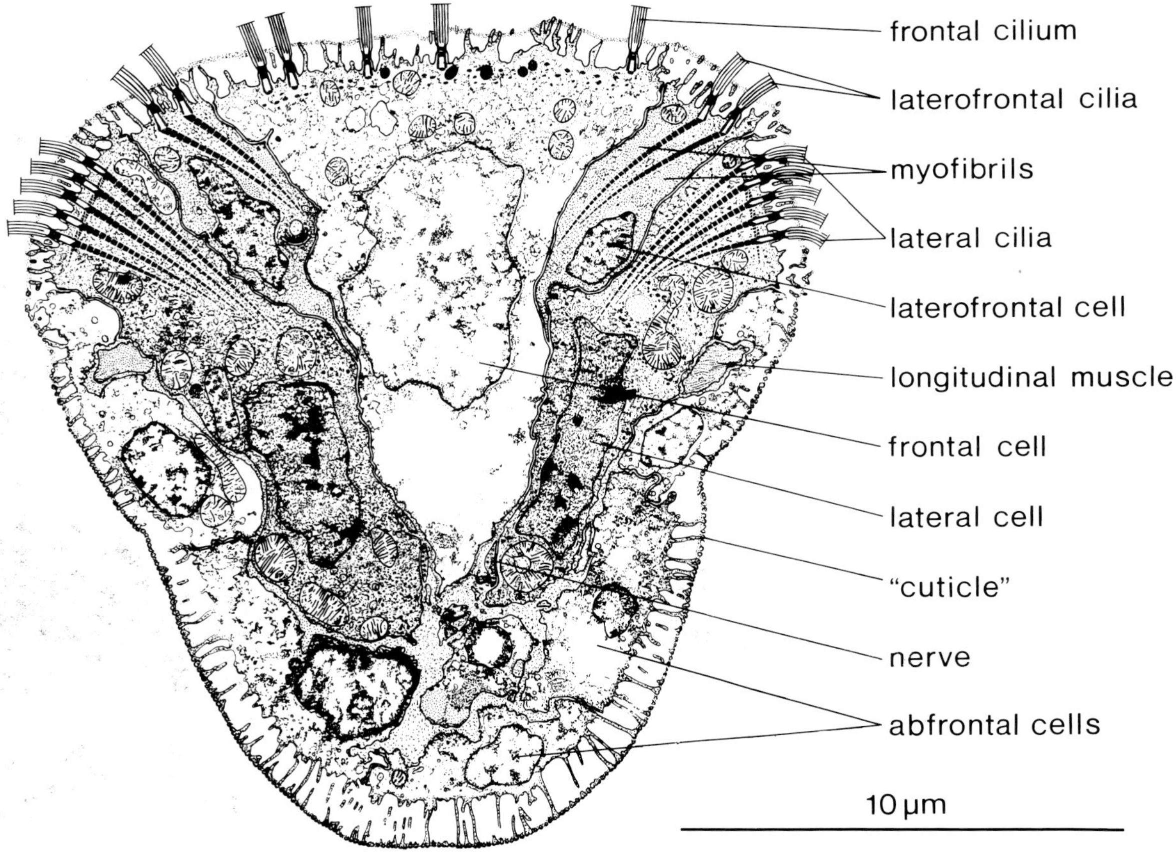

Fig. 3. Transverse section of a tentacle of *Loxosomella elegans;* composite drawing based on transmission electron micrographs.

Fig. 4. Syncilia of the lateral ciliated cells of a tentacle of *Loxosomella elegans.* Scanning electron micrograph.

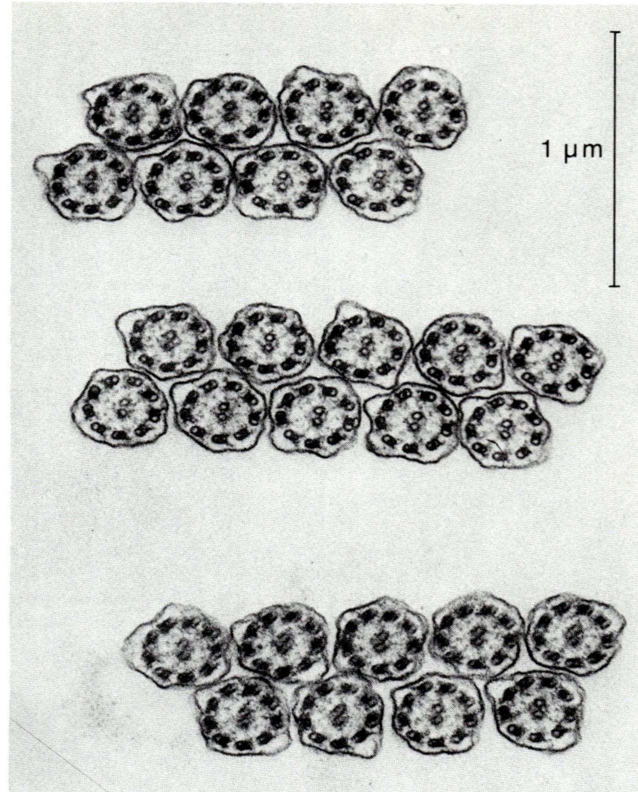

Fig. 5 Transverse sections of three syncilia of the lateral ciliated cells of a tentacle of *Loxosomella elegans.* Transmission electron micrograph.

J. Gonads

The gonads of the entoprocts are sac-shaped and have well-defined gonoducts. The gonads of the ectoprocts are merely clusters of germinal cells in or on the mesoderm of the body cavity; the sperm are shed through the tentacle tips, at least in a number of groups; the eggs or larvae are usually liberated through a temporary pore.

The spermatozoa of both groups are of a modified type with a pointed head with chromatin, a long middle piece with mitochondrial material, and a shorter or longer tail with a flagellum.

K. Summary

From the preceding sections it may be seen that several very important similarities exist between the ento- and ectoprocts: Unique features are found in the eyes of the larvae and in the metamorphosis of the larvae, and detailed similarities are observed in the budding and the formation of resting bodies. Important differences are found only in the cleavage and the ciliation. My conclusion is that the two groups must be related, and I find it natural to consider them as classes of a phylum called Bryozoa.

III. DISCUSSION OF POSSIBLE RELATIONSHIPS WITH THE PHORONIDS

The ectoprocts have often been regarded as derived from the phoronids; especially the similarities between the adult phylactolaemates and the adult phoronids have been stressed. It should, however, be kept in mind that entire life cycles ought to be compared instead of adults only. The very complex life cycle of the ectoprocts is very similar to that of *Pedicellina,* whereas it is very difficult to imagine the types "between" the phoronids and the ectoprocts.

A brief comparison of the more important features of the three groups in question may be instructive. The entoprocts have spiral cleavage whereas the two other groups have radial cleavage and full regulation after separation of the blastomeres of the 2-cell stage (Zimmer, 1973); the spiral cleavage tendencies described for some phoronids are questioned by Zimmer. The larvae of entoprocts and ectoprocts swim and in some cases feed by means of the prototroch (corona), whereas the actinotroch uses the metatroch. The entoproct and ectoproct larvae lack coeloms; these are well developed in the actinotroch. At metamorphosis a sac-shaped organ on the ventral side of the larva is everted or extended in representatives of all three groups. Budding in entoprocts and ectoprocts takes place from the episphere of the larva and involves ectoderm (+

mesoderm); in *Phoronis ovalis* budding takes place from the hyposphere and also involves the alimentary canal of the parent animal. The ganglion of the adult ectoprocts and entoprocts is formed as an invagination from the epithelium whereas it is intraepithelial in the phoronids. Both entoprocts and ectoprocts form thick-walled cysts from which new zooids are formed through a normal budding process; such cysts are unknown in phoronids. The phoronids have a well-developed coelom and a blood system, whereas ento- and ectoprocts have no coelom or a not very well-defined coelom, respectively, and lack blood vessels. The ciliation of the entoproct tentacles is clearly comparable to the larval prototroch + adoral ciliary zone (cilia of the food groove); the cilia of the phoronid tentacles are directly comparable to the metatroch (postoral ciliary band) + adoral ciliary zone of a trochophore; the ciliation of the ectoproct tentacle resembles that of the phoronids in structure and function, but its derivation is obscure. Metachronal waves of the cilia are present in ectoprocts and phoronids, but are lacking in entoprocts. The spermatozoa of entoprocts and ectoprocts are of a similar elongated type; those of *Phoronis pallida* and *Phoronopsis vancouverenis* are of a quite unusual V-shape (Franzén, 1956; Zimmer, 1967); but *Phoronis ovalis* seems to have a more normal, elongate type (Franzén, 1970). The three coelomic cavities of *Phoronis* are clearly absent in the entoprocts, but several authors have claimed that they can be found in the phylactolaemates. Brien (1960) has, however, made a detailed treatment of the anatomy of the phylactolaemates and concluded that they have no nephridia and that the "coelomic cavities" of cystid, lophophore, and epistome are confluent; the various parts of this cavity cannot be derived from corresponding larval cavities since the larva is highly specialized; all ectoproct larvae totally lack the tripartition characteristic of many deuterostomians.

It seems clear that the phoronids and the brachiopods are closely related and that they have many characteristics in common with the deuterostomians. The ectoprocts have usually been compared to the phoronids, and especially the comparisons of body "regions" have shown a number of tendentious interpretations. As Zimmer (1973, p. 594) expressed it (in a discussion of the relationships of the lophophorates): "Because a one-sided question has usually been asked, it is not surprising that a one-sided response has been given." In my opinion the above discussion has shown so many more important similarities between ectoprocts and entoprocts than between ectoprocts and phoronids that the ectoprocts must be regarded as derived from the entoprocts rather than from the phoronids. To derive the ectoprocts from the phoronids will need the speculation of a hologenetic scheme of an equal complexity as that given for the derivation of *Pedicellina* from a proentoproct (see Fig. 7), but of which we have none of the intermediate stages represented; although not totally impossible, this hypothesis seems very improbable.

IV. THE RELATIONSHIPS OF THE BRYOZOANS (ENTOPROCTS + ECTOPROCTS)

The entoprocts must clearly be the more primitive of the two bryozoan groups, as shown by the relatively uncomplicated life cycle of species such as *Loxosomella harmeri.* The entoproct larvae are easily comparable to the trochophores of annelids and mollusks. Their ciliation consists of an apical tuft, a prototroch, an adoral ciliary zone, and a metatroch; the gastrotroch (neurotroch) of the trochophore can probably be identified as the ciliation of the foot of the entoproct larva. The apical organ is clearly identical in all these larvae and the frontal ganglion of the entoproct larva perhaps corresponds to the cerebral

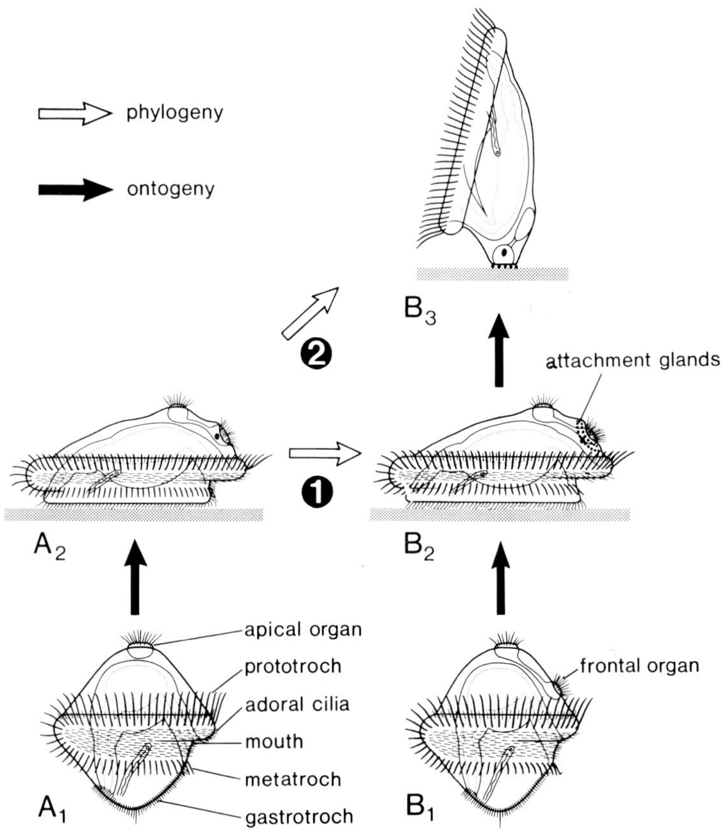

Fig. 6. A hologenetic schema of the evolution of a proentoproct from a creeping ancestor with a pelagobenthic life cycle.

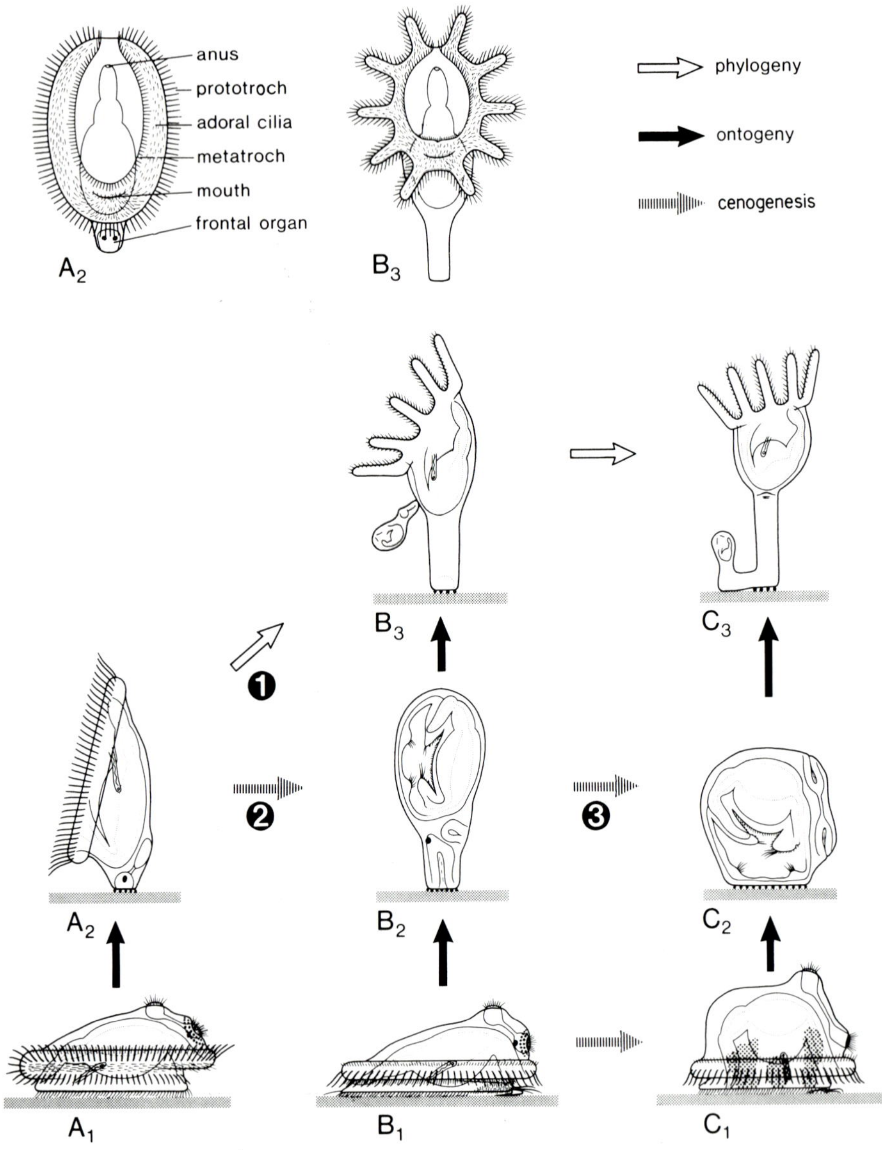

Fig. 7. A hologenetic schema of the evolution of loxosomatids and pedicellinids from a proentoproct resembling a sessile trochophore.

ganglion of larvae of polychaetes and mollusks. The gut has the same morphology and location, and a pair of protonephridia is also found.

It seems probable that an early ancestor of the entoprocts had a pelago-benthic life cycle such as that pictured in Fig. 6 (A) (see also Jägersten, 1972); an organism of this type may also have given rise to the earliest mollusks and with few modifications also to the annelids. The proentoproct had simply developed some gland cells around the frontal organ and the adult used these glands as an attachment organ [Fig. 6 (B_3)]; the foot and the apical organ may then have degenerated in the adult, and a new ganglion developed ventrally.

The evolution within the entoprocts may be followed in the hologenetic schema (Fig. 7). The loxosomatids [Fig. 7 (B)] with a simple life cycle (e.g., *Loxosomella harmeri*) have evolved directly from the proentoproct [Fig. 7 (A)] through the development of tentacles formed as protrusions carrying a row of the prototroch cilia and a narrow extension of the adoral ciliary zone [Fig. 7 (1)]. The closing of the atrium and the reorganization of the ciliation seen in extant loxosomatids must represent a cenogenetic specialization [Fig. 7 (2)]. Budding has also developed in the loxosomatids. The pedicellinids [Fig. 7 (C)] have become colonial and their larvae have developed adhesive glands on the foot and settle on the edge of the contracted prototroch; the change in the settling mechanism [Fig. 7 (3)] must likewise be regarded as a cenogenetic specialization.

Some of the ectoproct larvae settle in the same way as the *Pedicellina* larva, and the whole group has probably evolved from a *Pedicellina*-like ancestor which developed internal budding. It thus appears that the Bryozoa are related to Annelida and Mollusca and that hence they must be regarded as protostomians. The Phoronida and Brachiopoda, on the other hand, seem more closely related to some of the groups regarded as deuterostomians.

REFERENCES

Braem, F. (1951) Über *Victorella* und einige ihrer nächsten Verwandten sowie die Bryozoenfauna des Ryck bei Greifswald. *Zoologica (Stuttgart)* **102**, 1–59.

Brien, P. (1960). Classe des Bryozoaires. *In* "Traité de Zoologie" (P.-P. Grassé, ed.), Vol. 5, Part 2, pp. 1053–1355. Masson, Paris.

Franzén, Å. (1956). On spermatogenesis, morphology of the spermatozoon and the biology of fertilisation among invertebrates. *Zool. Bidr. Uppsala* **31**, 355–482.

Franzén, Å. (1970). Phylogenetic aspects of the morphology of spermatozoa and spermiogenesis. *Quad. Accad. Naz. Lincei* **137**, 29–46. [Also *in* "Comparative Spermatology" (B. Bacetti, ed.). Academic Press, New York.]

Jägersten, G. (1972). "Evolution of the Metazoan Life Cycle." Academic Press, New York.

Kupelwieser, H. (1905). Untersuchungen über den feineren Bau und die Metamorphose des Cyphonautes. *Zoologica (Stuttgart)* **47**, 1–50.

Lutaud, G. (1973). L'innervation du lophophore chez le Bryozoaire Chilostome *Electra pilosa* (L.). *Z. Zellforsch. Mikrosk. Anat.* **140**, 217–234.

Nielsen, C. (1971). Entoproct life-cycles and the entoproct/ectoproct relationship. *Ophelia* **9,** 209–341.

Nielsen, C. and Rostgaard, J. (1976). Structure and function of an entoproct tentacle with a discussion of ciliary feeding types. *Ophelia* **15,** 115–140.

Potts, E. (1891). Reproduction of *Urnatella. Am. Nat.* **25,** 380.

Smith, L. W. (1973). Ultrastructure of the tentacles of *Flustrellidra hispida* (Fabricius). *In* "Living and Fossil Bryozoa" (G. P. Larwood, ed.), pp. 335–342. Academic Press, New York.

Strathmann, R. (1973). Function of lateral cilia in suspension feeding of lophophorates (Brachiopoda, Phoronida, Ectoprocta). *Mar. Biol.* **23,** 129–136.

Woollacott, R. M., and Eakin, R. M. (1973). Ultrastructure of a potential photoreceptoral organ in the larva of an entoproct. *J. Ultrastruct. Res.* **43,** 412–425.

Woollacott, R. M., and Zimmer, R. L. (1972). Fine structure of a potential photoreceptor organ in the larva of *Bugula neritina* (Bryozoa). *Z. Zellforsch. Mikrosk. Anat.* **123,** 458–469.

Zimmer, R. L. (1967). The morphology and function of accessory reproductive glands in the lophophores of *Phoronis vancouverensis* and *Phoronopsis harmeri. J. Morphol.* **121,** 159–178.

Zimmer, R. L. (1973). Morphological and developmental affinities of the lophophorates. *In* "Living and Fossil Bryozoa" (G. P. Larwood, ed.), pp. 593–599. Academic Press, New York.

Appendix

Classification of Bryozoan Genera Cited in This Volume

In the following classification, all genera, families, and orders of Bryozoa (Ectoprocta) cited in this volume are listed. For consistency, the taxonomic heirarchy of each genus has been included whether or not its components were cited. The species considered for each genus are listed in the index under the generic name.

This appendix is provided for the convenience of the nonspecialist and should not be considered authoratative. The classification follows that of Bassler (1953)* as this currently is the most comprehensive and widely available single reference. Unfortunately, this bryozoan part of the *Treatise on Invertebrate Paleontology* is considerably out of date. An extensive revision of the treatise is near completion, and this volume should be consulted for a modern classification when it is available.

Following current convention, but in variance from Bassler (1953), the "cyclostomes" are placed in their own class Stenolaemata and the Ctenostome Gymnolaemata are organized as two suborders with two divisions each, not four equivalent suborders. Bassler's placement of all "ascophoran" cheilostomes in a single taxon is followed, despite the knowledge that this group is polyphyletic. Unfortunately, there is no generally accepted alternate classification, and for none of the revised groupings has a comprehensive listing of genera been published. An introduction to alternative classifications of ascophorans is available on pages 162–166 of this volume.

Class Gymnolaemata
  Order Ctenostomata
    Suborder Carnosa

*Bassler, R. S. (1953). Bryozoa. *In* "Treatise on Invertebrate Paleontology" (R. C. Moore, ed.), Vol. G., pp. 1–253. Geological Society of America, Lawrence, Kansas.

Division Halcyonelloida
Family Alcyonidiidae
Alcyonidium
Family Flustrellidridae
Flustrella (usually as *Flustrellidra*)
Flustrellidra
Family Pherusellidae
Pherusella
Division Paludicelloida
Family Nolellidae
Anguinella
Nolella
Family Victorellidae
Bulbella
Paludicella
Sundanella
Tanganella
Victorella
Suborder Stolonifera
Division Vesicularioida
Family Monobryozoontidae
Monobryozoon
Family Vesiculariidae
Amathia
Bowerbankia
Cryptopolyzoon
Hislopia
Vesicularia
Zoobotryon
Division Walkerioida
Family Buskiidae
Buskia
Family Immergentiidae
Family Mimosellidae
Farella
Hypophorella
Mimosella
Family Penetrantiidae
Penetrantia
Family Terebriporidae
Terebripora
Family Triticellidae
Triticella
Family Vinellidae
Marcusodictyon
Family Walkeriidae
Aeverrillia
Walkeria
Order Cheilostomata
Suborder Anasca

Division Inovicellata
Family Aeteidae
Aetea
Division Scrupariina (Scruparioida)
Family Scrupariidae
Eucratea
Scruparia
Family Labiostomellidae
Labiostomella
Division Malacostega (Malacostegoida)
Family Arachnopusiidae
Exochonella (or placed in own family Exochonellidae)
Family Calloporidae
Alderina
Callopora
Crassimarginatella
Parellisina
Tegella
Wilbertopora
Family Chaperiidae
Chaperia
Family Electridae
Aspidelectra
Electra
Pyripora
Tendra
Family Flustridae
Carbasea
Flustra
Securiflustra
Family Hiantoporidae
Tremogasterina
Family Hincksinidae
Antropora
Cauloramphus
Ellisina
Pseudolepralia
Family Membraniporidae
Conopeum
Cupuladria (or placed in own family Cupuladriidae)
Membranipora
Vincularia
Division Coilostega (Coelostegoida)
Family Alysidiidae
Alysidium
Catenicula
Family Aspidostomatidae
Megapora
Family Calpensiidae
Discoporella

Poricellaria
Family Onychocellidae
Onychocella
Family Steginoporellidae
Labioporella
Steginoporella
Family Thalamoporellidae
Thalamoporella
Division Pseudostega (Pseudostegoida)
Family Cellariidae
Cellaria
Euginoma
Melicerita (or placed in own family Meliceritidae)
Division Cellularina (Cellularioida)
Family Beaniidae
Beania
Family Bicellariellidae
Bicellariella
Family Bugulidae
Bugula
Camptoplites
Caulibugula
Dendrobeania
Kinetoskias
Semikinetoskias
Family Epistomiidae
Synnotum
Family Farciminariidae
Didymozoum
Nellia
Family Scrupocellariidae
Caberea
Menipea
Scrupocellaria
Tricellaria
Suborder Cribrimorpha
Family Cribrilinidae
Cribrilina
Figularia
Membraniporella (or placed in Division Malacostega, family Membraniporidae)
Puellina
Suborder Ascophora
Family Adeonidae
Adeona
Adeonella
Adeonellopsis
Metrarabdotos (or placed in own family Metrarabdotosidae)
Reptadeonella
Family Catenicellidae
Catenaria

Catenicella
Vittaticella
Family Celleporidae
Cellepora
Holoporella
Siniopelta
Trematooecia
Family Cheiloporinidae
Cryptosula
Hippopodinella
Tetraplaria
Veleroa
Watersipora
Family Conescharellinidae
Flabellopora
Family Crepidicanthidae
Crepidicantha
Mastigophora
Family Eurystomellidae
Eurystomella
Family Euthyrisellidae
Euthyrisella
Family Exochellidae
Escharoides
Family Gigantoporidae
Tessaradoma
Family Hippoporinidae
Aimulosia
Cleidochasma
Hippoporella
Hippoporidra
Hippoporina
Pentapora
Stephanosella
Family Hippothoidae
Celleporella (usually as *Hippothoa*)
Chorizopora
Haplopoma
Hippothoa
Tryptostega
Family Lekythoporidae
Turritigera
Family Mammiloporidae
Family Microporellidae
Fenestrulina
Microporella
Family Mucronellidae
Codonellina
Mucronella
Parsmittina

Porella
Smittia (usually as *Parasmittina, Smittina,* or *Smittoidea*)
Smittina
Smittoidea
Family Myriozoidae
Myriopora
Family Petraliidae
Hippopetraliella
Hippopodina
Petralia
Petraliella
Family Phylactelliporidae
Phylactellipora
Family Reteporidae
Reteporellina
Rhynchozoon
Family Savignyellidae
Savignyella
Family Schizoporellidae
Dakaria
Escharina
Gemelliporidra
Hippodiplosia
Schizobrachiella
Schizoporella
Stylapoma
Family Smittinidae (usually as Mucronellidae)
Family Stomachetosellidae
Posterula
Family Tubucellariidae
Tubiporella
Tubucellaria
Family Umbonulidae
Umbonula
Class Stenolaemata
Order Cryptostomata
Order Cyclostomata
Suborder Articulata
Family Crisiidae
Crisia
Crisidia
Crisiella
Filicrisia
Suborder Cerioporina
Family Heteroporidae
Heteropora
Suborder Rectangulata
Family Lichenoporidae
Disporella
Lichenopora

Suborder Tubulorina
Family Diastoporidae
Berenicea
Diastopora
Family Entalophoridae
Crisulipora
Diplosolen
Family Plagioeciidae
Plagioecia
Family Terviidae
Tervia
Family Tubuliporidae
Tubulipora
Order Cystoporata
Order Trepostomata
Suborder Amalgamata
Family Constellariidae
Nicholsonella
Class Phylactolaemata
Family Cristatellidae
Cristatella
Family Fredericellidae
Fredericella
Family Lophopodidae
Lophopus
Lophopodella
Pectinatella
Family Plumatellidae
Plumatella

Index

C

D

M

A 7
B 8
C 9
D 0
E 1
F 2
G 3
H 4
I 5
J 6